Ancestors in Evolutionary Biology

Linear Thinking about Branching Trees

Phylogenetics emerged in the second half of the nineteenth century as a speculative storytelling discipline dedicated to providing narrative explanations for the evolution of taxa and their traits. It coincided with lineage thinking, a process that mentally traces character evolution along lineages of hypothetical ancestors. *Ancestors in Evolutionary Biology* traces the history of narrative phylogenetics and lineage thinking to the present day, drawing on perspectives from the history of science, philosophy of science, and contemporary scientific debates. It shows how the power of phylogenetic hypotheses to explain evolution resides in the precursor traits of hypothetical ancestors. This book provides a comprehensive exploration of the topic of ancestors, which is central to modern biology and is therefore of interest to graduate students, researchers, and academics in evolutionary biology, palaeontology, philosophy of science, and the history of science.

RONALD A. JENNER is Principal Researcher in the Department of Life Sciences at the Natural History Museum, London. His research focuses on animal phylogenetics, body plan evolution, the evolution of venoms, and conceptual issues in systematics. He has published extensively in the primary literature and coauthored *Venom: The Secrets of Nature's Deadliest Weapon* (The Natural History Museum, 2017).

The Systematics Association Special Volume Series

SERIES EDITOR

GAVIN BROAD

Department of Life Sciences, The Natural History Museum, London, UK

The Systematics Association promotes all aspects of systematic biology by organizing conferences and workshops on key themes in systematics, running annual lecture series, publishing books and a newsletter, and awarding grants in support of systematics research. Membership of the Association is open globally to professionals and amateurs with an interest in any branch of biology, including palaeobiology. Members are entitled to attend conferences at discounted rates, to apply for grants and to receive the newsletter and mailed information; they also receive a generous discount on the purchase of all volumes produced by the Association.

The first of the Systematics Association's publications The New Systematics (1940) was a classic work edited by its then-president Sir Julian Huxley. Since then, more than 70 volumes have been published, often in rapidly expanding areas of science where a modern synthesis is required.

The Association encourages researchers to organize symposia that result in multi-authored volumes. In 1997 the Association organized the first of its international Biennial Conferences. This and subsequent Biennial Conferences, which are designed to provide for systematists of all kinds, included themed symposia that resulted in further publications. The Association also publishes volumes that are not specifically linked to meetings, and encourages new publications (including textbooks) in a broad range of systematics topics.

More information about the Systematics Association and its publications can be found at our website: www.systass.org.

Previous Systematics Association publications are listed after the index for this volume.

Systematics Association Special Volumes published by Cambridge University Press:

78. *Climate Change, Ecology and Systematics (2011)*
TREVOR R. HODKINSON, MICHAEL B. JONES, STEPHEN WALDREN AND JOHN A.N. PARNELL

79. *Biogeography of Microscopic Organisms: Is Everything Small Everywhere? (2011)*
DIEGO FONTANETO

80. *Flowers on the Tree of Life (2011)*
LIVIA WANNTORP AND LOUIS RONSE DE CRAENE

81. *Evolution of Plant–Pollinator Relationships (2011)*
SÉBASTIEN PATINY

82. *Biotic Evolution and Environmental Change in Southeast Asia (2012)*
DAVID J. GOWER, KENNETH G. JOHNSON, JAMES E. RICHARDSON, BRIAN R. ROSEN, LUKAS RÜBER AND SUZANNE T. WILLIAMS

83. *Early Events in Monocot Evolution (2013)*
PAUL WILKIN AND SIMON J. MAYO

84. *Descriptive Taxonomy: The Foundation of Biodiversity Research (2014)*
MARK F. WATSON, CHRIS LYAL AND COLIN PENDRY

85. *Next Generation Systematics (2016)*
PETER D. OLSON, JOSEPH HUGHES AND JAMES A. COTTON

86. *The Future of Phylogenetic Systematics: The Legacy of Willi Hennig (2016)*
DAVID WILLIAMS, MICHAEL SCHMITT, AND QUENTIN WHEELER

87. *Interactions in the Marine Benthos: Global Patterns and Processes (2019)*
STEPHEN J. HAWKINS, KATRIN BOHN, LOUISE B. FIRTH AND GRAY A. WILLIAMS

88. *Cladistics: A Guide to Biological Classification 3rd edition (2020)*
DAVID M. WILLIAMS AND MALTE C. EBACH

89. *Cryptic Species: Morphological Stasis, Circumscription, and Hidden Diversity (2022)*
ALEXANDRE K. MONRO AND SIMON J. MAYO

90. *The Origin and Early Evolutionary History of Snakes (2022)*
DAVID J. GOWER AND HUSSAM ZAHER

91. *Ancestors in Evolutionary Biology: Linear Thinking about Branching Trees (2022)*
RONALD A. JENNER

SYSTEMATICS ASSOCIATION SPECIAL VOLUME 91

Ancestors in Evolutionary Biology

Linear Thinking about Branching Trees

RONALD A. JENNER

Natural History Museum, London

CAMBRIDGE
UNIVERSITY PRESS

Shaftesbury Road, Cambridge CB2 8EA, United Kingdom

One Liberty Plaza, 20th Floor, New York, NY 10006, USA

477 Williamstown Road, Port Melbourne, VIC 3207, Australia

314–321, 3rd Floor, Plot 3, Splendor Forum, Jasola District Centre, New Delhi – 110025, India

103 Penang Road, #05–06/07, Visioncrest Commercial, Singapore 238467

Cambridge University Press is part of Cambridge University Press & Assessment,
a department of the University of Cambridge.

We share the University's mission to contribute to society through the pursuit of
education, learning and research at the highest international levels of excellence.

www.cambridge.org
Information on this title: www.cambridge.org/9781107105935

DOI: 10.1017/9781316226667

First published 2022

A catalogue record for this publication is available from the British Library

Library of Congress Cataloging-in-Publication data
Names: Jenner, Ronald A., author.
Title: Ancestors in evolutionary biology : linear thinking about branching trees / Ronald A. Jenner, Natural History
    Museum, London.
Description: First edition. | Cambridge, United Kingdom ; New York, NY : Cambridge University Press, 2022. | Series:
    The systematics association special volume series ; no. 91 | Includes bibliographical references and index.
Identifiers: LCCN 2022000801 (print) | LCCN 2022000802 (ebook) | ISBN 9781107105935 (hardback) |
    ISBN 9781316226667 (epub)
Subjects: LCSH: Phylogeny. | Evolution (Biology)
Classification: LCC QH367.5 .J46 2022 (print) | LCC QH367.5 (ebook) | DDC 576.8/8–dc23/eng/20220119
LC record available at https://lccn.loc.gov/2022000801
LC ebook record available at https://lccn.loc.gov/2022000802

ISBN   978-1-107-10593-5   Hardback

For Sarah, for everything, for always

In loving memory of Loeki Jenner-Fresen (1939–2019),
Frans Eijben (1931–2019),
Frank Webb (1949–2021)

Contents

Acknowledgments   x

1 A History of Narrative Phylogenetics   **1**
1.1 Why This Book?   1
1.2 Outline of the Book   4

2 From Archetypes to Ancestors   **8**
2.1 Ideal Beginnings?   8
2.2 Johann Wolfgang von Goethe: Imagining Archetypes   10
2.3 Geoffroy Saint-Hilaire and Unity of Plan   14
2.4 Richard Owen and the Vertebrate Archetype   29
2.5 Reifying Archetypes: The Birth of the Evolutionary Ancestor   35
2.6 The First Common Ancestors: A Coda on Lamarck   38
2.7 Epistemic Fallout from a Metaphysical Revolution   39

3 The Emergence of Lineage Thinking   **44**
3.1 Systematics and the Shape of Nature   45
3.2 Methods of Classification   46
3.3 The Shape of the Natural System   47
3.4 Toward a Realist Interpretation of the Natural System   55
3.5 The Temporal Reinterpretation of the Natural System   58
3.6 The Darwinian Revision of the Natural System   64
3.7 Crossing the Phylogenetic Frontier   78
3.8 The Need to Speculate   82

4 Ernst Haeckel's Evolutionary Storytelling   **86**
4.1 Ernst Haeckel and the First Phylogenetics   86
4.2 Haeckel's Phylogeny and Genealogy: Distinct but Inseparable   88
4.3 Are Haeckel's Trees Darwinian?   89
4.4 Haeckel's Ancestors   104
4.5 Haeckel's Phylogenetic Evidence   106
4.6 The Need for Imagination   108
4.7 A Portrait Gallery of Ancestors   109
4.8 Progressive Evolution   117
4.9 Jump-Starting the Post-Haeckelian Hunt for Ancestors   120

5 The Epistemic Rise of Hypothetical Ancestors   **122**
5.1 Hypothetical Ancestors as Central Subjects in Scenarios   122
5.2 Fritz Müller's Scenario for the Origin of Root-Heads   124
5.3 No *De Novo* Origins   127
5.4 The Precursor Potential of Hypothetical Ancestors   133
5.5 The Unequal Eye and the Perspectival Nature of Scenarios   140
5.6 When Imagination Falters   144

6 Intuiting Evolution   **150**
6.1 Intuiting the Arrow of Evolution   150
6.2 Morphological Seriation: The Arrow at the Heart of the Threefold Parallelism   152
6.3 Evolutionary Intuitions as Narrative Tools   160
6.4 The Flatlining of Evolutionary Intuitions   167
6.5 Unimaginable Evolution and the Denial of Common Descent   169

7 Telling Straight Stories with Fossils   **181**
7.1 Fossils and the Arrow of Stratigraphy   181
7.2 Hilgendorf, Hyatt, and the Steinheim Snails   185
7.3 Orthogenesis: A Straight Story from Beginning to End   191
7.4 Nonorthogenetic Linear Thinking with Fossils   199
7.5 Othenio Abel: A Straight-Thinking Orthogeneticist   206
7.6 Linear Thinking with Branching Evidence   208

8 Seeing Animal Ancestors in Embryos   **217**
8.1 Haeckel's Gastraea Theory   217
8.2 Lankester's Planula Theory   220
8.3 Balfour's Amphiblastula Theory   222
8.4 Metschnikoff's Parenchymella/Phagocytella Theory   225
8.5 Bütschli's Plakula Theory   231
8.6 Judging a Phylogenetic Face-Off   234
8.7 Modern Perspectives on Early Animal Evolution   239

9 Ancestral Attractions and Phylogenetic Folklore   **242**
9.1 Amphistomy, Anemones, and the Origin of Bilateria   243
9.2 A Dissenting View from America   253
9.3 The Ciliate Ancestry of Animals   257
9.4 Annelids as Ultimate Ancestors   263

10 Narrative Shortcuts and Phylogenetic Faux Pas   **282**
10.1 Narrative Ghosts in the Cladistic Machine   282
10.2 Cutting Corners with Pruned Trees   288
10.3 Wringing Water from Stones   290
10.4 Ancestral Adjectives and Using Trees as Predictors   291

11 Taxic Distortions of Lineage Thinking   **298**
11.1 Gould's Blind Spot and His Denial of Linear Descent   299
11.2 Tree Thinking and the Rejection of Linear Narratives   305
11.3 The Influence of Gould's Blind Spot and the Cladistic Blindfold   308
11.4 The Importance of Lineage Thinking   319
11.5 Paraphyly, Ancestors, and Taxic Extremism   320

12 Making Sense with Stories **327**

12.1 Evolutionary Storytelling Is Inescapable 327

12.2 Leveraging Stories with Stories 331

12.3 Optimality Criteria versus Storytelling 333

References 336

Index 379

Systematics Association Special Volumes 386

Acknowledgments

It is a pleasure to acknowledge friends and colleagues who provided inspiration and served as sounding boards for the ideas expressed in this book. Malte Ebach and Ariel Chipman offered the early encouragement that got this book started. Chapters were critically read by Patrick Laurenti, Sander Gliboff, Bernard Cohen, Wallace Arthur, Fred Schram, Mike Richardson, Rudolf Nieuwenhuys, David Williams, Greg Edgecombe, and Ivan Koludarov. Sarah Webb read the entire manuscript and found many rough edges that I had missed. Chuck Crumley, Dominic Lewis, and Olivia Boult provided editorial support.

1

A History of Narrative Phylogenetics

1.1 Why This Book?

In 2007 eminent population geneticist Michael Lynch took equally eminent evolutionary developmental biologist Sean Carroll to task over something he had written in his 2005 popular science book *Endless Forms Most Beautiful*. In the final chapter of his book, Carroll addressed the general public's abysmal understanding of evolutionary biology. To help remedy this problem, he suggested that the teaching of evolution could be made more inspiring by shifting the focus from the dry calculus of population genetics to the evolution of organismal form, because that "is the main drama of life's story... So, let's teach that story" (Carroll, 2005: 294). Lynch didn't like that. He accused Carroll of "two fundamental misunderstandings. Evolutionary biology is not a story-telling exercise, and the goal of population genetics is not to be inspiring, but to be explanatory" (Lynch, 2006: 8597). The story that I tell in this book shows why Lynch is wrong and why and how evolutionary biology is a storytelling discipline.

I have distilled many of my insights from the older phylogenetic literature, supplemented by the literature about the history and philosophy of biology. However, I am neither a historian nor a philosopher. I am a biologist with a long-standing interest in metazoan phylogeny and the evolution of animal body plans, and I have written this book primarily to better understand the conceptual contours and the history of the research agenda of this field. What I report in these pages is therefore my personal intellectual journey, but I believe the book will be of broader interest because it helps plug a gap in the literature. First, although our bookshelves sag with literature on fundamental evolutionary concepts like species and speciation, natural selection, and homology, no recent volume focuses on the concept of ancestors. The recognition and discovery of ancestors pose notorious epistemological challenges, but the power of phylogenetic hypotheses to explain the origin and evolution of organismal traits resides entirely in the traits attributed to hypothetical ancestors. This book shows how hypothetical ancestors came to function as central subjects in macroevolutionary narratives.

Second, much of the recent literature about the history of systematics and phylogenetics has been written with an emphasis on systematics, the science concerned with discovering and describing biodiversity and investigating the relationships between taxa (Willmann,

2003; Williams and Forey, 2004; Williams and Ebach, 2008; Hamilton, 2014; Rieppel, 2016; Williams et al., 2016). The destination of many of these works is Willi Hennig (1913–1976) and his role in the development of phylogenetic systematics and cladistics. Less attention has been paid to phylogenetics as the science of evolutionary storytelling.

Evolutionary storytelling emerged as a distinct discipline in the second half of the nineteenth century. In 1866 Ernst Haeckel called it *Phylogenie* (Haeckel, 1866a). The origin of phylogenetics ushered in a century of speculative storytelling about evolving lineages before, exactly a century later, Hennig's *Phylogenetic Systematics* appeared in English and transformed systematics and phylogenetics (Hennig, 1966). The ensuing conceptual revolution quashed speculative storytelling, enshrined the epistemological primacy of systematic pattern over evolutionary process, and helped turn phylogenetics into an objective historical science. This, in a nutshell, is the story that you are likely to encounter when you consult the literature. Often little ink is spent on the interval between Haeckel and Hennig.

Ward Wheeler's *Systematics. A Course of Lectures*, for instance, jumps straight from Haeckel to the evolutionary taxonomists of the 1930s and 1940s and thence to Hennig via three paragraphs on the numerical taxonomists, aka pheneticists (Wheeler, 2012: 13–14). In their *Tree Thinking. An Introduction to Phylogenetic Biology*, David Baum and Stacey Smith get from Haeckel to Hennig in just three sentences (Baum and Smith, 2013: 25). Even when the period between Haeckel and Hennig is fleshed out in more detail, these pioneers define the axis of progress, with post-Haeckelian thinkers and ideas chosen chiefly to highlight the road to Hennig (Willmann, 2003; Williams and Ebach, 2008; Rieppel, 2016). Hennig is less of a hero in Joe Felsenstein's *Inferring Phylogenies*, which instead focuses on the development of numerical methods, and reaches the pheneticists of the 1950s and 1960s two sentences after telling readers that phylogenies "were discussed by Darwin and Haeckel" (Felsenstein, 2004: 123). This is all we hear about Haeckel in Felsenstein's book. The relative neglect of work that is not part of the progressive historical arc of systematics and phylogenetics is understandable when the aim is to show where modern concepts and methods have come from. But it does leave one wondering how evolutionary inference worked in the century between Haeckel and Hennig.

Traditional systematic and phylogenetic practices are often pejoratively labelled as "imprecise, authoritarian, and unable to articulate a specific goal other than ill-defined 'naturalness' [of taxa]" (Wheeler, 2012: 14). They are said to rely on "intuition" (Wiley et al., 1991: 1) and "subjective 'artistic' interpretations" (Eldredge and Cracraft, 1980: 189) and to commit the sin of bringing "assumptions derived from process explanations to bear on pattern reconstruction" (Rieppel, 2010a: 487). Traditional practices "lacked optimality criteria for choosing among alternative hypotheses of relationships" (Chakrabarty, 2010: 513), so that "classification and the reconstruction of phylogeny were considered an 'art'" (Reif, 2002: 357). Although these statements ring true enough when one judges traditional approaches from our modern perspective, they don't tell us what our predecessors actually did. Were they just engaged in wild evolutionary arm waving, or did their research strategy have a certain logic and conceptual consistency that allows it to be recognized as a distinctive way of thinking? In this book I argue that it did.

When Haeckel founded phylogenetics in 1866, he set as its goal the reconstruction of the evolution of lineages. What was new was not the search for taxonomic relationships, which had long been the purview of systematics, but the possibility of tracing lineages of descent of traits and taxa. This required a new type of thinking, which I call *lineage thinking*, that remains at the heart of evolutionary thought today. It focuses the mind's eye on the epistemic hinterland of hypothetical ancestors and character transformations that flow along the branches of phylogenetic trees.[1] The realization that systematic pattern is the product of a phylogenetic process had an important epistemological consequence. The unobservable entities and events of evolutionary history are empirically inaccessible. The visible evidence of the morphologist and the systematist can only provide indirect and imperfect access to this invisible realm. To cross the phylogenetic frontier, biological practice had to float free from pure observation and embrace speculation, inference, and imagination to mentally recreate evolving lineages.

As a result, a new storytelling discipline was born that I propose to call *narrative phylogenetics*. Narrative phylogenetic hypotheses are driven by speculations about the evolutionary process, and they attempt to explain the origin and evolution of traits by linking them to precursors in hypothetical ancestors. The resulting lineage explanations are often little constrained by considerations of relevant systematic relationships. In this book I will outline the history of narrative phylogenetics, show that it continues to play an important role alongside its modern descendants, and reveal that an unrecognized failure to grasp lineage thinking is responsible for a family of conceptual flaws that afflict the professional, popular, and educational literature.

The story I tell in this book is not meant as an alternative to existing historical narratives. The histories of systematics and phylogenetics are intimately entwined, and much of what is relevant to the story told here is covered in the works mentioned earlier and in the expansive literature they represent. My intention is to provide a complementary story with a focus on the type of thinking that evolutionists use to visualize evolving lineages. From among the many sources that influenced my thinking, I want to mention four because together they provided much of the initial inspiration for writing this book.

First, Stephen Jay Gould's *Ontogeny and Phylogeny* (Gould, 1977b) was a challenging, but fascinating read for me as an undergraduate student about 25 years ago. It convinced me that spending a career thinking about evolution would be time well spent. It also taught me that research is greatly enriched when it is illuminated by the history of science, even though I have come to disagree fundamentally with some of Gould's conclusions.

Second, Peter Bowler's *Life's Splendid Drama* (Bowler, 1996) first showed me how evolutionists actually thought in the decades around the turn of the twentieth century in their efforts to understand the evolution of lineages. His study enticed me to engage with the historical literature myself, which set me on the path to writing this book. It remains one of the few modern books dealing with the early history of narrative phylogenetics.

---

[1] Lineage thinkers interpret cladograms as phylogenies or phylogenetic trees that depict the relationships of diverging lineages. See Chapter 11 for more discussion.

Third, when Michael Ghiselin kindly sent me a copy of his *Metaphysics and the Origin of Species* (Ghiselin, 1997), he introduced me to the individuality thesis, which holds that taxa and lineages are individuals, not classes. It greatly clarified my thinking, although readers of this book might find that opacities still remain.

Fourth, with *The Changing Role of the Embryo in Evolutionary Thought*, Ron Amundson (2005) excited me with his illuminating marriage of the history and philosophy of biology. He revealed that the explanatory ambition that ties nineteenth-century evolutionary morphology to modern phylogenetics and evo-devo research is the explanation of the evolution of form. With *The Changing Role of Ancestors in Evolutionary Thought*, I aped him for the initial working title of my book. Alas, like any evolving lineage, the book's title and flavor have changed substantially, while still maintaining a thread of conceptual continuity that ties what you are reading today to the first notes that I scribbled, now some years ago.

1.2 Outline of the Book

In Chapter 2 I trace the origin of the idea of evolutionary ancestors back to pre-evolutionary archetype concepts in the thinking of Johann Wolfgang von Goethe, Étienne Geoffroy Saint-Hilaire, and Richard Owen. Ancestors and archetypes were both used to explain unity of type, homology, and the origins of organismal form, but due to their metaphysical dissimilarity, they achieved this in fundamentally different ways. The archetypal thinking of these authors each illuminates a distinctive aspect of this explanatory strategy. I also diagnose and defuse a modern myth that has arisen about the views of Geoffroy Saint-Hilaire, which claims that he thought that the ventral surface of arthropods corresponds to the dorsal surface of vertebrates. When Darwin reinterpreted the archetype as an ancestor, evolutionary storytelling became possible, with hypothetical ancestors becoming the central subjects in phylogenetic narratives. But the emergence of ancestors as actors in evolutionary stories also required reconceptualizing the systematic relationships between taxa to provide pathways along which evolutionary stories could flow.

In Chapter 3 I outline the transformation of systematics into phylogenetics by tracing the emergence of lineage thinking. One of the routes to a realist interpretation of the natural system of systematic relationships was to temporalize it. Lineage thinking emerged when the previously atemporal and symmetrical affinity relationships between collateral relatives were replaced by asymmetrical ancestor–descendant relationships that tracked the arrow of time. This transition was accompanied by a rapid decrease in the diversity of shapes of affinity diagrams published in the systematic literature, and it marked a shift from predominantly reticulating or web-like systems to tree-like figures soon after the publication of Darwin's *On the Origin of Species* in 1859. I argue that this graphic revolution largely records the influence of evolutionary expectations, as biologists redrew their diagrams to fit the theoretical dictates of Darwinian descent with modification. Whereas previous reticulating diagrams had recorded tangles of character conflict, the new evolutionary trees depicted the flow of diverging lineages composed of ancestors and descendants. The current swell of enthusiasm for evolutionary networks has driven several recent authors to the peculiar argument that even Darwin disliked the tree of life as an evolutionary metaphor, an argument I will refute. Reconceiving the systematic relationships between taxa as phylogenetic pathways along which body plans evolve had an epistemic corollary as well. The empirical trinity of observation, description, and comparison that had been at the heart of natural history and systematics could not provide access to the invisible realm of ancestors. Speculation became a necessary tool for the evolutionary storyteller, and few employed it so deftly as the founder of phylogenetics.

In Chapter 4 I take a detailed look at the evolutionary storytelling of Ernst Haeckel. He founded phylogenetics as the science dedicated to tracing the evolution of lineages. Although Haeckel's phylogenetic scenarios were nourished from a broad buffet of evidence, the biogenetic law was his favorite shortcut to create lineages of hypothetical ancestors, most famously the tiny cup-shaped Gastraea. A recent consensus has emerged that stigmatizes Haeckel's phylogenies as un-Darwinian constructs that are conceptually stained by teleological thinking and the linearity of the *scala naturae*, the idea that all of God's creations can be arranged in a single, static, ascending chain of being.[2] Instead, I argue that his trees are fully Darwinian and that the linearity present in his trees and thinking is the linearity of evolving lineages that track the arrow of time. Lineage thinking was novel when Haeckel started writing, and his was marred by imperfections. It was up to the following generations of evolutionists to resolve the conceptual tension between the linear and branching aspects of evolution, which is a major theme of this book and an ongoing struggle in today's literature.

In Chapter 5 I discuss the anatomy of evolutionary storytelling. Historical narratives are woven around central subjects that lend them continuity through time. The central subjects of phylogenetic scenarios are lineages of hypothetical ancestors. These define the pathways of homology along which evolutionary change is reconstructed, and they root the power of phylogenetic hypotheses to explain the evolution of form by allowing characters in descendants to be traced back to ancestral precursors. *De novo* origins of novel traits are chinks in the explanatory armory of phylogenetic hypotheses. Hypothetical ancestors have therefore often been deliberately equipped with characters that provide suitable precursors of the traits that await evolutionary explanation. I will argue that the precursor potential of hypothetical ancestors functioned as an early phylogenetic optimality criterion used in the construction and judging of scenarios.

In Chapter 6 I survey how biologists and paleontologists have used their imagination and evolutionary intuitions to animate phylogenetic narratives. Before outgroup comparison and ancestral state reconstruction methods became available, many authors intuited the direction of character evolution from clues provided by the threefold parallelism. The form changes that can be observed during ontogeny, and those inferred from the stratigraphic sequence of fossils, have inbuilt time axes that can be used as shortcuts for proposing lineages of changing forms. But interpreting the polarity of character change suggested by

---

[2] Unless noted otherwise, all translations from non-English sources into English are my own.

the systematic leg of the threefold parallelism was less straightforward. Many researchers intuited or imposed the direction of evolutionary change by proposing a smoothly transitional linear series of observed or imagined forms. Some evolutionary intuitions date back to pre-evolutionary times, while others emerged from the search for evolutionary laws, such as Cope's rule of phyletic size increase. Unsurprisingly, evidence and imagination were typically unequal partners in the construction of scenarios, with the latter leading the former wherever it wanted. But when the first molecular phylogenies were published, researchers often muted their once vocal evolutionary intuitions to accept sometimes deeply puzzling results. The complete suppression of the evolutionary imagination is associated with the flawed, but widespread strategy of using observed differences between taxa as prima facie evidence against monophyly and homology. Not asking oneself if such differences could be due to descent with modification rather than convergent evolution signals the death of lineage thinking and diverts phylogenetic debates into fruitless avenues.

In Chapter 7 I take a close look at the early lineage thinking of paleontologists and their attempts to infer evolution from stratigraphic sequences of fossils. Linear thinking with fossils emerged as a core component of the traditional paleontological method in the second half of the nineteenth century. It was taken to extremes by orthogeneticists, and I use the late-nineteenth-century debate between Franz Hilgendorf and Alpheus Hyatt about the famous freshwater Miocene Steinheim snails as a case study for looking at competing forms of lineage thinking. It was in fact the straight-thinking orthogeneticist Othenio Abel who helped resolve the conundrum of how branching evidence can shed light on the evolution of linear lineages.

In Chapter 8 I examine one of the earliest debates about animal body plan evolution. Ernst Haeckel, E. Ray Lankester, Francis Maitland Balfour, Elie Metschnikoff, and Otto Bütschli were the main participants in an international debate about the origins of animals that was triggered by the publication of Haeckel's Gastraea theory in the 1870s. Each author proposed a different hypothetical animal ancestor, which with the exception of Bütschli's, were the products of recapitulationist reasoning. Each of these hypothetical creatures stood at the beginning of a unique scenario with a distinctive explanatory texture, and many of them can still be found in the pages of zoology textbooks today. This late-nineteenth-century clash of scenarios is representative of narrative phylogenetic debates generally. It shows how unique evolutionary stories are produced by authors wielding their personal evolutionary intuitions in the context of unequal attention to available evidence. Unsurprisingly, disagreements quickly became entrenched as dogma, but strikingly, several of these early scenarios, as well as their descendants, continue to inform debates today.

In Chapter 9 I chronicle the emergence of different phylogenetic traditions in invertebrate zoology that coalesced around attractive hypothetical ancestors. The first half of the chapter discusses scenarios for the origin of Bilateria that emerged in the late nineteenth and early twentieth centuries, including the enterocoel and archicoelomate theories, and the importance of the contested character of amphistomy for these scenarios. These scenarios and their associated hypothetical ancestors became phylogenetic totems in different parts of the world, with Gastraea at the core of the European zoological tradition, and Phagocytella rooting both the Russian and American traditions. In the mid-twentieth century, an alternative theory was proposed that derived bilaterians directly from ciliate ancestors. Remarkably, this attraction to ciliate ancestors emerged three times independently in quick succession, which illustrates the epistemic importance accorded to the precursor potential of hypothetical ancestors in narrative phylogenetic debates. The second half of the chapter discusses why many authors have felt so strongly attracted to annelid-like ancestors. These were proposed to have crawled at the cradles of many taxa, including molluscs, arthropods, vertebrates, and Bilateria. The arguments used to promote annelid-like ancestors form a conspicuous strand in the history of narrative phylogenetics that can still be traced today.

In Chapter 10 I show that although hypothetical ancestors largely lost their epistemic power as deliberately constructed phylogenetic tools after the spread of modern phylogenetics, narrative phylogenetic reasoning persists. A conspicuous and widespread example of employing narrative shortcuts in evolutionary storytelling today is the attempt to use the phylogenetic position of taxa to predict the presence of ancestral character states. In close analogy to the historical use of the label "lower" to designate taxa that are presumed primitive, modern authors use a range of evocative adjectives to promote the presumed presence of ancestral character states in their favored taxa. "Basal" is the most frequently used label, and many authors think that its link to the phylogenetic position of taxa gives it predictive power over whether these are likely to have retained ancestral states. I explain that phylogenetic and evolutionary theory provide no convincing rationale for this argument and argue instead that basal taxa can be especially misleading about the nature of their character states.

In Chapter 11 I address common flaws in lineage thinking that result from confusing the branching relationships between collateral relatives in the realm of systematics with the linear relationships between ancestors and descendants in the realm of evolutionary descent. The influential voices of the late Stephen Jay Gould and Robert O'Hara, who dubbed the now ubiquitous phrase "tree thinking," have warned readers for decades against the sins of linear evolutionary storytelling and the use of linear evolutionary imagery, with great success. Their works are widely cited and promoted in the professional, popular, and educational literature, but I argue that their impact has been deeply pernicious. Their writings fundamentally misconstrue the relationship between the branching realm of systematics and the linear realm of evolving lineages. Furthermore, I argue that a similar failure to understand lineage thinking is at the heart of the fruitless debate about paraphyletic ancestral taxa. It burned brightly for a while during the early history of cladistics, but it continues to throw out sparks occasionally today. I close with a discussion of the problem that, in the absence of a vocabulary designed to talk about lineages, we are forced to discuss them in the taxic language of systematics. This inevitably causes problems.

Finally, in Chapter 12 I offer a concluding meditation on the inescapability of evolutionary storytelling, if our goal is to go beyond the trees and try to understand what may have happened along life's myriad evolving lineages. As soon as you look at a tree and ask "What does this mean?," you become an evolutionary storyteller. Some dismiss such stories as pure fiction and would prefer not to go beyond the tree. But for those of us who wish to peer over the phylogenetic frontier, these stories are how we generate our understanding of evolution, or at least how we try to.

2

From Archetypes to Ancestors

2.1 Ideal Beginnings?

Every evolutionary story features ancestors. They are the sole actors in every evolutionary drama, but before they could assume their role in explaining the origin of biological diversity, the basic unity of biological organization itself had to be recognized. Although the notion of life's unity is now entrenched even in popular culture – at the time of writing you can find deep homology on Facebook and Urbilateria on Twitter – this insight was anything but straightforward prior to Darwin.[1] As Ospovat (1981: 97) noted, the intellectual effort required to formulate the idea of unity of type far outstripped that needed to make it an integral part of the new evolutionary worldview. And once common ancestry was established as a fundamental evolutionary principle, it could legitimize the further search for unity of type. Without the a priori context of common ancestry, however, unity of type was far from an obvious phenomenon.

Thinking about types and archetypes as models, patterns, standards, norms, and ordering categories goes back to antiquity (Van der Hammen, 1981). During the eighteenth and nineteenth centuries, such thinking assumed extraordinary importance at the disciplinary cradle of comparative morphology. Capturing the unity of plan that could be seen shimmering through nature's skin of diversity was a central preoccupation of what is variously called idealistic, philosophical, transcendental, or typological morphology or anatomy[2] (Russell, 1916; Farber, 1976; Rehbock, 1983; Amundson, 2005; Rieppel, 2016).

---

[1] Throughout this book, I use Darwin's evolutionary theory as a conceptual marker to distinguish pre- and post-Darwinian ideas and events. I often use it as shorthand for particular components of Darwin's theory, which the context of each reference should make clear. Here, as in many other places, the relevant idea is that of common descent. A reader of this chapter suggested I replace Darwin with Wallace-Darwin everywhere in the book because Wallace deserves recognition for his co-discovery of the basics of modern evolutionary theory. Indeed he does. But in the interest of brevity, I have decided against this. Wallace himself referred to their joint theory as Darwinism and its worldview as Darwinian (Wallace, 1891). I do discuss Wallace explicitly where necessary.

[2] Throughout this book, I use the term morphology as a general descriptor for the science of form and structure, irrespective of the precise disciplinary and institutional settings of the research, which can include zoology, botany, or anatomy.

This originally pre-evolutionary mode of thinking was most fully and most extremely developed as part of the *Naturphilosophie* movement in Germany, but types (archetypes, *Typen, Urtypen, Urformen, Haupttypen, Urbilden*, etc.) were also recognized by comparative anatomists working in France and England (Farber, 1976; Lenoir, 1981; Rehbock, 1983; Trienes, 1989; Richards, 2002; Gliboff, 2008).

Until the first decade of the new millennium, the dominant historiography labeled all or most pre-Darwinian idealistic morphologists as essentialists who considered species and other taxa to be immutable types, each defined by a set of necessary and sufficient characters. It was not until Darwin's wholesale transformation of the metaphysics of biology – a shift to population thinking that accorded importance only to the uniqueness of individuals, not the constancy of types – that comparative morphology could escape from the conceptual stranglehold of essentialism. Consequently, the label of essentialism has long tainted all typological thinking as deeply antithetical to the evolutionary worldview. This influential version of history, of which the late Ernst Mayr is the principal architect, has been called the "Essentialism Story" (Winsor, 2003, 2006; Amundson, 2005). Mayr didn't hide his low opinion of the accomplishments of pre-evolutionary comparative morphologists when he wrote that "[w]hat satisfaction idealistic morphology gave was primarily esthetic" (Mayr, 1982: 459).

However, recent work by historians and philosophers has exposed the Essentialism Story to be a self-serving myth used to promote the conceptual novelty and superiority of the Modern Synthesis, a theoretical edifice that Mayr helped to build (Amundson, 1998, 2005; Winsor, 2003, 2006; Levit and Meister, 2006; Wilkins, 2009; Witteveen, 2018). These revisionary works echo earlier historical scholarship (e.g. Farber, 1976; Lenoir, 1987), and offer more sympathetic readings of typological thinking by pointing out that there had been no single typical or dominant type concept, and that morphological types received a diversity of interpretations. Moreover, these authors argue that the typological thinking of idealistic morphologists didn't necessarily imply a commitment to an essentialist metaphysics that was fundamentally incompatible with, or even obstructed, the emergence of an evolutionary worldview. Instead, types functioned as heuristic devices for thinking about the nature and origin of organismic diversity, and this epistemic assistance helped pave the way for modern evolutionary thought. Although explaining form was not one of Darwin's main goals, the work of idealistic morphologists furnished him with invaluable evidence about the homologies and systematic relationships between organisms. Their discoveries provided the substrate that Darwin used for establishing the reality of evolutionary descent. Darwin's distinctive contribution was to transform the type into an ancestor through an act of metaphysical metamorphosis. In my view, this act, more than any other, planted the flag that demarcates the beginning of the Darwinian worldview. This transformation marks the symbolic, conceptual, and historical beginning of modern evolutionary thought.

To sketch the outlines of this metaphysical transformation, I will present the archetypal thinking of three morphologists from Germany, France, and England: Johan Wolfgang von Goethe, Étienne Geoffroy Saint-Hilaire (Geoffroy hereafter), and Richard Owen. They were not the first, and certainly not the only, morphologists whose work was part of the flow of ideas that fed into the development of the evolutionary worldview. However, their late-eighteenth- and early-nineteenth-century thinking about archetypes was extraordinarily influential in the development of comparative morphology, and their works continue to be cited today. My aim is to show how their search for the deep unity hidden underneath nature's diversity, and their different ways of conceiving of organismal relationships, illuminate important steps in the conceptual evolution of ancestors. All three conceived of archetypes in different ways as epistemic tools to help them discover and understand nature's unity in diversity. Goethe's approach was intuitive and dynamic. He imagined archetypes to be a continuous flow of forms in which all observable and imaginable diversity was embedded. Geoffroy's more analytical approach sought to discover the archetype in the unity of the structural composition of organisms, which he attempted to grasp with his principle of connections. Owen's archetype achieved a level of conceptual refinement that laid the perfect foundation for the wholesale reinterpretation of the archetype as ancestor. Taken together, these archetypal views provide complementary perspectives on the development of the explanatory apparatus that made evolutionary morphology the science of phylogenetics.

2.2 Johann Wolfgang von Goethe: Imagining Archetypes

When I started reading the pre-Darwinian literature to search for clues about the origin of the concept of evolutionary ancestors, I had no idea how challenging that task would turn out to be. The mind of the modern biologist so snugly fits the mold of Darwinian metaphysics that trying to understand the conceptual world of someone like Goethe or Geoffroy almost requires the mental equivalent of an out-of-body experience. The dangers of misapprehension lurk in every sentence, and even familiar, but heavily concept-laden words such as "evolution," "archetype," and "plan" can become traps of misunderstanding. Some concepts are so subtle, or so alien, that they could only be appropriately expressed in a specific language. Richard Owen dedicates the first four pages of *On the Nature of Limbs* to explain why his book should really have been titled *On the Bedeutung of Limbs*. And the holistic thinker Goethe mocked Geoffroy's use of the term "composition" because it reduced the structure of animals to being just the sum of its parts, "als ob es ein Stück Kuchen oder Biscuit ware, das man aus Eiern, Mehl und Zucker zusammenrührt!" ("as if it was a piece of cake or biscuit, mixed together from eggs, flower and sugar!") (Goethe in Eckermann, 1908: 329). I have found few things as challenging as properly understanding the biological thinking of Goethe.

Stephen Jay Gould included a section titled "Judgments of Importance" in his 2002 magnum opus that singled out some works of historical scholarship for their near oracular accuracy (Gould, 2002b: 163). For the science of morphology, Gould commended Russel's 1916 book *Form and Function* for its "brilliance and justice of its characterizations." If we were to heed Gould's words and consult Russell for the final verdict on Goethe's achievements in biology, we would have to conclude that they had not amounted to much. Russell sandwiches his short and scathing chapter on Goethe between two patronizing indictments of both the man – "little more than a dilettante" – and his work – "[s]uch an interpretation of the unity of plan reaches perhaps beyond the bounds of science" (Russell, 1916: 45, 51). Strictly speaking, neither of these judgments is inaccurate. Goethe was indeed an amateur scientist, and his methodology of gaining insight into nature's unity of plan certainly does not map neatly onto any cardboard version of the scientific method currently used in the twenty-first century.

2.2.1 An Aesthetic, Intuitive Appreciation of Nature

It may not be too surprising that nature's unity of plan first became most forcefully apparent to the expansive sensibilities of a poet. Biologists of course remember Johann Wolfgang von Goethe (1749–1832) for being a founder of the field of morphology and for giving the discipline its name. But Goethe did not think like a modern scientist and his insights about the unity of plan were not the result of a series of detailed, systematic, empirical studies of nature's diversity, even though he did do original research on both plant and animal morphology. Instead, he followed Spinoza's application of *scientia intuitiva*, the innate intuitive understanding, to grasp the essence of things (Richards, 2002: 379, 439). By applying his poetic and aesthetic sensibilities, Goethe reached his conclusions largely intuitively. At this point, we should note that it is difficult for modern readers – at least for me – to really understand Goethe's thinking. Given the complexities of the philosophical milieu of which Goethe was a part, the inconsistencies and ambiguities in Goethe's own published statements, and the enduring disagreements between modern commentators on central aspects of Goethe's thinking – Is Goethe a formalist? Is Goethe's conception of the archetype Platonic? Did Goethe invoke a *Bildungstrieb* to animate his archetypes? – my sketch of Goethe's thinking is inevitably somewhat crude. I advise the reader who wishes to know more to consult the secondary sources from which my own account draws so heavily. Robert Richard's wonderful book *The Romantic Conception of Life* (2002) is a good place to start.

Russell (1916: 46) and Rehbock (1983: 19) claim that Goethe merely aligned the facts of comparative anatomy with his preconceived conviction of unity of type. It is indeed true that Goethe did not convince himself about the unity of type until after a long process of empirical study. He did provide an empirical basis for his conclusions, but his epistemological route is distinctly foreign to modern scientists. Observing the variety of forms in nature, Goethe became intuitively aware of the deep unity underlying their differences. This intuitive perception, or *Anschauung* as Goethe called it (Ebach, 2005), reveals the hidden relationships between organisms in that we can conceive of different forms being able to metamorphose or transform into each other. The unity of ostensibly dissimilar forms is revealed in that hypothetical intermediate forms might be imagined that could connect them into a seamless series of forms, the extremes of which may look entirely different. Goethe did not conceive of such metamorphosis as a process of genealogical change over time, but as something that can only be seen by the mind's eye.

This mental experience, where unity is revealed to "der speculative Geist" ("the speculative mind") (Goethe in Lenoir, 1987: 23), and of becoming conscious of the existence of a pattern of relationships between forms, is what constitutes the archetype. The archetype is therefore not identifiable with any particular manifestation in nature, nor is it merely the sum of all the forms that belong to an archetype. The archetype should be seen holistically

Figure 2.1 A graded series of leaves taken from a single stem of the meadow buttercup *Ranunculus acris*. It illustrates Goethe's conception of the archetype, with each form representing a snapshot in a continuous flow of form
(after figure 1 on https://netfuture.org/2005/Jul0505_164.html, last accessed February 21, 2022, with permission from Stephen Talbott).

and dynamically: It includes all forms that belong to it that we can observe in nature, but it also includes all imaginable forms that could conceivably belong to it, plus the relationships between all these real and imagined forms. For Goethe the archetype was a mental flow of form (Figure 2.1). To use his own words, whereas an existing organism or part thereof exhibits a certain *Gestalt* – a static abstraction of the archetype – what he really aims to capture is *Bildung*. This designates both the dynamic process of mental transformation between gestalts and the products of this process. It is this putting of forms into motion via imagined metamorphoses that Goethe scholar Ronald Brady (1984: 340) labels as Goethe's "major contribution." For Goethe, morphology was the study of process rather than product (Brady, 1984: 334). Goethe admitted that this dynamic relationship between parts and wholes had a somewhat "esoteric quality" to it: "It can be expressed in its totality, but not proven, the singular can illustrate it but not completely capture it" (Goethe in Breidbach, 2006: 33).

A modern analogue that helps capture the essence of Goethe's dynamic conception of the archetype is a theoretical morphospace that encompasses all existing and imaginable gestalts, all of which are permutations of a basic form, such as Goethe's archetypal leaf (Riegner, 2013). This morphospace can be traversed via metamorphosis of a basic archetypal form (Figure 2.1).

This intuitive process of understanding the unity of nature's diversity was for Goethe intimately related to the intuitive, aesthetic appraisal of art forms. One's aesthetic intuition could reveal the essence of a work of art as surely as it could reveal the unity of forms in nature (Steigerwald, 2002; Richards, 2006). This intuitive experience of unity does not just become apparent to the mind's eye; it is in fact created by it. Goethe dealt with the apparent problem of the distinction between the thing as such and the mind's construct of it by denying there was a distinction. The observer is not just a reporter of knowledge, she is the creator of it. Things simply are as they appear to be.

2.2.2 The Ontology and Role of Goethe's Archetypes

The Goethean archetype is an intuited pattern of relationships between a diversity of forms that could logically metamorphose into each other. Since a Goethean archetype can encompass an infinity of forms, both actual and imagined, none is more typical than any other. Goethe did think that forms composed of more dissimilar parts were more complete or "ideal" than morphologically less differentiated ones, but all forms, irrespective of their degree of differentiation, were equally representative of the archetype – and, as gestalts, were equally unable to capture its totality. The essence of Goethe's dynamic perception of form is not a conserved morphological element that can be modified, but it is the continuity of the transformation of forms that comprises the archetype. Because the Goethean archetype is revealed only to the mind and not to the senses, it does not have a specific form or composition. You wouldn't be able to meet an archetype. This may be why Goethe mocked Geoffroy's use of the term "composition" in the latter's famous 1830 debate with Cuvier at the Académie des Sciences in Paris. For Goethe the archetype transcended any specific composition, while for Geoffroy a type was defined by a specific composition. Hence, it could not be represented or depicted in a drawing of a single object, although drawings could be used to illustrate the idea (e.g. Figure 2.1). Turpin's oft-reproduced drawing of Goethe's archetypal plant or *Urpflanze* (Breidbach, 2006: 107) cannot be a Goethean archetype because the very act of drawing it subtracts its universality by freezing it into a particular form (and as Breidbach points out, because it erroneously suggests that the archetype is the simple summation of all imaginable morphological forms, rather than the dynamic relationships that connect different forms via pathways of metamorphosis).

In his earlier thinking about archetypes, Goethe famously fantasized about finding the *Urpflanze* as a real plant in nature, but as his thinking evolved, he came to realize that he was trying to grasp an idea rather than a thing (Richards, 2002: 2, 395–396; Breidbach, 2006: 31–34). Similarly, Goethe's archetypal leaf is not a real leaf but an abstract form or primordial organ, without any definable composition.

Goethe's archetype – unlike the types of Geoffroy and Owen, as we shall see later – did not function in establishing a mature concept of homology. However, it did have heuristic value for interpreting the morphological variation observed in nature (Brady, 1984; Ebach, 2005; Breidbach, 2006). If a form can be imagined to fit into the series of forms comprising a certain archetype – defined by the flow of forms between the archetype's gestalts – it can be concluded to belong to that archetype. The archetype could thus function as an aid in classification. Accordingly, the archetype could be seen as a generating, productive unity, "a dynamic force actually resident in nature" (Richards, 2002: 440), a *Bildungstrieb* (Richards, 2002: 447) responsible for the lawful relationships between forms. Goethe thought that the actual forms of species resulted from the interaction between two kinds of forces: an internal *Bildungstrieb* and external, environmental forces (Richards, 2002: 447). The metaphysical status of the archetypal force remains somewhat vague. Brady (1984: 339) likens it to an Aristotelian *formal* cause, "like an *idea* in nature ... by which nature forms its productions," whereas Asma (1996: 44) sees it more as a conceptual blending of Aristotelian formal and efficient (the agent of change) causes. Yet, no matter how one interprets Goethe's dynamic archetype, it is clear that it was meant to represent more than a merely descriptive summary of the variety of form; it symbolized the lawlike relationships between forms, a "law-like pattern" (Richards, 2002: 450) capturing both their unity and endless variations.

Some see Goethe's morphology as the bona fide cradle of systematics (Ebach, 2005; Williams and Ebach, 2008). Others see it as a pseudoscience rooted in occult metaphysics that spawned little more than a "formalistic and sterile" morphological tradition (Ghiselin, 2000b: 46–47). What is beyond doubt is that Goethe was among the first to systematically ponder the unity of type hidden beneath the morphological diversity of forms in nature. He approached and solved the problem intuitively. My friend and Goethe enthusiast Malte Ebach sums up this approach: "Things are as we experience them, they need no further explanation" (Ebach, 2005: 266). Such an immediately intuitive and aesthetic approach to nature in a scientist is as uncommon today as it was when Russell penned his derisive chapter on Goethe. A real scientific approach, Russell expects us to think, should transcend intuition and instead apply explicit analytical thinking to a dense crop of empirical data. Indeed, Goethe thought more like a poet or artist than a modern biologist, but I think a strict dichotomy between intuitive and scientific thinking is fictional.

Goethe's intuitive approach to the study of unity of type was a precursor to more explicit scientific reasoning, both historically – as we shall see in the rest of this chapter – as well as mentally. Bowler (2000) reminds us that many of our deepest convictions, including our overall worldview or metaphysics, often originate as more or less fully formed intuitions that we only subject to explicit, rational reasoning when we are challenged to defend them. As all practicing scientists know perfectly well, this holds true with equal power for the many hunches or gut feelings that drive our daily research. In later chapters, we will see that much of what we think we know about the history of life today is still beholden to our often unexamined evolutionary intuitions. Without intuition, scientific research will grind to a halt. Fortunately, there was no danger of this happening in the career of Geoffroy.

2.3 Geoffroy Saint-Hilaire and Unity of Plan

To transform the unity of type into the foundation of an evolutionary worldview involved a radical ontological and epistemological revision. Unity of type, as Darwin and his intellectual descendants came to understand it, was a "fundamental agreement of structure, which we see in organic beings of the same class, and which is quite independent of their habits of life" (Darwin, 1859: 206). Hence, influential worldviews that considered form and function to be inseparably linked – such as the teleology of Cuvier and the nineteenth century British natural theologians (Ospovat, 1981; Amundson, 1998) – hardly favored the recognition of the primacy of structural unity. Moreover, it was equally challenging to bring unity of type down to earth, and to conceive of it in purely naturalistic and materialistic terms. Prevailing views often saw nature as being animated and regulated by God-given primary or final causes well into the nineteenth century. What was needed to effect such a radical conceptual shift was an iconoclastic thinker unafraid to oppose orthodoxy and willing to heed the famous future dictum of T. H. Huxley: "logical consequences are the scarecrows of fools and the beacons of wise men."

2.3.1 Unpromising Beginnings

In 1796 a young Étienne Geoffroy Saint-Hilaire (1772–1844) wrote in a memoir on lemurs that "[i]t seems that nature is confined within certain limits and has formed all living things on only one single plan" (Geoffroy in Le Guyader, 2004: 3). Although this idea was scarcely more than a fantasy at this point, it would quickly develop into the coordinating vision of Geoffroy's career. Given the inauspicious circumstances at the start of his career, it seemed distinctly unlikely that Geoffroy would be able to develop his intuitions into a powerful and innovative research program and the lasting pillar of his scientific fame. Geoffroy's tenure as a professor in vertebrate zoology at the Muséum d'Histoire Naturelle in Paris had started barely three years previous. In fact, he had initially declined the chair because he thought he lacked the qualifications needed for the post[3] (Appel, 1987: 11). He had no previous experience in either research or teaching in zoology – his only scientific training at the time was in mineralogy – and these deficiencies were anything but offset by the material resources made available to him upon accepting the post: "a few skeletons" and a "pitiful collection of poorly stuffed animals" (Appel 1987: 22). Luckily, empirical limitations didn't pose much of an obstacle to this philosophical anatomist.

In his book *The Philosophical Naturalists*, Rehbock (1983) summarized what he considered to be the characteristic features of pre-Darwinian philosophical naturalists in Britain and continental Europe: a search for the laws of nature, a concomitant dissatisfaction with teleology and the invocation of final causes, and a search for patterns that could be discovered by a "philosophical" or "transcendental" approach, defined by "intuitive leaps of the imagination, rather than by coaxing from impersonal heaps of data" (Rehbock, 1983: 9). Even though one cannot group all pre-Darwinians together on the basis of a shared set of epistemological and ontological commitments, these characteristics nonetheless capture the core of Geoffroy's scientific method. Geoffroy's idea of doing science was not to confine oneself to the generation of morasses of descriptive detail and

---

[3] Geoffroy was offered his post only because he had already been employed for a short time in a minor role at the Jardin des Plantes (Jardin du Roi) before the founding of the museum and because more suitable candidates had moved from Paris (Appel 1987: 21). The same was true for Jean Baptiste Lamarck, who was offered a chair in invertebrate zoology in the museum on the basis of little more than being friends with some experts on invertebrates and possessing a shell collection (Appel 1987: 19).

to keep theory so firmly tethered to "positive facts" (Appel 1987: 6) that it could never play a more prominent scientific role than providing the wavering outlines of inductive general-izations. He considered an "infatuation with details" (Geoffroy in Le Guyader, 2004: 29) and the slavish accumulation of facts to be the hallmarks of a stunted science. Unfortunately, the man whose scientific views came to represent "zoological orthodoxy" (Appel 1987: 51) during Geoffroy's career saw these supposed shortcomings as great virtues.

That man was, of course, Georges Cuvier (1769–1832), Geoffroy's erstwhile friend and collaborator, and later his fiercest intellectual adversary. Cuvier's scientific ideal, in both his zoological and geological work, was to adhere as strictly as possible to the straight and narrow of factual details, suppressing, whenever possible, the urge to speculate (Appel, 1987: 46–53). His contempt for unrestrained speculation in science was nowhere more clearly on display than in his "devastating critique" (Rudwick, 2008: 314) of geotheorizing. The genre of geotheory, which flourished in the seventeenth and eighteenth centuries, encompassed a diverse set of attempts to bring together all the Earth's physical and biotic phenomena under a single comprehensive explanatory system or theory, drawing only on natural causes (Rudwick, 2005). A successful geotheory was expected to provide satisfying explanations for both local and global phenomena – say the location of a particular mountain range in the Alps or the general origin of igneous rocks found anywhere in the world. Conspicuous among these were the two influential geotheoretical systems, pub-lished in 1749 and 1778, devised by Georges-Louis Leclerc, Comte de Buffon (1707–1788). Being of the opinion that good geology should adhere as closely as possible to "positive facts" and observations, Cuvier labelled such speculative systems as mere "castles in the air" (Rudwick, 2005: 459). He damned Buffon's and others' imaginative and conjectural systems with faint praise, calling their speculative tendencies a "mark of genius" (Rudwick, 1997: 46) and singling out the "eloquent manner" in which Buffon had presented his systems (Rudwick, 1997: 47), rather than their content. Far from sharing these sentiments, Geoffroy fully embraced the intellectual spirit of Buffon (Appel, 1987: 22), even though synthetic system building had fallen increasingly into disrepute from the late eighteenth century, in no small measure helped by Cuvier's disparaging and sneering remarks (Appel, 1987: 47, 168–169; Burkhardt, 1995: 39, 196–198; Rudwick, 1997: 46–47, 82–83, 123, 176–178; 2005).

2.3.2 Geoffroy's Radical Epistemic Orientation

Geoffroy did things differently than many of his predecessors and contemporaries. He focused not on the determination of differences between organisms – the chief modus operandi of descriptive morphologists and taxonomists since at least the Renaissance (Ogilvie, 2006: 191) – but instead on the similarities shared between species. This epistemic reorientation was crucial for attaining his goal of discerning the deep unity underlying nature's diversity, and later the establishment of common ancestry. However, the identifi-cation of shared structural similarities required the abstraction of taxon-specific functions. As pointed out by Amundson (2005: 57), this abstraction of function was regarded as "inferential and epistemologically more suspect" in the early nineteenth century. Since the functional individuation of structures was considered to be "empirically simple and direct" (Amundson 2005: 57), it was precisely this process of probing behind appearances that Geoffroy labeled "philosophical" or "transcendental," or what is called "idealistic" or "speculative" in the secondary literature (Rehbock, 1983; Appel, 1987; Asma, 1996; Amundson, 1998, 2005). Structural differences between taxa and traits needed to be abstracted as well to reveal the unity underlying diversity. Differences can be directly observed, but detecting similarities requires abstraction.

Geoffroy's conceptual innovations departed from the epistemological norms of the time, which were represented, and had in no small measure been constructed, by Cuvier. This was true not just on the continent, but also in England with its tradition of Paleyian natural theology focused on adaptations (Amundson, 1998). Yet, the central elements of Geoffroy's new epistemology were widely adopted by later anatomists and zoologists, notably through the influence of nineteenth-century radical thinkers in Edinburgh and London, such as Robert E. Grant (1793–1874) and Robert Knox (1793–1862) (Desmond, 1989). The search for homologues became the cornerstone of later comparative biology and phylogenetics. For Geoffroy himself, the focus on structural correspondences between species furnished the epistemological outlines within which he developed his influential and controversial research program.

The form-function[4] dialectic has occupied a central place in the historiography of morphology from Russell (1916) to Asma (1996), Gould (2002), and Amundson (2005), to name just a few broadly conceived studies spanning the last century. The most common historical tactic compares Geoffroy's method to that of Cuvier, a strategy used by Geoffroy himself in the opening pages of his 1830 *Principes de Philosophie Zoologique* (Le Guyader, 2004: 108–116; Le Guyader, 2004 provides an English translation of the full text of this work). It is also the focus of the excellent study by Appel (1987). Geoffroy throws his own views into sharp relief by labelling Cuvier an exemplar of the "old" and "ancient" method of comparative anatomy that emphasized the interrelationship of form and function. Indeed, Cuvier accorded explanatory primacy to function. This simultaneous consideration of form and function was typical of a long tradition of anatomical investigation going back at least to Aristotle (Russell, 1916), in whose thinking on final causes we find the seed for the influential concept of teleology.

Geoffroy, in contrast, tried to establish a science of "pure morphology" (Russell 1916: 52). He strove to "escape the seductive influence of forms and functions" (Geoffroy in Le Guyader, 2004: 111), judging both to be merely ephemeral adornments of an underlying fundamental structural unity expressed in the parts of organisms. His method consisted of abstracting the specific form (or shape) and function from structure to yield "an isolated object that the principle of connections illumines by its torch, and which invariably, despite its possible modifications, retains the fact of its primitive essence, its philosophical character of a uniform composition" (Geoffroy in Le Guyader, 2004: 111). In his *Principes de Philosophie Zoologique* Geoffroy was clear: "in the investigation of analogies [his term for

---

[4] In the context of Geoffroy's work, "form" should be understood as morphological structure rather than mere shape, the fundamental structural unity that underlies a variety of forms and functions.

what we now call homologues], I proposed rejecting considerations drawn from forms and from functions. Forms, I said, are ephemeral from one species to another; this view applies with further scope to functions" (Geoffroy in Le Guyader, 2004: 158).

To go beyond strict observation and to consider structure "in the abstract" (Geoffroy in Le Guyader, 2004: 112) was a radical departure from the norm of individuating parts of organisms along functional fault lines. Cuvier was never able to comprehend this. For him the vestigial remains of a pelvis in cetaceans merely signified a "slight resemblance" to animals with fully developed pelvises (Le Guyader, 2004: 142), not unity of composition, because the observed differences belied "identity." Cuvier's strict empirical conservatism would not allow the mental abstraction needed to document unity of composition. He thought that scientists should busy themselves primarily with the gathering of "positive facts" (Appel, 1987: 6, 138, 139) and should keep in check the urge to generalize and speculate beyond the boundaries of descriptive detail. It would become one of Geoffroy's signal successes to show that, far from fatally crippling the study of comparative anatomy, the process of mental abstraction actually allowed more penetrating insights into the structural unity of organisms than hitherto thought possible. He initially even managed to impress the skeptical Cuvier with the results of his philosophical anatomy (Appel, 1987: 90–91).

The second radical aspect of Geoffroy's epistemic orientation that I want to draw attention to is so deeply ingrained in the thinking of modern comparative biologists that it is odd to realize that there was once a time during which this perspective was far from obvious: an express focus on structural similarities shared between organisms. It is in large measure due to Geoffroy that we look to shared similarities or correspondences between taxa as the only available source of evidence for putative homologues, irrespective of whether "we"' denotes late-nineteenth-century evolutionary morphologists, such as Anton Dohrn and Carl Gegenbaur; early-twentieth-century idealistic morphologists, such as Adolf Naef and Rainer Zangerl; or modern phylogeneticists. He noted a widespread "infatuation with details" (Geoffroy in Le Guyader, 2004: 29) in the work of predecessors and contemporaries, which was in fact "a predilection for the study of differences" (Le Guyader, 2004: 158). This myopic focus on differences, Geoffroy claimed, had led to the unnecessary development of separate vocabularies for the veterinarian, the ichthyologist, and the surgeon. Trimming this nomenclatural excess, with identical names identifying different parts in different animals, and different names identifying corresponding parts in different animals, was also a driver of Richard Owen's work on the vertebrate archetype (see introduction in Owen, 1848). Instead of reveling in the fundamental correspondence of tetrapod limbs, for instance, "the monographic naturalists" (Geoffroy in Le Guyader, 2004: 29) chose to emphasize the distinctness of claws and paws, hoofs and hands, and wings and fins. Functional divergence trumps structural unity. Of course, a focus on differences makes absolute sense for the discovery and description of new taxa, which is a prime goal of natural history and taxonomy. But to trace the hidden bonds of diversity requires a different approach.

Remarking on the debates between Geoffroy and Cuvier before the Académie des Sciences in 1830, Goethe was struck by this dichotomy of viewpoint. He noted that

Cuvier worked as a "distinction maker" and "a separator," whereas Geoffroy was a seeker of "secret relationships" (Goethe in Le Guyader, 2004: 262). Geoffroy did not, however, consider the preoccupation with differences to be entirely useless, but rather providing an incomplete perspective. Before studying what is "dissimilar," he argued one must determine in detail, rather than merely assume, "the relations they have in common" (Le Guyader, 2004: 113). We might say that Geoffroy was applying an early version of Hennig's "auxiliary principle" (Hennig, 1966: 121) to the study of comparative morphology: assume homology of shared similarities in the absence of contrary evidence. This powerful heuristic principle indicates a remarkable epistemological unity shared between the otherwise very different conceptual worlds of Geoffroy and modern phylogeneticists. Differences have no value in the search for relationships and common origins. As Geoffroy wrote, "to infer the facts of relations from the observation of differences was to accept judgments that rested on perpetual contradictions in ideas" (Geoffroy in Le Guyader, 2004: 158–159). Yet, as we will see in Chapters 5 and 6, the misuse of differences to deny homology and monophyly remains disconcertingly common in modern evolutionary thinking. The tension between a focus on differences or similarities has, of course, been a major part both of the history of taxonomy and of the taxonomic history of many groups of organisms as well, as lumpers and splitters endlessly debate the relative merits of shared similarities and unique differences.

2.3.3 Documenting Unity of Composition: Establishing the Principles of Comparative Anatomy

Historians have suggested that Geoffroy's fascination with unity of composition may in part be the result of the influence of the mineralogist René-Just Haüy, Geoffroy's scientific mentor and friend (Appel, 1987: 90; Grene, 2001: 190; Rieppel, 2001a: 164; Le Guyader, 2004: 105). Haüy had been elected to the Académie des Sciences in 1783 for showing that the diversity of crystal form was underpinned by a constant geometrical pattern of elements. However, there is little doubt that a more pervasive and direct influence lay elsewhere, particularly in the writings of Buffon. Although Geoffroy had probably never met Buffon, his approach to research fitted a Buffonian mold perfectly: do not specialize too narrowly, distrust final causes, cast a sympathetic eye on materialism, think imaginatively, and theorize widely. Geoffroy adopted a set of methodological principles that allowed him to transform unity of plan[5] from an unverified intuition into an inductive generalization of

---

[5] Readers will have noticed that I use the expressions "unity of type" and "unity of plan" interchangeably. The most famous source of the latter expression is Darwin's *Origin*, and many modern writers also use this phrase to refer to pre-Darwinian ideas. Geoffroy used various expressions to represent his central idea that animals were constructed on the same plan (Rieppel, 2001a), including "unity of plan," "unity of organization," and "unity of (organic) composition," with the latter occurring commonly only from 1818 (Le Guyader, 2004). During their 1830 debate, Cuvier took Geoffroy to task for this terminological ambiguity since he thought it betrayed fuzzy thinking (Appel, 1987; Le Guyader, 2004), and even Goethe, who was otherwise sympathetic to Geoffroy's cause, carped that Geoffroy's cavalier use of language did his arguments no favors (Le Guyader, 2004: 269). In a sneering, if perhaps not altogether effective response, Geoffroy countered that this was nothing more than a "purely grammatical debate about words" (Le Guyader, 2004: 154–155).

sorts. He laid out these principles for the first time in detail in the first volume of his *Philosophie Anatomique* (1818), and he concisely summarized them in the second volume (1822). He listed "four rules or principles" that he thought were intimately related to each other: "The theory of analogues, the principle of connections, the elective affinities of organic elements, the balancement of organs" (Geoffroy in Le Guyader, 2004: 45). The "theory of analogues" referred for Geoffroy simply to the possibility that homologous organs[6] could in principle be identified in different organisms, or in his words the "philosophical thought of the analogy of organization" (Geoffroy in Le Guyader, 2004: 46). This denotes the absolute minimum requirement for the viability of his research program.

To fashion this "glimpse by the mind" into a practical tool, Geoffroy invoked the "principle" or "law of connections." This represented Geoffroy's sharpest epistemic tool. He thought it could reveal homologues in different organisms on the basis of their corresponding topological relationships. Geoffroy considered this to be a nearly infallible guide for identifying corresponding parts: "an organ is rather altered, atrophied, annihilated, then transposed" (Geoffroy in Le Guyader, 2004: 32). In fact, he considered his principle of connections to be so powerful that it could reduce the riotous diversity of animal form to its theoretical minimum: "[s]trictly speaking, there are not a number of animals, but one single animal, whose organs vary in form, use, and volume, but whose constitutive materials always remain the same in the midst of those surprising metamorphoses" (Le Guyader, 2004: 220).

The principle of the "elective affinities of organic elements" (from 1829 also known as the law of "soi pour soi," or "like for like") states that the "reciprocal suitability" of organic components determines the way they are put together to form the larger structure of organs. Finally, the "balancement of organs," also known as the principle of compensation (Panchen, 2001), states that the relative prominence of organs is balanced, such that a hypertrophy of one organ would be accompanied by a corresponding hypotrophy of another.

Geoffroy wielded these tools to great effect. The principle of connections allowed him to identify corresponding structures in widely different organisms, while the principle of compensation furnished an explanation for the sometimes extreme changes in relative proportions of structures with conserved relative positions, or even the complete loss of elements. After initial work documenting the unity of plan in tetrapods, Geoffroy extended his efforts to fishes. Although fishes are of course vertebrates, their skeletons show striking differences from those of tetrapods. In three memoirs published in 1807, Geoffroy presented the radical finding that his principle of connections could identify corresponding parts in fish and tetrapod skeletons that were previously hidden by their widely diverging shapes and functions.

He also showed that seemingly disparate adult anatomies could be bridged by careful study of developmental stages. Impressive demonstrations of the value of this approach can

---

[6] Geoffroy's analogous parts correspond to our modern understanding of homologous parts. Hossfeld and Olsson (2005: 247) note that Geoffroy was the first to use the term homologous when he wrote that "when the development of organs is analogous, they are called homologous."

be found in another 1807 memoir on the composition of the skulls of vertebrates. Here Geoffroy homologized the quadrate bone in birds (involved in the articulation of the jaw) with bones associated with the mammalian middle ear and concluded that fish and mammal skulls are composed of corresponding numbers of bones, deriving support for these findings from his studies of bird and mammal fetuses (Appel, 1987: 88; Le Guyader, 2004: 157). Cuvier, by then Europe's leading comparative anatomist, praised this early work (Appel, 1987: 88, 90–91; Le Guyader, 2004: 22–23), but as is well known, Geoffroy's attempt to follow his principles to their logical conclusion, and to establish unity of plan across different taxonomic *embranchements*, led to their famous public falling out in 1830.

2.3.4 The Ontology of Geoffroy's Type

Asma (1996) dedicates a substantial part of his book to establishing the nature of Geoffroy's archetype. He writes: "The basic options are that either the archetype is ideal and pseudo-Platonic, or it is an actual organism, or it is 'shorthand' for a combination of materialistic laws which form constraints on potential variation" (Asma, 1996: 16). He concludes, convincingly I think, that it is the latter. For Geoffroy the archetype was real but not concrete. More precisely, he did not conceive of it as a particular form, organism, or entity. Instead, the archetype represented the intersection of fundamental laws of organization, such that it could be said to have logical and temporal priority over any specific configuration of form and function (Appel, 1987: 88; Asma, 1996: 48; Ghiselin, 1999: 33; Gould, 2002: 299, 303). In this, Geoffroy's and Goethe's archetypes agree. It is thus inaccurate to say that Geoffroy believed that all animal body plans are variants of "a single fundamental form" (Raineri, 2009: 55) or "an archetypal form" (Quammen, 2003: 73). Geoffroy thought that different parts of the archetype were most accurately reflected in the different species in which they were most fully developed. For example, the opercular bones of fishes afforded a more revealing glimpse of one aspect of the archetype than what he presumed represented their diminutive homologues in mammals: the middle ear bones. Similarly, the nine bones that make up the tortoise plastron furnished Geoffroy with the context within which he evaluated the relatively small sternal bones of other vertebrates (Panchen, 2001: 42–43).

Historians and philosophers understand Geoffroy's unity of type either as a law, or as a consequence of the operation of a law or laws (Ospovat, 1981; Desmond, 1989; Asma, 1996; Amundson, 1998). Amundson (1998, 2005) calls unity of type a phenomenal (or geometrical) law to distinguish it from an ultimate or causal law. A phenomenal law refers to an observable pattern that results from some hidden fundamental cause(s). The phenomenal law that Geoffroy discovered was the pattern of homologues across diverse organisms. Evolutionary descent eventually furnished the ultimate causal law or explanation for this phenomenal law. Analogously, pre-evolutionary systematists discovered the outline of the natural system in the patterns of systematic relationships between taxa before evolutionary theory provided a causal explanation (see Chapter 3). Even though Geoffroy explicitly accepted evolution in some of his work, notably his research on fossil crocodiles, he never made it the causal foundation for the unity of type. Yet, he still attempted to pinpoint some causal principles. For him the operation of the same fundamental developmental and physiological processes across animals equally well explained unity of organization, in particular through the action of the principle of *soi pour soi* (similar parts attract), and the divergence between groups, which could result from differences in the flow of nutrients through the developing embryo. So he took at least a tentative step toward explaining unity of type through the operation of natural, materialistic, morphological laws.

Even though the archetype was abstract in the sense that it did not occur as a concrete organism in nature, it did not represent a supernatural, divine, or Platonic blueprint. Indeed, it was one of the major goals of Asma (1996) to show that Geoffroy's thinking should not be uncritically categorized with the quite different brand of idealistic morphology developed by the German *Naturphilosophen*. Asma blames Russell's classic 1916 study for having tainted Geoffroy's ideas with an inappropriate Platonic gloss, which was uncritically accepted by many later scholars. Asma (1996: 41) accuses Russell for having inappropriately "paired up" Geoffroy with Plato. Indeed, already in the preface to his book Russell labels Geoffroy a "typical representative" of transcendental morphology. However, despite classifying Geoffroy with the German *Naturphilosophen*, Russell (1916: 100) actually considered Geoffroy "vastly their superior in the matter of pure morphology... the Germans were transcendental philosophers first, and morphologists after." The current consensus is that Geoffroy's thinking doesn't align closely with the idealistic metaphysics of the German *Naturphilosophen* (Asma, 1996; Breidbach and Ghiselin, 2002).

Although Geoffroy's approach was capable of generating new insights into the deeper unity of nature, contemporaries and later commentators alike judged some of his comparisons to be far-fetched in the extreme. This was particularly true for his detailed comparisons of the external and internal anatomy of invertebrates and vertebrates. For example, his attempts to show that arthropods displayed the same fundamental organization as vertebrates, allowing a detailed part by part comparison, have been labeled as "fanciful" (Appel, 1987: 111; Russell, 1916: 64). This was also what many thought – Cuvier famously among them – about his attempt to fold molluscs into the vertebrate type, literally, by concluding that the internal anatomy of a cephalopod mollusc corresponded to that of a vertebrate folded back upon itself. Of course, Cuvier had little difficulty showing the "enormous differences" (Cuvier in Le Guyader, 2004: 144) that remained between the organizations of these animals, folded over or not. Yet, would it be fair to label Geoffroy's most daring comparisons as far-fetched? Geoffroy considered "unity of organic composition" to be a "fundamental truth" (Geoffroy in Le Guyader, 2004: 54). For Geoffroy the structural correspondences between organisms were the inevitable products of determinate morphological laws; they were not the accidental products of a contingent evolutionary history. Since he thought his principle of connections to be a near-infallible guide to those correspondences, any comparison would be mandated with equal validity, whether between highly similar or disturbingly dissimilar forms. Geoffroy was brave enough to follow his epistemological guide to its logical conclusion.

2.3.5 Geoffroy's Originality

For all his ingenuity and intellectual daring, Geoffroy could not escape charges that his thinking lacked originality. Geoffroy's ideas had not sprouted fully formed from nothing, and several conceptual umbilical cords root his thinking in ideas of both contemporaries and ancients. From the start of his career, Geoffroy's thinking had clearly been indebted to that of his "spiritual mentor" Buffon, from whom he may have gleaned the seed of the idea of unity of type (Appel, 1987: 22–24). It is also well-known that the transcendental morphology of the German *Naturphilosophen* and Geoffroy's philosophical anatomy were largely independent developments, at least prior to 1818 (Appel, 1987), and these systems of thought had independently incorporated some of the same ideas, such as the principle of connections and the law of compensation (Russell, 1916; Appel, 1987).

For example, the distinguished anatomist Johann Friedrich Meckel (1781–1833) – a German contemporary of Geoffroy who had done research at the Jardin des Plantes between 1804 and 1806 under the tutelage of Cuvier – disputed Geoffroy's claims of priority of having devised a new method for identifying homologues and unity of type (Göbbel and Schultka, 2003). Meckel also disagreed with Geoffroy that the principle of connections has general validity because nature shows too many exceptions, a charge to which Geoffroy responded in some length in his 1822 work on human monstrosities. Geoffroy's principle of the "balancement of organs," or compensation of parts, can also be traced back to an influential idea developed by the early *Naturphilosoph* Carl Friedrich Kielmeyer (1765–1844) (Rieppel, 2001a: 160). Kielmeyer summarized his ideas on the relationships between the organic forces in nature in one law: "The more one of these forces, from one side, increases, the more would those on the other side be reduced" (Kielmeyer in Richards, 2002: 244).

What was most exasperating for Geoffroy, however, was that the kernel of virtually every important idea in the natural sciences could be traced back to Aristotle. Pointing this out was precisely Cuvier's strategy for throwing doubt on Geoffroy's inventiveness during their 1830 debates. Cuvier flatly denied that Geoffroy's ideas had laid any "new foundations for zoology" because he considered them in essence to be nothing more than "the principal foundations" erected by Aristotle (Cuvier in Le Guyader, 2004: 141). According to Cuvier, ever since Aristotle had shone his light on comparative anatomy, "naturalists have [had] nothing else to do" (Cuvier in Le Guyader, 2004: 143). This blunt declaration of his irrelevance stung Geoffroy. In response, in the introductory essay to the debates, Geoffroy sarcastically asked the rhetorical question whether we should "religiously preserve an ancient method" or rather use a new method fit "to satisfy new needs" (Geoffroy in Le Guyader, 2004: 109). During the debates Geoffroy repeatedly emphasized how his theory of analogues was both different from, and superior to, the "Aristotelian doctrine" (e.g. Geoffroy Saint-Hilaire, 1830: 15–17, 88, 89, 91–94, 97–99, 109, 119–121). However, far from dismissing "Aristotelian principles"[7] of comparative anatomy as irrelevant to his thinking, Geoffroy considered his ideas to be a rigorous extension of them. He considered the organismal resemblances discovered on the basis of the application of Aristotelian principles to be so obvious as not to require more than "folk common sense" (Geoffroy in Le Guyader, 2004: 115).

---

[7] Aristotle's honorific label "father of the science of classification" rests on him using shared characteristics to group organisms, rather than classifying them via dichotomous division (Mayr, 1982: 150; Wilkins, 2009).

Reference to any methodological principles was scarcely required to become convinced of the homology of, for example, the eye of an ape and that of a human. These organs are identical in the "three givens that are always encountered in the species participating in natural families: *the anatomical element, the form, and the function*" [italics in original] (Geoffroy in Le Guyader, 2004: 115). Hence, Aristotelian principles could only lead to a "feeling," or "presentiment," of homologies (Geoffroy in Le Guyader, 2004: 45, 115). What was needed was a real "scientific method" based on explicit criteria for identifying homologues that did not reveal "themselves easily to the eyes of the mind" (Geoffroy in Le Guyader, 2004: 115). Geoffroy believed that he had found precisely the "instrument of discovery" (Geoffroy in Le Guyader, 2004: 45, 115) needed for this more challenging task. His principle of connections provided "Ariadne's thread, which keeps us to the true path, and which necessarily leads to a happy outcome" (Geoffroy in Le Guyader, 2004: 161). But not even this principle was original with Geoffroy.

Geoffroy's genius and the basis of his enduring fame is to have taken an idea that had lain dormant in the literature for several centuries and fashioning it into the most powerful method for identifying homologues that systematics has ever known. He recognized the importance of the earliest documentation of the principle of connectivity portrayed in the iconic 1555 illustration of Pierre Belon du Mans (1517–1564) (Geoffroy Saint-Hilaire, 1830: 89). Belon's famous drawing (Figure 2.2) juxtaposed the skeletons of a bird and a human, with topologically corresponding parts labeled with the same letters. By abstracting the specific shapes and functions of the equivalent bones, Belon revealed the fundamental unity hidden beneath the superficial overlay of diversity, and his insight has been part of the communal consciousness of systematics ever since (e.g. Haeckel, 1866b: 153; Russell, 1916: 18; Rieppel and Kearney, 2002: 71; Ghiselin, 2005b: 92; Williams and Ebach, 2008: 127).

Geoffroy's mentor Louis Jean Marie Daubenton (1716–1800) elaborated this principle into an important tool for comparative anatomical studies of skeletons. He determined the equivalence of skeletal elements on the basis of their relative structural positions and designated these with the same anatomical term. It was the anatomist Félix Vicq d'Azyr (1748-1794), who is probably best remembered today for his work on the serial homology of vertebrate fore- and hindlimbs, who subsequently used the notion of the unity of type systematically as a tool in the study of comparative anatomy. The predictive power of this idea led to his discovery of the human intermaxillary bone in 1784, around the same time as Goethe did[8] (Schmitt, 2009). But it was Geoffroy who came to use the principle of connections most penetratingly in his attempts to transform his intuition of the unity of animal organization into an inductive generalization.

---

[8] Secondary sources present various dates for Vicq d'Azyr's discovery of the human fetal intermaxillary bone, with the most common date being 1784, the same year Johann Wolfgang Goethe independently discovered it: Russel (1916), Rieppel (2001). Richards (2002: 374), however, cites 1779 as the date of Vicq d'Azyr's discovery, while Richards (1992a: 31) cites 1780.

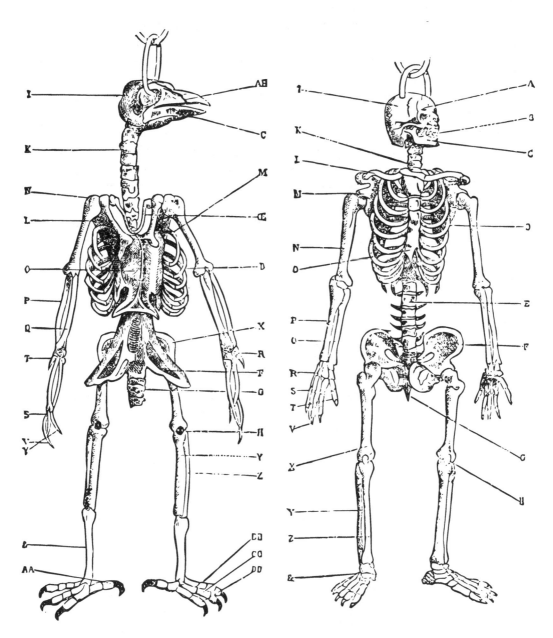

Figure 2.2 Pierre Belon's depiction of comparable parts of a bird and human skeleton (after figure 3 in Anonymous, 1889).

2.3.6 Twisting Geoffroy's Views: Creation of a Modern Myth

I doubt whether Geoffroy's accomplishments were widely appreciated in the twentieth century, outside the discipline of systematic biology. This situation changed abruptly, however, as a result of two spectacular discoveries in the mid-1980s and the mid-1990s. The first of these – the discovery that homologous homeobox genes functioned in the development of both frogs and flies – provoked Stephen Jay Gould to trumpet the apparent vindication of Geoffroy's dream that a deep unity of plan pervades the animal kingdom (Gould, 1986). Still more impressively, the later discovery that frogs and flies establish their dorsoventral body axes by expressing homologous transcription factors cemented this unity of body plans. Strikingly, because these developmental genes are expressed in an inverted pattern in the frog and fly, it was proposed that an actual evolutionary inversion of the dorsoventral body axis had occurred sometime after the fly and frog lineages split from their last common ancestor (Arendt and Nübler-Jung, 1994). Again, one of Geoffroy's ideas was exhumed. His 1822 work titled *Considérations Générales sur la Vertèbre* (*General Considerations on the Vertebra*) was cited as the earliest known source for the speculation that "the ventral side of insects would... correspond to the dorsal side of vertebrates" (Arendt and Nübler-Jung, 1994: 26), such that "arthropods chose to walk with their dorsal side oriented towards the ground" (Nübler-Jung and Arendt, 1994: 357). Later authors have been virtually unanimous in expressing their agreement with this terse assessment of Geoffroy's thinking, as I trust a selection of citations spanning a quarter century will attest:

"French naturalist E. Geoffroy Saint-Hilaire proposed that the ventral side of the arthropods was homologous to the dorsal side of the vertebrates" (De Robertis and Sasai, 1996: 37); "[w]e should note that St. Hilaire (1822) explained the differences by suggesting that vertebrates rolled over to an upside-down posture compared to protostomes (gastroneuralians)" (Bergström et al., 1998: 82); "in contrast to the vertebrates, the dorso-ventral axis of annelids and crustaceans would have had to be inverted such that what is the dorsal surface in deuterostomes is the ventral in protostomes" (Van den Biggelaar et al., 2002: 29); "Geoffroy resolved this striking difference by suggesting that the same groundplan underlay the development of both phyla, but that this common design appeared in reverse orientation, with arthropods interpreted as, essentially, vertebrates turned on their backs" (Gould, 2002b: 1119); "[t]hese and other details suggested in 1820 to the great French zoologist Geoffroy St Hilaire that a vertebrate could be thought of as an arthropod, or an earthworm, turned upside down" (Dawkins, 2004: 389); "Geoffroy produced a model that rationalized this dorsoventral difference, suggesting that the dorsoventral axis had simply been reversed" (Valentine, 2004: 421-422); "[t]he notion that the dorsal surface of the one [protostomes] corresponds to the ventral surface of the other [deuterostome] goes back to Étienne Geoffroy Saint-Hilaire" (Ghiselin, 2005b: 97); "the comparative zoologist Étienne Geoffroy de St Hilaire arrived at the surprising conclusion that chordates, such as fish and humans, are upside down relative to all other bilaterally symmetrical (bilaterians) animals" (Telford, 2007: 237); "Étienne Geoffroy Saint-Hilaire thought that the vertebrae of vertebrates correspond, through some odd topological inversion, to the exoskeletal body segments of arthropods" (Kuznetsov, 2012: 22); "[t]his finding leads to the suggestion that vertebrates evolved from protostomes by a dorsoventral inversion, resurrecting an earlier inversion hypothesis proposed in the work by Geoffroy Saint-Hilaire" (Northcutt, 2012: 10629); "This hypothesis proposed that the dorsoventral axis of vertebrates and arthropods are essentially the same if arthropods are flipped over on their back" (Hejnol and Lowe, 2015: 2); "[i]t was Geoffroy who adopted wild schemes in which lobsters could be transformed into vertebrates by turning them upside down" (Gee, 2018: 16); "which led the French naturalist Geoffroy Saint-Hilaire to propose the homology between the arthropod ventral side and the chordate dorsal side already on 1822" (Martín-Durán and Hejnol, 2021: 188); "[Geoffroy] proposed that vertebrates are, in a sense, upside-down arthropods" (Baum and Smith, 2013: xv).(Arthur, 2021: 24).

Alas, this unanimity of opinion consistently misrepresents Geoffroy's thinking. Geoffroy was indeed struck that

> all the soft organs, that is to say, the principal organs of life, are reproduced in the crustaceans, and consequently in the insects, in the same order, in the same relations, and with the same arrangement as their analogues in the higher vertebrates. (Geoffroy in Le Guyader, 2004: 77)

His often reproduced drawing of a longitudinal section of a lobster beautifully illustrates this point. So it is correct to conclude, as some authors indeed explicitly do, that Geoffroy proposed that crustaceans and insects share the body plan of vertebrates, but in "an inverse state" (Geoffroy in Le Guyader, 2004: 78). Geoffroy, however, was explicit that this conclusion only applied to the orientation of the internal organs. He did *not*, as virtually all modern commentators claim, suggest that arthropods and vertebrates inverted their ventral and dorsal sides or surfaces or their dorsoventral axes. Instead, he concluded, explicitly and repeatedly, that compared with fishes, crustaceans "really swim on their sides" (Geoffroy in Le Guyader, 2004: 77)! To miss this point is to miss the entire purpose of his 1822 paper.

The consensus view is that Geoffroy did not propose a real phylogenetic transformation of arthropods and vertebrates,[9] but it is clearly difficult to properly grasp Geoffroy's thinking. As reflected in the title of the work, *General Considerations on the Vertebra*, Geoffroy's purpose was to describe in detail the general structure of the vertebra in vertebrates and to determine its homologous parts in the exoskeleton of arthropods. The motivation for this was provided by his previous conclusion that insects are vertebrates that live within their vertebral columns. After first describing the vertebrae of a young flatfish, Geoffroy then describes and illustrates the corresponding parts of a lobster's exoskeleton (Figure 2.3). He concludes that the dorsal and ventral vertebral processes of the flatfish correspond to the legs on the thorax and pleon of a lobster. He realizes, though, that some may claim a "serious objection" (Geoffroy in Le Guyader, 2004: 74) to this interpretation in that it requires the lobster to be oriented on its side. He devotes about a quarter of the entire text of his paper to refute this objection, concluding that, indeed, like flatfish, crustaceans move around on their sides. Geoffroy applied his principle of connections to conclude that "the crustacean makes of one of its sides its dorsal face and of the other side

---

[9] Galera (2021) has recently argued that in addition to proposing that the animal kingdom was tied together by a fundamental unity of composition, Geoffroy also proposed an actual phylogenetic transformation of invertebrates into vertebrates. I am unable to judge this conclusion because I haven't read all of Geoffroy's relevant work, but Galera (2021: 237, 240–241) perpetuates the misunderstanding that Geoffroy thought this involved a dorsoventral inversion of the body.

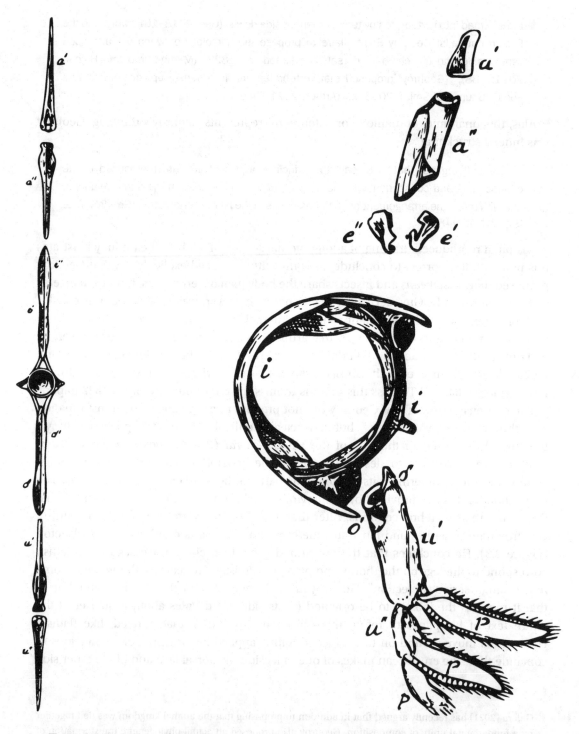

Figure 2.3 Details from plates 5 and 6 in Geoffroy's 1822 *General Considerations on the Vertebra*. On the left is a vertebra of a young plaice, and on the right is a pleon segment of a lobster placed on its side, with appendage segments oriented vertically. Geoffroy indicated the homologous parts with the same letters (modified after figure 2.3 in Ochoa and Barahona, 2014).

its ventral face" (Geoffroy in Le Guyader, 2004: 75), with the result that they "really swim on their sides" (Geoffroy in Le Guyader, 2004: 77).

Geoffroy explains the fact that the internal organs of crustaceans are completely inversed with respect to those of the vertebrates by claiming that they had rotated within the exoskeleton. The dorsoventral axis was thus not inverted, it was twisted, with the arthropod exoskeleton and appendages on the one hand, and its internal organs on the other hand, shifting relative to each other compared to vertebrates.

Recent authors have created an historical myth about Geoffroy's views on the unity of plan between arthropods and vertebrates by failing to appreciate that his primary framework for comparison was the vertebra. It was present as an internal vertebral column in vertebrates and as an exoskeleton in arthropods. According to Geoffroy, arthropods are vertebrates oriented on their sides, with the mass of internal organs rotated to assume an inverted orientation with respect to "higher" vertebrates. As a final irony, on the last page of his fine book that brings Geoffroy's ideas to English readers for the first time, Le Guyader (2004: 254) twice approvingly cites De Robertis and Sasai (1996) for attributing the idea of dorsoventral axis inversion to Geoffroy: "What an acknowledgment, more than a century and a half later!" A fine acknowledgment indeed, but alas, only of the unacknowledged persistence of a modern myth.

2.4 Richard Owen and the Vertebrate Archetype

It was not that long ago that Richard Owen (1804–1892) was still stigmatized as being little more than an inferior yin to Darwin's yang. However, through the efforts of historians over the last several decades – with Nicolaas Rupke's *Richard Owen: Biology without Darwin* paving the way – Owen has emerged as an important and complex thinker whose work provided a valuable springboard for some of Darwin's central insights. Although this intellectual rehabilitation has gone a bit too far for some – "Owen has become a sort of culture-hero among contemporary anti-evolutionists (in the broad sense)" (Ghiselin, 1997: 205) – Owen's work on the vertebrate archetype is of special significance for understanding the origin of ancestors. As we will see in the next section, Owen's 1849 monograph *On the Nature of Limbs* provided the archetypal roots from which Darwin grew his concept of evolutionary ancestors. In order to be able to judge the nature of this accomplishment, we first need to grasp the essence of Owen's archetype.

In the early 1840s, as Geoffroy's health deteriorated in the last few years before his death in 1844, Owen was busy developing his own brand of archetypal thinking. The results were to become an enduring pillar of Owen's fame. His theory of the vertebrate archetype established the concepts of special, serial, and general homology and separated homology from analogy. The pictorial representation of the vertebrate archetype as a fish-like diagram became one of evolutionary biology's iconic images. Appel (1987: 228) concludes that "Owen's was a far more flexible and sophisticated treatment of philosophical anatomy than Geoffroy's," and there is much truth in this characterization. Whereas Cuvier's teleology and Geoffroy's structuralism formerly seemed to be located on opposite sides of an unbridgeable chasm between incompatible worldviews, Owen incorporated both in his theory of the vertebrate archetype. In fact, while Owen was dubbed the "British Cuvier" early in his career, at least in part to reflect his status as "the custodian of the functionalist design argument" (Rupke, 2009: 66), his work on the vertebrate archetype moved him close to a position of pure formalism, also known as structuralism, which explains the unity and diversity of form primarily in terms of features intrinsic to the morphology of organisms. In contrast, functionalist or adaptationist theories locate most of their explanatory power in factors extrinsic to organisms, such as natural selection. Owen's strict focus on form secured his enduring reputation as that "rare anglophonic exponent of a predominantly formalist theory" (Gould, 2002b: 65).

Owen's archetype theory was also built upon a much richer empirical basis than Geoffroy's. The vertebrate archetype was the unique distillation of Owen's deep reading of the continental literature on vertebrate comparative anatomy, combined with his own extensive first-hand knowledge of vertebrate skeletal structure. Who else but Britain's foremost naturalist would choose to illustrate the common loss of the first digit (thumb) in mammals with the spider-monkey and the hyena, and the loss of the fifth digit (little finger) with the "marsupial bandicoot" (Owen, 1849: 30)? And who else would have felt confident that he had studied enough orangutan and wombat skeletons to be able to conclude that the former "commonly" and the latter "constantly" has only one phalanx on the first metatarsal bone of the foot (Owen, 1849: 35)? His unmatched expertise even drove him to criticize classical painters and sculptors for failing to properly depict the relative sizes of human toes in their artworks (Owen, 1849: 39).

Unlike Geoffroy, Owen refused to taint the veracity of his conclusions by stretching his comparisons beyond the boundaries of types. He constrained his archetypal theorizing safely within the boundaries of the vertebrate *embranchement*, and even within those limits he did not accept Geoffroy's version of the unity of composition that prescribed that homologues parts in different organisms should be composed of the same number of bones. This was why, for example, Owen labeled Geoffroy's theory that the vertebrate skull was fundamentally composed of seven modified vertebrae as being "based less on observation than on purely *à priori* views" (Owen, 1848: 79). Yet, despite being rich on empirical detail and light on speculative excess, Owen's archetypal theory still managed to rile his contemporaries and challenge later generations of scholars. To understand why, we need to confront Owen's metaphysics.

2.4.1 A Metaphysical Mystery

For more than a century and a half, Owen's vertebrate archetype has presented historians and philosophers with a refractory metaphysical riddle. Paradoxically, however, for most of this time few seemed to realize or care there even was a riddle. The difficulty of pinning down the precise nature of what has been labeled "one of the most important theoretical developments in comparative anatomy in the mid nineteenth century" (Sloan, 2003: 53) is primarily due to the striking metaphysical inconsistency of Owen's own statements in the two works in which he presented the vertebrate archetype in detail: *On the Archetype and Homologies of the Vertebrate Skeleton* from 1848 and *On the Nature of Limbs. A Discourse* from 1849 – the latter reprinted in a facsimile edition with new introductory essays in 2007. Both Owen's contemporaries and many modern scholars chose to focus almost exclusively on his 1849 presentation of the archetype theory. The result was a tradition of agreement, but unfortunately for Owen, of near universal condemnation as well.

In his 1848 book, Owen introduced the archetype – the word, the concept, and the iconic diagram – as a generalization of the endoskeletal elements shared among vertebrates. Owen considered it the "fundamental or general type" of the vertebrates, comprising "a series of essentially similar segments succeeding each other longitudinally from one end of the body to the other, such segments being for the most part composed of pieces similar in number and arrangement" (Owen, 1848: 7). These repeated segments were vertebrae with their associated processes. He defined the "ideal typical vertebra" as "one of those segments of the endo-skeleton which constitute the axis of the body, and the protecting canals of the nervous and vascular trunks: such a segment may also support diverging appendages" (Owen, 1848: 81).

The vertebrate archetype does the explanatory work for Owen's homology concepts. He defined serial homology as homology between serially repeated skeletal elements within the body of one animal. Special homology occurs when structures in different animals are diagnosed, pace Geoffroy's criteria, as having the same relative position and connections. These two kinds of homology are presided over by a "higher relation of homology" that is defined with respect to the archetype, "in which a part or series of parts stands to the fundamental or general type on which a natural group of animals, the vertebrate for example, is constructed" (Owen, 1848: 7). It is this general homology that allows the morphologist to detect special homology between such dissimilarly looking structures as the shoulder blade of a human and a little rod-shaped bone in a fish, as both are modified versions of thoracic vertebral processes present in the archetype. It was this rooting of the concept of general homology in the archetype that Owen understood as the *Bedeutung*, or essential nature, of a part of an animal body. He borrowed this word from the German morphological literature and eventually settled on titling his 1849 book *On the Nature of Limbs*, after rejecting the "signification," the "meaning," and the "idea" of limbs as terms that were too theoretically feeble.

The vertebrate archetype's segmental composition reflects the operation of a "law of vegetative or irrelative repetition," which is produced by a "polarizing force," and which also operates in the formation of inorganic structures in nature, such as crystals (Owen, 1848: 171). Owen thought a second "specific organizing force" or "adaptive force" worked in opposition to the polarizing force to produce the "diversity of form belonging to living bodies of the same materials" (Owen, 1848: 172). Unopposed by this adaptive force, the polarizing force alone would produce the purest manifestation of the archetype as a series of identical parts or homonomous segments. The famous diagram of the archetype, however, already betrays unmistakable signs of activity of the adaptive force, for example, producing definite axial differentiation (e.g. anterior sense organs). Owen believed that the relative strength of these opposing forces determines for each species "the index of the grade of such species," as well as "its ascent in the scale of being" (Owen, 1848: 172). Vertebrates are located relatively high on the scale of being because in them the adaptive force represses the effects of the polarizing force to a larger extent than lower down the scale. The more the polarizing force is antagonized by the adaptive force, the more the contours of the archetype are obscured. Hence, repetition of similar segments "is so much more conspicuously manifested by the... exoskeleton of the invertebrata" located lower on the scale of being, such as arthropods, annelids, and echinoderms (Owen, 1848: 171).

The polarizing force is least encumbered by the adaptive force at the inorganic bottom of the scale. Because Owen thought that most inorganic components in the vertebrate body were concentrated in the endoskeleton, it was here that the operation of the polarizing force was most visible (Owen, 1848: 172). The ultimate expression of the omnipresence of this force came later in Owen's career when he implicated its operation in the origin of life by spontaneous generation: "at the period when life became possible on the earth's surface the conditions were sufficiently varied to permit the conversion of the general polaric [sic] into a special organic mode of force to operate under circumstances resulting in a variety of the simplest forms of life" (Owen in Rupke, 2009: 174). In short, for Owen the archetype reflected the existence of a natural law of vegetative repetition that resulted from the operation of a material, developmental force that was active in both the animate and inanimate worlds. It was to this force that "the similarity of forms, the repetition of parts, the signs of the unity of organization may be mainly ascribed" (Owen, 1848: 172).

In his 1849 book *On the Nature of Limbs*, however, Owen painted over the archetype with an idealistic, Platonic gloss. The polarizing force is still operative, ready to "shoot out particle upon particle" (Owen, 1849: 69) to form the basic, repetitive structural patterns that characterize the simpler life forms. But the archetype it produces has now taken on another meaning: "the recognition of an ideal Exemplar for the Vertebrated animals proves that the knowledge of such a being as Man must have existed before Man appeared. For the Divine mind which planned the Archetype also foreknew all its modifications" (Owen, 1849: 86). The archetype came to represent a "predetermined pattern, answering to *the 'idea'* of the Archetypal World in the Platonic cosmogony" (Owen, 1849: 2). Unfortunately for Owen, this divine elevation of the archetype caused a precipitous decline in its scientific reputation, especially in the eyes of those, like T. H. Huxley, who were at the forefront of the democratization and professionalization of science (Desmond, 1982, 1989, 1994, 1997; White, 2003; Rupke, 2009).

2.4.2 The Metaphysics of Owen's Archetype under Siege

When after Owen's death in 1892 his grandson naively invited Owen's archenemy Huxley to write an overview of his grandfather's work, he invited disaster. Although Huxley was not averse to the use of archetypes as purely heuristic devices – simple summaries of traits shared between organisms – he ridiculed Owen's idealization of the vertebrate archetype as an idea in the mind of God. Huxley invited those who admired Owen for his "exact observations, the clear-headed and sagacious interpretations, of the anatomical and palae-ontological memoirs" to ponder the travesty of distorting "essentially naturalistic abstractions," such as the polarizing force, and turning them into personified agents (Huxley in Owen, 1894: 318). A divinely inspired archetype was equally distasteful to materialists such as Darwin and the medical radicals of London. Again, they saw nothing wrong with hypothesizing archetypes – the medical radicals were ardent followers of Geoffroyan philosophical anatomy (Desmond, 1989) – but imagining the inductively inferred archetype to be a predetermined plan was tantamount to mistaking a result for a cause.

Some of the sharpest criticism came from former admirers. The polymath George Lewes, who is perhaps most famous today for having been George Eliot's long-term partner, was an erstwhile fan of the homologizing vanguard of Goethe, Geoffroy, and Owen. Yet, the man who so memorably wrote of the necessary role of imagination in both literature and science – "it is because we see little, that we have to imagine much" (Lewes in Carignan, 2003: 464) – later accused those who thought plan preexisted pattern of "infirmity of thought" (Rupke, 2009: 138). Presumably, then, Lewes must have similarly considered Darwin's later refashioning of Owen's archetype as a concrete ancestor a tragedy rather than a triumph because converting "the formal into material elements" and to "realise abstractions" was an equally damnable infirmity of thought (Lewes, 1871: lxxviii). Interestingly, as we will see later, some modern biologists actually agree with such an assessment of Darwin's accomplishment.

Why did Owen change his interpretation of the vertebrate archetype so dramatically from one year to the next? Rupke (1993, 2009), in a careful historical reconstruction of the event, suggests that Owen bowed to peer, or rather patron, pressure. Owen's initial presentation of the archetype as a reflection of the operation of nothing but a pervasive natural cause (the polarizing force) upset his conservative Oxbridge patrons who figured prominently on the boards of trustees of the Hunterian and British Museums, where Owen worked before and after 1856, respectively. A material archetypal force would erode too much of the Creator's omnipotence and open the door to the pantheistic worldview that Owen had imbibed from the German literature on *Naturphilosophy*. Platonizing the archetype would allow God to remain in his customary role as the "traditional British craftsman working to a blueprint" (Desmond, 1982: 48). Rupke argues that because Owen changed his mind merely to placate his patrons, we should not take his Platonization of the archetype too seriously. Moreover, the interpretation of the archetype as a Platonic blueprint also loses force, Rupke argues, because "a Platonic idea is the highest, most perfect reality," whereas "the vertebrate archetype represented the opposite: the simplest and least perfected conception of a vertebrate" (Rupke, 1993: 243).

2.4.3 Owen's Archetype as a Nomological Construct

I agree with Rupke (1993, 2009) and Amundson (2005) that historians have unfairly exaggerated the Platonic nature of Owen's archetype. Yet, I do think that Owen's work betrays a commitment to idealistic metaphysics that is more than just "harmlessly rhetorical" (Camardi, 2001: 507). This leaves me with a degree of ambiguity about Owen's thinking about the archetype. Sloan (2003) considers this uncertainty to be inescapable. According to Sloan, Owen was heavily influenced by strands of German philosophy that reached him via the work of his philosopher friend William Whewell and Owen's colleague and erstwhile mentor Joseph Henry Green. They emphasized the interplay of a priori ideas and empirical data. Sloan concludes that Owen intended the archetype to be regarded simultaneously as "both an empirical and a transcendental concept" [italics in original] (Sloan, 2003: 58). I do not know whether Owen consciously promulgated a deliberate dialectic of Platonic ideas and empirics, or simply allowed a degree of philosophical inconsistency in his work. What is more important in our quest to understand the relationship between archetypes and ancestors is that Owen's archetype is a nomological construct.

The distinction between nomothetics and idiographics goes back to an 1894 work of the philosopher Wilhelm Windelband (see Ghiselin, 1997 and Rieppel, 2006 for discussions of the significance of this distinction for evolutionary biology and systematics). Nomothetic sciences, or aspects of sciences, aim to discover laws of nature and law-like generalizations and are concerned with repeatable, predictable, law-like (nomological) phenomena. Idiographic science, in contrast, concerns itself with the description and understanding of unique particulars, individuals, the historically contingent. Put another way, idiographics is concerned with the study of individuals, philosophically defined, such as individual organisms, species, or lineages and their traits, whereas nomothetics deals with generalizations for classes of individuals that are defined on the basis of shared traits. Even though this conceptual framework postdates Owen's work, I think it illuminates the difference between Owen's archetype and Darwinian ancestors.

It is clear that irrespective of whether the ultimate cause of the archetype was thought to be a divine plan or not, Owen considered it to be the direct, predictable, law-like outcome of the operation of the polarizing force that pervaded both the inorganic and organic worlds. Animal structure has no choice but to be "obedient to the archetypal law" (Owen, 1849: 27) that created the "law of vegetative or irrelative repetition" (Owen, 1848: 171). In actual species, parts of the archetype may of course be "extremely modified for special functions, yet never so as to wholly mask their typical character" (Owen, 1848: 7). The inescapability of the archetypal law is strikingly illustrated by a paragraph included in both *On the Archetype* (pp. 102–103) and *On the Nature of Limbs* (pp. 83–84). In the latter, we read that if life were to exist on other planets it may look surprisingly familiar:

> But the laws of light, as of gravitation, being the same in Jupiter as here, the eyes of such creatures as may disport in the soft reflected beams of its moons will probably be organized on the same dioptric principles as those of the animals of a like grade of organization on this earth. And the inference as to the possibility of the vertebrate type being the basis of the organization of some of the inhabitants of other planets will not appear so hazardous, when it is remembered that the orbits or protective cavities of the eyes of the Vertebrata of this planet are constructed of modified vertebrae.

This paragraph establishes beyond doubt that the archetype is the predictable outcome of the operation of a law of nature. Incidentally, I can't help but notice a remarkably similar strand of archetypal thinking a century and a half later in the work of University of Cambridge paleobiologist Simon Conway Morris, although I doubt that he would characterize his own thinking as such. In 2005 he published the first of a number of articles dealing with extraterrestrials, titled "Aliens like us?" In it he asks, "Could it be that biologists have been lulled into believing that evolution is really free of those general laws and principles that physicists, chemists and indeed astronomers, take for granted and indeed without which simply could not operate?" (Conway Morris, 2005: 4.25). He argues that, yes, we have been missing a law of biology that has been hiding in plain sight: "Convergence confers, therefore, a predictability to evolution and as a general law what applies here should apply on any heritable planet, in this galaxy and beyond" (Conway Morris, 2005: 4.26). Conway Morris's is a world but little buffeted by the unpredictable impacts of historical contingency, such that even the overall "configuration of the Tree of Life is pre-determined" (Conway Morris, 2009: 1323). The subtitle of his 2003 book *Life's Solution* reveals where the law-like momentum and sharp directionality of convergent evolution has been aiming: *"Inevitable humans in a lonely universe"* (Conway Morris, 2003).

Owen's vertebrate archetype is the set of traits necessarily shared by all vertebrates as a result of the operation of an archetypal law – Vertebrata is defined as a class united by the shared possession of inerasable archetypal traits. It is this interpretation of Owen's archetype as a class concept that led Asma (1996: 108–112) to conclude that Owen's Platonism cannot be dismissed entirely, let alone be called "a complete red herring" (Amundson, 2005: 93). Asma points out that Platonic Ideas can be understood as class concepts: "The vertebrate archetype allows us to relate—to classify together—the dog, the fish, and the man, just as Plato's Ideal 'state' allows us to unite a democracy, an oligarchy, and a timocracy (*Rep* bk. 8) under the Idea 'government'" (Asma, 1996: 110). Because this implies that "essence precedes existence," Owen's "is just the reverse of Darwin's metaphysics" (Asma, 1996: 112). Michael Ghiselin likewise remained unimpressed by recent reimaginings of Owen's archetype as something other than Platonic or Neoplatonic. He surmises that within the context of Owen's Christianity his archetype's fishy form "makes a fine symbol of both Man and the Son of Man" (Ghiselin, 2000b: 45).

It will be clear by now that Owen's archetype is not idiographic; it cannot be interpreted as an individual organism or a particular species. We must disagree with Ospovat (1981: 133) when he writes that Owen "had in mind also a real, vaguely amphioxus-like animal from which the vertebrates might actually have developed through time." Perhaps Owen did have an ancestral creature like this in mind, but it could not have been the archetype. Ospovat is right that the vertebrate archetype betrays some influence of the adaptive force in having several adaptations that would be needed if it were to live in the real world. However, there is no doubt that the archetype is an abstraction, for it lacks most soft anatomy needed to be a living organism. Similarly, to describe the archetype as being "rather like a bony amphioxus" (Padian, 2007: lxii) or to note the "striking similarity" (Raineri, 2008: 106) of the archetype and amphioxus stretches the concept of resemblance to breaking point. Owen's archetype depicts a smorgasbord of anatomical specializations found in different animals, including the median horn of the rhino, the lateral horns of ruminants, the dorsal humps of camels, and the anal fin of fish. It fell to Darwin to prune Owen's chimera down to form a credible and concrete creature.

2.5 Reifying Archetypes: The Birth of the Evolutionary Ancestor

A pencil and something to write on was all that Charles Darwin (1809-1882) needed for laying a new metaphysical foundation for biology. He substituted fact for fantasy by collapsing *Owen's "ideal pattern"* into a concrete ancestor. He famously recorded this metaphysical metamorphosis on the back flyleaf of his copy of Owen's *On the Nature of Limbs*: "I look at Owens Archetypus as more than ideal, as a real representation as far as the most consummate skill & loftiest generalization can represent the parent form of the Vertebrata."[10] Irrespective of the amount of transcendental patina that adhered to Owen's construct, Darwin's reconceptualization brought it down to Earth to function as a materialist component of a new, truly historical, evolutionary biology. Ghiselin (2005a: 127) aptly summarizes the essence of Darwin's metaphysical revolution as follows: "[m]uch as Copernicus had moved the sun to the center of the physical world, Darwin moved concrete particular things to the center of the metaphysical world."

Darwin's radical materialism immediately cast a profoundly new light on the mysterious causes of the unity of type that had been revealed by comparative anatomists: "[o]n my theory, unity of type is explained by unity of descent" (Darwin, 1859: 233). Homologous or "ideally similar" (Darwin, 1859: 221) characters now reflected and were explained by common ancestry. The ancestor replaced the archetype in the explanation of homology: "[t]he naturalist, thus guided, sees that all homological parts or organs, however much diversified, are modifications of one and the same ancestral organ" (Darwin, 1862: 288). Far from rejecting the results of the transcendental or idealistic morphologists, Darwin embraced their methods, such as the principle of connections, as well as their data to detect homologues and infer unity of type. His contribution was to offer what we today still deem to be the correct interpretation and explanation of the results obtained by the labors of his predecessors and contemporaries.

The mysterious "force by which all the modifications of the vertebrate skeleton. . . are still subordinated to a common type" (Owen, 1848: 171) is simply inheritance from common ancestors. Darwin transformed Owen's nomological archetype, which reflected the universal action of enigmatic laws of organization, into the idiographic pinpoint of an ancestor. Homology was reduced to mere historical inertia. The evolutionary ancestor emasculated the transcendental type, and purloined its explanatory power. No longer did homology relations reflect the law-like force of an active archetype. Instead, they were the fading traces of shared ancestry. Bonds of special homology could still be detected between structures in different species in the manner laid out by Owen, but their general homology, or *Bedeutung* (essential meaning) as Owen preferred to put it, would no longer be anchored by the archetype. Instead, general homology now revealed the material traces of ancestry.

A mild irony shouldn't escape us here. In his 1969 book *The Triumph of the Darwinian Method*, Ghiselin wrote: "the worst defect of the old morphology was that it concerned itself with a search for something [the archetype] which does not exist" (p. 107). This, of course, is true. Yet, the new morphology's triumph was to find something that does exist – the ancestor – by searching in the same way in the same place. Such is the nature of a metaphysical revolution.

---

[10] Darwin's handwriting has been deciphered differently by different authors. I base mine on a scan of the relevant page available via the online Biodiversity Heritage Library (www.biodiversitylibrary.org/collection/darwinlibrary; last accessed February 8, 2022).

Darwin thought that his materialist reimagining of the archetype and homology had the virtue of making the merely metaphorical real: "the science of Homology clears away the mist from such terms as the scheme of nature, ideal types, archetypal patterns or ideas, &c.; for these terms come to express real facts" (Darwin, 1862: 288). Darwin used the last paragraph of the section on morphology in Chapter 13 of the *Origin* to annex the metaphorical language of pre-evolutionary morphologists for his materialist worldview in order to express the reality of phylogenetic, historical transformations of organisms:

> Naturalists frequently speak of the skull as formed of metamorphosed vertebrae: the jaws of crabs as metamorphosed legs; the stamens and pistils of flowers as metamorphosed leaves; but it would in these cases probably be more correct, as Professor Huxley has remarked, to speak of both skull and vertebrae, both jaws and legs, etc.,–as having been metamorphosed, not one from the other, but from some common element. Naturalists, however, use such language only in a metaphorical sense: they are far from meaning that during a long course of descent, primordial organs of any kind–vertebrae in the one case and legs in the other–have actually been modified into skulls or jaws. Yet so strong is the appearance of a modification of this nature having occurred, that naturalists can hardly avoid employing language having this plain signification. On my view these terms may be used literally; and the wonderful fact of the jaws, for instance, of a crab retaining numerous characters, which they would probably have retained through inheritance, if they had really been metamorphosed during a long course of descent from true legs, or from some simple appendage, is explained. (Darwin, 1859: 418–419)

In this paragraph idealistic comparative biology has become historical. Geoffroy's principle of connections guides the evolutionist to the patterns of homologues that form the blueprints of body plans. Goethean mental metamorphoses of form trace the bonds of homology between living species and outline the possible pathways of descent by which they connect to Owenian archetypes made flesh. The tools morphologists had used to trace the flow of form on the synchronic surface of nature now afforded a view into its diachronic depths. For comparative morphologists like Carl Gegenbaur, "[f]rom the standpoint of descent theory the 'relatedness' of organisms loses its figurative meaning," allowing "what previously had been designated as *Bauplan* or *Typus* to appear as the sum of the structural elements of animal organization which are propagated by means of inheritance" (Gegenbaur in Coleman, 1976: 162). Gegenbaur was never the dogged Darwinian disciple that his colleague Haeckel was, but as he started to come to terms with the explanatory merit of Darwin's theory, he came to understand the type as the roster of homologues that traces the ties of genealogical continuity within evolutionary lineages. Viewed through the lens of descent theory,

> we should imagine the type as a developmental series of organisms that, departing from an ancestor, differentiates into many branches and twigs during geological development... That which is continuous in these multiple differentiations – that which is inherited with modification from an ancestor, is what is typical in their organization. Every type therefore comprises a total of genealogical connections, i.e. by descent of related organisms it forms an animal lineage [Thierstamm] (Phylum, Haeckel). (Gegenbaur, 1870: 74)

What Gegenbaur shows here is what I call lineage thinking: the realization that the systematic and morphological relationships between collateral relatives are expressions of a more fundamental type of relationship – namely that between ancestors and descendants. As we will see in the next chapter, the spread of lineage thinking had a profound impact on how evolutionists depicted and understood the shape of the natural system.

2.6 The First Common Ancestors: A Coda on Lamarck

Darwin was not the first materialist or evolutionist, nor was he the first to propose the existence of common ancestors in the context of an evolutionary genealogy. The work of Jean Baptiste Lamarck (1744–1829) deserves a few words. Although organismal transformation was clearly at the center of Lamarck's evolutionary theory, the status of common descent was much more ambiguous. This might be surprising to modern biologists because common ancestry is a *sine qua non* of Darwinian evolution. After all, for Darwin life had "originally been breathed into a few forms or into one" (Darwin, 1859: 459–460), and universal common ancestry is regarded a "central pillar of modern evolutionary theory" (Theobald, 2010: 219). For Lamarck, however, spontaneous generation of microorganisms continuously established evolutionary lineages unbound by common ancestry. Such lineages would then progress along independent evolutionary trajectories of increasing morphological complexity, with the complexity of extant organisms functioning as an index of the relative age of their lineages. More complex organisms simply had ridden their evolutionary escalators for longer along tracks that paralleled parts of a temporalized *scala naturae*. Modern commentators have taken this to mean that in Lamarck's system different Classes (Lefèvre, 2001: 198), or species (Wilkins, 2009: 108), or lineages (Ghiselin, 1997: 247), or each "point on the scale of being" (Bowler, 1989: 85) arose by separate events of spontaneous generation, never joining into a trunk of common ancestry.

This common interpretation of Lamarck's phylogenetic views, however, has been judged inaccurate by historian Richard Burkhardt. Although Burkhardt agrees that Lamarck's theory "should not be seen as a theory of common descent" (Burkhardt, 1995: xxii), Lamarck's evolving phylogenetic views nevertheless moved toward a branching view of evolution. Several authors have rediscovered Burkhardt and Mayr's (1972) earlier insights and have anointed Lamarck's first tree-like branching diagram for the animal kingdom in his 1809 *Philosophie Zoologique* (*Zoological Philosophy*) as possibly the first genuine phylogenetic tree (Gould, 2000; Archibald, 2009; Tassy, 2011). It combines evolution and lineage branching, which implies the presence of common ancestors located at the branching points of the tree. As pointed out by Gould (2000), Lamarck envisioned in his 1809 and 1815 (*Histoire Naturelle des Animaux sans Vertèbres*, or *National History of Invertebrates*) works two independent evolutionary lineages of animals, rooted in the spontaneous generation of infusorians and parasitic worms, respectively. One encounters these diagrams frequently in the secondary literature, but Gould also draws attention to Lamarck's last book from 1820 (*Système analytique des connaissances positives de l'homme*, or *Analytical System of Positive Knowledge of Man*), which he claims has been neglected by historians interested in Lamarck's thinking about animal evolution. In this book Lamarck conceives, but does not illustrate, a branching evolutionary tree rooted in a single common ancestor of all animals, called a monad, which gave rise to the infusorians and the more complex animals. So, common ancestry was undeniably part of the first great evolutionary theorizer's thinking well before Darwin.[11]

Lamarck's evolutionary theory was influential and polarized opinions both inside and outside France, and some found Lamarck's ideas so similar to Darwin's that they felt they had merely to dust off their earlier criticisms of Lamarck to slay Darwinism as well (Farley, 1974). But the importance of common ancestry for Lamarck's and Darwin's theories is very different. The essentials of Lamarck's two-factor theory (Burkhardt, 1995; Gould, 2002b), encompassing forces driving increases in complexity in lineages on the one hand, and forces producing diversity by causing lateral deflections from the main thrust on the other hand, remain fully intact whether evolution starts with one or more spontaneously generated lineages. In contrast, common ancestry is the absolute conceptual bedrock for Darwin's evolutionary theory, so much so, argues philosopher Elliot Sober (2009), that one can ask whether the *Origin* has actually been written backward. Sober argues that although Darwin put natural selection first and foremost in the *Origin*, common ancestry has evidential priority. Common ancestry is the necessary framework within which the hypothesis of natural selection can be tested, whereas common ancestry does not require the existence of natural selection.

2.7 Epistemic Fallout from a Metaphysical Revolution

By tracing steps in the origin of the concept of ancestors like I did in this chapter, I do not imply that it was a simple linear story. We cannot trace a direct conceptual umbilical cord from Goethe through Geoffroy and Owen to Darwin. The beginning of modern evolutionary biology is rooted in a metaphysical discontinuity. Michael Ghiselin has written that "archetypes... must be dismissed as metaphysical delusions – vestiges of Platonism which cannot be reconciled with what happened in 1859" (Ghiselin, 1976: 122). Indeed, the explanatory apparatus of Darwin's evolutionary theory drew its legitimacy from a materialistic worldview that left no room for transcendental interpretations of types and relationships between organisms. But several of the ideas, discoveries, and methods used by the idealistic morphologists are nevertheless reflected, in modified form, in the evolutionary biology that came after them. The order of nature that was discovered by pre-evolutionary comparative morphologists and systematists did not collapse when Darwin transformed its meaning. For example, the main animal types, such as vertebrates, arthropods, and molluscs, were still recognized, but they could now be interpreted as lineages arising from common ancestors. The sort of dynamic thinking that Goethe used to understand the relationship between archetypes and their real-world manifestations, although based in a

---

[11] In his new book *Progress Unchained*, Peter Bowler argues that Lamarck's branching diagrams were not meant as phylogenies. Irrespective, Lamarck was undeniably an early lineage thinker.

completely different metaphysics, was conceptually similar enough to the Darwinian idea of real evolutionary transformations to allow Gegenbaur to reinterpret his typological comparative morphology along evolutionary lines without difficulty. The identification of homologous characters between taxa on the basis of similarities in their topological relations and connectivity was a tool that retained its sharpness when it was wielded by evolutionary morphologists. And the ties of homology that were discovered between species no longer received their meaning from reference to a type, but instead to an ancestor.

The Darwinian revolution was an intellectual phase transition in which everything took on a new meaning and significance. But not everyone welcomed this. Unsurprisingly, the biggest hurdle to acceptance was religion. Throughout the first half of the nineteenth century, comparative morphologists of a materialist bent had already found their worldview sharply at odds with the theism of those in power (Desmond, 1989). As we saw earlier in this chapter, even Richard Owen felt compelled to give his vertebrate archetype a metaphysical makeover in order to placate the theistic beliefs of his patrons. The pantheistic aroma of continental archetypal theories wrinkled many noses as effectively as the atheism of the most radical thinkers. Darwin's achievement certainly did not eliminate the distasteful connotations of a materialist worldview overnight. The initial reception of his ideas in France, for example, was strongly colored by a persistent biological tradition that was "still theoscientific in nature" (Farley, 1974: 275), according to which change in nature, if such was imagined at all, was channeled along pathways reflecting the Creator's purposes. Of course, Darwinism had to overcome such theistic evolutionism also in England (Bowler, 1989), and it was the unmistakable sound of disapproval that echoed closer to home that was much more upsetting to Darwin.

The *Origin* opens with a quote from the philosopher William Whewell stating that events in the "material world" are ruled by "general laws," while the first paragraph of the first chapter refers to the origin of species as "that mystery of mysteries, as it has been called by one of our greatest philosophers," the polymath John Herschel. Darwin was strongly influenced by the writings of these two pillars of the newly emerging philosophy of science (Hull, 1989; Ruse, 2000), and he arranged for both of them to receive an advance copy of the *Origin* accompanied by a personal note. Unsurprisingly, Darwin was dismayed when he learned that both Herschel and Whewell remained fundamentally unconvinced by his theory, with Herschel contemptuously referring to it as the "law of higgledy-piggledy." A theory that was cleansed of divinely ordained natural laws, and animated by blind selection of fortuitously favorable variants, was philosophically repulsive to both Herschel and Whewell. Of course, Darwin could have expected this reaction, at least from Whewell. He had lifted the opening quote on the title page of the *Origin* from Whewell's *Bridgewater Treatise* on astronomy, which he full well knew embodied a metaphysics almost perfectly antithetical to his own. The quote reads: "But with regard to the material world, we can at least go so far as this—we can perceive that events are brought about, not by insulated interpositions of Divine power, exerted in each particular case, but by the establishment of general laws." For Whewell this meant that although God didn't deign to be burdened with the daily running of his creation, he was still its ultimate architect. Darwin's subversive use of the quote meant that the God hypothesis was strictly unnecessary to make his own theory work.

But it wasn't just the religious implications of evolutionary descent that rankled. A small but vocal group of comparative biologists in the twentieth century judged the materialization of comparative biology to be a rather unwelcome change. Instead of trying to trace the contours of shapeshifting archetypes, comparative biologists in Darwin's world could now hunt for more concrete quarry. In his book on the fertilization of orchids, Darwin (1862: 288) wrote:

> [the naturalist] may feel assured that, whether he follows embryological development, or searches for the merest rudiments, or traces gradations between the most different beings, he is pursuing the same object by different routes, and is tending towards the knowledge of the actual progenitor of the group, as it once grew and lived. Thus the subject of Homology gains largely in interest.

Not everyone agreed. As plant morphologist Agnes Arber (1879–1960) phrased it,

> To many workers of the time, the diversion of biology into historical channels was a welcome relief, since it transformed theoretical botany into something material, amenable to picture-thinking, and not demanding difficult mental activity of a metaphysical kind. Thus, by a feat of legerdemain, which seems to have passed almost unnoticed, the Ancestral Plant was substituted for the Archetypal Plant, and those characters which had, with reason, been attributed to the mental conception of the archetype, were, without further justification, assumed to have been proved for an actual historically existent ancestor. (Arber, 1950: 63–64)

To suggest that Darwin's metaphysical revolution was nothing more than a cheap conjuring trick is an interesting, if perhaps also misguided, conclusion. A more charitable and accurate reading explains the adoption of Darwin's perspective because it made immediate and overwhelming sense of what was previously stupendously puzzling. Common ancestry and ancestors are the general principle and idiographic expression of the foundation of a new and consistent view of the natural world that for the first time *explained* the existence of unity of type. It was time for metaphysical gymnastics to come down to earth. The material bonds of ancestry and descent set new boundaries for morphologists' mental manipulations of organismal form. And within those boundaries comparative morphologists could reconstruct real history.

What, then, was Arber's problem? Why would a biologist writing in the middle of the twentieth century deplore that "[t]o evolutionary schemes, the type concept fell an immediate victim" (Arber, 1950: 63)? The short answer is that Arber was concerned with the epistemological purity of comparative morphology. Post-Darwinian idealistic morphologists didn't want to worry what their morphological comparisons could reveal about the real world. For them comparative morphology was a game of logic in the realm of pure theory that was played with pawns made of pure observations. Who knew if relations of homology between taxa traced lines of phyletic descent from common ancestors? Evolutionary descent exists in a realm inaccessible to empirical investigation. The observations of comparative morphologists and systematists can only offer speculative and indirect access to this realm. Besides, Arber noted that instead of discovering tree-like relationships between species, as one would expect in a Darwinian world, taxonomists and morphologists instead often found "a tangled reticulum in more than two dimensions [which] involves a loss of faith in phylogenetics" (Arber, 1950: 60). If purely empirical research, unsullied by any constraints from evolutionary theory, doesn't produce the type of relationships that evolutionists expect to find, perhaps it's best to steer clear of phylogenetics. Indeed, Arber comes to this conclusion with obvious relief, writing that the loss of faith in phylogenetics "has been a loss of shackles, and has given freedom for a revival of comparative morphology, studied for its own sake, and not in subservience to any evolutionary theory" (Arber, 1950: 66).

Twentieth-century idealistic morphologists like Agnes Arber, Adolf Naef, Sinai Tschulok, and Rainer Zangerl and pattern cladists of today adopt a stance of more or less strict historical agnosticism that tries to preserve the independence and epistemic priority of the study of comparative morphology over phylogenetic speculations (e.g. Zangerl, 1948; Rieppel, 2010b, 2011a, 2011c, 2016). Their work is firmly tethered to the empirical trinity of observation, description, and comparison, and they consider phylogenetic speculation an optional extra or derivative from what Naef called systematic morphology. These post-Darwinian idealistic morphologists have done useful work delineating the boundaries between facts and phylogenetic fantasy (see Chapters 6 and 7), but their unwillingness to dip their toes into the empirically inaccessible realm of phylogenetic transformations was frustrating to those who wanted their morphological research to be a window into the real world.

Willi Hennig developed his views on phylogenetic systematics in explicit opposition to the tenets of idealistic morphology, but thinking about types or ancestors are not necessarily sharply different activities. Adolf Remane defined the "systematic type" of a taxon as that constellation of traits that could be attributed to it based on the distribution of character states revealed by the relevant systematic relationships (Remane, 1956: 140–142). If those systematic relationships had been established correctly, they would allow the inference of a systematic type that could also be interpreted as a more or less accurate rendition of a once-living ancestor. The epistemic proximity of archetypes and ancestors facilitated the evolutionary reinterpretation of the insights of pre-evolutionary morphologists and systematists. It also provided flexibility for those who needed some time to adjust to the new world order. In his treatise on the morphology of invertebrates, T. H. Huxley wrote:

> Whatever speculative views may be held or rejected as to the origin of the diversities of animal form, the facts of anatomy and development compel the morphologist to regard the whole of the Metazoa as modifications of one actual or ideal primitive type, which is a sac with a double cellular wall, inclosing a central cavity and open at one end. This is what Haeckel terms a Gastraea. (Huxley, 1888: 50–51)

Some cladists have carried the torch of idealistic morphology into the new millennium but without wishing to shine a light on evolution. For them Darwin's theory of common descent just provides an excuse for naively literal thinking about imagined transformations of homologous characters (Brady, 1994; Williams and Ebach, 2008). These authors tell us that "the notion of transformation is really a myth" (Williams and Ebach, 2008: 259), that considering "common ancestors as real things" (Brower, 2014: 110) is a conceptual flaw, and even if they do believe in the reality of evolutionary transformations, hypotheses of common ancestry are said to have zero explanatory power (Patterson, 2002: 16, 20). This general outlook will receive a critical probing in Chapter 11, but first we need to get a better understanding of the epistemological transformation of comparative biology that accompanied the Darwinian revolution.

3

The Emergence of Lineage Thinking

In Chapter 2 we saw that comparative morphologists in the eighteenth and nineteenth centuries sought and found the unity of type in different ways. Their work culminated in the insight that unity of type is held together by the relationship of homology between corresponding parts of organisms. The insight by Darwin and Alfred Russell Wallace that these ties reflected descent from common ancestors was the decisive step in transforming comparative biology into a historical science. But this did not mean that relationships of homology between taxa could simply be reinterpreted as ancestor–descendant relationships. The atemporal ties of homology between contemporaneous taxa do not map neatly onto the temporal ties of ancestry. One of the tasks of Darwinian evolutionary theory was therefore to "harmonise many incongruities in the excessively complex affinities and relations of living things," as Wallace (1891: 8) put it. Moreover, application of the hallowed empirical trinity of observation, description, and comparison could not give morphologists and systematists direct access to the realm of evolutionary time. To transform comparative morphology into evolutionary morphology, and systematics into phylogenetics, biologists had to adopt a new way of thinking that I call lineage thinking.

Lineage thinking focuses on the ties of homology and the transformations of traits between ancestors and descendants. It is a fundamentally linear type of thinking that tracks evolution along lineages across phylogenetic trees. It is the type of thinking one uses when trying to understand the evolution of any trait, clade, or taxon of interest. Every evolutionist is intimately familiar with lineage thinking, but as far as I know, no one has proposed a specific term for it. Lineage thinking isn't simply an unnecessary synonym for evolutionary thinking. Similar to what Ernst Mayr called population thinking, it is a fundamental type of evolutionary thinking, but there are others. One, which may be called taxic thinking, is concerned with the systematic relationships between collateral relatives and the origins of biodiversity. Another has come to be called tree thinking (O'Hara, 1988), and it incorporates aspects of both lineage and taxic thinking (see Chapter 11 for further discussion). An important reason why we need a new term is that many of the conceptual flaws that I discuss in this book derive from improper lineage thinking. Examples are the accusation that Haeckel's trees are thinly veiled *scala naturae* (Chapter 4), that differences between taxa have phylogenetic value (Chapter 6), that the phylogenetic position of taxa can predict how likely it is that they have retained ancestral character states (Chapter 10), that evolutionary stories and imagery shouldn't be linear, that evolution doesn't happen by anagenesis, and that birds didn't evolve from dinosaurs (all discussed in Chapter 11). This chapter traces the historical emergence of lineage thinking.

3.1 Systematics and the Shape of Nature

As we saw in Chapter 2, Étienne Geoffroy Saint-Hilaire noticed during his debates with Georges Cuvier in the 1830s that a frustratingly large number of ideas in the history of biology had sprouted from one place: the fertile mind of Aristotle. It is no different with the historical roots of natural history and taxonomy. These trace back at least two millennia to the zoological and botanical works of Aristotle (384–322 BC), and his student and collaborator Theophrastus (ca. 371–287 BC) (Lovejoy, 1936; Mayr, 1982; Ogilvie, 2006; Grant, 2007; Wilkins, 2009; Leroi, 2014). In their hands, natural history arose as an empirical enterprise aimed at generating a philosophical understanding of nature. However, although a continuous textual tradition links the founding works of Aristotle and Theophrastus to modern times, they are not at the core of an unbroken research tradition.

Natural history was reinvented in the sixteenth century as an autonomous discipline dedicated to the exacting empirical study of nature (Ogilvie, 2006). Breaking with the tradition of natural philosophy, which traces back to even before the ancient Greeks (Grant, 2007), Renaissance natural historians no longer primarily collected facts to underpin a philosophical understanding of nature. Instead, they preoccupied themselves with painstakingly observing, describing, and comparing nature's particulars, propelled by a strong aesthetic interest. They were more interested in nature's productions for their own sake than seeing them as murky windows into nature's essences. Similarly, the natural historians' pursuit of what Ogilvie (2006: 8) calls a "fine-grained empiricism" was sharply at odds with the preceding medieval practice of using plants and animals as vehicles for moral and spiritual instruction. As Ogilvie (2006: 102) writes, if medieval "bestiaries' readers learned a few facts about animals, that was incidental."

This fixation on accurately rendering Mother Nature's dress, instead of trying to peek under it, also betrays a myopia in the Renaissance naturalists' gaze. For instance, naturalists had long noticed and described the dew-like droplets on the leaves of the sundew (genus *Drosera*), and they must have seen the insects that were frequently stuck to them. Yet, it wasn't until 1623 that Caspar Bauhin (1560–1624), a Swiss botanist and then Europe's most famous naturalist (Ogilvie, 2006: 267), recorded the observation that after touching these sticky droplets, they could be drawn out into hardening filaments. But this then did not lead Renaissance naturalists to write about plant carnivory. They strictly used their observations for descriptive purposes, not for any theoretically guided penetration of nature's deeper mysteries. Naturalists were forced to weave a philosophical strand back into the fabric of their investigations from the late sixteenth century onward in order to deal with the increasingly cumbersome catalogue of life.

The corpus of natural history was becoming unwieldy by the end of the sixteenth century. Whereas several hundred species of vascular plants had been included in the herbals during the 1530s and 1540s, Bauhin had to fit more than 6,000 species into his *Pinax Theatri Botanici* (*Illustrated Exposition of Plants*) of 1623 (Ogilvie, 2006: 139). Similarly, when the sheer diversity of the animal world started to sink in, it signaled the end of the era of comprehensive, loosely structured animal compendia, such as Conrad Gessner's alphabetically organized *Historia Animalium* (*History of Animals*) from the second half of the sixteenth century, which even reserved space for fable animals. Species-rich groups, such as insects, clearly required specialist attention. Paralleling the situation in botany, Renaissance work on insects had focused on producing detailed observations and descriptions. But from around the turn of the seventeenth century, authors started to devote much more of their attention to classification (Engel and Kristensen, 2013). This is when natural history gradually gave birth to systematics.

An interest in classification was not completely absent in the early Renaissance, but pragmatic concerns were initially the strongest drivers of taxonomic work (Winsor, 1976; Mayr, 1982; Stevens, 1994; Ogilvie, 2006). One of the things early naturalists needed most were identification keys for common flora and fauna to aid medical practitioners and teachers, as well as effective information storage and retrieval systems for plants with desirable medicinal properties. The result was a preponderance of classifications tailored to specific practical needs. But naturalists recognized such classifications for what they were: artificial. They did not try to impose a realist interpretation upon these schemes. The resulting groupings received a strictly nominalist interpretation – the names of groups and taxa were not meant to imply that these really existed as things in the world. They were convenient mental constructs erected for purely utilitarian ends. But most naturalists also intuited that the organisms they studied were not just isolated leaves on an endless lawn, swept into arbitrary piles by their attempts at classification. Mapping out the structure of what came to be called the "natural system" became a prime concern for systematists, especially from the mid-eighteenth century onward (Winsor, 1976; Mayr, 1982; Stevens, 1994; Amundson, 2005; Pietsch, 2012; Archibald, 2014). Yet it wasn't until well into the nineteenth century that it became clear from what kind of superstructure the leaves had dropped.

The flux of conceptual and empirical developments in eighteenth- and nineteenth-century systematics was great and complex. Outlining all the relevant epistemological, ontological, and disciplinary details associated with the fitful transition from a pre-evolutionary to an evolutionary systematic biology is beyond the scope of this book. For our purpose of understanding the emergence of lineage thinking, we will limit ourselves to a consideration of several interrelated issues: methods of classification and the geometry and interpretation of the natural system.

3.2 Methods of Classification

Many naturalists understood that their utilitarian classifications were unlikely to reflect the natural order, if they even believed that a verifiable natural order existed. Artificial classifications, such as identification keys, were often produced by logically dividing larger groups into smaller ones. Called downward or analytical classifications (Mayr, 1982: 158–166; Stevens, 1994: 5), such utilitarian schemes were generally not serious attempts to carve nature at her joints. Although rooted in natural history's empirical trinity of observation, description, and comparison, utilitarian classifications were deliberately not based on a comprehensive consideration of all available evidence. For instance, Linnaeus' sexual system of plant classification was rooted in flower anatomy, and he acknowledged it was artificial for being so narrowly restricted. Linnaeus thought that his classifications also revealed real fragments of the structure of the natural system – he was a realist about the existence of genera, for instance (Stevens, 1994: 177; Stamos, 2005: 86, 93; Breidbach and Ghiselin, 2006: 21–22; Ragan, 2009) – but naturalists began to understand that the contours of the natural system were more likely to emerge when one applied a more densely empirical approach to classification.

Two pioneers who developed a more empirical approach to classification were the French botanists Antoine-Laurent de Jussieu (1748–1836) and Michel Adanson (1727–1806) (Mayr, 1982; Stevens, 1994). Although Adanson's taxonomic work had little influence on the course of systematics (Mayr, 1982: 195; Stevens, 1994: 95), de Jussieu's approach to classification was profoundly influential in both botany and zoology. Adanson and de Jussieu advocated an upward or synthetic approach to classification, which involved grouping species together into ever more encompassing higher taxa on the basis of information gleaned from a diversity of character systems (Mayr, 1982: 190; Stevens, 1994: 3). In their view, the strength of a natural classification depended on the breadth of its empirical foundation. However, the fact that botanists built their classifications on a broad empirical basis did not mean that they considered all available evidence. Characters derived from the internal anatomy of plants were generally disregarded until well into the nineteenth century, a situation in striking contrast to zoology where investigations of internal anatomy were important for classification much earlier (Stevens, 1994).

A comprehensive approach to data gathering did not mean that characters were selected indiscriminately or weighted equally. The application of a deliberately unequal eye to available evidence was assumed to result in artificial classifications. A richly empirical approach to classification increasingly came to replace the downward or analytical classification by logical division from the eighteenth century onward. Still, an empirical approach to classification did not guarantee a consensus of results. Incomplete evidence and unequal attention to available data, as well as different aesthetics, conspired to produce a kaleidoscope of systems. In botany, for instance, dozens of configurations of the higher-level relationships of plants emerged from the late eighteenth to the mid-nineteenth centuries (Stevens, 1994: 64). Such classifications came in a great variety of shapes.

3.3 The Shape of the Natural System

Eighteenth- and nineteenth-century naturalists explored a great many classificatory geometries, including linear series, branching trees, reticulating networks or webs, maps, tables, and other geometric forms such as circles, stars, triangles, and polyhedrons (O'Hara, 1991;

Stevens, 1994; Ragan, 2009; Rieppel, 2010a; Tassy, 2011; Pietsch, 2012; Archibald, 2014). The philosophical basis of the simplest and oldest of these iconographies – the linear *scala naturae* or Great Chain of Being – can be traced back two millennia to Aristotle's thinking about the principles of plenitude (every conceivable form does in fact exist) and continuity (nature's productions form a graduated series without sharp qualitative breaks), although Aristotle himself never drew an actual *scala naturae* (Lovejoy, 1936; Wilkins, 2009; Archibald, 2014; Leroi, 2014). Yet, despite its antiquity, or perhaps because of it, the practical role that the idea of the *scala naturae* played in day-to-day systematic work in the eighteenth century was limited. The concept of the *scala naturae* was tied to the idea of man as the measure of all things. While this anthropocentrism appealed to natural philosophers who pondered reality at the grandest of scales (Lovejoy, 1936), it was of less theoretical and practical value to natural historians engaged in more mundane activities, such as trying to work out the relationships of thistles or periwinkles.[1]

Even the most enthusiastic advocates of the *scala naturae* found it hard to compress all of nature's diversity into a single chain. Charles Bonnet (1720–1793), the Swiss naturalist philosopher who is perhaps the best known advocate of life's linearity, asked himself whether he could justify cramming shellfish and insects into a single linear chain running from the elements to man (Ragan, 2009). The famous and frequently reproduced ladder in his 1745 treatise on insects shows that he answered in the affirmative, but he saw his effort "only as a trial" and "perhaps foolhardily" (Archibald, 2014: 7). Indeed, naturalists in the eighteenth century increasingly rejected the *scala naturae.* If one didn't pay too much attention to individual characters, one could line up organisms along a vague gradient of complexity or perfection. But such a scale was scarcely a useful taxonomic principle if you worked on organisms with body plans of broadly comparable complexity. Also, the limitations of a linear system quickly became clear if you studied more than a single character system.

A single, linear order of nature was increasingly seen as the imposition of an idea rather than the discovery of a pattern. This was certainly true for one of the last comprehensively linear schemes of animal relationships that was advocated by a contemporary of Geoffroy and Cuvier during the first half of the nineteenth century in France. Henri de Blainville (1777–1850) was initially on friendly terms with both Geoffroy and Cuvier, but eventually he fell out with both of them and their ideas, although he succeeded Cuvier in the chair of comparative anatomy after his death at the Muséum d'Histoire Naturelle. He rejected Geoffroy's unity of plan, but like Geoffroy, he saw no unbridgeable gaps between the main animal types. Blainville strung together Cuvier's *embranchements* (taxonomic branches)

---

[1] After completing my manuscript, I became aware of Peter Bowler's 2021 book *Progress Unchained*. In Chapter 2 of his book, he traces the history of the idea of progress by discussing many of the same thinkers as I do here in Chapter 2 and especially Chapter 3. However, our perspectives differ, as he traces the continuity of the idea of progress from the *scala naturae* to temporalized versions of systematic relationships, whereas I focus on the emergence of lineage thinking and how that changed the meaning of relationships between taxa. Readers might find it instructive to compare our accounts, especially our disagreements about Ernst Haeckel's phylogenetic thinking (see Chapter 4).

along a linear continuum, from sponges to mammals. Yet, no matter how much, or how little, this arrangement reflects the empirical results of morphological research, Blainville's worldview as a devout Catholic simply demanded a progressive animal series. His system was a static taxonomic arrow pointing through man to God (Appel, 1980).

3.3.1 Character Conflict

Character conflict forced systematists to contemplate nonlinear iconographies such as parallel, branching, or reticulating schemes (Stevens, 1994; Rieppel, 2010a). This was especially true in botany, where pervasive character conflict led to a preponderance of reticulating and map-like diagrams of relationships, right up to the end of the nineteenth century when the depiction of the natural system as an evolutionary tree or bush became increasingly popular (Stevens, 1994; Morrison, 2014, 2015). For instance, one of Europe's leading botanists of the early nineteenth century, the Swiss naturalist Augustin-Pyramus de Candolle (1778–1841), noted in his major work *Théorie Élémentaire de la Botanique* (*Elementary Theory of Botany*, in which he coined the word "taxonomy") that a linear series could not accurately depict affinities because "each genus, or each family, would resemble not only the group that precedes it and the group that follows it, but would have multiple affinities with many other groups" (de Candolle in Nelson and Platnick, 1981: 103). De Candolle therefore preferred, following Linnaeus, the image of a geographical map for depicting plant relationships, a map with territories that would be progressively filled in as taxonomists circumscribed more and larger taxa.

The ability of character conflict to collapse linear classifications was not restricted to particular times, taxa, or systematic philosophies. This is nicely illustrated by Lamarck's classification of animals (Burkhardt, 1995: 160–164; Gould, 2000: 124–143). Whereas de Candolle rejected the *scala naturae* and saw nature as a landscape or archipelago of discrete higher taxa separated by real gaps, Lamarck perceived higher taxa to be arbitrary segments delineated along a linear, evolutionary continuum. Starting in 1797 with a single animal series containing five classes of invertebrates, Lamarck inserted new classes into various positions in this chain in subsequent years. However, the chain fractured because he could not smoothly accommodate parasitic worms. Lamarck had long suspected that Linnaeus' Vermes was a can of worms. It grouped together animals with vastly different body plans, including complex, free-living annelids and morphologically simpler parasitic worms. Lamarck placed annelids even higher than insects in his animal scale, but although parasitic worms did show advanced traits, like bilateral symmetry, that would place them above animals such as echinoderms, their lack of internal complexity suggested they should actually be lower on the scale.

To resolve this conflict, Lamarck arranged the diversity of the animal kingdom in two parallel series. He established the class Annelida to house the morphologically complex, free-living worms, grouping them together with other segmented invertebrates in a series that originated in the morphologically simple parasitic worms. He then placed nonsegmented animals, such as molluscs, ascidians, cnidarians, and echinoderms, in a parallel series that began in an infusorian cradle. In his last book of 1820, titled *Analytic System of*

Positive Knowledge about Man, Lamarck integrated these parallel series into a branching, monophyletic tree of the animal kingdom.

Lamarck's revision shows that a move away from strict linearity does not have to sever the threads of taxonomic continuity. Branching and reticulating schemes could still accommodate the continuity of nature that had been at the core of the *scala naturae* and which had functioned as a leitmotif of systematic practice throughout the eighteenth and nineteenth centuries (Stevens, 1994). For de Jussieu and many of his successors, the need for a synthetic, empirical approach to systematics was rooted in the very conviction that nature was continuous. If nature is continuous, the limits of higher taxa cannot be recognized by the presence of conveniently placed gaps. They can then only be circumscribed by a catholic synthesis of empirical evidence. But when character conflict is rampant, a dispassionate empirical probing of nature's variety is unlikely to reveal an orderly geometry. This explains in part why Cuvier rejected Lamarck's linear animal series (Eigen, 1997: 206). Influenced by de Jussieu's thinking about continuity, Cuvier saw species ensnared in a network of relationships that radiated in multiple directions, depending on which character you followed. Lamarck's linear series were the result of applying an unequal eye to available evidence. Overly neat and aesthetically pleasing classifications clearly risked being seen as the imprints of preconceived ideas rather than as discovered patterns, even if they contained the seeds for conceptual progress. This is strikingly illustrated by the ingenious and infamous productions of the quinarians.

3.3.2 Quinarianism and the Distinction of Affinity and Analogy

Quinarianism is a remarkable taxonomic philosophy that flourished briefly during the early decades of the nineteenth century, especially among members of the Zoological Club of the Linnean Society of London (Desmond, 1985; O'Hara, 1991; Ragan, 2009; Pietsch, 2012; Archibald, 2014). The principles of quinarianism were established by entomologist William Sharp MacLeay (1792–1865) in his two part work *Horae Entomologicae, or Essays on the Annulose Animals* (1819–1821). Looking at the diversity of dung beetles, MacLeay saw circles of five. Then, broadening his view to encompass the entire animal kingdom, he saw nature revealed in a vista of circles. Each quinarian circle arranged five taxa along a continuous line of affinity (Figure 3.1). According to MacLeay a "natural series of affinity is such as, taking the majority of characters for our guide, shall be found uninterrupted by any thing known, although possibly broken by chasms occasioned by the absence of things unknown" (MacLeay, 1819–1821: 401).

MacLeay constructed – although he would say discovered – his circles by connecting the beginning and end points of linear series of affinity. His circle of vertebrate relationships, for instance, was unexceptional in its progression from fishes through amphibians, reptiles, and birds to mammals, but MacLeay bent the line into a circle by connecting mammals, via cetaceans, back to fishes: "the *Cetacea* lead us by a very distinct and natural transition from the *Mammalia* to Fishes" (MacLeay, 1819–1821: 272). Indeed, MacLeay considered his ability to convert linear classification schemes into circular ones to be a sure sign of the strength of his approach. He thought that one of his signal achievements was his transformation of Lamarck's two parallel animal series of 1815 into a single circular system.

(a)

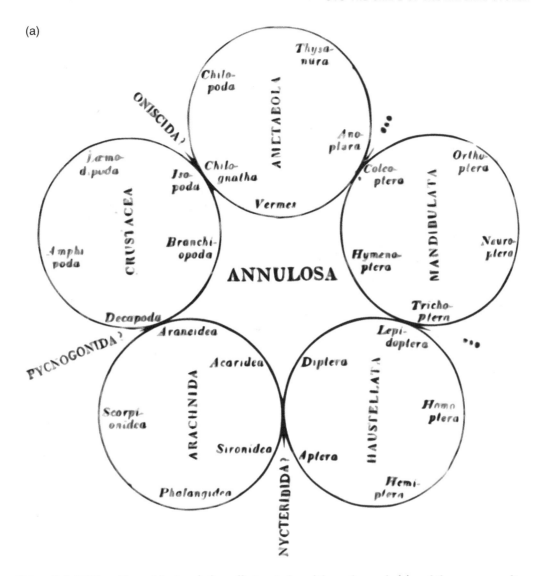

Figure 3.1 William Sharp MacLeay's five affinity circles of the arthropods (a) and the corresponding (b) pattern of analogies
(after pages 390 and 395 in MacLeay, 1919–1921).

"We have only to join the *Radiata* to the *Cirripeda,* and the *Annelides* to Fishes" to convert what "was little less than inspiration on the part of Lamarck" into convincing proof of the veracity of MacLeay's system (MacLeay, 1819–1821: 332–333).

Even seemingly extreme chasms between body plans were no match for the continuity of MacLeay's circles of affinity. He pressed some of the most idiosyncratic animals into service as intermediates between highly dissimilar taxa. For example, although "no two animals can differ so much from each other as a bird and a tortoise" (MacLeay, 1819–1821: 264), the

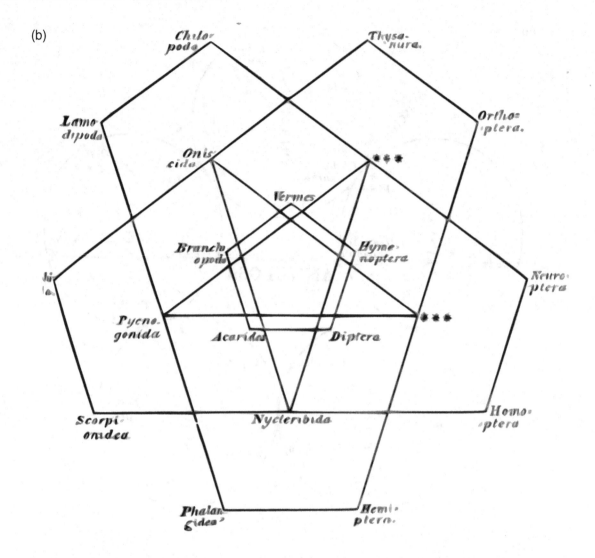

Figure 3.1 (*cont.*)

horny beak of turtles provided MacLeay with a strong clue that chelonians were the reptile group most likely to reveal the transition to birds:

> [T]he Hawksbill turtle puts the matter beyond doubt. This animal exhibits to the view a rude sketch of the form of a Bird, so distinctly that we can hardly refrain from supposing that Nature must, in a sportive mood, have intended to show us by the union of animals totally dissimilar in habit, what wonders she can perform in the prosecution of her favourite principle of affinity. (MacLeay, 1819–1821: 263)

As to where this chelonian bridge would make landfall,

> the birds which come the nearest to this animal in structure must be aquatic; that they ought to be covered with scales rather than with plumes; that their sternum ought to be very large, protecting all their viscera; their wings short, of no use for flight, but serving rather as fins to swim with; finally, that their legs ought to be placed so far behind as to render the bird almost incapable of walking. They ought in short to be true reptiles with respect to locomotion. If such then be the sort of bird we are to look for, who does not see the Patagonian Penguin or the genus *Aptenodytes* of Forster in the above description. (MacLeay, 1819–1821: 264)

Although MacLeay was convinced that "[t]hose birds which appear in their internal as well as external structure to approach the nearest to the *Mammalia* belong undoubtedly to the Ostrich family" (MacLeay, 1819-1821: 267), he couldn't locate the transition from birds to mammals in any particular group of birds. This was not because it didn't exist, but because this link had not yet been discovered among living or fossil animals. For MacLeay the principle of plenitude was threaded through nature, so that the natural system, when it had been fully reconstructed, would be extremely regular and devoid of gaps: "the test of a relation of affinity is its forming part of a transition continued from one structure to another by nearly equal intervals" (MacLeay, 1819-1821: 364). If taxa in a circle could not be arranged in a smoothly regular series, MacLeay inserted placeholders, confident that gaps would be filled by taxa yet to be discovered. The quinarian carousel allowed no empty seats.

Just as character conflict had forced Lamarck to move to a nonlinear and increasingly branching arrangement of taxa, so too did MacLeay realize that animal diversity could not be arranged in a single affinity circle. If a taxon could not be inserted into a circle without disrupting the regularity of the intervals separating its constituent taxa, he placed it into a new circle, or as an intermediate taxon between circles. At each supraspecific taxonomic level, MacLeay therefore constructed five circles of affinity that together formed a seamless ring, with each of the five taxa in each circle themselves containing five lower-level taxa arranged in a circle. So earwigs came to glue together the rings of beetles and orthopterans, cephalopods joined molluscs to vertebrates, and barnacles cemented the affinity of Annulosa (segmented invertebrates) and Radiata (echinoderms and jellyfish) on the strength of what MacLeay saw as similarities between barnacle valves and the mouthparts of sea urchins (MacLeay, 1819-1821: 206, 313, 436-437).

The unbroken continuity of affinity between circles was essential. Quinarian aficionado Nicholas Vigors (1785-1840), cofounder of both the Zoological Club of the Linnean Society and the Zoological Society, was distressed that a printer error had resulted in one of the two quinarian diagrams in his 1825 monograph on bird classification being printed with the circles not touching. This suggested the "erroneous idea of the series of affinity being incontinuous" (Vigors, 1825: 509). For quinarians, nature's continuity did not just run through ties of affinity. It was also expressed in relations of analogy, which "consist in a correspondence between certain insulated parts of the organization of two animals which differ in their general structure" and which may reflect "barely an evident similarity in some one or two remarkable points of formation" (MacLeay, 1819–1821: 363–364). MacLeay pinpointed analogies across nature's taxonomic tableau, from the "obvious analogy" of the jaws of centipedes and carnivorous neuropteran larvae, to the similarities of fly maggots and parasitic worms "in form as well as manners" (MacLeay, 1819–1821: 397–398). Analogies were especially conspicuous when relations of affinity were least obvious. In an example of a kingdom-spanning analogy, MacLeay labelled dicotyledonous plants the "*Vertebrata* of the vegetable kingdom" because of "their hard or osseous parts being as it were in the middle, and thus affording the most perfect and intricate plan of vegetable construction" (MacLeay, 1819–1821: 211).

The distinction between affinity and analogy was crucial for MacLeay. Only relations of affinity tied together natural groups. Analogies on their own demarcated artificial groups. Those who mistook analogies for affinities were fated to produce false taxonomies, a danger to which, MacLeay felt, not even the most illustrious naturalists were immune (MacLeay, 1819–1821: 365, 392, 453). Although analogies are misleading when considered in isolation, they provided MacLeay with an independent strut for his quinarian classifications, especially where affinities on their own provided less than compelling evidence (MacLeay, 1819–1821: 401). Once the naturalist had puzzled out the correct pattern of affinities he could use "the study of relations of analogy... to throw light on almost every general and specific notion, that may have been or can be entertained on the nature of these animals" (MacLeay, 1819–1821: 396). By connecting taxa in corresponding positions in different circles of affinity, relations of analogy completed the harmony, symmetry, and regularity of the quinarian system, thereby lending support to the whole edifice. Tracing analogies within segmented invertebrates, MacLeay perceived a pattern so stunningly geometrical that he wrote: "Nothing in Natural History is, perhaps, more curious than that these analogies should be represented by a figure so strictly geometrical. One is almost tempted to believe that the science of the variation of animal structures may, in the end, come within the province of the mathematician" (MacLeay, 1819–1821: 211, 397–398) (Figure 3.1b).

Several influential naturalists, including Louis Agassiz (1807–1873) and Thomas Henry Huxley (1825–1895), similarly used analogies to prop up affinity relationships of animals depicted in parallel series or circles (MacLeay, 1819–1821: 363, 380, 401). While toiling aboard the *HMS Rattlesnake*, the young Huxley developed ideas about the systematic relationships of polyps and jellyfish that bore some striking similarities to a quinarian system (Huxley, 1850; Winsor, 1976: 81–94, 136–138). Huxley met the older MacLeay, who had emigrated to Australia, in Sydney, and they struck up a correspondence that continued after Huxley had returned to England. Huxley suspected that there was "a great law hidden in the 'Circular system' if one could but get at it" (Huxley, 1850: 67). But when he attempted to demystify the difference between affinities and analogies in terms of embryological development – by proposing that affinities are rooted in shared developmental processes, while analogical traits have different kinds of development – MacLeay demurred (Winsor, 1976: 92). For him affinities and analogies were distinct. Affinities tied together organisms of broadly similar organization into natural series, whereas analogies were isolated correspondences between fundamentally dissimilar organisms, but these differences were not explicable in terms of materialistic causes. For MacLeay this distinction was more epistemological than ontological. Unexpected ties of affinity could be discovered when a sufficient number of what at first appeared to be analogies tied together two taxa into a common type of organization. For instance, noting the striking analogies between the aquatic larvae of caddisflies and certain lepidopterans, including their gills and protective casings, MacLeay concluded that "the particulars of analogy become here sufficiently numerous to compose an affinity" (MacLeay, 1819–1821: 366), so cementing together the different circles housing these taxa.

MacLeay's insights into the concepts of affinity and analogy were modified and refined by others (Amundson, 2005: 45–50; Williams and Ebach, 2008: 155–167). This ensured that his ideas were incorporated into the canon of systematics. However, these other naturalists were not quinarians. Their revisions fundamentally changed the systematic significance of MacLeay's two types of correspondence, they rejected analogies as taxonomically impotent, and they unwittingly readied affinities for a metaphysical makeover that would leave the rationale of the quinarian system in tatters.

3.4 Toward a Realist Interpretation of the Natural System

Before the early nineteenth century, few naturalists were committed to a realist interpretation of the natural system (Stevens, 1994; Amundson, 2005; Stamos, 2005). The idea of species realism – that biological species are not just imagined or arbitrary chunks of nature's continuous fabric, but are instead real segments of natural diversity – had already spread widely from the mid-seventeenth century into the eighteenth century, largely piggybacking on the notion of species fixism (Amundson, 2005; Stamos, 2005; Wilkins, 2009). The fixist interpretation of species replaced previous beliefs that species could transmutate haphazardly across generations, for instance through hybridization (giraffes, *Giraffa camelopardalis*, being leopard-camel hybrids) or spontaneous generation (barnacle geese sprouting from rotting logs via stalked barnacles – the scientific name of the gooseneck barnacle, *Lepas anatifera*, still betrays its etymological origin in the Renaissance word "anatiferous," meaning "producing duck or geese'"). As Amundson (2005: 34–39) has argued, the idea that species are fixed or permanent was a necessary precondition for building the natural system, because without the existence of discoverable, more or less stable patterns of species relationships, the natural system would dissolve into a taxonomic quicksand. The move toward a realist interpretation of supraspecific taxa, however, did not gain serious momentum until the nineteenth century.

Before then, most classifications were constructed to fulfil the pragmatic need of ordering the chaos of species diversity into a workable constellation of higher taxa. Most such classifications were acknowledged to be artificial. They were mental, nominalist constructs. Their order was manufactured, not discovered. Even richly empirical taxonomic work that aspired to transcend the artificial systems was not guaranteed a realist interpretation. The nominalist interpretation of the natural system was entwined with the pervasive acceptance of the principle of continuity in the eighteenth and early nineteenth centuries (Lovejoy, 1936; Rieppel, 1988a; Stevens, 1994; Amundson, 2005; Stamos, 2005). If the fabric of nature is continuous, higher taxa can only be demarcated arbitrarily. That is why de Jussieu, who was an important pioneer of the synthetic empirical approach to systematics, was not a taxonomic realist. He considered his classifications to be natural because they were based upon a broad selection of characters, but the limits of his higher taxa were arbitrary. For de Jussieu, "natural" did not equate with "having a real existence independent of mind" (Stevens, 1994).

The same is thought to be true for Lamarck, who considered supraspecific taxa to be convenient constructs, without any real existence. After he became an evolutionist, he expanded this nominalist interpretation to the species level as well (Stafleu, 1971: 424; Wilkins, 2009: 104-108). When authors did suspect that their taxonomic efforts revealed something real about nature, although many didn't state their beliefs one way or the other, theirs was typically a cautious realism, uncommitted to any specific ontological nature or ultimate reality (Amundson, 2005: 34). This situation was to change dramatically in the nineteenth century. And among the revolutionary vanguard that advanced a realist inter-pretation of the natural system, we again find William MacLeay.

3.4.1 Quinarianism Revisited

MacLeay did not hide his disdain for architects of artificial systems. In the preface to the *Horae Entomologicae*, he noted that these could be constructed with "the exercise of no other faculty than that of vision," requiring "neither talent nor ingenuity" (MacLeay, 1819-1821: xiii). Besides cautious inductive reasoning, those in search of the natural system also needed courage and conviction: "it is said that every naturalist who has hitherto proposed a natural system, has thereby only deceived himself and others with an illusive structure, which, like the castle in a fairy tale, falls to pieces on being tried by the talisman of truth." (MacLeay, 1819-1821: xviii) Worse, anybody presumptuous enough to think that they had seen the shimmering outlines of the natural system risked being labelled a "false naturalist" (MacLeay, 1819-1821: xviii). Still, MacLeay was convinced he possessed the talisman of truth: the distinction between relationships of affinity and analogy. The fact that these types of correspondence tallied so neatly to produce pentamerous circles bridged by parallel links, whether they arranged beetles or the kingdoms of life, was for him incontrovertible evidence that his system captured something real in nature.

Moreover, MacLeay managed to escape the taxonomic nominalism inherent in accepting the indivisible continuity of life by making his affinity series circular. I don't know if this was a deliberate part of his strategy, but in this way he ensured that his natural groups were at the same time self-contained and part of a seamless continuum of affinity by virtue of the circles touching one another. By placing his system on a divinely ordained basis, MacLeay left his readers in no doubt about the ultimate reality of his natural groupings: "the harmony it may display is the work of God, not man" (MacLeay, 1819-1821: 324).

MacLeay's realism about his affinity circles was a major step toward taxonomic realism in systematics (Amundson, 2005: 45-50; Williams and Ebach, 2008: 155-167). Others followed his lead, but with the exception of a few, most infamously his fanatical acolyte William Swainson (1789-1855) (see Desmond, 1985: 168-178 for some biographical details about this tragic character), they were not aiming for the same destination. Entomologist John

Obadiah Westwood (1805–1893); ornithologist, geologist, and Victorian dodo expert Hugh Edwin Strickland (1811–1853); and comparative anatomist Richard Owen further developed the conceptual distinction between affinity and analogy. Their work ensured that MacLeay's insight that natural groups are tied together by bonds of affinity came to be widely accepted.[2] Yet after the 1840s, few naturalists who were not card-carrying quinarians followed MacLeay in interpreting relationships of affinity and analogy as complementary pillars of the natural system. Instead, analogies were damned as treacherous sirens that lured naturalists into taxonomic quagmires.

After seeing "affinities and their attendant analogies all combining to one sublime effect" in his studies of the animal kingdom, MacLeay asked: "Is it then possible to suppose that the rest of organized beings constitute a chance-medley map of reticulation, as some seem to think, or offer to the view a few scattered fragments of a temple now in ruins, as others have esteemed them to be?" (MacLeay, 1819–1821: 460). This was precisely the conclusion that many naturalists felt forced to swallow when they attempted to tame morasses of character conflict into intelligible systems.

Corresponding with Swainson, Zoological Club cofounder Adrian Haworth wrote: "I'm afraid they [analogies] are merely the shadows of a shade: and of no importance whatever... For, were they otherwise, whatever should we do in Botany? where so many 1000ds of Plants occur clothed in the foliage (and even flowers) of other groups" (Desmond, 1985: 163). It was hopeless to try and incorporate into classifications the isolated and incidental similarities between fundamentally different taxa. Even if taxonomists only focused on taxa with comparable body plans, different character systems often threw up confusingly complex relationships. Rather than seeing the symmetry and regularity of quinarian classifications as a talisman of truth, most mid-nineteenth century naturalists dismissed these graceful geometries as manmade illusions, MacLeay's passionate protestations to the contrary notwithstanding (MacLeay, 1819–1821: 322, 324, 363). Similarly, the empty seats reserved in the quinarian carousels for future discoveries could easily be dismissed as self-fulfilling prophecies rather than hallmarks of the method's heuristic value. Strickland captured this growing consensus in 1840 when he wrote that one possessing a "healthy tone of mind" would "Instead of assuming an *à priori* system of his own, and then twisting the facts into a partial coincidence with that system, he is content to take Nature as he finds her, and not the less to admire her luxuriant variety because she refuses to marshal her irregular troops into straight lines, circles, or pentagons" (Strickland, 1840: 219–220).

It doesn't become easier to distinguish affinities and analogies when a priori taxonomic groupings are rejected. As Strickland (1840: 220) observed, "[i]n order to arrive at their definitions, we must first prove the existence of a real *natural system,* a subject which

---

[2] In discussing how Strickland's work on affinity and analogy built upon and departed from that of MacLeay, Amundson (2005: 47–48) and Williams and Ebach (2008: 162) correctly conclude that for Strickland, in contrast to MacLeay, analogies were taxonomically impotent. Their discussions also create the misleading impression that it was Strickland's idea that only ties of affinity unite natural groups. This was already at the heart of MacLeay's thinking. Where Strickland differed was his denial that analogies indicated any meaningful relationships in addition to affinities.

involves an enquiry into the designs of creative power, one of the most awful themes which the human intellect can attempt" [italics in original]. This presents a thorny problem. If the natural system must first be known before affinities and analogies can be distinguished, how does one even start? Strickland's starting point is the observation that organisms resemble each other to different degrees, and it is so-called essential characters that mark out ties of affinity. These essential characters are those that mark out larger taxa of obviously related species, such as beetles, or vertebrates. But to recognize the essential characters in each particular case "must, in great measure, be left to the determination of what Linnaeus called a 'latent instinct' which Professor Whewell defines to be 'an unformed and undeveloped apprehension of physiological functions'" (Strickland, 1840: 221).

The recognition that natural groups are rooted in relationships of affinity and that analogies should be discarded was not the only route toward increased taxonomic realism. Taxonomic nominalism could also be avoided by breaking nature's continuum, as MacLeay had achieved, wittingly or unwittingly, by making his affinity series circular. Cuvier, who did not distinguish analogies and affinities, cut through nature's continuity by basing his four animal *embranchements*—Radiata, Mollusca, Articulata, and Vertebrata—on different patterns of functional organization, separated from each other by nonfunctional gaps (Eigen, 1997; Amundson, 2005: 42). Cuvier rejected any linear continuum of groups. But he accompanied his realism about *embranchements* and species (he was a species fixist) with a nominalism about intermediate level taxa. Species in each *embranchement* were part of a continuous fabric, but since each species was perched upon its own isolated pinnacle of functionality, there were for Cuvier no objective relationships between species that demarcated real supraspecific taxa such as genera or families.

Karl Ernst von Baer (1792–1876) likewise conceived of four types in the animal kingdom, which coincided with Cuvier's *embranchements*, even though his taxonomy was based on shared patterns of development rather than functional organization. Louis Agassiz (1807–1873), who like MacLeay used both affinities and analogies as props for the natural system, also ascribed an objective reality to the supraspecific groups captured by Cuvier's *embranchements* and von Baer's types (Rieppel, 1988b: 34; Wilkins, 2009: 113). Like MacLeay, he thought that natural groups reflected God's creative plan, but in contrast to Cuvier, he was also a realist about taxa on intermediate levels (Winsor, 1991: 24–26).

This illustrates that steps toward a realist interpretation of the natural system in the nineteenth century took place in different epistemological and ontological contexts. The taxonomic coincidence of different systems added support to the notion that classifications could capture an objective reality that existed independent of the human mind (Amundson, 2005: 42, 60, 133). Still, there was another, more ancient, route to realism that was poised to take center stage.

3.5 The Temporal Reinterpretation of the Natural System

For all their geometrical, epistemological, and ontological diversity, most schemes of the natural order conceptualized and drawn before the mid-nineteenth century had one thing in common: they were atemporal. Nothing moved along the affinity lines drawn between taxa. Nothing happened there, not even along the most linear of them. The linear order of increasing complexity or perfection in early depictions of the great chain, ladder, or staircase of being was one of divinely ordained permanence. It did not trace a time axis. Likewise, some of the first tree-like taxonomic diagrams, such as those published by the German botanist Johann Philipp Rühling in 1774 and the French botanist Augustin Augier in 1801 (reproduced as figures 18 and 21 in Pietsch, 2012, respectively), were similarly static structures (Lovejoy, 1936; Stevens, 1994; Fisler and Lecointre, 2013; Archibald, 2014). The taxonomic networks, webs, and maps that resulted from systematists' efforts to deal with rampant character conflict often had no discernible direction at all. From the very infancy of natural history, many had sensed a broad-brush directional order in nature's diversity. It was especially thinkers who were not themselves practicing systematists obsessed with taxonomic trivia whose speculations about this order furnished the seeds from which would eventually grow the most important foundation for a realist interpretation of the natural system.

As we saw earlier with the example of the carnivorous sundew, the flip side of Renaissance naturalists' obsession with objectively describing nature's details was that anything not subservient to this goal fell on their blind spot. This doesn't necessarily imply that these naturalists weren't interested in gaining a deeper understanding of nature, but it does illustrate that they could and did carry out their descriptive and systematic work perfectly well without simultaneously attempting to formulate causal explanations for their observations. Renaissance botanists, whose publications were an important component of medical teaching, considered the position of a species in a system as "a pedagogical, not a philosophical question" (Ogilvie, 2006: 182). Although the efforts of later naturalists increasingly shifted away from artificial and purely pragmatic systems toward reconstructing the natural system, like their predecessors, they generally did this without also pursuing a causal explanation of their schemes.

This separation of descriptive and causal investigations is of course not unique to systematics. It has long legitimated the split between natural history and natural philosophy as separate intellectual disciplines, and observations and descriptions logically have to precede attempts at causal explanation in all empirical sciences. In geology, for example, descriptive research into the deep history of the Earth and life preceded, and long progressed on a parallel track, to studies into the mechanisms of change, before they eventually fused (Rudwick, 2005, 2008). The history of systematics itself contains a long-lasting tension between the epistemic roles of description and explanation. As Stevens (1994: 255) put it for botanical systematics, from the late eighteenth to the beginning of the twentieth century there was "no transition in epistemes, no transformation of natural history into a biology that looked at the workings of life and living organisms. In particular, there was no change from *Naturbeschreibung* [description of nature] to *Naturgeschichte* [history of nature]." While systematists were absorbed in their morphological minutiae, other more philosophically minded naturalists engaged their speculative faculties in the search for an explanation of the natural order.

Although originally conceived as a static hierarchy, the directionality of the *scala naturae* contained a tantalizing clue about its own possible explanation. The concept of an arrow of time has long been deeply embedded within Western culture (Gould, 1991b). Given the *scala*'s linear aspect, no great intellectual feat is required to intuit the possibility that it could parallel the passage of time. The French philosopher Jean-Baptiste Robinet (1735–1820) did precisely that when he proposed that the *scala* was a pathway for the march of nature. In his writings on the Great Chain of Being, Robinet displayed an unwavering commitment to the principle of plenitude. He was convinced that in the beginning God had made every conceivable thing, from monkeys to mermaids and mermen (Lovejoy, 1936: 271–272). But for Robinet this didn't suggest the crystalized perpetuity of a divine design. Rather than thinking that Mother Nature wore a permanent mask, he was convinced that she continuously and successively actualizes new things by unfolding them from latent seeds or germs, not letting "any *nuance*, any variation, go unrealized" (Robinet in Lovejoy, 1936: 280; italics in original).

The germs from which everything in nature sprouted were indestructible, and as the history of the Earth unfolded, they came to form new beings. Robinet was convinced that there was a definite directionality to these transformations, leading to ever greater complexity and perfection. The infinite variety of natural productions should be seen, he thought, as "so many essays of Nature" as she advanced "fumblingly towards that excellent being who crowns her work" (Robinet in Lovejoy, 1936: 280). All that is in nature that is not man is to be regarded as "an innumerable series of sketches. I think we may call the collection of the preliminary studies the apprenticeship of Nature in learning to make a man" (Robinet in Lovejoy, 1936: 280). For Robinet, the chain of being was an anthropoid tractor beam.

While it may be amusing that Robinet saw semblances of human faces and limbs in radishes as signs of a divine sculptress in training, his attempt to temporalize the scale of being is nevertheless a notable early step toward biology becoming historical (Burkhardt, 1995: 83; Ruse, 1996: 45; Wilkins, 2009: 52). For Robinet the scale of being was not just a convenient arrangement of nature's productions. Its reality was rooted in a process of historical change, and for Robinet, historical change worked upon a single primitive substrate. He conceived of a prototype, which he envisioned as an elongated and hollow cylinder, that formed the basis of all variations observed in nature (Lovejoy, 1936: 277–280). This universal element of diversity, which he at times thought of as concrete, an ideal plan or pattern, model, or intellectual principle (Lovejoy, 1936: 279; Van der Hammen, 1981: 8), was most perfectly elaborated in man.

A contemporary of Robinet, the philosophical naturalist Charles Bonnet (1720-1793), held somewhat similar views. Bonnet is the best-known exponent of the *scala naturae* in the eighteenth century, and his 1745 and 1764 human-crowned scales of being are still frequently reproduced (Ruse, 1996: 44; Wilkins, 2009: 83; Kutschera, 2011: 4; Archibald, 2014: 8). But although Bonnet did important empirical work when he was young – he was an avid student of insects and discovered parthenogenesis in aphids, for instance – his failing eyesight increasingly forced him to turn to more philosophical matters. Deaf and practically blind from his mid-twenties, Bonnet examined the world with his mind's eye, spinning a fantastic embryological and historical theory to explain the origins of organized beings (Lovejoy, 1936: 283–286; Gould, 1977b: 19–28; Richards, 1992a: 10–12; Pinto-Correia, 1997: 56–58).

Like Robinet's theory, Bonnet's was rooted in elementary and indestructible particles he called *germs*. Bonnet suggested that in the beginning God had filled the world with germs, which swarmed, unseen, everywhere in the environment, in water, air, earth, and organisms. Each germ encapsulated an organized being, which in turn encapsulated a smaller being, so forming telescopic lineages of individuals of species. This preformationist theory of *emboîtement* or encapsulation suggested that new individuals would unfold from their preexisting germs, both in embryological and in historical time. When an organism died, its germs were released to join the hordes of free germs, ready to be recycled into new organized bodies. Over time, Bonnet envisioned new and more perfect organisms would rise from the immortal germs.

This progressive perfection of organisms did not unfold within continuous lineages, however. Instead, as the Earth went through a series of life-extinguishing cataclysms, it was filled with new cohorts of improved organisms. The germs that once made up a plant, for example, could in the future end up unfolding into a more perfect being that was unrecognizably different. Thus, while humans are currently the crown of creation, in the future nature's creative force would repopulate the rungs of the ladder of life with a finely graded series of newly evolved beings that would be strikingly different from those living today. In Bonnet's vision, our current world is sandwiched between alien pasts and futures thronging with strange creatures. In the less perfect past,

> a Réaumur [an entomologist that Bonnet struck up a correspondence with as a teenager], a Jussieu, a Linnaeus would be lost. There we would look for our quadrupeds, our birds, our reptiles, our insects, and we would see in their place only bizarre figures, whose irregular and incomplete traits would leave us uncertain that we were looking at what would become quadrupeds and birds. (Bonnet in Gould, 1977b: 25)

The future would be scarcely less fantastic:

> Man—who will then have been transported to another dwelling-place, more suitable to the superiority of his faculties—will leave to the monkey or the elephant that primacy which he, at present, holds among the animals of our planet. In this universal restoration of animals, there may be found a Leibniz or a Newton among the monkeys or the elephants, a Perrault or a Vauban among the beavers. (Bonnet in Lovejoy, 1936: 286)

Robinet's and Bonnet's were early attempts to tie the reality of the natural order to a historical process of change. Their systems unfolded from indestructible germs that provided filaments of historical continuity, even though these did not extend across all of deep time. Moreover, for Robinet the existence of the tubular prototype, whether ideal or material, was a logical consequence of his adherence to the principle of continuity. For him continuity could not exist without an element that was manifested in various forms throughout the system. In this we can glimpse a foreshadowing of the concept of ancestry, the grounding of evolutionary continuity in a common root.

As other thinkers pondered the natural system and offered their own interpretations of its structure during the late eighteenth and early nineteenth centuries, its reality became increasingly anchored in a consideration of natural processes. In Germany the philosophical naturalist Carl Friedrich Kielmeyer (1765-1844) and the philosopher Friedrich Wilhelm Joseph Schelling (1775-1854) stood at the crucible of an influential intellectual tradition that rooted the individuality, and thus the reality, of species, higher taxa, and their relationships in a dynamic conception of matter (Rieppel, 2011d, 2013a).[3] For Kielmeyer the natural system not only needed to be founded on a comprehensive selection of taxonomic characters, it also had to be cemented together by a causal theory for the generation of diversity (Coleman, 1973; Lenoir, 1981). He took his cue from embryology. Kielmeyer thought that the causal laws governing embryological change were identical to those ruling the transformation of species in geological time (Lenoir, 1981: 161–169; Richards, 2002: 245–248; Gliboff, 2008: 45; Rieppel, 2011d: 631). He also detected suggestive parallels between the systematic order of organisms and the developmental stages of embryos. He thought, for instance, that the order in which sense organs develop in chickens, geese, and ducks paralleled the sequence of appearance of these faculties in the animal scale.

Kielmeyer seemed to have been more circumspect about the existence of a third parallel – the sequence of forms preserved in the fossil record – which while theoretically mandated, was at the time sadly lacking in empirical support (Coleman, 1973). Although he didn't think that embryos faithfully recapitulated a single continuous and linear evolutionary chain of forms, he did propose that new species could emerge from ancestral ones, which may now be extinct, thus forming phylogenetic lineages within each epoch of the Earth's history (Coleman, 1973: 345; Lenoir, 1981: 163–164). Kielmeyer's phylogenetic ties did not necessarily stretch across the entirety of the Earth's history. He thought that gaps in the series of forms could indicate that some organisms did not trace back to ancestors because they were "original products of the fertile earth," or even alien intruders (Coleman, 1973: 348). Nevertheless, the reality of Kielmeyer's natural system was clearly anchored in natural historical processes.

Kielmeyer's thinking greatly influenced Friedrich Schelling, "philosopher king of the Romantic circle" (Richards, 2002: 214), and one of the most important architects of *Naturphilosophie*. At the heart of Schelling's *Naturphilosophie* was a nature in continuous flux, where every hint of stasis was only a temporary lull. Like Kielmeyer, Schelling accepted

---

[3] The individuality thesis in the Anglo-American literature was chiefly developed by Michael Ghiselin and David Hull in the 1970s. According to the individuality thesis, species, lineages, and supraspecific taxa are composite wholes composed of lower-level parts. They are concrete, spatiotemporally delimited individuals that can take part in natural processes. The Anglo-American development of this idea proceeded partly in parallel to its development in Germany, much to the frustration of Hull, who wrote in a paper published in the year of his death: "Why did I have to start from scratch? Why did Ghiselin and I have to rediscover the wheel?" (Rieppel, 2013a: 164). Richards (2002: 302) wryly notes that Hull and Ghiselin "would shudder at the association with Schelling."

the deep causal connection between embryological and species change. He considered the embryological and historical genesis of species to be steps in the actualization of an ideal type, with less developed forms preceding more developed ones. But Schelling rejected genealogical descent as an explanation of nature's unity in diversity (Richards, 2002: 301–306; Rieppel, 2011d: 631). In Schelling's teleological universe, nature strove to achieve the full potential of a perfect, presiding archetype by the gradual emergence of new species. A flesh-and-blood ancestor simply couldn't contain the infinite potential of an ethereal archetype. In Schelling's world, new species were born from Mother Nature's ideal womb. Thus, although Schelling's temporalization of nature was naturalistic, it wasn't materialistic. But the tide of materialistic genealogical thinking was on the rise elsewhere.

In the course of the eighteenth and nineteenth centuries, genealogical and evolutionary thinking became an increasingly well-traveled route to a realist interpretation of the natural system. Pre-evolutionary beliefs did not yield to evolutionary thinking in a single, smooth transition, however. Naturalists, medical doctors, philosophers, authors, anatomists, geologists, and physiologists across Europe entertained in various guises, to different degrees, and in different contexts the idea of naturalistic and materialistic genealogical relationships between organisms produced by historical processes. In France, Buffon became convinced that the Earth had a long history and that natural history should be placed on a historical footing. Although he did not accept the transmutation of species, Buffon did draw genealogical ties between dog breeds and traced their origin to the shepherd dog (figure 10 in Ragan, 2009; Fisler and Lecointre, 2013). His fellow countryman Geoffroy traced genealogy back into deep time by supposing that species of extinct and extant crocodilians may be bound together by real ancestor-descendant relationships (Panchen, 2001). But whereas Geoffroy never drew an evolutionary tree, Lamarck placed an increasingly tree-like natural system at the heart of his evolutionary worldview.

Meanwhile in Germany, and following on from Kielmeyer, thinking about the natural system had acquired a developmental flavor. Gottfried Reinhold Treviranus (1776–1837), who used the word *Biologie* in its modern sense for the first time in a book, grounded his radiating, web-like natural system in an elaborate causal theory that explained the diversity of organisms on Earth as the result of historical transformations (Lenoir, 1981). Like Robinet, Treviranus envisioned a teleological world with man as the natural prototype that was imperfectly reflected by other organisms. And like his near contemporaries Kielmeyer, Friedrich Tiedemann (1781–1861), and Johann Friedrich Meckel (1781–1833), Treviranus believed that the development of individuals and species was orchestrated by forces governed by the same natural laws (Gould, 1977b: 37, 46–47; Richards, 1992a: 42–55; Jahn, 2002; Göbbel and Schultka, 2003). In contrast, for someone like Louis Agassiz (1807–1873), the threefold parallelism of embryological, systematic, and fossil correspondences indicated the necessary harmony of divine thought (see Chapter 6 for further discussion). Together their works laid the foundation for recapitulationist transformationist thinking by pointing out parallels between the embryos of higher forms and the adult stages of lower forms. But the most influential reinterpretation of the natural system took place in England.

3.6 The Darwinian Revision of the Natural System

Charles Darwin wasn't unaware of these conceptual developments on the continent. The Scottish radical evolutionist Robert Edmond Grant (1793–1874) had been a favorite mentor when he was a student in Edinburgh, although their period of intense contact lasted only for several months (Browne, 1995: 80–88). Grant was a great admirer of Lamarck and Geoffroy, and on their coastal walks together in search of marine invertebrates, Darwin must have imbibed a generous swig of the philosophical outlook of these Frenchmen, seeing the natural system glued together by unity of type and genealogical descent (Desmond and Moore, 2009: 34–40). Darwin and Grant had also both enjoyed reading Erasmus Darwin's (1731–1802) *Zoonomia*, in which Charles' grandfather had speculated about the "living filament" that might have sired all life. But whatever formative influence these thinkers may have had on the young Darwin, they are conspicuously absent from the first two editions of the *Origin*. Only in the third edition did Darwin add his unambitiously titled *Historical Sketch*, where he presents an uninspired chronological list of thinkers who had published transmutationist views prior to 1859. He was struck, though, by the fact that Lamarck, Goethe, and his grandfather had, seemingly independently, come to similar conclusions about the transformation of species in the years 1794 to 1795. Still, Darwin's perfunctory historical sketch was too little, too late, for one of his fiercest critics.

The English writer Samuel Butler (1835–1902) – today still famous for his novel, or at least its title, *The Way of All Flesh* – penned four books that present lengthy and not infrequently vitriolic critiques of Darwin's evolutionary theory: *Life and Habit* (1878), *Evolution Old & New* (1879), *Unconscious Memory* (1880), and *Luck or Cunning?* (1887). After initially admiring Darwin's work, Butler eventually came to rail mightily against Darwin's dissing of design. He had no time at all for natural selection as an evolutionary mechanism. Shorn of its distinctively Darwinian characteristics, to Butler Darwin's theory looked no better than a reheated version of the ideas of Buffon, Lamarck, and Darwin's grandfather. Had Darwin not "been born into the household of one of the prophets of evolution," Butler surmised, he would never have concluded that species are not fixed (Butler, 1922: 246). Charles Darwin's greatest achievement, in Butler's mind, was nothing but a contemptible case of intellectual theft, for which he had been undeservedly crowned "with leaves that have been filched from the brows of the great dead who went before him" (Butler, 1922: 250). Butler (1922: 247) wrote: "Buffon planted, Erasmus Darwin and Lamarck watered, but it was Mr. Darwin who said, 'That fruit is ripe,' and shook it into his lap." If there was anything at all to admire about the younger Darwin's accomplishments, it was that he had been lucky, so that "Mr. Darwin with one of the worst styles imaginable did all that the clearest, tersest writer could have done" (Butler, 1922: 247).

Darwin evidently did not think that he had erected his theoretical edifice from the rubble of older systems. He believed he had built his insights on the basis of a comprehensive and penetrating interrogation of empirical evidence. He held his evolutionary synthesis to be distinctly his own, not just a modified descendant or a scraggly hybrid of the views of his predecessors. His and Wallace's new world order was not just transmutationist. It was

Darwinian. Natural selection was the new ruler of a natural world with an almost inconceivably long history. By adapting organisms to their local environments, it produced the mind-blowing diversity of life. Even though lineages followed their own autonomous evolutionary trajectories, in a Darwinian world they traced back to shared starting points. Darwin's was a world of common descent. This was his radical explanation for the puzzling phenomenon of the general resemblance between species that naturalists called affinity. Darwin (1859: 404) wrote:

> if I do not greatly deceive myself... the natural system is founded on descent with modification; that the characters which naturalists consider as showing true affinity between any two or more species, are those which have been inherited from a common parent, and, in so far, all true classification is genealogical; that community of descent is the hidden bond which naturalists have been unconsciously seeking, and not some unknown plan of creation, or the enunciation of general propositions, and the mere putting together and separating objects more or less alike.

Darwin's predecessors had ignited the first flames of a new historical biology, but their efforts had not illuminated the full shape of nature. Ideas had been developed, disseminated, and debated in different intellectual milieus, but their influence had been limited. It was Darwin who eventually succeeded in fanning these flames into an all-consuming conceptual conflagration. The theory of descent with modification provided a naturalistic and materialistic explanation of systematic relationships. In a Darwinian world, taxa are not mere mental abstractions. They are real natural groups, diagnosable by common evolutionary origins and tied together by the material bonds of genealogy. Terms such as "varieties, species, genera, families, orders, and classes" are real segments of the natural system that just differ in marking varying "grades of acquired difference" (Darwin, 1859: 433). Darwin realized that his theory wouldn't extinguish the debates about taxonomic nominalism because how to assign these ranks to taxa and where to draw their boundaries would remain largely subjective. At least systematists could finally stop worrying about the reality of the entities in their systems, which he felt sure "will be no slight relief" (Darwin, 1859: 455).[4]

Darwin understood the significance of his achievements well enough to predict "that there will be a considerable revolution in natural history," but he thought that "[s]ystematists will be able to pursue their labours as at present" (Darwin, 1859: 455). This sentiment was still very much reflected by the leading systematist of the first half of the twentieth century: "Fortunately, accepting evolution did not necessitate any change in the taxonomic technique. No longer did the taxonomist have to 'make' taxa, evolution had done this for him. All he needed to do was to discover these groups" (Mayr, 1969: 61). The empirical trinity of observation, description, and comparison served systematists just as well as before.

---

[4] John Wilkins rejects, convincingly I think, the common conclusion that Darwin was a nominalist with respect to species and higher-level taxa (Wilkins, 2009: 155–159).

One can give different interpretations to Darwin's admission that systematists could go about their business as if his explanatory theory didn't exist. Some confirm that systematic practice has indeed stayed essentially the same (Engel and Kristensen, 2013: 591). Others consider this methodological stasis ironic given the immense intellectual effort involved in Darwin's wholesale reconceptualization of nature (Gould, 1977b: 69). Others, who like to enliven their phylogenetic thinking with speculations about the evolutionary process, claim that Darwin's prediction is, or at least should be, flatly contradicted by post-Darwinian systematic practice (Ghiselin, 1969, 2000a). And yet others, who prefer to keep process speculations under wraps during systematic research, feel that Darwin's insight should have been heeded more widely (Nelson and Platnick, 1981; Patterson, 2002: 31).

These disagreements reveal the issue of whether Darwin's evolutionary theory can and should play a role in discovering patterns of systematic relationships or be used only to attempt their explanation. Darwin was right that systematists could retain the empirical core procedures that had been at the heart of natural history for centuries. Indeed, systematists have done so up to the present day. To some factions, like the early-twentieth-century idealistic morphologists and pattern cladists of today, this is, or should be, all there is to systematics. On the other hand, the impact of evolutionary thinking on systematic practice is there for all to see. It is a willingness to go beyond observation and to allow inferences, assumptions, and intuitions about evolutionary events to influence the reconstruction of patterns of relationship and the discovery of homologues. It is a willingness to inform the narrative reconstruction of phylogenetic lineages with evolutionary process assumptions, whether these sprout from researchers' minds or reside in the substitution models of probabilistic phylogenetic methods. One of the first telltale signs of Darwin's impact on systematics can be found in the geometry of systematic diagrams published after 1859.

3.6.1 Changing the Shape of Nature

Before 1859 most systematic schemes were either highly regular, like those produced by the quinarians, or characterized by reticulate relationships, like networks, webs, and maps (Stevens, 1994; Ragan, 2009; Pietsch, 2012; Morrison, 2014, 2015). Instead, Darwin envisioned the natural system to be tree-like,[5] with ever-diverging branches. Darwin promoted this shape because he understood that systematic pattern is the product of a process. His was a world of descent with modification. The natural system therefore has to follow the geometric dictates of the evolutionary process. Tree-like diagrams can conveniently express the temporal direction of evolution. Linear schemes are directional too, but they can't explain the origins of biological diversity. They can depict pathways of transformation, but not common descent. This was the Achilles' heel of the evolutionism espoused by the younger Lamarck and by Darwin's erstwhile mentor Robert Grant, for instance.

---

[5] The essential aspect of Darwin's evolutionary metaphor is that it is branching, showing the divergence of lineages. In his earlies notes, he thought about a coral-like metaphor. When I refer to trees or tree-like diagrams, I do not imply that they should have a central trunk, like some of Haeckel's trees famously did.

Many of the reticulating schemes popular before the *Origin* had no discernible direction whatsoever. Darwin understood that the radiating and crossing tendrils of affinity depicted in reticulating diagrams were a graphical representation of character conflict. He suspected that systematists "shall never, probably, disentangle the inextricable web of the affinities between the members of any one class" (Darwin, 1859: 415), but he knew that this was only an epistemological problem, albeit a serious one. Beneath the tangled web of systematic evidence hid a genealogical tree. Only a branching natural system could produce, and thus explain, natural history's "wonderful fact" (Darwin, 1859: 170): the hierarchical nesting of taxonomic groups. Darwin had the key to understanding the puzzle of "the excessively complex and radiating affinities by which all the members of the same family or higher group are connected together" (Darwin, 1859: 413). It was caused by the different degrees by which descendants had diverged from their last common ancestor, the extinction of intermediate taxa, and the convergent evolution of adaptive analogies between distant evolutionary relatives.

Darwinian naturalists had to expect that the shape of the systematic patterns they sought to discover would bear the imprint of the process that had created them. To a Darwinian systematist, success would look like a tree. But living up to this theoretical expectation was far from straightforward. One of T. H. Huxley's protégés offers a good illustration.

After enthusiastically embracing Darwinian evolution, and feeding Darwin information about all manner of zoological arcana, anatomist St. George Mivart (1817–1900) became one of Darwin's most perceptive critics. This earned him a star role, albeit as a villain, in the only new chapter that Darwin ever added to the *Origin*, titled "Miscellaneous objections to the theory of natural selection." In 1874 Mivart published a book titled *Man and Apes* that addressed a question then at the center of a lively debate in the popular press. What was the significance of "the semi-humanity of the Gorilla... [h]ighest of the apes—close ally of the Negro" (Mivart, 1874: 5)? Could this enormous ape be man's closest evolutionary cousin? Could it illuminate aspects of our supposed bestial origins? To investigate this, Mivart started from the premise that anatomical resemblance should scale linearly with propinquity of descent (Mivart, 1874: 6–7). If the gorilla really was man's closest relative, it should have the greatest overall resemblance to man. In 1865 and 1867, Mivart had already published two osteological studies of primates that, disconcertingly, inferred different patterns of affinities depending on whether he looked at spines or limbs. Then, when Mivart took readers of his 1874 book on a broader survey of primate anatomy, he led them into chaos.

One message quickly emerged. The gorilla was not most similar to humans when deconstructed into separate characters. This analytical procedure was utterly unable to reveal a clear genealogical path to man. A bewildering mosaic of affinities emerged, buttressed by conflicting characters. The length of the hand of the gorilla, measured against the length of its spine, is the most man-like of the apes. But the relative length of the lower and upper arm bones of man are more like those found in chimpanzees. Continuing upward, the shoulder blade of the gorilla has a shape most similar to that of man, but the relative length of the shoulder blade and collar bones are most similar in man and the orangutan. How could independently evolved analogies be distinguished from true signs of evolutionary affinity? Why would siamangs and humans have a "more or less developed chin," lacking in other primates? Why would the "slightly aquiline nose" of the hoolock gibbon be common in humans too (Mivart, 1874: 179)? Could natural or sexual selection explain these features? "Can either character be thought to have preserved either species in the struggle for life, or have persistently gained the hearts of successive generations of female Gibbons?" (Mivart, 1874: 179).

Darwin's evolutionary mechanisms were a poor tool for the systematist, Mivart thought. They were unlikely to be able to assist in explaining why chimps, orangutans, and humans had similar livers, while gorillas were stuck with "a very degraded liver" that was more similar to those of monkeys (Mivart, 1874: 155). Mivart's verdict was clear. The affinities between primates were a "tangled web" (Mivart, 1874: 176). "[I]t is to no one kind of ape that man has any special or exclusive affinities... [and] the much vaunted Gorilla... is essentially no less a brute and no more a man than is the humblest member of the family to which it belongs" (Mivart, 1874: 193).

Retrieving trees from tangles of evidence became the defining challenge for Darwinian systematics. The speed with which tree-like diagrams became the dominant systematic icon showed that systematists embraced this challenge with relish (Ragan, 2009; Pietsch, 2012; Archibald, 2014; Morrison, 2014). The changes that Carl Gegenbaur made to his *Grundzüge der vergleichenden Anatomie* (*Principles of Comparative Anatomy*) between the first edition of 1859 and the second edition of 1870 provide an informative example of this transformation (Rieppel, 2011c). Alongside the introduction of evolutionary terms, such as *Abstammung* (descent), *Stämme* (lineages), and *Ahnen* (ancestors), there was a conspicuous pictorial change. Gegenbaur replaced the diagram that depicted the relationships between the main animal types as a network with a tree (Figure 3.2).

This radical shift in iconography in the decades following the publication of the *Origin* was abrupt. This is nicely illustrated by David Morrison (2014) (Figure 3.3), who tabulated whether systematic relationships were depicted as networks or trees in 124 works published between 1750 and 1900 (I will explain shortly why I disagree fundamentally with Morrison what this pictorial shift means). This shift was clearly less the result of a sudden improvement of methods that better revealed trees hidden in tangles of systematic data, than a reflection of systematists falling in line with new theoretical expectations. For the first researchers who did their work explicitly under the banner of Darwinism, such expectations could weigh heavily. The German paleontologist Franz Hilgendorf (1839–1904) published a series of papers between 1866 and 1901 on the evolution of fossil freshwater gastropods from the Steinheim basin in southern Germany (see Chapter 7 for further discussion of the Steinheim snails). In the first paper on these snails, which was based on his 1863 dissertation thesis, Hilgendorf summarized his ideas in a branching diagram, showing several diverging lineages (Figure 7.1). However, as part of his unpublished thesis, he also drew another phylogeny that, alongside lineage splitting, showed a fusion of two lineages (Reif, 1983; Janz, 1999; Glaubrecht, 2012). This may well be the earliest known fossil-based phylogeny, but an interesting question is why Hilgendorf didn't publish it. The text of his thesis provides the answer. According to Hilgendorf, the fusion of lineages contradicts "the beautiful image that Darwin presents of the relationships of species in a tree rich in twigs;

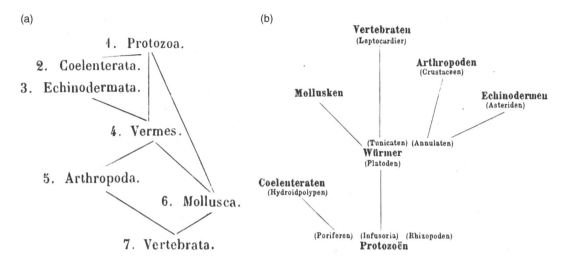

Figure 3.2 Carl Gegenbaur's reticulated diagram of the relationships between the major animal types in the 1859 edition of his *Grundzüge der vergleichenden Anatomie* (a) and depicted as a phylogeny in the 1870 edition (b) (a after the figure on page 38 in Gegenbaur, 1859, and b after the figure on page 77 in Gegenbaur, 1870).

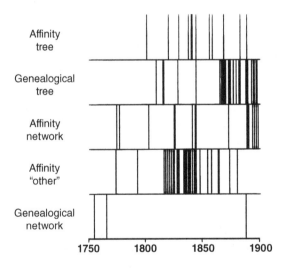

Figure 3.3 A profound shift in how systematic relationships were depicted happened in the second half of the nineteenth century
(after figure 1 in Morrison, 2014, by permission of Oxford University Press).

the twigs of a tree never grow together again" (Hilgendorf in Glaubrecht, 2012: 255). Such is the power of theoretical expectations.

The more thoroughly systematists harvest characters from their organisms, the more conspicuous character conflict becomes. As a devout Christian, Mivart may not have been too disappointed that his work on primate skeletons didn't recover a single evolutionary tree. Darwinian evolution clashed fundamentally with his ideological and metaphysical view of the world. But nature is equally well able to trip up those eagerly searching for evolutionary trees. In his monograph on American characid fishes, ichthyologist Carl Eigenmann (1863–1927) admitted that he would have loved to have summarized the systematic relationships of tetragonopterinid fishes (a group including popular aquarium fishes such as tetras) "in the form of the conventional phylogenic tree," but instead they seemed "to form an interlacing fabric rather than a branching tree" (Eigenmann, 1917–1929: 48). Sadly, taxa that seemed to be a "paradise for the student of divergent evolution" are precisely those that lead to "the dispair of the systematist" (Eigenmann, 1917–1929: 47). Eigenmann was unable to disentangle the many strands of convergent evolution in his fishes, and he summarized their relationships in a unique diagram very unlike a tree (Figure 3.4).

Although Eigenmann was a morphologist writing a century ago, trees don't emerge from systematic data any more readily today. Complex substitution models are used to infer bifurcating trees from increasingly large and complex molecular datasets, with the aim of separating tree-like phylogenetic from nonphylogenetic signals that can be caused by things like heterogeneous substitution rates across sites in alignments or between taxa, and compositional biases across sites or between taxa. Moreover, since Darwin there has been a quantum leap in our understanding of processes that actually cause nontree-like phylogenetic signals. Processes such as lateral gene transfer between distantly related taxa and introgressive hybridization, particularly of newly diverged lineages, are fundamentally incompatible with the existence of a neat inclusive taxonomic hierarchy that would be produced were evolution strictly tree-like and diverging. We now understand that the detection of nontree-like signals in our data is no longer just an epistemological problem that could, at least in principle, be solved by more detailed analysis of characters. The nexus of life is now more accurately seen as a genealogical network that is more or less tree-like depending on where you zoom in (Doolittle and Bapteste, 2007). But surprisingly, some recent commentators have turned this new understanding into an anachronistic critique of Darwin's championing of the tree of life.

3.6.2 Is Darwin's Tree of Life a Flawed Metaphor?

David Penny (2011: 3) makes the peculiar claim that "Darwin basically rejected the Tree of Life phrase in favor of the theory of descent with modification." He claims that "Darwin did not really like the tree of life concept" (Penny, 2011: 3) and that the tree simile was "not a central part of his theory" (Penny, 2011: 3) because the tree analogy cannot accommodate hybridization, which would require reticulate relationships. Penny speculates that Darwin labeled his theory "descent with modification" because this formulation is less prescriptive about the geometry of evolution. Darwin supposedly disliked the tree of life metaphor because "scientists at the time already knew about hybrids," (Penny, 2011: 3), and this Penny argues, is why Darwin "deliberately chose not to use the phrase Tree of Life" (Penny, 2011: 3). Except, of course, he did. Still, Penny (2011: 2) claims that the only time in the *Origin* where Darwin does present the full biblical borrowing, with capital T and capital L,

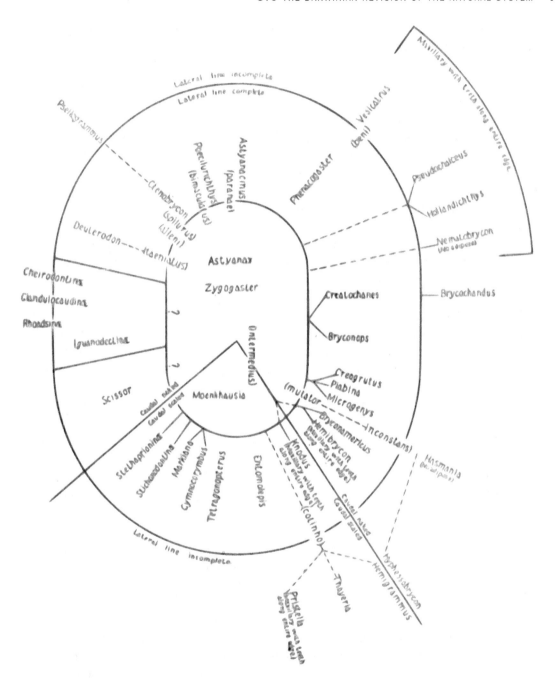

Figure 3.4 Relationships between the genera of tetragonopterinid fishes as envisioned by Carl Eigenmann, with extensive convergent and parallel evolution obscuring a neat phylogenetic tree (after figure 1 in Eigenmann, 1917–1929).

"it is <u>not</u> a description of the relationship between taxa" (underlined in the original). These are puzzling conclusions.

The relevant paragraph in the *Origin* that evokes the Tree of Life starts like this: "The affinities of all the beings of the same class have sometimes been represented by a great tree. I believe this simile largely speaks the truth" (Darwin, 1859: 171). How Penny concludes that this is not a statement about relationships between taxa is a mystery to me. As Penny knows, the only diagram in the *Origin* that Darwin drew to illustrate descent with modification is tree-like. However, David Morrison (2016: 460) agrees with Penny and writes that Darwin included the tree diagram in the *Origin* to "illustrate continuity of evolutionary descent and the processes of extinction and diversification, rather than strictly as representing a phylogeny—his Tree of Life metaphor had nothing to do with the diagram." I disagree.

Darwin included both his diagram and the Tree of Life metaphor precisely to convey the branching nature of phylogeny. What else do diversifying lines of descent and extinction represent if not phylogeny? Why else would Darwin label the diverging lines in his diagram "lines of descent," the tracing of which he called "phylogeny" from the fifth edition of the *Origin* onward, after he had read Haeckel's works? Because Darwin knew about hybridization, Morrison (2016: 459) claims that his Tree of Life "must therefore have reticulations," which he claims Darwin failed to note. Indeed, Darwin did know about hybrids. He devoted an entire chapter to hybrids in the *Origin*. But Darwin apparently thought that hybridization wasn't an important enough factor in the production of new species to dismiss the tree simile as a metaphor. Darwin didn't draw much, but when he doodled in his unpublished notebooks and letters, he only produced branching tree-like diagrams, not networks, not even when he explored hybridization between species (Archibald, 2014: 80–112; Putten, 2020: 36–39).

Morrison (2014: 629) suggests that "the tree is a historical artifact" as a metaphor for relationships between taxa: "the network metaphor precedes the tree in both theory and practice. It has, however, been marginalized for some time, and is only now starting to move back into the spotlight." He therefore labels evolutionists' long preoccupation with trees "the Darwinian Digression," (Morrison, 2014: 630) and claims it is good that "systematists have started seriously questioning this change of metaphor" (Morrison, 2014: 630). I find this argument perplexing. As Morrison admits in his article, Darwin transformed systematics by moving away from affinity relationships, based on the overall resemblance of taxa, to genealogical relationships that trace lines of descent. Darwin looked beyond the chaos of character conflict summarized in systematic networks and intuited the presence of a hidden evolutionary tree. In the twentieth century, evolutionists discovered that the key to dissolving the chaos of conflicting characters was to distinguish primitive and derived character states (see Chapter 6). Darwin saw systematic pattern through the lens of evolutionary descent. Although it isn't untrue that Darwin used the tree as "a rhetorical device" (Morrison, 2014: 629), it was also a concept deeply rooted in his understanding of the evolutionary process.

As discussed earlier, the tree-like conception of evolutionary divergence is central to Darwin's theory. Even today, despite the increasing prominence of reticulating lineages, the tree of life metaphor hasn't run out of heuristic juice (Blais and Archibald, 2021). It cannot be dismissed as a digression in the history of science. As Morrison acknowledges, pre-evolutionary systematic networks are a different beast from their post-Darwinian successors. It makes meaningless his claim that modern insights are "a return to the original metaphor of 250 years ago" (Morrison, 2014: 631). The conceptual dichotomy clearly falls between unrooted pre-evolutionary affinity networks on the one hand and rooted phylogenetic trees and networks on the other. If Darwin's tree-like metaphor was a mere digression, as Morrison would have us believe, one would not expect genealogical (as opposed to affinity) networks to be as rare as they are – Morrison knows of only nine published genealogical networks between 1900 and 1990, all of them botanical.

Darwin wasn't the only nineteenth-century biologist whose image of the natural system was fundamentally shaped by his thinking about the historical process. Nor was he the first to publish his views on this topic. Morrison (2016) reminds us that the French botanist Charles Naudin published an article in 1852 in which he explicitly rejected linear and reticulated affinities between species and instead favored a branching tree-like system. On his phylogenetic blog (http://phylonetworks.blogspot.com/2015/05/naudin-wallace-and-darwin-tree-idea.html; last accessed July 23, 2021), Morrison writes that it is "rather a pity that [Naudin] explicitly rejects a network." Maybe so, but the reason why he did should be clear. Naudin adopted a branching image of affinity relationships precisely because he considered them to reflect the descent of diverging lineages from shared common ancestors. He was not an evolutionist in a modern sense (Farley, 1974), but Naudin clearly thought that the systematic relationships between taxa should have the shape of a tree because of the branching historical process that produced them.

The codiscoverer of natural selection, Alfred Russell Wallace (1823–1913), adopted the same reasoning as Darwin in a 1855 paper where he discusses how biogeographical and geological evidence support what he considered to be a natural law: "Every species has come into existence coincident both in time and space with a pre-existing closely allied species" (Wallace, 1855: 186, 196). If this law is true then "the natural series of affinities will also represent the order in which the several species came into existence, each one having had for its immediate antitype a closely allied species existing at the time of its origin" (Wallace, 1855: 186). Here Wallace makes the inference that MacLeay castigated Lamarck: for perhaps there is a direction to the affinities between species that coincides with their evolutionary descent. If each ancestral species gives rise to only a single descendant species, a linear series of affinities would result. "But if two or more species have been independently formed on the plan of a common antitype, then the series of affinities will be compound, and can only be represented by a forked or many-branched line" (Wallace, 1855: 186). He thus recommends "a branching tree, as the best mode of representing the natural arrangement of species and their successive creation" (Wallace, 1855: 191).

Wallace tried to put this theoretical insight into practice in a paper published a year later, where he drew two affinity diagrams for groups of birds (Wallace, 1856) (Figure 3.5). These diagrams are branching and the lengths of the lines connecting the taxa roughly reflect the number of affinities that separate them. But Wallace found it difficult to give the diagrams an unambiguous reading direction due to his ignorance about the fossil record. He writes:

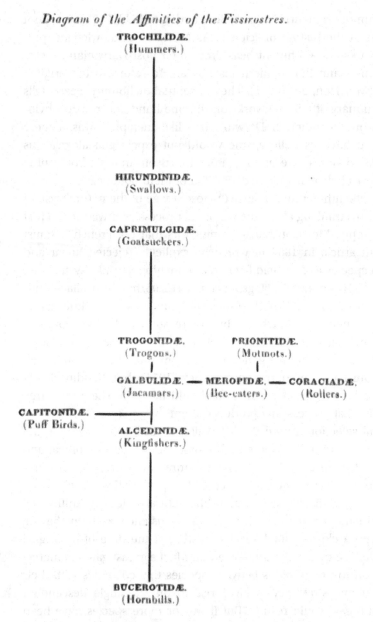

Figure 3.5 An affinity diagram of birds drawn by Alfred Russell Wallace (after the figure on page 205 in Wallace, 1856).

It is probable that in very few cases is there a direct affinity between two groups, each being more or less distantly related to some common extinct group, so that we should represent their connexion more accurately by making our central line a blank, for the extinct portion of the group, and placing our families right and left, at different distances from it. (Wallace, 1856: 214)

What Wallace demonstrates here is lineage thinking: the understanding that the systematic relationships between taxa are genealogical relationships, with lineages diverging from shared common ancestors. Wallace (1856: 212) also echoes Strickland's earlier dismissal of systematic systems that obey numerical and geometrical regularities. In his paper titled "On the true method of discovering the natural system in zoology and botany," Strickland (1841: 185–186) writes "how erroneous must be all those methods which commence by assuming an a priori system, and then attempt to classify all created organisms in conformity with that system." Strickland rejected the regularity of linear, circular, and numerical systems as well as networks with regular reticulations. Instead, the natural system should be irregularly branching because organisms vary irregularly, so that it "may, perhaps, be most truly compared to an irregularly branching tree" (Strickland, 1841: 190). He thought that branches could sometimes fuse and form circles, but not in regular patterns.

There is an irony here. Evolutionists rejected systems shaped according to a priori beliefs. However, their preference for irregularly branching tree-like diagrams was also influenced by theoretical assumptions. The natural system was best captured in a tree-like diagram because the evolutionary theory developed by Wallace and Darwin stipulated that evolutionary lineages diverged from common ancestors. In nonevolutionary systematic diagrams, irrespective of their shape, affinity relationships were represented by lines drawn directly between collateral relatives. Evolutionary trees connected taxa differently. They replaced the abstract, symmetrical affinity relationships of taxa with the concrete, asymmetrical, and temporal genealogical relationships of ancestors and descendants. But schemes of affinities couldn't simply be transformed into genealogical diagrams by the imposition of a time axis. This would imply the direct evolution of one extant taxon into another, which Darwin thought would be an erroneous inference in most cases (Darwin, 1859: 292–293, 405–406, 438). Nature's continuity wasn't located on the surface of systematics. It was hidden in the depths of time.

In affinity diagrams, taxa can be arranged like beads on a string. But in evolutionary trees, observable taxa are relegated to the periphery. Evolutionary connections are indirect. They run backward in time through unknown common ancestors, which are represented by the internal nodes of trees. The synchronic ties of homology between collateral relatives (taxic homology) are orthogonal to the diachronic ties of homology between ancestors to descendants (transformational homology). The taxonomic gaps that mark the discreteness of taxa and had helped to establish the reality of the natural system (Stevens, 1994; Amundson, 2005) were now explained by the continuity of common descent. This was the new systematic realism.

Not everyone was happy to be swept up in the advancing tide of Darwinian evolutionary theory. As Darwin had already prophesied in the *Origin*, his evolutionary theory did not necessarily require systematists to change their ways. And many didn't. The eminent English botanist George Bentham (1800–1884) used the occasion of his 1862 presidential address to the Linnean Society of London to laud Darwin's accomplishments. Although he didn't deny the value of Darwinian theory as a spur for new empirical research, he "certainly should be sorry to see our time taken up by theoretical arguments not accompanied by the disclosure of new facts or observations" (Bentham, 1862: lxxxi). A year later, he reinforced his position and that of the Society by adding that discussions of Darwin's theory did not fall "within the legitimate scope of our Society" (Bentham, 1863: xvi), and "we cannot allow the time of our own meetings nor the pages of our publication to be devoted to the abstract discussion of any such theories" (Bentham, 1863: xiii). Interestingly, more than 70 years later, botanist members of the Systematics Association remained equally reticent to marry taxonomy and phylogenetic speculation, in sharp contrast to their zoologist colleagues (Winsor, 1995).

Bentham's stern remarks seem to be born from a mounting concern for the autonomy and purity of systematics. All the way back to its roots in Renaissance natural history, systematics had been proudly empirical. Its methodological tripod – observation, description, and comparison – was tailored perfectly to the study of the static aspects of nature. But these tools cannot penetrate the synchronic surface of biology. Biology's new realism introduced a serious epistemological headache. The evolutionary past lies beyond the phylogenetic frontier, which can only be crossed by using imaginative inference. Bentham feared that allowing speculation free reign could corrupt systematics. He was disconcerted that popular scientific lecturers and writers, such as T. H. Huxley, were using the term "biology" to refer to anything relating to living things and subsuming "the classificatory sciences" (Bentham, 1863: xii). To Bentham this suggested the unwelcome prospect that, under biology's broad umbrella, systematists might feel justified to engage in the speculative pursuit of such inscrutable topics as the origin and evolution of species, thereby diverting resources away from the core mission of systematics: the observational quest for facts.

Good systematists should keep speculation strictly at bay, an attitude Bentham duly exemplified in his monograph on Compositae published a decade later in the society's botany journal. He singled out for a salvo of snipes two of the "numerous more speculative naturalists whom the promulgation of the Darwinian theories have called into action" (Bentham, 1873: 342). One senses Bentham's relish in labelling these men mere "theorists" (Bentham, 1873: 342) given to propounding evolutionary hypotheses "for which the data at hand are wonderfully few" (Bentham, 1873: 342), supplementing solid observations "rather largely from the sources of imagination" (Bentham, 1873: 342), and drawing conclusions that are scarcely more than "mere conjectures" (Bentham, 1873: 342) and "a play upon words" (Bentham, 1873: 343). Not surprisingly, Bentham didn't draw a tree to depict the evolutionary relationships between his taxa, but instead produced a reticulated affinity diagram that was but little modified from an original published in 1826 by a French botanist (Cassini, 1826) (Figure 3.6).

This desire to stay as close as possible to observational data rather than follow the siren song of evolutionary speculation wasn't a short-lived mindset that petered out soon after the *Origin* appeared. As we saw in the previous chapter, a number of post-Darwinian idealistic morphologists had similar epistemological qualms about tree building well into the twentieth century. Evolution may be true, but why shackle the study of comparative morphology to unprovable speculations about evolutionary history?

Bentham needn't have worried that systematics would necessarily be contaminated by evolutionary speculations when Huxley classified systematics as a subdiscipline of biology.

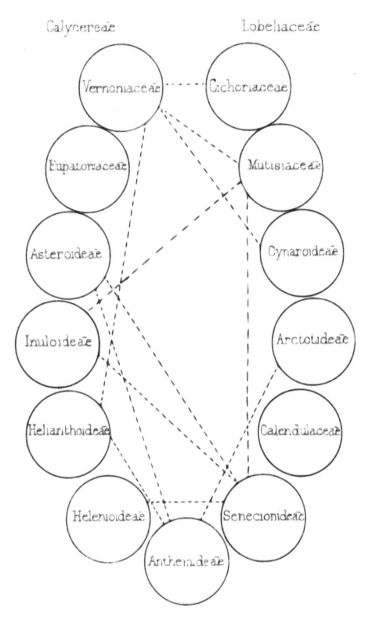

Figure 3.6 Affinity diagram drawn by George Bentham for Compositae (after plate X by Bentham, 1873).

Huxley couldn't have been clearer on the need to separate research into nature's patterns and the processes that cause them:

> Darwin, by laying a novel and solid foundation for the theory of Evolution, introduced a new element into Taxonomy. If a species, like an individual, is the product of a process of development, its mode of evolution must be taken into account in determining its likeness

or unlikeness to other species; and thus "phylogeny" becomes not less important than embryogeny to the taxonomist. But while the logical value of phylogeny must be fully admitted, it is to be recollected that, in the present state of science, absolutely nothing is positively known respecting the phylogeny of any of the larger groups of animals. Valuable and important as phylogenic speculations are, as guides to, and suggestions of, investigation, they are pure hypotheses incapable of any objective test; and there is no little danger of introducing confusions into science by mixing up such hypotheses with Taxonomy, which should be a precise and logical arrangement of verifiable facts. (Huxley, 1875: 52–53)

Huxley diagnoses here what was to become a chronic issue in evolutionary biology: the nature of the relationship between taxonomy and phylogeny. Huxley held fast to the ideals of empiricist science throughout his career, but there are conspicuous differences in the extent to which he was willing to broaden his epistemological horizon in his professional, popular, and private output (Ruse, 1996: 118–122; Desmond, 1997: 109–110). It was primarily in his popular and private writing that he explored the diachronic depths of geological time, sketching phylogenies and speculating about ancestors and evolutionary processes. Michael Ruse argues that Huxley was uneasy about the role of evolutionary speculation in professional science, even though he did value its heuristic value. But even researchers who used the mortar of their imagination to build intricate air castles could be reluctant to speculate in full view.

As we saw earlier, by avowing to be not the fanciful architect but the discoverer of taxonomy's divine carousel, William MacLeay revealed his own capacious talent for self-deception. Yet, he made up for his lack of self-scrutiny with his willingness to criticize others. After awarding Lamarck the back-handed compliment of being on the right track in as much as his work agreed with his own, MacLeay then condemned him for daring to suggest that taxonomic order might reflect the temporal sequence in which taxa had arisen during geological time. To MacLeay, this was an unwarranted mixing of "effects of which we are sure, with the means by which these effects have been produced, of which we known [sic] nothing" (MacLeay, 1819–1821: 330). Desirable or not, it was the insertion of time into comparative biology that led to the founding of a new discipline.

3.7 Crossing the Phylogenetic Frontier

Phylogenetics was born when the arrow of time penetrated the study of systematics and morphology. The introduction of a time axis gave systematic diagrams a reading direction that allowed relationships to be specified with unprecedented precision. It established relative recency of common ancestry as a new arbiter of relationships. Without a reading direction, ambiguity about which taxa are most closely related remains even in nonreticulated branching diagrams, which is why modern phylogeneticists, whenever possible, root their trees. The defining difference between trees depicting evolutionary relationships and trees showing affinity or homology relationships is that only the former have a reading direction.

It is this, as Nelson and Platnick (1981: Chapter 2) point out, that limits the information content of pre-evolutionary branching diagrams. Two taxa are each other's closest relatives if they share a more recent common ancestor than either does with any other taxon. This criterion transformed systematic into phylogenetic relationships. It explains the synchronic arrangement of collateral relatives with the diachronic ties of genealogy. The expression "phylogenetic relationships" is therefore far more common in the evolutionary literature than "systematic relationships." Although the explication of this concept of relationships is often mentioned in connection with twentieth-century works by the zoologists Daniele Rosa and Willi Hennig and the botanist Walter Zimmermann (Nelson and Platnick, 1981: 135–136; Williams and Ebach, 2008: 118; Wheeler, 2012: 16), it was already understood and used in the nineteenth century.

Fritz Müller's (1821–1897) name rings familiar today mostly for his 1878 explanation of the type of mimicry that bears his name, but this remarkable naturalist's short 1864 book on crustaceans, titled *Für Darwin*, has a small but important place in the history of phylogenetics. In it Müller developed the concept of recapitulation that Haeckel later elaborated into his biogenetic law, and he described the new phylogenetic criterion of systematic relationships based on relative recency of common ancestry, which he illustrated with two of the first post-Darwinian evolutionary trees. Although he was born and educated in Germany, Müller did the work that he is remembered for today in Brazil. The eldest son of a pastor, Müller grew into a radical atheist who, after studying medicine for four years, refused to take the religious oath necessary to be able to practice as a doctor. The negative repercussions this had for his private and professional life – a situation that wasn't helped by having a child out of wedlock with his partner Carolina – were so serious that Müller decided to emigrate to Brazil (West, 2003).

His 1864 book earned him Charles Darwin's patronage and initiated a lifelong correspondence with the English sage. Darwin was so pleased with Müller's application of his theory in reconstructing the evolutionary history of crustaceans that he arranged and helped finance an expanded English translation of the book, which was titled, after a suggestion by Charles Lyell, *Facts and Arguments for Darwin*. Müller clearly understood that relationships "in the sense of Darwinian theory" were based on the exclusive sharing of common ancestors, which he explained with an example involving amphipod crustaceans and illustrated with two figures reminiscent of cladograms (Müller, 1869: 10--11).

But phylogenetic precision comes at an epistemic cost. Common ancestry cannot be observed. Ancestors are the after-images of an evolutionary process that we can only ever hope to glimpse with our mind's eye. In order to use common ancestry as a criterion to adjudicate taxonomic relationships, one has to be willing to supplement observation, description, and comparison with imaginative inference. To turn systematic into phylogenetic relationships, the focus shifts from the taxa that form the terminal twigs of trees to the lineages that form its internal branches. To infer the direction of the evolutionary process not only requires the systematist to cut through knots of conflicting characters to remove reticulating affinities at odds with the direction of descent. He also has to pin down the axis of evolutionary time by divining ancestors and designating ancestral character states, a task that cannot be tackled without wielding one's speculative faculties.

The reluctance of post-Darwinian systematists, such as George Bentham, to engage with this challenge is understandable enough. The use of speculation had attained more than a whiff of epistemological rot in the hands of the transcendental morphologists and *Naturphilosophen* on the European continent and their acolytes in Britain. Their attempts to explain the static with the dynamic, to root pattern in process, to relate organismal variations to law-like archetypes, and to try and grasp the ethereal behind the observable had come to be widely disparaged by the early nineteenth century. No wonder many systematists were reluctant to enter this speculative arena, even if they were evolutionists. Systematists whose interests were limited to classification could stay true to systematic tradition and renounce speculative theorizing. But those whose curiosity stretched beyond the tree had to be willing to allow their methodology to float free from pure observation. This turned out to be exactly to the taste of a new breed of comparative biologists.

During the decades spanning the late eighteenth and early nineteenth centuries, geologists and paleontologists established the immense antiquity of the Earth (Rudwick, 2005, 2008). The directionality of the rock record was widely accepted in the early nineteenth century, and strongly directional geotheories, such as Abraham Gottlob Werner's Neptunist model of the history of the Earth, were linked to emerging transformationist interpretations of the history of life (Jenkins, 2016). Molluscs were a particular focus for fossil-based lineage thinking in early-nineteenth-century research by Lamarck and the Italian Giambattista Brocchi (Eldredge, 2010), as well as research later in the century (see Chapter 7). When it became clear that the modern biota is merely the icing on an ancient cake, a new generation of biologists and paleontologists set out to discover how Mother Nature had managed to bake it, using evolutionary theory as a recipe book. Edmund Beecher Wilson (1856–1939) nicely captured the exciting prospects:

> The central question in every morphological investigation became twofold: it was no longer simply *what is?* it was also *how came it to be?* And this second question, be it observed, is not properly a speculative matter at all, but an historical one; it related not to an ideal or hypothetical mode of origin, but to a real process that has actually taken place in the past and is to be determined like any other historical event. "Speculative zoology" thus, by slow degrees, became the guide and leader of research, and every morphological inquiry became, in the last analysis, a genealogical one. (Wilson, 1891: 54; italics in original)

Ultimate understanding for an evolutionist means understanding origins. This was the founding insight of the influential research program of evolutionary morphology that Ernst Haeckel and Carl Gegenbaur established at the University of Jena in the 1860s. Evolutionary morphologists retained the traditional interests of systematists and morphologists in identifying homologues, inferring systematic relationships and documenting unity of type, but these objectives became subservient to a new goal: tracing the phylogenetic history of organismal form (Nyhart, 1995; Bowler, 1996; Amundson, 2005). The interests of the earlier idealistic morphologists and *Naturphilosophen* acquired a new legitimacy after passing through the purifying filter of evolutionary metaphysics. The once mysterious ties of affinity were explained by the material bonds of genealogy, and timeless archetypes collapsed into the historical pinpoints of ancestors, reducing the laws of form to historical residue. But this ontological cleansing left an epistemological challenge.

The phylogenetic frontier couldn't be crossed with empirical evidence alone. Evolution happens in an epistemic hinterland of unobservable entities and invisible processes that are located in the crannies of the natural system. No matter how focused the phylogeneticist's observations are, evolution will only reveal itself to his peripheral vision. The phylogeneticist has to use inference, extrapolation, generalization, and guesswork to triangulate ancestors. Since it is highly unlikely that any known living or extinct organism is the de facto ancestor of a taxon of interest, the phylogeneticist has no choice but to paint ancestral portraits with a palette of descendants' traits. As Darwin expressed it in the *Origin*,

> [i]n looking for the gradations by which an organ in any species has been perfected, we ought to look exclusively to its lineal ancestors; but this is scarcely ever possible, and we are forced in each case to look to species of the same group, that is to the collateral descendants from the same original parent-form, in order to see what gradations are possible, and for the chance of some gradations having been transmitted from the earlier stages of descent, in an unaltered or little altered condition. (Darwin, 1859: 218)

Phylogeneticists must play a game of systematic slalom. In order to home in on distant ancestors, they have to zigzag between species extant and extinct, gleaning observations and braiding strands of indirect evidence into "how possibly" scenarios of evolutionary change. Phylogeneticists must shift their focus from observable taxa to the gaps in the natural system where only the mind's eye can see. When the imagination is allowed free reign, even the peculiar traits of highly modified organisms can become ancestral beacons. The simple photoreceptors of starfish come to exemplify a possible step in the evolution of complex image-forming eyes; the swim bladders of teleosts presage the evolution of the tetrapod lung (Darwin, 1876: 144, 148). The mind's eye works just as well as a crystal ball for the future as a mirror of the past. Were black bears to specialize in catching insects in the water by swimming around with open mouths, "I can see no difficulty," Darwin writes, in believing an animal "as monstrous as a whale" would ultimately evolve (Darwin, 1859: 215). Since flying reptiles, birds, mammals, and insects have evolved, why wouldn't flying fish evolve "into perfectly winged animals" (Darwin, 1859: 214)? But the challenges of inferring what has, in fact, happened are daunting.

The common ancestor of two taxa cannot be triangulated by simply envisioning a form intermediate between them (Darwin, 1859: 292, 438). Worse still, as Darwin noted with frustration both in his chapter on the imperfection of the geological record and in his concluding chapter, ancestors will almost always lie beyond our empirical reach (Darwin, 1859: 293, 439).

> We should not be able to recognise a species as the parent of any one or more species if we were to examine them ever so closely, unless we likewise possessed many of the intermediate links between their past or parent and present states; and these many links we could hardly ever expect to discover, owing to the imperfection of the geological record. (Darwin, 1859: 439)

The phylogeneticist has no choice but to surrender to speculation. The scarcer the evidence, the more the touchstone of phylogenetic truth resides in the eye of the beholder.

3.8 The Need to Speculate

Darwin received a blessing to speculate from a physicist. John Tyndall (1820–1893) was an intimate friend of T. H. Huxley and a fellow member of the X-Club, a scientific dining club-cum-pressure group that Huxley had stocked with evolutionary enthusiasts. Tyndall delivered an address, titled "The Scientific Use of the Imagination," at the 1870 meeting of the British Association for the Advancement of Science, the year that Huxley assumed its presidency. Tyndall's lecture gave wings to that "soaring speculator" from Down House (Tyndall, 1870: 32) who, although he had "drawn heavily upon the scientific tolerance of his age" (Tyndall, 1870: 31), had a mind that "can never sin wittingly against either fact or law" (Tyndall, 1870: 32). If the scientific imagination of a Darwin could lift transforming theories from the mute realm of facts, then "a good deal of wildness on the part of weaker brethren may be overlooked" (Tyndall, 1870: 31). But would Tyndall's imprimatur extend to phylogenetic theorizing as well? The two Darwinian theories that Tyndall mentioned in his address, species evolution and the theory of particulate inheritance known as pangenesis, were nomothetic or law-giving theories. Although no theory has ever originated without the animating spark of a scientist's imagination, once erected, nomothetic theories are expected to support their own weight with a scaffold of deductive logic. Wherever and whenever a specific set of initial and boundary conditions prevails, nomothetic theories, when correct, provide their explanations with law-like inescapability, offering what are generally referred to as deductive-nomological or nomological-deductive explanations.

In contrast, phylogenetic theories are idiographic. The explanatory reach of phylogenetic theories is limited to a unique address in space and time. They are balanced on a knife edge of incomplete and historically contingent evidence, and their detailed shape typically depends in no small measure on the inspired imagination of their creators. Moreover, phylogenetic theories do not describe and explain evolutionary transformations with the inevitability of deductive logic. They can only delineate historical events with more or less probability. More or less informed inference and speculation necessarily substitute for deductive certainty in any historical hypothesis. The scarcer and older the available evidence is, the more prominent the role of the imagination becomes in animating historical hypotheses. It is therefore not surprising to see that the early phylogeneticists explicitly defended their need to speculate.

If Darwin had Tyndall's blessing, evolutionary morphologists had Karl Ernst von Baer's, or so some of them claimed. Von Baer subtitled the celebrated 1828 embryological work in which he presented his four embryological laws *Über Entwicklungsgeschichte der Thiere "Beobachtung und Reflexion"* (*On the Developmental History of Animals "Observation and Reflection"*). Haeckel read into this exactly what he needed: an endorsement of philosophical speculation as the necessary yin to natural science's empirical yang. He let von Baer's phrase echo through many of his works, no less than five times on the page on which he introduced it in the *Generelle Morphologie* (*General Morphology*). This is not without irony given that in the same work von Baer mercilessly ridicules transmutationist thinking:

> Still recently we read in a paper about the embryonic circulatory system that not a single animal form is excluded from the human embryo. One gradually learned to think of the different animal forms as if they had developed one from another—and then at least some seemed to want to forget that this metamorphosis was merely a way of conceiving things. Supported by the finding that remains of vertebrates are absent from the older layers of the earth, one thought one could prove that such a transformation of the different animal forms was really historically grounded, and finally told in all seriousness and in detail how they had evolved from each other. Nothing was easier. A fish swimming up to the land wanted to go for a walk, for which its fins couldn't be used. Their width shrank through lack of exercise, and they grew longer. This was passed on to children and grandchildren for several millennia. It is therefore no wonder that fins finally became feet. It is even more natural that the fish in this way, because it doesn't find water, gasps for air. Consequently, over the same period it will sprout lungs, which meanwhile only requires that several generations cope without being able to breathe. (von Baer, 1828: 200)

The first generations of phylogenetic biologists were well aware that excessive speculation could lure them into treacherous terrain. At the end of *Facts and Arguments for Darwin*, Fritz Müller was tempted to offer a speculative recapitulationist reading of the evolutionary history of crustaceans. Yet, he didn't do it because he feared that the uninitiated would mistake the weakest parts of his thinking for established fact, while crustacean aficionados might dismiss his ideas in toto once they had spotted the inevitable weaker points.

In her book *Biology Takes Form*, Lynn Nyhart claims that these younger researchers "almost inadvertently" (Nyhart, 1995: 195) adopted Haeckel's speculative tendencies in their own phylogenetic research. I think there was nothing inadvertent about this. Nyhart offers three reasons why speculation could thrive to an unprecedented degree in late-nineteenth-century evolutionary morphology. First, the great merit of phylogenetic theories, like Haeckel's Gastraea theory, was their synthetic nature and broad explanatory scope, which offset their speculative nature. Second, speculative papers may have been quicker to produce than new empirical works, which was a benefit in the increasingly competitive academic arena. Third, the line of demarcation between restrained theorizing and outright fantasizing is subjective. As long as speculations are couched in appropriately tentative language, they can lend richness to empirical studies and provide guidelines for future work (Nyhart, 1995: 199–200).

I suspect that all three reasons played a role in fostering a more speculative approach in evolutionary morphology at the time. But I think that the most important reason for the rise of the speculative approach was that the younger evolutionists understood that it was an epistemological *sine qua non* of phylogenetic research. They were intensely aware of the central role of speculation in their work, and unlike much of the older generation, they were willing to explicitly justify using it. This is illustrated by the leading evolutionary morphologists of the late nineteenth century, including E. Ray Lankester in Britain, Ernst Haeckel and

Carl Gegenbaur in Germany, William Keith Brooks in America, and Anton Dohrn and Nikolai Kleinenberg in Italy (Lankester in Gegenbaur, 1878: viii; Lankester, 1880: 3; Brooks, 1893a: 179; Kleinenberg in Müller, 1973: 131; Dohrn in 1900 in Groeben, 1985: 16; Hatschek in 1877 in Nyhart, 1995: 198–199). As Anton Dohrn put it in a letter to E. B. Wilson, "Phylogeny is a subtle thing, it wants not only the analytical powers of the 'Forscher' [researcher], but also the constructive imagination of the 'Künstler' [artist],— and both must balance each other, which they rarely do,—otherwise the thing does not succeed" (Dohrn in Groeben, 1985: 16).

Not all researchers embraced speculation with equal relish. While a Haeckel would unhesitatingly and unapologetically lay "claim to that liberty of natural philosophical speculation" (Haeckel, 1874: 153), a Huxley would only grudgingly feel "reduced to speculation" (Huxley, 1875: 201) because of the poor quality of the fossil record. He would see the value of phylogenetic speculations primarily in their ability to guide future investigations. Others separated the descriptive/comparative and evolutionary sections of their papers, a strategy commonly adopted in early studies of phylogenetic morphology (Nyhart, 1995: 196). Yet others, such as Ambrosius Hubrecht (1883: 368), implored readers not to judge all aspects of his phylogenetic scenarios equally harshly and not to criticize the whole edifice for speculations about character systems that the author admitted he had not studied in any great depth. But few hedged their bets as much as William Bateson.

Bateson introduced his famous paper "The ancestry of the Chordata" by noting that "[o]f late the attempt to arrange genealogical trees involving hypothetical groups has come to be the subject of some ridicule, perhaps deserved. But since this is what modern morphological criticism in great measure aims at doing, it cannot be altogether profitless to follow this method to its logical conclusions" (Bateson, 1886). For three dozen pages, Bateson delves deeply into a discussion of deuterostome morphology, presumably to shed light on chordate origins. Then in the final two paragraphs, instead of rewarding readers with a crowning phylogenetic insight, he screeches to an abrupt halt. After claiming to have analyzed "the facts as they stood... [while] avoiding as far as possible a reliance upon phylogenetic changes of whose occurrence we have no evidence" (Bateson, 1886: 571), he finishes with a sentence that is a complete cop-out: "The foregoing views are, perhaps, more clearly expressed in the following table, which is not meant so much as a genealogical tree as to serve as an exhibition of the logical relation of the various forms, showing their points of divergence" (Bateson, 1886: 571) (Figure 3.7). The paper was his swansong in evolutionary morphology, and he devoted the rest of his career to the study of variation and heredity (Hall, 2005).

Bateson wasn't the only young biologist – he was 25 when his paper on the ancestry of chordates appeared – who bid evolutionary storytelling farewell in the late nineteenth and early twentieth centuries (Allen, 1981; Nyhart, 1995). Barely two decades after phylogenetics had gotten out of the starting blocks, Alexander Agassiz, son of Louis Agassiz and curator at Harvard's Museum of Comparative Zoology, declared in an address published in *Science* that the "time for genealogical trees is passed" (Agassiz, 1880: 148). This paper provided a more visible platform for critical comments that he had published a few years earlier in a massive monograph on sea urchins. There he lambasted evolutionists who "outdarwin Darwin himself" (Agassiz, 1874: 754) by building "castles in the air, which they have been

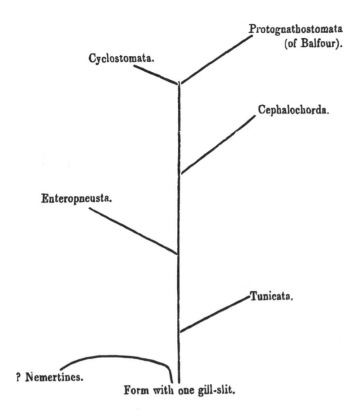

Figure 3.7 Relationship diagram of chordates and near relatives drawn by William Bateson (from Bateson, 1886).

obliged to pull down in rapid succession" (Agassiz, 1874: 753). Unsurprisingly, these critical sounds from an internationally respected scientist in a leading journal caused ripples. They triggered invertebrate zoologist William Keith Brooks, who had done graduate research with Agassiz, to pen a two-part defense of phylogenetic speculation in a popular science magazine (Brooks, 1882b, 1883). Brooks admitted that tracing ancestral lineages was an exercise fraught with difficulties, but that "the only way to avoid danger will be to stick to facts, and, stripping our science of all that renders it worthy of thinking men, to become mere observing machines" (Brooks, 1883).

Brooks clearly considered himself a thinking man. He was irresistibly beckoned by phylogenetic adventures that lay just beyond the empirical horizon, and he was not afraid to lace phylogenetic speculations through his densely empirical studies of the morphology of invertebrates (Brooks, 1882a, 1893a). The emergence of biological novelties was one of the most captivating mysteries that science had to offer to evolutionary morphologists like Brooks. All one needed to do to be transported to distant evolutionary vistas was to exercise one's more or less informed imagination. The mind's eye could watch evolution unfold from the safety of an armchair without even so much as a glance through a microscope. As we will see in the next chapter, the biologist who was accused of making this his main modus operandi led the phylogenizing vanguard of the new evolutionary biology.

Ernst Haeckel's Evolutionary Storytelling

The theory of common descent stoked the imagination of the first generation of evolutionary biologists (Nyhart, 1995; Bowler, 1996). As we saw in the previous chapter, the *Origin* dispelled the idea that the diversity of life somehow reflected the existence of a set of ethereal archetypes in a timeless realm of pure morphology and replaced it with the image of evolving lineages of flesh-and-blood ancestors linked by ties of homology and stretching across the immensity of geological time. Yet, making tangible the formerly figurative turned out to be a challenging and controversial task. Careful observation of organisms cannot make the invisible visible. Neither does the theory of common descent allow the simple deduction of the nature of distant ancestors and phylogenetic relationships. Instead, a complex mixture of observation, comparison, synthesis, inference, and speculation can hold hope for illuminating the unique attributes of each specific section of life's enveloping genealogical nexus. It was a bold young German researcher who, in the years immediately following the *Origin*, outlined the new science of phylogenetics. He started to sketch with characteristic panache the first broad phylogenetic hypotheses, animated by daring imaginings of long-lost ancestors. The result was an illustrious scientific career that established its author as the indisputable founder of the science of phylogenetics, as well as the coiner of the discipline's name. However, his life's work also became an object lesson about the dangers of wielding an unbridled imagination in the pursuit of scientific facts.

4.1 Ernst Haeckel and the First Phylogenetics

The paleontologist Heinrich Georg Bronn (1800–1862) prepared the first foreign translation of the *Origin* in 1860. He translated the second edition of the *Origin* into German just a few months after it was published in Britain, with a second posthumous German edition (of the third edition of the *Origin*) appearing in 1863. Bronn's rendering of Darwin's thinking was extraordinarily influential in the German-speaking world (Gliboff, 2007, 2008), even though it went significantly beyond being a verbatim translation of Darwin's words. Darwin's apprehension about Bronn's liberal editing of his text – including deliberate omissions and the addition of a whole chapter critiquing Darwin's ideas – was such that he arranged for a new German translation to be prepared by Bronn's friend Victor Carus, who eventually translated several of Darwin's books. Nevertheless, Bronn's original translation had a widespread influence.

Most famously it awakened the scientific imagination of a young researcher of radiolarians. From the moment he read the *Origin*, Ernst Haeckel (1834–1919) fell completely under the spell of the Darwinian worldview. He confided in a 1864 letter to Darwin that of "all the books that I have read, none has made so powerful and marked an impression on me as your theory of the origin of species" (Haeckel in Richards, 2008: 168). He added that "[I] only have one goal in life, namely, to work for your descent theory to support it and perfect it" (Haeckel in Richards, 2008: 170). Haeckel's scientific zeal was in no small measure the result of having to cope with the tragic loss of his wife Anna, who died on his 30th birthday. Although Haeckel had started planning the *Generelle Morphologie* during the summer of 1864, the two-volume, more than 1,000-page work was published in 1866, after just one year of monomaniacal writing. Whereas Darwin had established common descent as evolution's general principle in the *Origin*, he had not, and would not, devote his life to illustrating the plethora of phylogenetic links that his theory dictated must bind together all living and extinct organisms. The *Generelle Morphologie* was a bold first attempt to accomplish this ultimate goal.

The full title of Haeckel's book leaves no doubt about his esteem both for Darwin and for common descent as the metaphysical center of the new evolutionary theory: *Generelle Morphologie der Organismen. Allgemeine Grundzüge durch die von Charles Darwin reformierte Descendenz-Theorie* (*General Morphology of Organisms. General Outline of the Descent Theory Reformed by Charles Darwin*). *Generelle Morphologie* is rightly considered the founding document of evolutionary morphology. It represents Haeckel's translation of comparative idealistic morphology into a science cleansed of the empty vitalistic and teleological explanations that he found contaminated much contemporary morphological work. Haeckel was a morphologist and his founding definition of phylogenetics bears the distinctive stamp of that science. Haeckel considered *Phylogenie* to be a subdiscipline of *Morphologie*, the scientific study of form. He distinguished two branches in the science of morphology. *Anatomie* was the science of the *vollendenten* (completed) form of organisms, while *Morphogenie* was the science of the *werdenden* (becoming, developing) form of organisms. Within *Morphogenie* Haeckel defined two further subdisciplines, *Ontogenie* (or *Embryologie*), the science of the embryonic development of organismal form, and *Phylogenie*, the science of the evolutionary history of organismal form.

The goal of phylogeny was to "investigate the connected chain of forms of all those organic individuals that have branched off from one and the same shared stemform" (Haeckel, 1866a: 30). Thus, for Haeckel the sciences of ontogeny and phylogeny were two intimately related counterparts aimed at uncovering the changes in organismal form that occurred along the successive developmental stages of an individual over embryonic time and along the series of ancestors that comprises that organism's lineage over evolutionary time. Of course, as we saw in the previous chapter, Haeckel was by no means the first to find meaning in the parallels of the progression of developmental stages and the arrangement of taxa in the natural system, nor was he the first to give these a recapitulationist interpretation (he was clearly indebted to Fritz Müller's 1864 *Für Darwin*). Nevertheless,

Haeckel captured the intimacy of these relationships in his biogenetic law, according to which the series of morphological changes that can be observed during the ontogeny of a species recapitulate those undergone by its evolving lineage during phylogeny. This theory became widely known and much discussed. In Haeckel's words, if ontogeny is the "developmental history of concrete morphological individuals," phylogeny is the "evolutionary history of abstract genealogical individuals" (Haeckel, 1866a: 60).

This conceptual affinity between ontogeny and phylogeny is further expressed in Haeckel's use of the term *Entwicklungsgeschichte* for both developmental and evolutionary history. Yet, although conceptually united, ontogeny and phylogeny presented completely different epistemological hurdles (Haeckel, 1866b: 306–307). Whereas careful empirical observations, either descriptive or experimental, can provide insights into ontogeny, Haeckel realized that this approach was altogether impotent in the realm of phylogeny. Only by clothing observations in a cloak of imaginative inference could phylogeneticists have any hope of revealing evolution's unseen mysteries. Before having a more detailed look at how Haeckel attempted to reveal phylogeny, we need to get a clearer grasp of what Haeckel's phylogeny is and isn't.

4.2 Haeckel's Phylogeny and Genealogy: Distinct but Inseparable

For modern biologists, "phylogeny" explicitly refers to the evolutionary relationships of organisms. For Haeckel it did so implicitly. Haeckel captured the evolutionary relationships between organisms in *Stammbäume*, what we would now call evolutionary or phylogenetic trees. Haeckel called this science of systematic relationships *Genealogie*. The evolutionary history of morphology inferred along individual lineages was what Haeckel referred to as *Phylogenie*. In other words, his *Phylogenie* refers to the linear realm of evolutionary descent, whereas his *Genealogie* refers to the branching realm of collateral relatives. Despite this conceptual distinction, he saw these realms as indissolubly linked. In the first volume of *Generelle Morphologie*, Haeckel presents a lengthy discussion of the relationship between the disciplines of systematics and morphology. He strongly disagreed with biologists, such as the aforementioned Victor Carus, who considered these disciplines to be fundamentally distinct, with morphologists studying form for its own sake and systematists studying it only as a means to classify species. Haeckel concludes that systematics and comparative morphology are intimately related disciplines: "the concentrated extract of all comparative morphology is: the various relationships of all forms that have descended from one and the same stemform" (Haeckel, 1866a: 38). The resulting "genealogical systematics" will then be "the essential core of the entire morphology of organisms" (Haeckel, 1866a: 38). If we summarize Haeckel's concepts of genealogy and phylogeny in modern terms, the first is represented by trees that depict systematic relationships between collateral relatives, whereas the latter is represented by the series of character state transformations reconstructed along lineages in the trees.

Even though Haeckel considered genealogy and phylogeny to be theoretically distinct, he considered them inseparable in practice. Because "the discovery of the relationships between organic forms is so fundamentally the task [of phylogeny], she could also be called genealogy of organisms" (Haeckel, 1866a: 30). In the second volume of *Generelle Morphologie*, Haeckel expresses the intimate relationship between genealogy and phylogeny somewhat differently, stating that phylogeny encompasses both the genealogy and the paleontology of organisms (Haeckel, 1866b: 4), with genealogy concerning the branching evolutionary relationships of organisms and paleontology concerning the linear evolutionary histories of their morphology. For Haeckel, as for modern phylogeneticists, insights into the evolutionary relationships of organisms are the end product of phylogenetic research. In this important sense, Haeckel's phylogenetics was fully modern. However, the original meaning of his term "phylogeny" has subsequently shifted from the linear aspect of evolutionary descent to the branching aspect of systematic relationships. Confusing these aspects has led recent authors to misjudge Haeckel's phylogenetics.

4.3 Are Haeckel's Trees Darwinian?

Despite correctly emphasizing the close epistemological ties between genealogy and phylogeny in Haeckel's thinking, Benoît Dayrat (2003) has advanced the peculiar argument that the phylogenetic forest planted by Haeckel was nevertheless distinctly non-Darwinian. Haeckel biographer Mario Di Gregorio (2005: 187) echoes this view and concludes that Haeckel's "trees only had a superficially Darwinian gloss." These opinions align these authors with numerous historians and other commentators who have long denied Haeckel the honorific, if myopic, label of "Darwinian" (Gliboff, 2008). Peter Bowler's characterization of Haeckel as "a Darwinian only by the loosest definition" (Bowler, 1992a: 33) and Michael Ruse's even stronger assertion that Haeckel "was not a Darwinian scientist" (Richards and Ruse, 2016: 77) are emblematic of this widespread view. According to Ruse, Haeckel's science wasn't Darwinian because he was more interested in tracing phylogenetic lineages than studying evolutionary mechanisms. But that's just a matter of emphasis and focus. This facile criterion would remove virtually all evolutionary morphologists from the Darwinian camp. Haeckel's focus on homology and unity of plan can easily, and rightly, be seen as a modified outgrowth of the continental European tradition of transcendental comparative morphology, but Darwinian evolutionary theory infused this type of research with renewed legitimacy. Denying Haeckel's science the Darwinian label on this basis makes about as much sense as denying that a curry is Indian because it contains tomatoes and chilies from the Americas. Haeckel certainly developed his own ideas, but he consciously worked within the Darwinian tradition.

To get a clear view of what the phylogenetics of Haeckel represents, it is instructive to discuss it in the context of Dayrat's 2003 paper in *Systematic Biology*, in which he tried to demarcate a Darwinian and a Haeckelian phylogenetics. This paper represents views that are widespread in the contemporary literature on Haeckel. It notes that in contrast to real Darwinian trees, Haeckel's trees are only superficially branching, with branches being mere ornaments diverging from a major trunk; that his trees are an evolutionary incarnation of the *scala naturae* by representing evolution as a linear process; and that Haeckel's phylogeny doesn't refer to an actual series of ancestors, but rather an evolutionary march of morphology that is decoupled from evolving lineages. How accurate are these views?

Haeckel thought that he was fulfilling his Darwinian duty by stocking the literature with phylogenetic trees, but many modern commentators judge his achievements differently. They insist that just underneath the bark of his trees lurks the specter of the *scala naturae*. The Great Chain of Being was originally conceived as a static, eternal, anthropocentric, and teleological hierarchy along which all worldly and heavenly productions could be arranged according to vague and variable notions of increasing complexity or perfection. Within the later context of evolutionary worldviews, this notion then transmogrified into the idea of inevitable, inherent, goal-directed evolutionary progress, with organisms being driven to or striving for ever greater complexity (Lovejoy, 1936). In a Darwinian world such a view is justly berated, but Haeckel unfortunately didn't manage to escape the lure of the linear ladder of progress. Or so it is widely claimed in textbooks, historical works written by historians and scientists, as well as research and review articles (Panchen, 1992; Bowler, 1996, 2021; Breidbach, 2003; Dayrat, 2003; Bowler and Morus, 2005; Catley and Novick, 2008; Omland et al., 2008; Reynolds and Hülsmann, 2008; Arthur, 2011; Baum and Smith, 2013; Casane and Laurenti, 2013; Archibald, 2014).

For example, in the online addendum to his popular textbook *Developmental Biology*, Scott Gilbert writes that "Haeckel brought the Western notion of the Great Chain of Being into evolutionary thought... [I]t celebrated the ascent of Man" (https://iws.oupsupport .com/static/5c7cbefacde8ba00107ba35c/c25_01_2.html; last accessed August 5, 2021). Consequently, the "ladder of progress view is most clearly documented in Haeckel's phylogenetic trees" (Omland et al., 2008: 854), which represent "the epitome of a teleological view of evolution" (Catley and Novick, 2008: 983) and reflect the idea that "evolution is striving towards a perfect being" (Breidbach, 2003: 178). His *Stammbaum des Menschen* (*Pedigree of Man*) in particular showcases the "inexorable central upward thrust towards humans" (Arthur, 2011: Fig. 18.3), in a world where the evolutionary process has foresight. In drawing up his trees, Haeckel "was mainly inspired by the idea of the *scala naturae*" (Dayrat, 2003: 515) and "faithfully followed Lamarck instead of Darwin" in arranging "living beings along a linear series" (Dayrat, 2003: 525). Lateral branches were "mere ornament" (Dayrat, 2003: 515) and devoid of any phylogenetic information because "Haeckel assumed that the organisms of a given lateral branch would share the same general morphology as the stem group from which it would originate" (Dayrat, 2003: 524). As Bowler and Morus (2005: 152) put it, Haeckel treated "all those creatures that do not lie on the 'main line' as side branches leading off to stagnation." In other words, body plan evolution would more or less stop in its tracks once an evolutionary lineage had split away from the tree's trunk. This interpretation is echoed in Omland et al.'s (2008) paper on "Tree thinking for all biology" in which they single out Haeckel's *Pedigree of Man* as the clearest example of improper tree thinking. Haeckel's tree shows "extant groups such as 'amoeba' and 'primitive worms' low down and embedded within the main trunk [whereas] [f]urther up, evolution passes through 'jawless animals', 'pouched animals', 'apes' and finally, to the apparent pinnacle of evolution – 'man.'" Perhaps not wishing to pass an all too patronizing judgement on the founder of phylogenetics, they write that "with the benefit of nearly 150 years of evolutionary research... we know that evolution generally has not stopped in any lineages. Thus it is misleading to think of an extant amoeba species as ancestral to humans, or an extant amphibian as ancestral to snakes" (Omland et al., 2008: 854).

The problem with these views is that Haeckel thought no such things. He didn't need 150 years of new research to know that phylogenetic lineages could not be recreated by lining up living taxa like beads on a string. Stamping his trees with the stigma of the *scala naturae* is a tactic that grossly distorts Haeckel's thinking. In fact, accusing Haeckel of believing in the Great Chain of Being is the conceptual equivalent of accusing Darwin of believing in archetypes. The ideas of both men had undeniable links and similarities to previous ideas, but in both instances deep metaphysical metamorphoses had produced entirely novel concepts that were as distinct from their predecessors as butterflies are from caterpillars. So yes, Haeckel was convinced of the unity of the inorganic and organic worlds, just like in the *scala naturae*. And yes, Haeckel was convinced that phylogenetic continuity was unbroken and that gaps are only apparent, just like in the *scala naturae*. And yes again, Haeckel was anthropocentric at least in his *Pedigree of Man*, just like the *scala naturae*. And finally, yes, Haeckel's evolutionary world was suffused with progress, just like the *scala naturae*. But is this convincing evidence that the *scala naturae* represents the essential scaffolding of Haeckel's trees? I think not.

Haeckel was fully Darwinian in realizing that evolutionary trees branch throughout, from root to tips (Figure 4.1). He knew that terminal twigs were no less significant than deeper branches because evolutionary trees were produced by a continuous process of lineage splitting and transformation. He considered taxonomic categories to be convenient, but artificial constructs for the ordering of systematic knowledge (Haeckel, 1866a: 195). They reflected the continuity of phylogeny that was expressed in the blood relationships of species on different scales (Haeckel, 1894: 28). Haeckel drew most of his trees as bushes, rather than trees, lacking conspicuous trunks, with some even resembling hedgerows with many twigs diverging in parallel, such as the echinoderm and vertebrate trees in *Generelle Morphologie*, which were drawn against the background of a geological timescale. In fact, as Dayrat admits, Haeckel drew only one tree with a conspicuously thick trunk: his famous *Pedigree of Man* published in *Anthropogenie* (Figure 4.2), which is his most frequently reprinted tree. Dayrat nevertheless claims that all of Haeckel's trees are fundamentally linear and that "according to Haeckel the information conveyed by trees was entirely contained in the trunk of the tree and not in the lateral branches" (Dayrat, 2003: 524).

Modern criticisms that Haeckel's diagrams are merely ladders masquerading as trees are never supported by detailed arguments. The "evidence" that is presented is always the same: a picture of Haeckel's *Pedigree of Man*, which is often wrongly attributed to the *Generelle Morphologie* from 1866 (Gould, 1995a: 65; Omland et al., 2008: 855; Wheeler, 2012: 13). This is as convenient as it is misleading because it is Haeckel's only tree that perches humans on top of a massive trunk, with the rest of animal diversity arranged as a series of side branches. What these modern commentators fail to realize is that this anthropocentric emphasis is intentional, as the aim of the tree is to visualize the

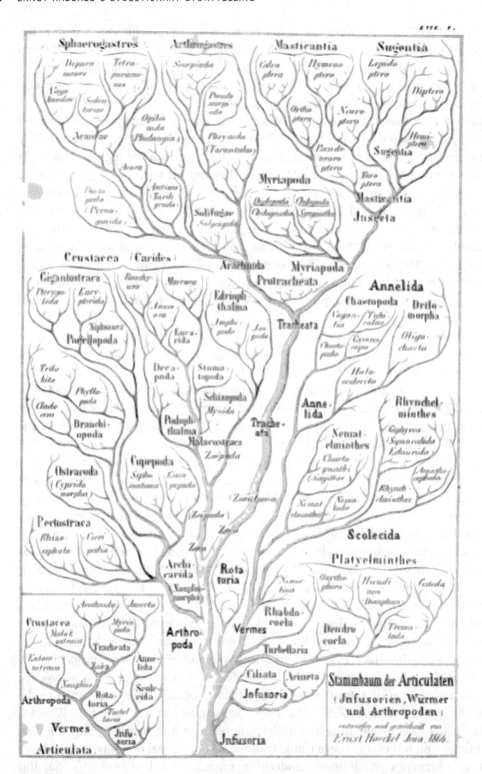

Figure 4.1 Haeckel's phylogeny of Articulata, including platyhelminths, annelids, and arthropods (after plate 5 in Haeckel, 1866b).

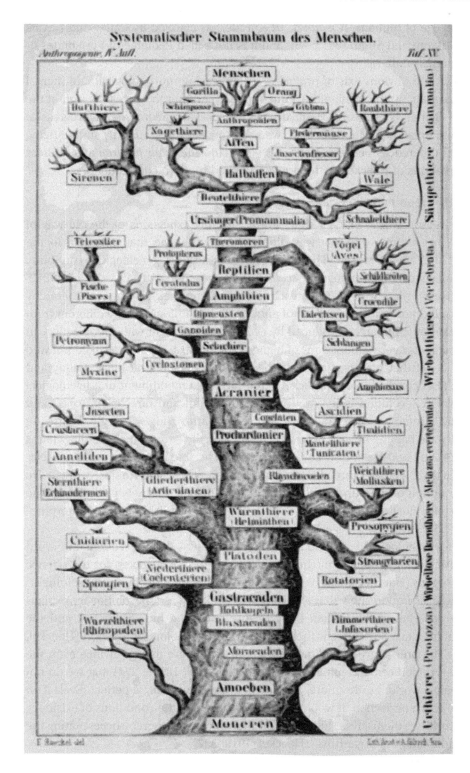

Figure 4.2 Haeckel's artistic *Stammbaum des Menschen* or *Pedigree of Man*
(after plate 15 in Haeckel, 1891).

phylogenetic lineage of humans. It is *not* a "tree of life," as claimed by Catley and Novick (2008: 983) and Baum and Smith (2013: 26), and it is not called Pedigree of Vertebrates or Pedigree of Mammals either.

But what about the obvious axis of progress – of increasing morphological and mental complexity – that runs up the trunk of his human pedigree? Isn't that a telltale sign of the Great Chain of Being? No, it isn't. I will discuss Haeckel's thinking about progressive evolution in more detail in Section 4.8. First, we need to clear the ground by summarizing the fundamental differences between Haeckel's views and the *scala naturae* and address the misrepresentations of Haeckel by the previously cited modern commentators.

4.3.1 Haeckel's Trees and the *Scala Naturae*

The most basic reason why Haeckel's views are not simply extensions of the old way of thinking is that his entire worldview was different from the one that gave rise to the assumptions of the *scala naturae*. There was no historical or genealogical continuity between the objects arranged along the *scala naturae*. As noted by Ernst Mayr (1982: 205–206), they were ordered into a static exclusive hierarchy that, just like the hierarchy of military ranks, did not form a nested set of entities. This is fundamentally different from the inclusive taxonomic hierarchy produced by evolutionary descent with modification, in which lower-ranking taxa are subdivisions of higher-ranking taxa. Haeckel also did not wed his undeniable belief that man was a special beast to any teleological ideas. Catley and Novick's (2008: 984) conclusion that Haeckel's trees are the embodiment of a teleological worldview in which "evolution is in some way a purposeful and directed process" and Breidbach's (2003) inference that Haeckel's *Pedigree of Man* shows evolution's striving toward humanity are baffling in view of what Haeckel actually wrote. So are similar claims by Bryant (1995: 201) and Fisler and Lecointre (2013). The intellectual fight against teleology is a massive golden thread running through Haeckel's major works. He denounced any notion of the purposeful design of the universe that saw evolution as the mere unfurling of a goal-directed process. He called this "childishly anthropomorphic" (Haeckel, 1891: 659).

His disdain for teleological explanations was so complete that he coined the word *Dysteleologie* (dysteleology), tongue in cheek at least in part one suspects, for the science of rudimentary organs (Haeckel, 1866a: 99–100). He used this "theory of unpurposefulness" (*Unzweckmässigkeitslehre*) to show that teleology was bankrupt as an evolutionary explanation. The thin covering of the human body in tiny hairs, the human coccyx, and the appendix are rudimentary structures without any obvious physiological function (Haeckel, 1891: 837–838). The prevalence of rudimentary organs in nature must convince even the "eyes of the most prejudiced and idiotic naturalists" (Haeckel, 1866b: 274) that we do not live in a teleological world. Furthermore, no creature, not even man, is perfect. Even if an organism would reach perfection, it wouldn't last long because "the conditions of existence in the outside world are themselves subject to continuous change, thereby necessitating the uninterrupted adaptation of organisms" (Haeckel, 1919: 337).

Second, as I argued earlier, Dayrat's (2003: 526) conclusion that "Haeckel's trees were not branching diagrams" is simply nonsense. Haeckel was explicit in his rejection of a single scale that includes "all animals from the lowest to the highest, from infusorian to man" (Haeckel, 1891: 47). For him there were as many chains of organisms as there were lineages. The expression of linearity that is a real feature of Haeckel's trees is the result of lineage thinking and of linearly arranging evolutionary sequences of paraphyletic groups, as also noted by Williams and Ebach (2008: 52). Paraphyletic groups are supraspecific taxa with a unique common ancestry, but which deliberately exclude some taxa that share that common ancestry. Examples are dinosaurs, excluding birds; apes, excluding humans; and invertebrates, excluding vertebrates. Many supraspecific taxa in Haeckel's trees are logically paraphyletic, whether or not Haeckel understood what this meant, and these taxa include, or even primarily comprise, living taxa.[1] He invented many hypothetical supraspecific ancestral groups, from stem-genera to stem-classes, all of which were logically paraphyletic. Linear arrangements of paraphyletic groups form the backbones of the major branches of many of his trees. But De Queiroz (1988: 252) and Panchen (1992: 31–34, 40) consider the very recognition of paraphyletic grades to be convincing evidence of the lingering stranglehold of the *scala naturae*.

Panchen (1992: 31, 34) concludes that Haeckel's linear arrangements of taxa don't follow a time axis and cannot represent ancestor–descendant series, even though Haeckel clearly intended them as such, because our current understanding is that supraspecific taxa cannot be ancestral because paraphyletic taxa are abstractions without any real existence. The problem of paraphyly will be discussed in more detail in Chapter 11. For now, I will just point out that paraphyletic taxa indeed aren't real because they lack individuality, in contrast to monophyletic clades. But you can't indict a nineteenth-century biologist for theoretical crimes that weren't recognized as such for another century. When Haeckel or any other biologists or paleontologists writing in the late nineteenth and early twentieth centuries drew phylogenies that lined up paraphyletic taxa, their aim was to express the linearity of descent. Judged by our modern understanding, this is indeed bad taxonomic bookkeeping because paraphyletic taxa are abstractions. To taxic thinkers today, the formal recognition of paraphyletic taxa is a conceptual sin. But these early evolutionary thinkers were primarily lineage thinkers. What they tried to convey with their trees was the linearity of lineages, not the linearity of the *scala naturae* (Figure 4.3). This was challenging because they could only express themselves in the taxic language of taxonomy, which inevitably causes problems. As we will see in Chapter 11, a substantial number of researchers, science writers, and educators today still fail to get the distinction between taxic and lineage thinking right.

Third, Haeckel knew that in a Darwinian world of open-ended evolution, phylogenetic lineages were not made up of and could not be reconstructed by strings of living species.

---

[1] The problem is that paraphyly, as argued in the text of Haeckel's works, although not using this term, is by no means always obviously indicated in his trees. For example, Haeckel argued in *Generelle Morphologie* that the cnidarian group Stauromedusae, which he called Calycozoa, had evolved from a group called Elasmorchida. Yet, in his *Stammbaum der Coelenteraten* (*Pedigree of Coelenterates*) he depicts them as sister groups. In contrast, he argued that sea stars are a paraphyletic group. He thought that the fossil taxon Crinastra represented transitional forms to crinoids, and his *Stammbaum der Echinodermen* (*Pedigree of Echinoderms*) in *Generelle Morphologie* indeed depicts asteroid paraphyly.

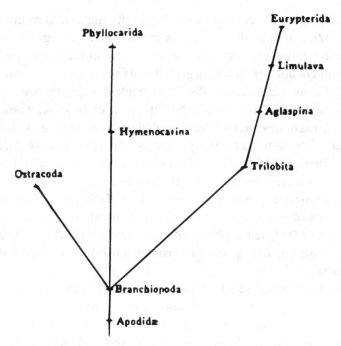

T HEURETICAL EVOLUTION UF CAMBRIAN CRUSTACEA FROM THE BRANCHIOPODA

Figure 4.3 The attempt of Charles Doolittle Walcott, discoverer of the Burgess Shale fauna, to trace the "lines of descent" of several major groups of arthropods. He lined up paraphyletic fossil taxa to draw these lineages, an image that Stephen Jay Gould, in his book *Wonderful Life*, incorrectly rejected as flawed
(after the figure on page 164 in Walcott, 1912).

Yet, Bowler (1996: 60–61) claimed that a "potentially dangerous aspect of Haeckel's reconstructions was his tendency to look for living equivalents of the ancestral form of each class. In theory it was most unlikely that the common ancestor could have survived unchanged while its descendants had radiated out into various lines of specialization." The correct impression that "Haeckel seems to have gone out of his way to provide a living analog of each node in the tree of life" (Bowler, 1996: 60) suggests, incorrectly I think, to Bowler that Haeckel succumbed "to the lure of the old chain of being. He wanted to believe that all stages in the ascent of life were still represented in the modern world" (Bowler, 1996: 60–61). Haeckel believed this no more or less than modern biologists. If the traits of ancestors aren't retained in more or less modified form in descendants, evolutionary inference becomes impossible. Haeckel understood well that all terminal twigs on the tree of life represent body plans more or less modified from those of their ancestors:

> If we now try to find the most important among the old palingenetic forms from the morphologically diverse groups of the lower animals and worms, it is obvious that not a single one of these can be seen as the unchanged or even just little changed image of a long extinct stemform. (Haeckel, 1891: 509)

He took care in making this explicit throughout his major works, even when his adjectival flourishes would seem to suggest that some animals were veritable phylogenetic oracles. For instance, "Therefore we cannot say: 'amphioxus is the ancestor of the vertebrates'; but we can say: 'of all animals known to us, amphioxus is the nearest relative of this ancestor'" (Haeckel, 1891: 530). Haeckel thought that the lineage of amphioxus had degenerated to a degree, with loss or simplification of its sensory organs and brain in connection with living submerged in sand (Haeckel, 1895: 208). Similarly, "Of course, no living fish can be regarded as the direct stemform of the higher vertebrates" (Haeckel, 1891: 535). Around the time that Haeckel published his *Anthropogenie* and *Systematische Phylogenie* (*Systematic Phylogeny*) in the 1890s, biologists were no longer expected to fall into this trap. As Edmund B. Wilson put it, "I scarcely need to add that no zoologist would look for the actual progenitor of vertebrates among existing invertebrates" (Wilson, 1891: 55). As we will see in Section 4.7, Haeckel also did not uncritically consider fossils as direct lineal ancestors, not even the most suggestive so-called missing links, such as *Archaeopteryx lithographica* and *Pithecanthropus erectus*.

Haeckel had the same hope as evolutionists today that living organisms would capture salient aspects of ancestral body plans. The conviction that some surviving taxa are passable models of ancient progenitors is still present in our modern literature. Amphioxus hasn't lost its ancestral badge of honor in a century and a half, despite debates about its degree of degeneracy. Researchers continue to express this view in various ways and with varying degrees of accuracy. For instance, according to Garcia-Fernàndez and Benito-Gutiérrez (2009: 665) "amphioxus (lancelet) is now recognized as the closest extant relative to the stem chordate." This is inaccurate as all extant chordates are equally related to their stem-relatives. At different times Simon Conway Morris (2000: 1) called amphioxus "the most primitive living chordate, widely regarded as a sort of pre-fish" or inaccurately labeled it "the predecessors [sic] of the vertebrates" (Conway Morris, 2003: 180). According to Peter Holland (2010: 85), "the latest common ancestor of all chordates was a motile animal of similar body form to amphioxus," while Vopalensky et al. (2012: 15383) call "amphioxus, our most primitive chordate relative." Mallatt and Chen (2003: 2) express the ancestral gleam of amphioxus by writing that "cephalochordates... closely resemble the ancestor of craniates," while Chen et al. (2009: R78.2) conclude that "amphioxus affords the best available glimpse of a proximate invertebrate ancestor of the vertebrates and is likely to exemplify many of the starting conditions at the dawn of vertebrate evolution." Indeed, amphioxus experts Nick, Linda, and Peter Holland summarize the current consensus as the "transition from ancestral invertebrate chordates (similar to amphioxus and tunicates) to vertebrates is well accepted" (Holland et al., 2015: 450).

A particularly convincing early illustration of Haeckel's understanding of the relationship between the linear and branching aspects of evolution can be found in the second volume of *Generelle Morphologie* when he discusses human evolution. After praising Huxley's *Evidence as to Man's Place in Nature* for having embedded us so snugly in our beastly cradle, he then takes him to task for coming too close to suggesting that the living groups of primates can be arranged in a linear series leading up to man (Figure 4.4). Haeckel (1866b: cliii) writes:

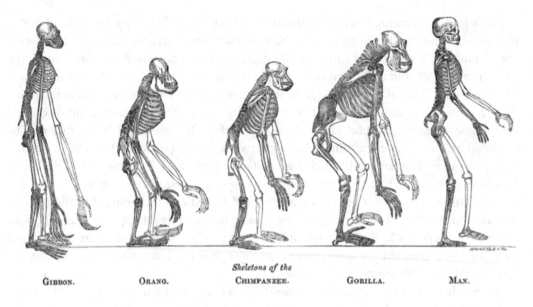

GIBBON.   ORANG.   *Sheletons of the*
                             CHIMPANZEE.   GORILLA.   MAN.

Figure 4.4 The famous frontispiece of Huxley's 1863 *Evidence as to Man's Place in Nature* (from Huxley, 1863).

> So, even though this unbroken gradation from the lowest rodent-like discoplacentalian to the highest, up to man, is of the highest importance, however, the true genealogical relationships of the different groups to each other are not simply expressed by the image of a ladder, but much more by that of a group of branches, as is suggested in the right upper corner of the pedigree in Table VIII.

On the next page he writes:

> We cannot simply describe the platyrrhines [New World monkeys] as lower and less perfect, and the catarrhines [Old World monkeys] as higher and more perfect monkeys, because both groups have independently from each other, as two divergent, coordinated branches, developed their own typical perfection separately; and among the platyrrhines in America, the most improved monkeys (*Callithrix*, *Cebus* [a genus of marmosets and capuchin monkeys, respectively]), have through progressive perfection of the skull and brain, and correspondingly their intellect, raised themselves above the lower platyrrhines, just as man has above the monkeys of the Old World. (Haeckel, 1866b: cliv)

Haeckel clearly understood the distinction between the evolution of body plans captured in a linear series of hypothetical human ancestors (phylogenesis) and the branching order of the living taxa (genealogy) that forms the necessary context within which this evolution is inferred. At this point, I should note that Haeckel's trees are frequently not an unambiguous guide to his thinking as captured in his texts. Different trees in the same work can conflict with each other and with the accompanying text. For instance, and in contrast to the earlier citation, in the third volume of *Systematische Phylogenie* Haeckel argues that New World monkeys are an older, less highly organized group than the Old World monkeys and that they have given rise to the latter. This is reflected in the tree of placental mammals (Haeckel, 1895: 491) in which Haeckel drew a linear series of paraphyletic primate taxa. But in the tree of primates (Haeckel, 1895: 601), Old and New World monkeys seem to have a sister group relationship. The same is true for *Anthropogenie*, where the tree of primates (Haeckel, 1891: 584) depicts the paraphyly of New World monkeys with respect to Old World monkeys, in line with the text, but where the tree of mammals (Haeckel, 1891: 577) seems to indicate these groups as each other's closest relatives. More pronounced conflicts between trees and text, however, also occur in places, which makes it challenging to determine what Haeckel really thought at any one time and cautions against any literal interpretation of his trees.

Finally, Haeckel knew that phylogenesis was not an escalator that dropped species off at their allotted branching points, leaving them to linger forever after in evolutionary stasis. He knew that lineages evolve indefinitely. For this reason, living as well as most extinct species could only ever be expected to partly lift the veil on ancestors. For Haeckel, the basic body plan of Gastraea, for example, was only partly revealed in its least modified descendants among the living *Gastraeaden. Trichoplax* had become flattened as an adaptation to its creeping lifestyle, flattening out its archenteron. Mesozoans had lost the archenteron entirely as a result of their parasitic habits. And even though *Hydra* exemplified the basic structure of Gastraea fairly well, its body plan reflected the adoption of a sessile life, with tentacles and cnidocytes added as cnidarian novelties. Similarly, although the fossil triconodonts evoked a more or less accurate impression of the overall morphology of Architherium, the last common ancestor of mammals, they did not retain its undifferentiated reptile-like dentition. Haeckel realized that even taxa that he labeled "living fossils," such as echidnas and the platypus, had evolved many new specializations so that they only reflected certain aspects of the body plan of Architherium. The same was true even of that "real Urvertebrate" (Haeckel, 1891: 254) amphioxus, which Haeckel thought had lost a clearly demarcated and complex head with sensory organs, such as paired eyes and otic vesicles, and a well-defined brain after its split from Prospondylus, the hypothetical vertebrate ancestor. Haeckel spelled out this important general insight in the *Generelle Morphologie* (Haeckel, 1866b: 265–266): "Therefore we find that very many simultaneously existing organisms, although they have descended from one and the same stemform, nevertheless show very different degrees of perfection and differentiation."

Dayrat (2003: 525) concludes that "[a]lthough Haeckel wrote many pages on the history of natural history, he never mentioned the *scala naturae.* He surely wanted to preserve his reputation as the German Darwin." I suppose this is possible. I offer a simpler explanation. Haeckel rejected the Great Chain of Being. The widespread myth that he didn't results from a superficial reading of his work, or perhaps from just looking at some of his pictures.

4.3.2 Haeckel's Lineage Thinking

The linearity in Haeckel's trees is the linearity of descent. It is the result of lineage thinking and the foregrounding of specific lineages in the interest of evolutionary storytelling. As Dayrat explains in detail, this linearity resulted from how he reconstructed phylogeny. Mayr and Bock (2002: 175) aptly characterized Haeckel's approach to phylogeny as "a backward looking endeavor, the search for and study of common ancestors." Haeckel's primary goal was to reveal how morphology had evolved along the lineages of focal groups, whether these were species, such as *Homo sapiens*, or larger taxa, such as angiosperms. Haeckel therefore focused on lineages, the linear series of ancestors that stretch backward in time from every tip on the tree of life to its root. By foregrounding a single lineage, most famously humans in his *Pedigree of Man* (Figure 4.2), Haeckel created the impression of a main trunk in his trees, even when the tree was drawn as a simple stick diagram. But there is nothing non-Darwinian about such linear reconstructions of evolution.

In his popular science book *The Ancestor's Tale* (2004), Richard Dawkins adopts precisely the same backward-looking perspective in tracing the evolutionary lineage of humans back to the origin of life. The impression of a central trunk is simply the result of emphasizing body plan evolution along a focal lineage, which for understandable reasons is that of *Homo sapiens* for both Haeckel and Dawkins. Indeed, of the 31 human ancestors that we encounter in Dawkins's narrative before we reach the origin of the animal kingdom, 20 were already present in the same sequence in Haeckel's story of human origins, published 96 years earlier (Haeckel, 1908).

Foregrounding the evolution of a single lineage doesn't mean that the other lineages are "mere ornament" or that the "[p]osition and number of lateral branches had no real meaning in Haeckel's trees" (Dayrat, 2003: 515, 524). Visually singling out one lineage for emphasis has been a common narrative device in phylogenetics ever since Haeckel (O'Hara, 1992). Modern phylogeneticists accomplish this by judiciously pruning or collapsing side branches to home in on one lineage. This is perfectly reasonable. One can easily see that the thick-trunked *Pedigree of Man* in *Anthropogenie* is the result of Haeckel's deliberate attempt to highlight the human lineage by comparing it with trees that contain *Homo sapiens* as one twig among many in his works not specifically dedicated to human phylogeny. For instance, his *Stammbaum der Säugethiere mit Inbegriff des Menschen* (*Pedigree of Mammals, Including Humans*) in *Generelle Morphologie*, and his *Stammbaum der Primaten* (*Pedigree of Primates*) in *Systematische Phylogenie* do not elevate humans above the apes, refuting claims (Pietsch, 2012: 100) that humans were always placed at the apex in all of Haeckel's trees.

Admittedly, as Dayrat points out, Haeckel was not always consistent when placing the same taxa on different trees, but it is clear that the proximity of groups in his trees is a measure of their evolutionary relatedness, with the number of branches reflecting the number of lineages with distinct ancestors. Indeed, Haeckel intended his trees to be succinct genealogical summaries (Haeckel, 1866a: 88). Nor is it true that Haeckel "always selected a single species whose phylogeny would provide a trunk" (Dayrat, 2003: 524). Many of Haeckel's trees have no discernable trunk at all, nor do their tips always represent single species. Haeckel's trees trace lineages from their roots to their tips, exactly like modern trees do. All lineages in his trees are equivalent in the sense that they depict descent from common ancestors, irrespective of their relative thickness.

The linearity in Haeckel's diagrams is not a symptom of conceptual flaws, but results from his adoption of lineage thinking. Stephen Jay Gould's (1995b: 41) assertion that Haeckel's *Pedigree of Man* is a failure in both a"logical and geometrical sense... though

Haeckel seems unaware of the difficulty" is off the mark. Gould claims that Haeckel molded his human pedigree according to the unspoken, yet pervasive, dictates of what Gould called the "cone of increasing diversity," according to which the horizontal dimension of conventional phylogenetic trees records the evolutionary spread of diversity, while time runs along the vertical axis. According to Gould (1995b: 41), Haeckel could only maintain the impression of the joint unfolding of morphology and diversity by "spreading the measly four thousand species of mammals across the entire topmost level of the tree, and relegating all arthropods (more than a million species) to a single twig near the tree's base." But Haeckel did not intend the *Pedigree of Man* to reflect the diversity of taxa. The tree emphasizes the chain of human ancestors. Diverging lineages provide phylogenetic context, but they are otherwise literally beside the point. Although there is no single time axis in this tree, with lineages sprouting in all directions, the foregrounding of the human lineage cannot qualify as a logical and geometrical failure. The shape of the *Pedigree of Man* serves one overriding purpose: highlighting the evolutionary lineage of humans. Interestingly, Gould is right that Haeckel's tree conflates the passage of time with evolutionary progress (see discussion in Section 4.8), but that is independent of the precise shape of his trees.

As we will see in Chapters 10 and 11, a surprising number of modern authors, Gould prominently among them, don't seem to understand lineage thinking. This has resulted in confusing and misleading claims, especially in the popular science and educational literature. I believe Haeckel knew perfectly well what he was doing. In his *Pedigree of Man*, he foregrounded the evolutionary lineage of humans because that was the central topic of the entire work.

Dayrat advances another surprising argument why Haeckel's trees should not be considered Darwinian. He claims that "according to Haeckel, phylogeny referred to morphological changes and not to genealogical lineages themselves," (Dayrat, 2003: 520). Haeckel's phylogenies do not illustrate "a series of actual ancestral organisms or species but rather. . . phylogeny refers to a succession of morphotypes in geological time" (Dayrat, 2003: 521). This, Dayrat claims, fundamentally contrasts with Darwinian trees, which do illustrate "actual series of ancestors" (Dayrat, 2003: 526). I think Dayrat misunderstands Haeckel. His interpretation wrongly implies that Haeckel viewed the evolution of body plans (his *Phylogenie*) as being detached from the lineages that are their material substrate. Haeckel was an uncompromising materialist for whom the evolution of body plans could not be disconnected from the series of ancestors that made them concrete.

In the second chapter of *Generelle Morphologie*, Haeckel points out that the study of morphology, including phylogeny, only deals with real, concrete material entities. Morphology does not deal with the "insubstantial bodies of mathematics" ("den stofflosen Körpern der Mathematik") or "natural bodies without substance" ("Naturkörper ohne Materie") (Haeckel, 1866a: 13). Hence, for Haeckel the series of morphologies that characterize different segments of a lineage – his phylogeny – cannot exist separate from the lineages themselves. Haeckel considered the lineages that he depicted in his trees, and which he called *Stämme*, *Typen*, or *Phyla* (also *Phylen*; singular *Phylon*, hence *Phylogenie*), as real and concrete genealogical units bound together by ties of common ancestry (Haeckel, 1866a: 28, 206). He was a concrete lineage thinker. Many nineteenth-century thinkers chewed over materialism, but not everyone swallowed it. Haeckel did. In his influential history of biology, E. S. Russell (1916: 260) still judged Haeckel as "too trenchantly materialistic" more than half a century after the publication of *Generelle Morphologie*.

For Haeckel, as for modern phylogeneticists, it was metaphysical anathema to describe phylogeny as a march of ethereal, disembodied morphotypes through time. It is only by denying or failing to appreciate this that I can understand Rieppel's (2011b: 2, 3) puzzling statements that ancestral forms, such as Gastraea, were not actual "ancestral species," but merely ancestral forms, or morphological stages characteristic of some portion of the evolutionary lineage of Metazoa. Since any real organism is part of an evolving lineage and must be assigned to a species, it makes little sense for a materialist evolutionist to separate ancestral forms from ancestral species. Yet, Rieppel (2011b: 4) echoes the interpretation of Dayrat (2003) that "[m]orphological phylogeny was decoupled in Haeckel's thinking from the evolutionary history of taxa." Rieppel (2011b: 4) finds it "astonishing with which laxity Haeckel used the concepts of monophyly [shared evolutionary origin] and polyphyly [separate evolutionary origins]" because Haeckel could talk about the monophyletic origin of traits, while not at the same time relating this "to an ancestral species, nor to ancestral stages of morphological differentiation" (Rieppel, 2011b: 3). My interpretation is very different. I agree that Haeckel doesn't provide an explicit and strict definition for his concept of monophyly, but I don't see him using it inconsistently either, as Rieppel claims. As Vanderlaan et al. (2013) point out, all Haeckel's uses of the concept of monophyly are genealogically rooted. For Haeckel the meaning of monophyly referred fundamentally to a single phylogenetic origin. It is this interpretation that unites Haeckel's various uses of monophyletic, whether it refers to taxa or traits.

Rieppel's strongest condemnation of Haeckel's conceptual slips is Haeckel's claim that the mono- or polyphyly of morphological characters needn't be coupled to the mono- or polyphyly of the taxa involved. Rieppel (2011b: 4) writes that Haeckel "sketched a monophyletic tree for protophytes and specified: 'these phylogenetic relationships do not lose their morphological significance even if we assume a polyphyletic origin of protophytes.' This means that the actual history of protophyte taxa may differ from the monophyletic differentiation of their morphology from a common primitive form." But this is not a straightforward interpretation of Haeckel's words. In the text immediately following, Haeckel explains that there are three main groups of protophytes, each characterized by a distinct morphology. The most logical interpretation of Haeckel's words is therefore that irrespective of whether these three groups have mono- or polyphyletic origins, their morphological distinctness remains the same.

For Haeckel, as for Darwinians of every other variety, then or since, the strictly materialistic underpinnings of evolutionary theory simply do not allow form to float free from substance. Darwinian evolutionary trees, including Haeckel's, can only indirectly represent series of real ancestors. Phylogeneticists can try to infer the nature of ancestors by tracing organismic traits on trees, and this is precisely what Haeckel's phylogeny attempted to achieve. Although such inferred ancestors are hypothetical, they are believed to point to the existence of real, concrete, once-living creatures. Irrespective of whether one starts with a modern cladogram or a Haeckelian *Stammbaum*, ancestors, as hypotheses about real organisms, only emerge through the reification of characters plotted along trees.

If for the sake of argument, we agree with Dayrat that the depiction of actual ancestors is the defining hallmark of Darwinian trees, then we can celebrate Haeckel's diagrams as some of the most boldly Darwinian trees ever published. This was certainly how Darwin himself seemed to have assessed Haeckel's work. After reading *Natürliche Schöpfungsgeschichte* (*Natural History of Creation*) in 1868, Darwin wrote Haeckel an appreciative letter: "Your chapters on the affinities & genealogy of the animal kingdom strike me as admirable & full of original thought. Your boldness however sometimes makes me tremble, but as Huxley remarks some one must be bold enough to make a beginning in drawing up tables of descent" (www.darwinproject.ac.uk/entry-6466; last accessed July 2, 2021). Darwin considered Haeckel's phylogenetics sufficiently exemplary to close the section on classification in his chapter on the "Mutual affinities of organic beings" in the fifth and sixth editions of the *Origin* with this paragraph:

> Professor Häckel in his "Generelle Morphologie" and in several other works, has recently brought his great knowledge and abilities to bear on what he calls phylogeny, or the lines of descent of all organic beings. In drawing up the several series he trusts chiefly to embryological characters, but draws aid from homologous and rudimentary organs, as well as from the successive periods at which the various forms of life first appeared in our geological formations. He has thus boldly made a great beginning, and shows us how classification will in the future be treated.. Darwin (1876: 381)

Dayrat offers one final argument for his shaky thesis that Haeckel's phylogenetics is fundamentally non-Darwinian. He states that while Darwin thought that progress could in principle characterize any lineage, Haeckel considered progress to be limited to only "a few lines... This definition of progress constitutes a major difference between the Darwinian and Haeckelian evolutionary theories" (Dayrat, 2003: 525). This conclusion clashes with Haeckel's own words. He dedicated section nine of Chapter 19 of his *Generelle Morphologie* to a discussion of the idea of evolutionary progress (see Section 4.8 for further discussion). He was convinced that natural selection resulted in the increasing improvement or perfection of organisms, a process he called "Vervollkommnung"[2] or "Teleosis" (not to be confused with goal-directed teleology) (Haeckel, 1866b: 257). Indeed, he considered this to be one of the most important laws of nature ("eines der obersten organischen Grundgesetze" ["one of the supreme organic laws"], Haeckel, 1866b: 257), a "logical necessity" (Haeckel, 1866b: 443) of the operation of natural selection. This "general and universal" process of evolutionary progress would on average, and despite the opposing effects of the "special and local process" of regression, impart a broad direction upon evolution in most evolving lineages (Haeckel, 1866b: 263).

---

[2] Gliboff (2008) offers a sensitive discussion of the meaning of the German concept of *Vervollkommnung*, or perfection, in the work of paleontologist Heinrich Georg Bronn and its relation to Darwin's thinking.

For Haeckel tracing the degrees of "differentiation and perfection" of taxa was one of the "most important and rewarding" tasks of phylogenetics (Haeckel, 1894: 11). Hence, as far as progress can characterize any evolving lineage in a Darwinian world, without this being an inevitable outcome, Haeckel's views certainly do not qualify him as a non-Darwinian. Dayrat is right, of course, that Haeckel's belief in the prevalence of progress strongly infused his general thinking about evolution – for Haeckel the eight phylogenetic trees in his *Generelle Morphologie* potently illustrated "this great law" (Haeckel, 1866b: 443). As we will see, Haeckel also used this belief in progress as a guide for inferring the nature of distant ancestors.

4.4 Haeckel's Ancestors

Ahn, Ahn-Form, Ausgangsform, ancestrale Gruppe, Grundform, Progonen, Stammform, Stammeltern, Stammart, Stammgruppe, Stammvater, Stockform, Urahn, Urart, Urbild, Urform, Urspecies, Urtypus, Urmutter, Urvater, Vorfahr, Wurzelform, Wurzelgruppe – Haeckel used a surfeit of terms to designate ancestors or ancestor-like organisms. Not all of these terms are identical in meaning as some refer to a particular ancestral species or higher-level taxon, whereas others refer to aspects of their abstracted body plans. For example, in the first volume of *Anthropogenie* Haeckel includes a drawing of the "ideal Urvertebrate," (Haeckel, 1891: 256) (Figure 4.5), which represents the body plan (*Urbild* or *Urtypus*; Haeckel, 1891: 254) of an actual extinct ancestral vertebrate (*Stammform*; Haeckel, 1891: 254). In contrast, the word *Grundform* has a more specific meaning for Haeckel, referring to the basic external form and symmetry characteristics of an organism's organization.

Haeckel generally shunned the term *Bauplan* (blueprint), which he considered to be outdated and saturated with the teleological stench of a divinely ordained universe. I find Breidbach's (2003: 178) claims that the human pedigree in *Anthropogenie* reflects "the idea that evolution is striving towards a perfect being" and that for Haeckel the diversity of nature is merely "an expression of something already designed" to be completely incomprehensible in view of Haeckel's outspoken disdain for teleology. As far as Haeckel was concerned, teleology had died the "definitive death" (Haeckel, 1866a: 100) under the explanatory weight of the new Darwinian worldview. The charges by Breidbach and others (Bryant, 1995: 201; Fisler and Lecointre, 2013: 11) that Haeckel was a teleologist are nonsense.

Similarly, it is important to understand that for Haeckel all these terms unequivocally referred to real, concrete, once living organisms and their inferred characteristics. When Haeckel used terms such as *Urtypus* (archetype) or *Grundform* (basic form), he did not advocate a typological interpretation that implied the existence of an independent metaphysical realm populated by ethereal types. He clearly understood these terms as abstractions useful for designating particular aspects of body plans of real organisms. Yet, as Sander Gliboff argues in his illuminating book on the origins of German Darwinism (Gliboff, 2008), there is a long tradition of mislabeling Haeckel as an unimaginative

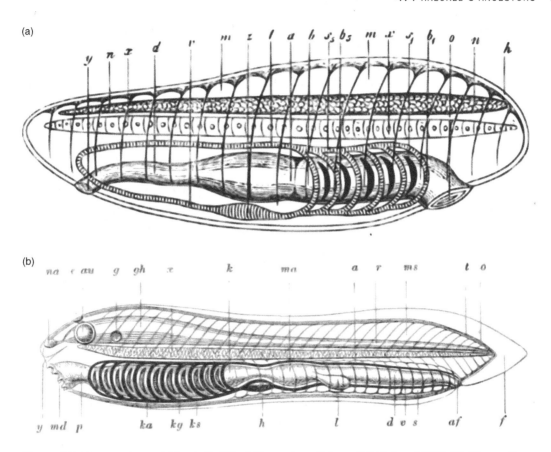

Figure 4.5 Two versions of Haeckel's "ideal Urvertebrate" from 1874 (a) and 1891 (b). The earlier version is more amphioxus-like, and the later version is somewhat more cyclostome-like (a after figure 115 in Haeckel, 1874, and b after figure 95 in Haeckel, 1891).

typologist. Gliboff argues that interpretations of Haeckel as a typologically tainted not-quite-Darwinist have become self-confirming: "Any time Haeckel uses the word 'type' he is read as a reactionary, and any time he uses 'common ancestor,' 'phylum,' or 'heredity' he is presumed actually to mean 'type'" (Gliboff, 2008: 163). I agree with Gliboff that Haeckel's thinking is definitely not perniciously typological. The way in which Haeckel adopted a typological perspective agrees closely with what Amundson (2005) called the explanatory rather than the metaphysical concept of type. Haeckel employed types to gain descriptive and explanatory purchase on nature's chaotic diversity, without committing to an indefensible typological metaphysics. The ultimate cause of unity of type was undeniably naturalistic: "There is no other explanation for the morphological appearances than the true blood relationship of organisms" (Haeckel, 1866b: 290). "Every type with its specific body plan [*Specialbauplan*] is for us a single, independent lineage [*Stamm (Phylum)*], with its own pedigree [*Stammbaum*]." (Haeckel, 1866b: 388).

If one wishes to summarize what Haeckel meant his trees to illustrate, one could do worse than state the precise opposite of Dayrat's (2003: 526) conclusion that "Haeckel's genealogical trees did not illustrate the actual series of ancestors." Throughout his publications, from *Generelle Morphologie* onward, Haeckel adorned his trees with hypothetical ancestors, such as his famous Gastraea on the metazoan tree in *Systematische Phylogenie* and the series of human ancestral groups labeled on the trunk of his grand, and often reproduced, human pedigree in *Anthropogenie*. Even where ancestors were not indicated on the trees themselves, they figured prominently in the texts of his phylogenetic works. Haeckel considered the extent of our acquaintance with evolution's portrait gallery of ancestors as an overall yardstick of progress in phylogenetics. This focus on ancestors was already obvious in *Generelle Morphologie*, where he stated that "the theory of descent can only advance insofar as the descent of individual groups of organisms from shared stem-forms is determined more nearly, and the number and nature of the latter is established" (Haeckel, 1866b: 290). The formidable challenge of this task is perhaps most clearly indicated by the paradoxical fact that the most compelling "evidence" for the existence of ancestors in Haeckel's time consisted of the glaring gaps in organization that separated the body plans of related taxa. How Haeckel attempted to plug these gaps with hypothetical ancestors is the topic we turn to next.

4.5 Haeckel's Phylogenetic Evidence

Haeckel drew on several sources of evidence to erect his trees and to infer the ancestors that articulated their branches. For him the threefold parallelism of ontogeny, paleontology, and morphology had established the theory of evolutionary descent as an irrefutable scientific fact (Haeckel, 1866b: 31, 266). At the same time, it provided the three primary empirical foundations of phylogenetics. In addition, Haeckel thought that several ancillary sources of evidence could benefit the phylogenetic reconstruction of particular taxa. For instance, ecological evidence – *Oecologie* (ecology) being one of Haeckel's most enduringly successful neologisms – could alert phylogeneticists to misleading cases of convergent evolution, such as the simplified morphologies of distantly related parasites caused by similar lifestyles, whereas geographical and topographical evidence could be used to pinpoint where on Earth a particular taxon may have originated and to provide clues about the relative age of the group. For example, the disjunct distribution of living lungfishes in tropical Africa, South America, and Australia suggested to Haeckel that they represented the last relics of a formerly much more diverse group (Haeckel, 1891: 547). Yet he considered the value of these additional sources of evidence to be generally restricted to plugging some of the holes left in the phylogenetic tapestry woven by the other three strands of evidence (Haeckel, 1894: 386).

The main sources of phylogenetic evidence did not exist in a state of equal partnership. Paleontology ruled this unholy trinity, although only in principle. Because Haeckel thought that phylogenetic history was most literally inscribed into the fossil record, phylogeny could also be called "scientific paleontology" (Haeckel, 1866a: 30). Yet, because this evolutionary record is a "completely gappy and torn up patchwork" ("vollständig lückenhaftes und zerrissenes Flickwerk") (Haeckel, 1866b: 307), phylogenetics can only be viable by drawing on the empirically superior evolutionary records contained within the embryology and morphology of living organisms. As is well-known, ontogeny was Haeckel's special weapon for resolving phylogenetic riddles.

For Haeckel ontogeny and phylogeny were intimately related as sister sciences within the discipline of *Morphogenie* or *Entwicklungsgeschichte* (developmental history), with a respective focus on individual organisms and evolutionary lineages. But the epistemological challenges of tracing ontogeny and phylogeny differed starkly. Ontogeny yielded its secrets to simple observation (*Beobachtung*). In contrast, the "much darker and more difficult" (Haeckel, 1919: 101) questions of phylogeny required "critical reflection through the comparative use of empirical sources" (Haeckel, 1919: 101). But Haeckel's (in)famous biogenetic law seemed to offer a shortcut. If ontogenesis was just a "short and rapid recapitulation of phylogenesis" (Haeckel, 1919: 111), then substantial chunks of phylogenetic history could be read off more or less directly from the development of individuals. The biogenetic law could function as Haeckel's thread of Ariadne (Haeckel, 1891: 391) in the labyrinth of phylogenetic history as long as the process of adaptation did not completely erase the signal of evolutionary inheritance – i.e. if cenogenesis (disruption of the ontogenetic recapitulation of phylogeny) did not overwhelm palingenesis (repetition of phylogenetic stages in ontogeny) by disrupting the parallel order in which traits emerged in ontogeny and phylogeny. As we will see, this law was particularly useful for conjuring images of life's most ancient ancestors. Among these, Gastraea, Haeckel's conception of the last common ancestor of all Metazoa, was the most conspicuous, the most controversial, as well as the most enduring (see Chapter 8 for a fuller discussion of the Gastraea theory).

Revealing the phylogenetic history of an evolutionary lineage in full detail required the combined strength of all strands of the threefold parallelism. The phylogenetic history of vertebrates best illustrated this optimal interlocking of the tendrils of evidence. Although dwarfed by the swarming masses of invertebrates, it was the depauperate group of vertebrates, and one bipedal species above all, that had been awarded a disproportionate amount of developmental and anatomical study by the time Haeckel penned his *Generelle Morphologie*. Their durable skeletons had even provided for an unmatched fossil record, so that the vertebrate evolutionary tree was only one of two in the *Generelle Morphologie* that bore the proud label of being "based on palaeontology" (in diagram seven in Haeckel, 1866b).

However, this ideal of a mutually reinforcing trinity of evidence was frequently thwarted by either the simple lack of knowledge or by the many cenogenetic distortions of the palingenetic record. For example, Haeckel considered the evolutionary relationships within the great group of Articulata (within which he classified arthropods, annelids, various other "worm" groups, and infusorians) to be the most complex of all the main animal groups. Unfortunately, in view of its sparse and biased fossil record, and the limited knowledge about the ontogeny of most articulates, Haeckel was forced to sketch the outlines of their phylogenetic history principally on the basis of comparative anatomy (Haeckel, 1866b: lxxvii–lxxviii). Yet for other groups the lack of one or more of the main sources of phylogenetic evidence was scarcely a problem. Although the fossil record was singularly uninformative about the origin and evolution of calcareous sponges, comparative study of the development and anatomy of living forms yielded a picture of their last common ancestor that Haeckel thought was as convincing as any phylogenetic conclusion could ever hope to be (Haeckel, 1872: 76). Even with just ontogenetic data at hand, the discerning naturalist could still trace the origins of the whole group as well as its main forms "in the most satisfying way" (Haeckel, 1872: 342). But even an imaginative zoologist had to throw down his pen when all sources of evidence remained resolutely silent. Hence, in the echinoderm tree in *Generelle Morphologie*, the sea cucumbers are precariously poised atop the barest branch in the book, enveloped by a cloud of question marks.

4.6 The Need for Imagination

Haeckel conceived of phylogenetics as the descriptive and explanatory science of evolutionary origins. He summarized what could be his science's motto: "Jedes Sein wird nur durch sein Werden erkannt" ("Every being can only be understood by its becoming") (Haeckel, 1866a: 23). Establishing the evolutionary relationships between organisms was only a first necessary step toward fulfilling the ultimate goal of understanding the evolution of morphology. Haeckel was not content with just showing that it was possible to trace different groups of organisms back to their various common origins. He wanted to understand in as much detail as possible the nature of the ancestors that connected distant dots of diversity. The branching patterns in Haeckel's trees only outlined the diverging avenues of phylogenetic history. Tracing the march of evolving body plans along these paths was his ultimate aim.

Haeckel needed two things to color in the empty spaces located at the intersections of the branches in his trees: data and imagination. He felt that imagination deserved an explicit defense. The reason for this was that the memory of the fantastic excesses of the erstwhile *Naturphilosophen* was still fresh when Haeckel set to work. He was concerned that many of the real achievements of the *Naturphilosophen* had been buried alongside the corpses of their "daydreams and fantasies" (Haeckel, 1866a: 70). He considered as tragic and deplorable the common view that philosophy and natural science were antagonistic. Haeckel even thought he could sketch the recent history of his science as an ongoing battle between strictly empirical and philosophical approaches to morphology.

Haeckel divided the history of the science of morphology into four periods, each of which was characterized by a different relative emphasis on empirics and philosophy (Haeckel, 1866a: 71–74). The first period, which started in the early eighteenth century with the work of Linnaeus, was empirical and showed the "rule of external morphology (systematics)." This was followed by the second, philosophical, period in the first third of the nineteenth century, during which the work of Lamarck and Goethe established the "rule of fantastic-philosophical morphology (older Naturphilosophie)." The third period, during the second third of the nineteenth century, saw Cuvier's work reassert the dominance of empirical approaches to morphology, which Haeckel called the "rule of the empirical internal morphology (anatomy)." Finally, starting in 1859 with Darwin's *Origin*, the fourth stage in the history of morphology saw the pendulum of fashion swing back to a renewed appreciation of philosophical approaches, representing the "rule of empirical-philosophical morphology (newer Naturphilosophie)."

Haeckel, who had jump-started his career with prize-winning and densely empirical work on radiolarians, felt it was the natural philosopher's obligation to also exercise his critical imagination. He approvingly quoted his friend and scientific idol Johannes Müller: to the nature researcher "fantasy is an indispensable good" (Haeckel, 1866a: 74). Pure empiricists could only hope to create "disordered heaps of stones," while pure philosophers could only erect "air castles." A true edifice of learning can only result from the intimate entwining of data and imagination. It was imagined ancestors that animated Haeckel's phylogenetic scenarios. The conspicuous gaps between the body plans of groups on adjoining branches in his trees provided Haeckel with the empty canvases on which he painted their silhouettes.

Many of Haeckel's phylogenetic speculations in *Generelle Morphologie* were tentative sketches, often coated with only the thinnest veneer of evidence. For example, how could one make a phylogenetic connection between the highly unusual body plan of echinoderms and the rest of the animal kingdom? Haeckel presented a fantastic, yet ingenuous speculation that drew heavily on analogies about the biology of other taxa (Haeckel, 1866b: lxii–lxiii). Imagine, wrote Haeckel, a segmented worm-like animal much like an annelid that reproduces either like a cecidomyid fly, in which the larva produces other larvae, or like a parasitic nematode in which the progeny remains inside the body of the mother animal. Further imagine that the progeny would fuse together with their anterior ends in groups of five so that they would come to share a single mouth opening, much like colonial botryllid ascidians share an exhalant opening. The result would be a composite pentaradial individual, representing the shared common ancestor of echinoderms. Haeckel licensed this seemingly far-fetched comparison by accepting the homology of the ciliated larvae of worms and echinoderms as well as the homology of the tubular excretory system of worms and the water-vascular system of echinoderms.

4.7 A Portrait Gallery of Ancestors

By the time that Haeckel's *Systematische Phylogenie* and *Anthropogenie* appeared in the 1890s, the sketchy scenarios of his earlier works had matured into fully fledged phylogenetic vistas, populated with veritable zoos of hypothetical ancestors. Some of these received descriptions longer than those of real species in the taxonomic literature. The urfish *Ichthygonus primordialis*, hypothetical ancestor of gnathostomes, was introduced in a two-page profile, describing its morphology and development in exquisite detail (Haeckel, 1895: 224-225). The length of its description was proportional to the ancestral burden it had to bear – Haeckel created it to fill the large gap between its own ancestor, which was cyclostome-like, and its proximate descendant, which was shark-like. The urvertebrate Prospondylus even received its own schematic anatomical drawings, outlining its basic morphology and internal anatomy (see Figure 4.5; figures 95–99 in Haeckel, 1891).

In sharp contrast to modern phylogeneticists, who generally restrict their ancestral inferences to only the nodes in their trees, Haeckel also speculated enthusiastically about the ancestors located along the unbranched lineages of his trees to bridge the gaps between disparate taxa and spin seamless scenarios of body plan evolution. This is on display at the base of his metazoan tree, where he individuated ancestors on the basis of ontogenetic stages observed during the development of living organisms. For example, in his *Die Kalkschwämme* (1872), Haeckel identified six ancestors at the base of Metazoa, each corresponding to a developmental stage: Monera (no longer represented in development) → Amoeba (fertilized ovum) → Synamoeba (morula) → Planulata (planula) → Planogastraea (planogastrula, with gut cavity but no mouth) ⇒ Gastraea (gastrula) (Figure 4.6). Whereas the first two of these hypothetical ancestors are shared between Protozoa and Metazoa, the last four are successive ancestors unique to the lineage of Metazoa. These early ancestors remained the foundation of Haeckel's metazoan trees in later works, although he sometimes used different names to indicate the same ancestor – Planulata became Planaea or Blastaea, for example – and he only mentioned Planogastraea in his earlier works focusing on sponges and the Gastraea theory.

The Gastraea theory was the Holy Grail in Haeckel's hunt for ancestors. As we will see in Chapter 8, this fertile and controversial theory has endured for more than a century and a half. Gastraea itself is the purest product of Haeckel's application of the biogenetic law. It is the hypothetical last common ancestor of all animals, revealed in the gastrula stage of extant organisms. For Haeckel the presence of a gastrula stage, with an invaginated archenteron, in the life cycle of disparate phyla did not just indicate the previous existence of a common ancestor with such a character. The biogenetic law stipulated that this ancestor *was* a gastrula. Haeckel employed this long-extinct animal as a potent piece of phylogenetic Velcro. Whereas the previous von Baerian and Cuvierian types of animals – Vertebrata, Radiata, Articulata, and Mollusca – had stood in eternal separation, Haeckel united their stems into the single cradle of Gastraea. This all-encompassing entwining of evolutionary lineages and the concomitant implications for the homology of the gut and germ layers across all animals threw open wide avenues for phylogenetic research into all aspects of animal body plans.

Among late-nineteenth-century explorers of life's ancient ancestry, Haeckel was somewhat of an Indiana Jones, and not just because of the adventurous attire he sports in some of his most famous photos (figure 1 in Nyhart, 2002, and photo on page 133 in Haeckel, 2008). As a charismatic lecturer and prolific writer, he took his audiences on spellbinding trips into the deep past, conjuring up long lost worlds. None of his contemporaries matched Haeckel's drive, daring, and panoramic vision in the quest for a fully connected phylogeny of life. Haeckel's imagination breathed life into the merest scraps of evidence to spin bold phylogenetic scenarios and to rouse imagined ancestors from the ashes of history. He scoffed at paleontological textbooks that offered a literal reading of the fossil record by labeling the trilobites entombed in Cambrian rocks as some of the oldest organisms on Earth. He thought they couldn't be more wrong. The morphological complexity of Cambrian fossils could only be explained by considering them the products of hidden phylogenetic fuses "many hundreds of thousands, if not millions of years" long (Haeckel, 1896: 15). Laying bare these fuses constituted a major goal of Haeckel's research.

The phylogenetic lineage to which Haeckel devoted by far the most attention was, of course, that of humans. He imagined it included at least two dozen ancestors[3]

---

[3] The precise number of ancestors that Haeckel identified in each phylogenetic lineage is best interpreted as a minimum number depending on how he plugged the gaps between body plans in the context of the differing branching patterns of his trees. For the phylogenetic lineage of humans, this minimum number was around two dozen – 25 human ancestors in *Anthropogenie* and 24 in *Systematische Phylogenie*.

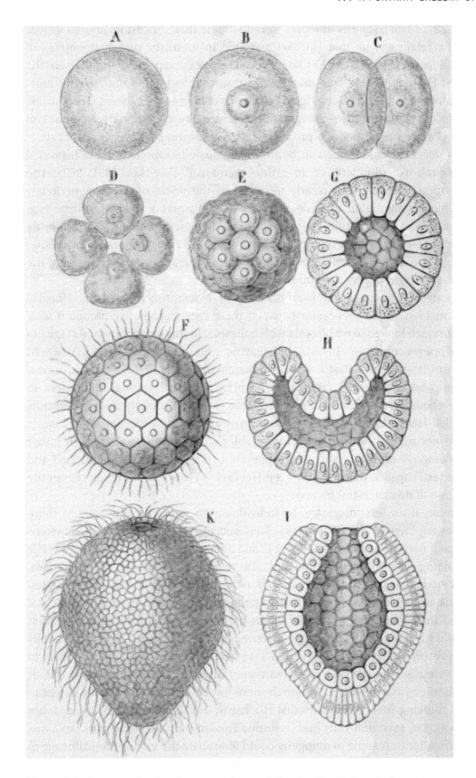

Figure 4.6 Stages in the development of a coral that for Haeckel evoked ancient ancestors, including cytula (B; evoking Cytaea), morula (E; evoking Moraea), blastula (F and G; evoking Blastaea), depula (H; evoking Depaea), and gastrula (I and K; evoking Gastraea)
(after figure 4 in Reynolds and Hülsmann, 2008, based on figure 20 in Haeckel, 1889).

(Haeckel, 1891: 524; 1895: 631). His attempts to individuate these ancestors affords a clear illustration of the relative value that Haeckel assigned to the three principal sources of phylogenetic evidence. Haeckel thought that he could replay the tape of evolution from the reel of ontogeny. He judged developmental stages observed during the ontogeny of morphologically complex animals to be convincing glimpses of long-lost forebears. The succession of stages in development also conveniently revealed the evolutionary sequence of these ancestors (Figure 4.6). Especially promising were the various "lower invertebrates," for which "each developmental stage can be unambiguously interpreted as the historical repetition or portrait-like silhouette of an extinct stemform" (Haeckel, 1891: 387). The earliest human ancestors were thus clearly refracted by the prism of the biogenetic law and correspondingly named after their developmental reincarnations. The fertilized egg, which Haeckel named Cytula, recalls the oldest eukaryotic ancestor Cytaea. The morula stage (a compact mass of blastomeres) in its turn evokes the next and first multicellular ancestor, Moraea. The hollow blastula points to the former existence of Blastaea, and the cup-shaped gastrula stage mirrors the famous Gastraea.

However, although Gastraea evolved from Blastaea, in *Systematische Phylogenie* Haeckel posits an additional hypothetical ancestor between these two: Depaea. He named it after the depula stage, which looked like a blastula with a dimple. In this developmental stage the archenteron has not yet completely invaginated, leaving a substantial blastocoel in place. In the following gastrula stage, the blastocoel is completely obliterated by the tight juxtaposition of the ecto- and endodermal cell layers. If such slight and gradual differences in morphology are enough for imagining discrete ancestors, one can easily understand how ontogeny provided Haeckel with a nearly endless trove of ancestral treasure. But the fact that Depaea did not make it into the summary tables of mankind's ancestors in either *Systematische Phylogenie* or *Anthropogenie*, despite the depula stage being mentioned and illustrated in the text, suggests that perhaps even Haeckel felt less than confident about the empirical substrate of this ancestral fantasy.

It is much easier, if no less imaginary, to individuate ancestors on the basis of more conspicuous developmental discontinuities. The successive appearance of coeloms (Coelomula stage), notochord (Chordula stage), and metamery (Spondula or Vertebrella stage) in the ontogeny of chordates heralds the erstwhile existence of Coelomaea, Chordaea, and Vertebraea (aka Prospondylus) early in the evolutionary lineage of vertebrates.

In his textbook on evolution, Wallace Arthur poses the question of whether Haeckel actually believed that embryos go through stages that are similar to the adults of their ancestors: "should we accept that a thorough student of embryology, such as Haeckel, really believed this? Probably not" (Arthur, 2011: 16). As we have seen, and as we will see again in Chapter 8 when we discuss Haeckel's Gastraea theory in detail, he most certainly did, at least for the earliest embryonic stages. Although there has been debate about the extent to which Darwin's thinking was recapitulationist (Richards, 1992a, 2000), there is no doubt that Haeckel's was. Yet, even ontogeny had its limits. The lamprey's ammocoetes larva was as far as a recapitulationist reading of ontogeny could illuminate the phylogenetic lineage of vertebrates. The Archicranula stage of ammocoetes showed the developing subdivisions of the brain, sensory organs, pharyngeal slits, and heart. It recapitulated Archicrania, a hypothetical beast with a cyclostome grade of organization, which Haeckel thought human embryos passed through at about three weeks of development (Haeckel, 1891: 877).

Early developmental stages could not always put detailed flesh on the bones of his progenitors. Developmental stages evoked adult ancestors, but failed to put them into sharp focus. Comparative morphology allowed Haeckel to supplement these glimpses of the basic body plans of long-lost ancestors. For example, to bridge the gap between the body plans of the simple Gastraea and the more complex Coelomaea, Haeckel consulted the adult morphology of invertebrates. Various turbellarians suggested the gradual evolutionary elaboration of morphological complexity. Closest in structure to the hypothetical ancestral group Archicoela were the acoels with their simple nervous systems and their lack of nephridia. The fact that these tiny carnivorous worms were free-living suggested to Haeckel that their morphological simplicity was primitive and not the product of degeneration, as had been claimed earlier (Haeckel, 1896: 241, 251). Acoel-like ancestors then gave rise to more complex forms, with differentiated mesoderm, a more complex brain, mesodermal gonads, and nephridia, characters shared by most nonparasitic turbellarians. The tiny gastrotrichs suggested that a through-gut with anus had also evolved before the origin of coeloms. Subsequently, the evolution of Coelomaea signaled the origin of a blood vascular system and coelomic body cavities (via elaboration of the gonad lumen), which Haeckel believed the morphology of nemerteans and enteropneusts indicated in their earliest and least modified incarnations. Finally, Haeckel inferred the presence of a pharynx with gill slits and a dorsal nerve cord in Coelomaea from their presence in enteropneusts.

Wherever possible, Haeckel also employed the fossil record to paint ancestral portraits, and nowhere more enthusiastically than for man's most recent vertebrate ancestors. He realized that the ontogenetic trajectories of most craniates, including man, had become distorted by later evolutionary changes. The right fossils therefore presented less murky pictures of ancestors in the final segment of the evolutionary lineage of humans. Architherium, the last common ancestor of all mammals, for instance, Haeckel thought looked quite similar to triconodonts (Haeckel, 1895: 475), an extinct group of generally furry, mouse-to-cat-sized, long-tailed insectivores or carnivores. Triconodonts have retained their ancestral gloss remarkably well, as they are still thought to "offer a close look at common ancestors [of] modern mammals" (Pietsch, 2012: 200) in the twenty-first century. Haeckel thought the nature of Architherium was also captured by the living monotremes, albeit less accurately. Although he labeled echidnas and the platypus as "living fossils" (Haeckel, 1895: 469), he understood they were "strongly specialized" (Haeckel, 1895: 472) in many of their traits, such as their reduced dentition and the aquatic adaptations of the platypus. However, he thought the permanent cloaca and egg-laying habits of these animals did reveal parts of Architherium's body plan.

More fossil treasure could be unearthed further along the lineage toward modern humans. Some fossils, Haeckel thought, could possibly be direct human ancestors. For instance, he thought that the fossil ape species *Dryopithecus fontani*, which was known from Miocene remains found in France, might be in the direct lineage of humans (Haeckel, 1866b: clvi). He similarly believed that *Pithecanthropus erectus* could be a possible direct human ancestor (Haeckel, 1895: 634). When in 1894 Eugène Dubois christened the remains of a fossil femur, skullcap, and a tooth found in Java as *Pithecanthropus erectus,* he honored Haeckel's prediction that upright walking ape-men must once have existed to connect humans to apes. Java Man, aka *Homo erectus*, has of course since become the virtual flagbearer of the concept of missing links (see also Chapter 11).

We should note here that Haeckel's willingness to entertain the notion that *Dryopithecus* and *Pithecanthropus* might be actual lineal ancestors of man is more the exception than the rule in his treatment of fossils. He generally treated fossils as he did extant taxa: they afforded more or less accurate pictures of ancestors depending on how much their body plans were thought to have diverged. Although he frequently placed supraspecific taxa in the direct ancestry of living groups (a common strategy of biologists and paleontologists that stretches into the twenty-first century with the formal recognition of paraphyletic groups), he was much more careful in doing the same with fossil species. Even with so-called missing links or transitional forms, such as *Pithecanthropus* or *Archaeopteryx lithographica*, he was careful to point out that although they were hugely informative about the ancestors of humans and birds, they were not necessarily, or even likely, to be themselves direct ancestors of these taxa. Haeckel thought it was more likely that they were little modified descendants of the ancestors he was trying to reconstruct (Haeckel, 1889: 642; 1895: 412). Haeckel used a discussion of the evolution of languages to highlight the importance of not confusing ancestral lineages with side branches. He wrote "you should never confuse the direct descendants with the side lineages, and likewise never the extinct forms with the living. These confusions occur very often, and our opponents very often use the misunderstandings resulting from this confusion to criticize the theory of descent" (Haeckel, 1891: 466). In the same way that languages evolve, "not a single ancestral form has reproduced itself to the present time without changes" (Haeckel, 1891: 467). Haeckel's Jena colleague Carl Gegenbaur cautioned readers in exactly the same way. He pointed out that you should no more expect to find living animal ancestors in today's fauna as expect to encounter an ancient ancestor of your family or your people or your race on the street (Gegenbaur, 1870: 75).

As is clear from this discussion, Haeckel knew that he didn't live in a phylogenetically Panglossian world in which distant ancestors are perfectly preserved in the flesh of living species and the remains of fossils. Most ancestral stepping stones have sunk irretrievably into the depths of time, but some species and groups retained the stamp of their ancestry better than others. Some clues, Haeckel thought, relate to the habits and habitats in which organisms lived. For instance, because they were free-living, the morphological simplicity of acoel flatworms was a sure sign that they were not the simplified descendants of more complex ancestors (Haeckel, 1896: 241, 251). The same was not true for parasitic platyhelminths, whose morphology could be expected to have been simplified in connection with their evolutionarily derived habits. Similarly, the free-living amphioxus was more likely to retain the imprint of its ancestry than the sessile adults of ascidians. The same assumption that sessile taxa are generally derived from free-living taxa led Haeckel to propose a heterodox hypothesis for the evolution of echinoderms. The sessile crinoids (sea lilies and their relatives) were, and are, generally considered to be the most primitive group of living echinoderms. Haeckel instead proposed that they were derived from free-living sea star-like ancestors and that their overall structure reflected adaptations to a sessile life. Seemingly grasping at straws to find support for this idea, he saw the "most glowing confirmations" arising from paleontology (Haeckel, 1866b: lxiv). New discoveries of fossil sea stars suggested that they might be the oldest echinoderms known. A strict reading of the threefold parallelism then of course suggested that sea stars must also be the most primitive echinoderms.

As we will see in Chapter 6, evolutionary rules of thumb, such as that sessile taxa are unlikely to be primitive, have been widely used since the nineteenth century to provide scenarios with an evolutionary direction. Haeckel also thought that some environments were more likely than others to be havens that could shelter species from evolutionary change. Freshwater taxa, such as *Hydra*, naidid oligochaetes, and sturgeons were more likely to have conserved body plans than their relatives living in other environments because he thought the struggle for existence was less severe in freshwater (Haeckel, 1866b: li, lxxviii, xxviii).

Yet sometimes little more than wishful thinking was the basis for promoting organisms to the status of ancestral totems. Even when the evidence for such phylogenetic prestige turned out to be paper thin, Haeckel was loath to let go. He was infamously reluctant to accept the demise of *Bathybius haeckelii,* which T. H. Huxley had described in his honor in 1868 as the living representative of the primordial *Urschleim* (Figure 4.7). Haeckel took the pure protoplasmic simplicity of *Bathybius* and its widespread distribution on the bottom of

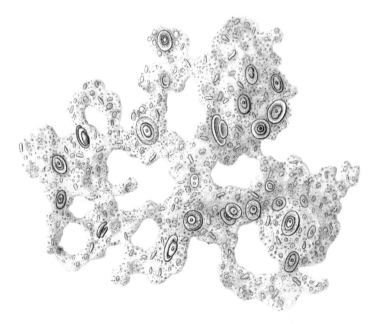

Figure 4.7 Haeckel's drawing of *Bathybius haeckelii*, life's ancient ancestor that turned out to be a chemical precipitate
(from https://commons.wikimedia.org/wiki/File:Bathybius_haeckelii_Haeckel_1870.png, last accessed January 12, 2022).

the Atlantic Ocean to be spectacular evidence that life was the product of spontaneous generation (Rehbock, 1975). He was convinced throughout his career that spontaneous generation was the link that united the living and nonliving worlds, and with *Bathybius* now spanning this chasm, he saw the foundation of his monistic worldview empirically vindicated. Interestingly, rather than interpreting *Bathybius* as a frozen remnant of the primordial ooze, he instead thought it more likely that this submarine protoplasmic blanket represented a vast phylogenetic placenta, continuously spawning new life and forever being replenished by abiogenesis (Haeckel, 1870: 182). Considering the theoretical significance of *Bathybius* for Haeckel's overall worldview, his reticence to acknowledge its death once it was exposed as a chemical precipitate (Rehbock, 1975) is at least somewhat understandable. Huxley's own immediate recantation upon learning in 1875 that *Bathybius* was a chemical chimera stands in sharp contrast.

Similarly, Reynolds and Hülsmann (2008) have recently argued that Haeckel's use of *Magosphaera planula* (the "magician's ball") to illustrate the nature of the ancestor Blastaea (the blastula-like ancestor of Gastraea) was essentially wishful thinking. Haeckel observed this small organism in 1869 along the Norwegian coast, and he described it as a protist that formed hollow multicellular colonies. However, he later promoted *Magosphaera* to a protozoan status more closely related to animals, reinterpreting it as an "Uranimal" (Haeckel, 1891: 492), without, Reynolds and Hülsmann claim, providing any explicit rationale for this phylogenetic revision. For this reason, they label *Magosphaera* a *Wunschbild*, a product of wishful thinking. Haeckel indeed diligently sought living analogues for his hypothetical ancestors, right down to the simplest and most distant forms. The summary tables of humankind's inferred ancestors in both *Anthropogenie* and *Systematische Phylogenie*, for example, list the "living nearest relatives" of these forebears,[4] those living organisms that in their adult stage evoked ancestral body plans most faithfully (Haeckel, 1891: 524; 1895: 631).

In this respect, amphioxus was a real trophy animal. Haeckel thought amphioxus, considered the most primitive vertebrate, to be a slightly modified descendant of the last common ancestor of the vertebrates. Even when he was writing *Generelle Morphologie*, Haeckel labeled the one species of amphioxus then known as "the last of the Mohicans" (Haeckel, 1866b: cxix) of a formerly more diverse group of animals that had diverged from the rest of the vertebrates early in their evolutionary history. The "invaluable amphioxus" (Haeckel, 1891: 242) provided such an accurate picture of the morphology of the hypothetical Urvertebrate (Haeckel, 1891: 256) that the headers of pages 254 and 255 in *Anthropogenie* read "Amphioxus, the real Urvertebrate," and the hypothetical ancestor "Prospondylus, the ideal Urvertebrate," respectively.

---

[4] It is clear that Haeckel's notion of phylogenetic relationships differs from that of our modern conception by mixing a purely cladogenetic component with one based on morphological similarity. For example, although all modern descendants of the vertebrate ancestor Prospondylus are equally related to it if only branching order is used as a criterion, Haeckel singled out amphioxus as the "nearest relative" of this ancestor because it most faithfully reflected its morphological and developmental body plan.

The ontogeny of amphioxus was equally pregnant with ancestral clues, revealing some of the most conserved versions of phylogenetically telltale developmental stages in the animal kingdom. Amphioxus boasted not just a nearly perfectly primitive gastrula stage, called Archigastrula, but also Archiblastula, Archidepula, Archicoelomula, and Archispondula stages. For Haeckel this made its development a dependable script of the main acts in the evolutionary drama of vertebrate origins. Compared to craniate vertebrates, amphioxus could be considered "in a certain sense a persistent embryo, a permanent developmental stage of craniates" (Haeckel, 1891: 395). Amphioxus was for Haeckel a paragon of primitiveness, so much so that he made it symbolic – *Hydra* became the amphioxus of cnidarians, the platypus and echidna the amphioxuses of mammals (Haeckel, 1866b: lii, cxlii).

4.8 Progressive Evolution

The *scala naturae* arranged both living and nonliving aspects of a static world in a single ascending series, culminating in a human or divine apex (Lovejoy, 1936; Kutschera, 2011; Archibald, 2014; Bowler, 2021). Through the centuries, different authors created different ladders, and although they didn't track a single explicit and precisely defined measure, they followed an arrow of assumed worth, such as increasing complexity, perfection, or excellence. Is this set of ideas present in Haeckel's work? As discussed earlier, I don't think it is. Whatever the various reasons and intuitions may have been for pre-evolutionists to believe that the productions of the natural world could be organized along a static scale of progress, Haeckel's basis for believing evolution to be, on average, progressive was firmly rooted in a plausible causal mechanism, explicit criteria for assessing relative degrees of morphological complexity and differentiation, and it was in line with a reasonable intuitive logic of change, be it developmental or evolutionary.

Haeckel set out his "law of progress" in the *Generelle Morphologie* (Fortschritts-Gesetz: Haeckel, 1866b: 257–266). It is immediately clear that the basis of his belief in evolutionary progress was fundamentally the same as Darwin's: progress is the expected result of the operation of natural selection, but it is not the inevitable result of an inherent drive toward a predefined goal, human or otherwise. Moreover, progress was not inevitable. Haeckel's understanding was that, on average, natural selection would cause the progressive differentiation and improvement or perfection of organisms. Certain aspects of morphology were of such general adaptive significance that they came to characterize many evolutionary lineages. For instance, he thought that the relentless struggle for survival would favor the centralization of organ systems, the increasing concentration of organs and their shift from the periphery to more internal locations in the body, reduction of the number of serially repeated structures and their attendant differentiation in form and function, and the increase in size of structures (see Chapter 6 for a fuller discussion of such evolutionary intuitions). In this he largely followed the ideas developed by the paleontologist Heinrich Bronn, whom we encountered earlier in this chapter as the man who had first translated the *Origin* into German. Haeckel considered such evolutionary changes to be progressive because he thought they conferred adaptive benefits upon the affected organisms, giving them an advantage over relatives lacking these features.

The two sets of criteria that Haeckel thought were particularly important in assessing an organism's degree of morphological perfection were the amount of axial differentiation they showed and the level of individuality they had achieved (summarized on pages 370–374 and 550–551 of the first volume of *Generelle Morphologie*). For instance, the greater the number of different axes an organism has and the more different the poles of each axis are, the more perfect it is. Hence, by this criterion bilaterally symmetrical animals had reached a higher level of perfection than radially symmetrical animals. By this criterion alone, the most perfect organism of all would be something like a flounder (Haeckel, 1866a: 524).

Haeckel based a related criterion of organismal perfection on distinguishing six levels of morphological individuality (Haeckel, 1866a: 266). These levels in order of ascending perfection are cells (first order), organs (second order), antimeres or bilaterally organized counterparts (third order), metameres or anteroposteriorly organized segments (fourth order), persons (fifth order), and colonies (sixth order). Individuals on all levels could in principle be autonomous living units, although individuals on lower levels also illustrate parts of the structure of individuals on higher levels. For instance, antimeres are illustrated by the left and right sides of bilaterally symmetrical organisms such as humans, but also by the ambulacra (radial axes) of echinoderms. Molluscs, nematodes, and other nonsegmented invertebrates are individuals of the fourth order and are equivalent to individual segments of animals such as annelids and arthropods. Most of what Haeckel called the higher animals were on the fifth level of individuality and composed of at least two metameres. This is true for arthropods and vertebrates, but also for many bryozoan and tunicate colonies. Haeckel called the former "chain-persons" and the latter "bush-persons," reflecting the different developmental modes of the formation of their basic units (terminally budded segments and laterally budded zooids).

The apex of individuality was reserved for true colonies composed of metameric individuals of the fifth order. Although common in plants, animals on this level of individuality are distinctly rare. Despite being completely unknown among the morphologically complex arthropods and vertebrates, real colonies on this level were especially common in the lowly coelenterates, such as siphonophores, and a handful of other invertebrates, including compound ascidians, some bryozoans, and echinoderms. Interpreting each ambulacrum as the equivalent of a segmented worm, Haeckel concluded that by his criterion of individuality alone, echinoderms were the most perfect of all animals (Haeckel, 1866a: 329).

In development as well as evolution, Haeckel thought that organisms could only reach higher levels of individuality by traversing most or all of the lower levels (Haeckel, 1866a: 267). This is, of course, precisely the arrow of progress that many modern commentators think corrupts Haeckel's trees as a conceptual relic of the *scala naturae*. Although Haeckel's belief in progressive evolution does show some conceptual similarity to the same notion embedded in the Great Chain of Being, however, I would argue that it is in fact a reasonable intuition rooted in the understanding of organismal development in his time. The direction of organismal development is overwhelmingly from simple or homogeneous beginnings to more differentiated, heterogeneous, and complex stages. Applying Haeckel's explicit criteria of morphological perfection, development does indeed generally proceed toward stages of greater axial differentiation and higher levels of individuality. With Haeckel's reliance on ontogeny as a phylogenetic guide, it is only logical that he thought evolutionary change showed a similar progression. After all, just imagine the implausibility of the alternative – morphological complexity emerging in a big bang without a fuse. Unless complexity can spring forth fully formed without any antecedent stages, the only reasonable inference is that complexity evolves from simpler beginnings (i.e. following a path of increasing perfection by Haeckel's criteria). Moreover, Haeckel knew full well that the arrow of morphological progress was often thwarted or reversed, for instance in many parasitic, subterranean, and sessile lineages. This was true, for instance, for two of the four major groups of lower worms (*Vermalien*) that he distinguished: Strongylaria (Nemathelminthes), with its many parasitic forms, including acanthocephalans and nematodes, and Prosopygia (Brachelminthes), with its many sessile forms (brachiopods, phoronids, bryozoans).

Peter Bowler and others have argued that Haeckel's view of evolution was much more progressive than that of Darwin (e.g. Bowler, 2013: 131–133). Bowler thinks that this is due to Haeckel having adopted a developmental model of evolution, "in which life advances toward higher levels of organization almost automatically" (Bowler, 2013: 133). While Haeckel certainly thought that evolution by natural selection was broadly progressive, I remain unconvinced by Bowler's argument that this was due to teleological thinking. One reason why I think progressive evolution is on more prominent display in Haeckel's writings than Darwin's is that he was tracing the macroevolution of animal body plans, rather than explicating the microevolutionary mechanisms of species evolution, as Darwin was. Conspicuous phenotypic disparities between taxa don't generally come into play when thinking about evolution in microevolutionary time. But when thinking about the evolution of the animal phyla, you cannot help but be struck by striking differences in their morphological complexity. In this context, the impression that morphological complexity has increased in many lineages naturally presented itself to Haeckel, as it continues to do, unsurprisingly, to modern authors (Valentine, 2004: 73–75, 494).

I will close this section with a remaining riddle. Combining Haeckel's criteria of morphological progress, the most perfect or complex organisms were those that had reached the highest levels of individuality, contained the largest number of morphologically and functionally differentiated components, and showed the most pronounced degree of axial differentiation. So, which organisms were the "most perfect and highest" of all? Perhaps surprisingly, these criteria would anoint the humble sea pens and sea pansies, cnidarians whose more or less feather-shaped colonies tend to bilateral symmetry (Haeckel, 1866a: 539). But Haeckel wouldn't allow them to be the real winners, for he awarded the medal to none other than man herself (Haeckel, 1866b: cxvi). Why? Neither criterion by itself would select humans as the crown of complexity, as flounders and sea pens display superior axial and hierarchical complexity, respectively. The key is function. Although the flounder's axial differentiation is unparalleled, its departure from pure bilateral symmetry impedes its ability for free and directed movement (Haeckel, 1866a: 524). It compromised its functional perfection.

For Haeckel, humankind's elevated status similarly resides in an argument of superior function. Humans possess a unique combination of four morphological traits associated with important functions: a larynx (speech), a large and complex brain (thought), sophisticated extremities (manipulation), and an upright posture (locomotion) (Haeckel, 1866b: 430–431). The morphological complexity of each of these traits is limited, and none is unique or necessarily developed to their fullest extent in humans – he notes that parrots can talk, penguins can walk upright, and many animals have more perfect modes of locomotion. However, what lifts humans to a level unattained by any animal is the emergent functional synergy of these traits. I find it remarkable that although Haeckel has long been thought of as a comparative morphologist with little interest in function (Russell, 1916; Bowler, 1996), he used functional complexity to place man at the summit of organismal perfection. Our unparalleled capacity for thought, which Haeckel considered the "highest and most perfect of animal functions" (Haeckel, 1866a: 214), allows us to make exceptional use of the functional versatility of these traits. Haeckel thought this "fortuitously lucky combination" of traits had opened for our "Pithecanthropus ancestors the temple of mankind" (Haeckel, 1895: 624). So, it seems that in the end, and despite his deep antipathy against anthropocentrism, Haeckel couldn't entirely escape its clutches. Yet, he saw our evolutionary ascent as a lucky break, not a prophesied outcome.

4.9 Jump-Starting the Post-Haeckelian Hunt for Ancestors

The fruits of Haeckel's pursuit of "speculative Stammesgeschichte" (Haeckel, 1894: 13) have been variously judged. Depending on whom you asked, when you asked, and about what bit of his work you asked an opinion, answers span the full spectrum from "ingenious" (Agassiz, 1876: 73) to "disastrous" (Gilbert, 1991: 843). Among his contemporaries, the reception of his science suffered from the vehemence with which he presented his views, accusations that he deliberately exaggerated the similarity of embryo pictures in his works (Hopwood, 2006; Richards, 2008), his willingness to gloss over details and conflicting evidence, especially in his more popular works, and his willingness to theorize on the basis of fragmentary evidence to reach conclusions that he presented with the confidence of revealed truth. Haeckel's zealous advocacy of the biogenetic law raised epistemological hackles far and wide. On the other hand, some of his ideas, notably his Gastraea theory, laid the basis for newer theories of animal body plan evolution. These are still cited in research papers today, and some of his ancestral portraits are likewise still present in recent works (see Chapter 8).

Although unanimity about the value of Haeckel's work will never be reached, we can be unanimous about the influence of his work: it has been enormous. For the purposes of this book, Haeckel's signal achievement was to place ancestors at the center of phylogenetics. Gone were the literally undepictable Goetheian archetypes immanent in nature. Gone also was the intangible and inviable taxonomic medley of Owen's vertebrate archetype that could "exist" only in the mind of the Creator. Phylogenetic continuity runs through flesh and blood ancestors, who in life are indistinguishable from living organisms. Hence,

Haeckel could sculpt ancestors by mining the whole of extant and extinct biodiversity for suitable models. However, this places a constraint on imagination as well. Hypothetical ancestors have to be viable organisms in principle. This requirement was a fulcrum for many debates about the nature of ancestors. Among a variety of conceivable ancestors for a given group, which was the most plausible in functional (e.g. physiological, ecological) terms? Haeckel's Gastraea was viable insofar as the gastrula stage upon which it was based was an obviously functioning part of the ontogeny of real organisms. But how well would it have been able to fare as a free-living creature? Not very well, some thought. Élie Metschnikoff conceived his hypothetical Phagocytella as a more likely metazoan ancestor because it made better functional sense to him as a free-living organism (Metschnikoff, 1886) (see Chapter 8).

Haeckel's phylogenetic work has been hugely influential as a driver of debates and research for more than a century. He showed how the biogenetic law in combination with evidence from adult anatomy and the fossil record could be used to infer ancestors to tie together the branches of the tree of life, even though this phylogenetic strategy is rejected today. The wealth of new research this inspired, both in support of and against Haeckel's views, makes up a significant chapter in the history of phylogenetics, right up to the present day. As the first full-time phylogeneticist, Haeckel did by no means get everything right, but we should not judge his achievements too anachronistically. I doubt that phylogenetics could have been off to a more stimulating start than through the work and advocacy of Haeckel and his disciples. Some of the weaknesses of Haeckel's theorizing, such as his authoritarianism, his flights of fancy, and his unequal eye for evidence, are less flaws unique to him than inherent features of what I call narrative phylogenetics. Haeckel was the first full-time evolutionary storyteller and his speculations inspired generations of evolutionists to do the same. The next chapters will explore some of these intellectual adventures.

5

The Epistemic Rise of Hypothetical Ancestors

5.1 Hypothetical Ancestors as Central Subjects in Scenarios

The metaphysical makeover of biology that accompanied the understanding that evolution is a process of descent with modification made it clear that the observable skin of biodiversity represents just the tips of countless lineages that are composed entirely of ancestors. But whatever impact this insight had on the field of systematics, systematists did not suddenly start using hypotheses about ancestors as epistemic tools in the elucidation of systematic relationships. It is in this sense that Darwin was right when he noted in the final chapter of the *Origin* that "Systematists will be able to pursue their labours as at present" (Darwin, 1859: 455). As before, systematists could stay on the empirical straight and narrow and base their hypotheses about systematic relationships exclusively on observable traits of living and extinct taxa. However, if they wanted to understand how those systematic relationships had been produced, they had no choice but to enter the realm of evolutionary processes and hypothetical ancestors.

As we saw in Chapter 4, Haeckel defined phylogeny as the "evolutionary history of abstract genealogical individuals" (Haeckel, 1866a: 60). This doesn't mean that Haeckel considered species and supraspecific lineages as abstract concepts without any real material existence. Haeckel thought that the delimitation of species boundaries could not escape an element of arbitrariness (Rieppel, 2011d) and that the erection of genera, families, orders, and classes would be "equally artificial" (Haeckel, 1866b: 393). But he did consider the higher-level lineages that he called *Phylum* or *Stamm* to be concrete entities. He defined a phylum or stem as the sum total of all species that had evolved from a shared common ancestor, such as the vertebrates, and he considered each to be a "real and complete, closed unit" (Haeckel, 1866a: 28). Each was "a natural group, a concrete unit. . . [that] through the continuous bond of common descent is tied together into an inseparable, real whole, by homology" (Haeckel, 1866a: 206). In other words, Haeckel considered evolutionary lineages to be real and concrete units of evolution that possessed an individuality bestowed upon them by the exclusive bond of common descent.

Haeckel's understanding of evolution's genealogical nexus is remarkably in line with our modern philosophical understanding of phylogenetic lineages as concrete individuals, irrespective of whether they are best seen as spatio-temporally extended entities (Hull,

1975, 1989; Ghiselin, 1997, 2005b, 2016; De Queiroz, 1999), or processes (Dupré, 2017; Dupré and Nicholson, 2018). When looked at from the perspective of the ontology of individuals or processes, lineages evolve through time in fundamentally the same manner as individuals develop throughout their lives. This realization makes phylogeneticists the biographers of evolutionary lineages. Like biographers, phylogeneticists seek answers to the same questions that mark the main events of their subjects' lifetimes: who, where, when, what, how, and why. Phylogeneticists sketch scenarios about the evolution of lineages (who) that seek to describe and explain the evolutionary origins of organismal traits (what, where, and when) by tracing them back to precursors in ancestors and, by reconstructing the relevant character transformations (how), where possible in the context of the relevant conditions of existence and selection pressures (why). Hence, like biographies, phylogenetic scenarios are factually informed stories, and their narrative nature marks them out as ingredients of a truly historical science (for accessible discussions of the characteristics of historical narratives and their relevance to evolutionary biology see Hull, 1975; O'Hara, 1988; Richards, 1992b; Griesemer, 1996; Ghiselin, 1997; Bock, 2007; Winther, 2008, 2011; Currie and Sterelny, 2017; Ereshefsky and Turner, 2020).

Let's elaborate these points a little more. First, historical narratives are woven around what David Hull (1975) called "central subjects" that provide them with cohesion and continuity through time. As will be clear by now, the central subjects of phylogenetic scenarios are evolutionary lineages. Lineages are linear segments of life's genealogical nexus, such as the lineage leading to flowering plants, or the lineage leading to *Homo sapiens*. They are linear in the sense of tracking the arrow of time to a single endpoint without splitting or fusing with other lineages, although the splitting and fusion of lower-level lineages can occur within higher-level lineages. Lineages comprise unbroken strings of hypothetical ancestors. Hence, scenarios are an example of what Calcott (2009) has aptly named lineage explanations. Claims that evolutionary narratives are fundamentally different from literary or historical narratives because the evolutionary chronicle is branching (O'Hara, 1988, 1992) or because evolutionary narratives lack central subjects (Oikkonen, 2009) are mistaken. Lineages are the central subjects in evolutionary stories. They can diverge and fuse, and time-extended populations, species, and higher-level taxa are lineages or components of lineages.

Second, every phylogenetic scenario has a specific spatiotemporal address as it reconstructs unique events that happened at the leading edge of an evolving lineage. No two scenarios can be identical, not even in the case of parallel evolution, because each has a unique starting point in one or more (in the case of the fusion of lineages) ancestors. Third, although Mayr and Bock (2002: 175) aptly called phylogenetics "a backward looking endeavour, the search for and study of common ancestors," most of the intellectual effort that goes into devising scenarios actually involves forward thinking and trying to imagine how and why descendants could have evolved from ancestors. Fourth, evolutionary narratives may reference natural laws and must not contradict any, but they are not derivable from, or reducible to, them.

Finally, although each scenario traces unique events, evolutionists have long tried to animate and lend force to their narratives with what I have called *evolutionary intuitions*

(Jenner, 2014, 2019) (see Chapter 6). These are assumptions or generalizations about how evolution is generally thought to occur and which are thought to be applicable across taxa. One example is Cope's rule of phyletic size increase, which holds that organisms tend to get larger because large body size has a competitive advantage (Kingsolver and Pfennig, 2004). Another is the assumption that the relative morphological simplicity of parasitic and sessile animals is typically the result of morphological degeneration. As we will see, evolutionary intuitions such as these press their mark on the nature of hypothetical ancestors, give scenarios a direction, and provide clues about the plausibility of evolutionary events. Let's have a look at the anatomy of an early phylogenetic scenario to illustrate these points.

5.2 Fritz Müller's Scenario for the Origin of Root-Heads

Fritz Müller sketched the first scenario for the evolutionary origin of the parasitic rhizoce-phalan barnacles commonly known as root-heads. First published in German in 1864, and later in an emended English translation as *Facts and Arguments for Darwin* (Müller, 1869), it is one of the earliest phylogenetic scenarios to explain the origin of any group of animals. Müller's scenario can stand as a prototype for the structure of scenarios, as well as for how they are constructed.

The origin of root-heads was an attractive topic for evolutionary biologists because of their bizarre morphology and their unusual mode of feeding. The only external part of adult root-heads is the sack-like reproductive externa of the female that connects via a narrow stalk to a root system that extends deep into the host's body and which the parasite uses to absorb nutrients. Because they lack any adult traces of arthropod characters such as segmentation or limbs, reconstructing their evolutionary origins is a fascinating challenge. The first step that Müller took was to determine the nearest relatives of root-heads. Because they share similar larval forms with barnacles (Cirripedia) (nauplius and cypris larvae), he concluded that they were each other's closest relatives, and he placed them "in opposition as equivalent" (Müller, 1869: 89). In modern parlance, he considered them sister taxa.

Second, based on the recapitulationist expectation that earlier developmental stages are likely to retain clues about the ancestry of species whose adults have become greatly modified, Müller looked at root-head larvae.[1] This allowed him to identify the barnacle homologue of the most peculiar character of root-heads: the roots with which they infest their hosts. Müller noted that the cypris larvae of barnacles and root-heads have similar prehensile antennules, which in barnacles house the cement ducts with which they attach themselves to substrates. He posited that root-head antennules would perform the same function, although this had never been observed. He proposed that structures he observed on the tips of root-head antennules were homologous to the cement ducts of barnacles and

---

[1] Müller's little book is widely credited to be the first to illustrate in detail how a recapitulationist reading of animal development could produce ancestral insights, but Müller's priority is overshadowed by Haeckel's subsequent voluminous writings on the topic.

that they represent the nascent roots of the parasite.[2] Having thus identified possible precursor structures for root-head roots, he next developed his scenario to answer the question: "How were Cirripedia converted by natural selection into Rhizocephala?" (Müller, 1869: 137).

He first notes that many barnacle species settle on living animals, making it a plausible assumption that they did so in the past as well. He then proposes the first evolutionary step in his scenario. In order to gain a more secure attachment, as well as to prevent being dislodged when the host molts, it would be selectively advantageous for the barnacle not to cement itself just on the surface of the host, but to penetrate its skin. Once the incipient roots are bathed in the body fluids of its host, they may passively take up nutrients via osmosis, and this reliable food source would provide a selective advantage over barnacles that obtain nutrients solely by filter feeding. The next step in Müller's scenario involves the development of a more intimate contact between the invading roots and the organs of the host to facilitate a more effective uptake of nutrients. Concomitant with the elaboration of the evolving parasite's root system, the functional importance of its original food procuring organs, such as the cirri, mouth parts, and intestine, will decrease, gradually resulting in their atrophy and eventually their complete disappearance. Finally, with the parasite protected by the abdomen of a crab or the shell of a hermit crab host, it would lose its own shell plates, leaving just a sack of gametes anchored by a root system. At this point, root-heads had evolved.

But Müller didn't stop there. What, he asked, may our parasite "have looked like when halfway in its progress from the one form to the other" (Müller, 1869: 139)? To help visualize this ancestor he enlisted *Anelasma squalicola*, a peculiar shark-dwelling stalked barnacle. Referring to the figures of *Anelasma* in Darwin's barnacle monograph, Müller noted that this animal is anchored in the skin of sharks with a thick stalk that sprouts ramifying, root-like outgrowths, and that it lacks shell plates, has very small mouth parts, and has thick inarticulate cirri. The animal examined by Darwin had a completely empty stomach. One could scarcely hope for a better snapshot of the possible body plan of a missing link. In 1875 Anton Dohrn pressed *Anelasma* into the same role in his own scenario for the origin of root-heads (Dohrn in Ghiselin, 1994: 79–86). The surprising fact that Dohrn published this scenario of crustacean evolution in an appendix to his work on the origin of vertebrates makes sense because the evolutionary transformation of *Anelasma*'s attachment stalk into a feeding organ beautifully illustrates the principle of the succession of functions around which Dohrn had organized his scenario of vertebrate origins from annelid ancestors. Remarkably, the significance of *Anelasma squalicola* in shedding light on the evolution of parasitism in barnacles remains virtually unchanged a century and a half later (Rees et al., 2014; Ommundsen et al., 2016).

When we summarize the steps that Müller followed when constructing his historical narrative, we see that he started out like any systematist would. He used morphological

---

[2] We now know that barnacle cyprids have cement glands that secrete the substance by which they attach themselves to the substrate or a host, but these glands do not give rise to the rhizocephalan roots. The adult rhizocephalan develops entirely inside the host from an instar that is injected into it.

observations to identify homologous characters between the larvae of barnacles and root-heads, and he concluded that they are each other's closest relatives. He further observed that barnacle larvae attach themselves to substrates, and he knew how the sessile adults feed themselves. Müller then supplemented these observations with three straightforward inferences, which could be tested by new observations. First, given the similarity of the anatomy of their cypris larvae, he inferred that root-heads attach themselves to substrates in the same manner as barnacles. Second, since extant barnacles may attach themselves to other animals, he inferred that their extinct relatives from the lineage leading to root-heads did likewise. And third, he proposed that the processes he observed on the antennules of root-head cypris larvae are the ontogenetic precursors of the adult roots. But this is as far as observational evidence will go. To construct his scenario, Müller now needed to change his epistemic approach. The systematist needed to give way to the phylogeneticist who supplements observations with inference and informed speculation to gain access to the empirically inaccessible realm of ancestors and evolutionary events.

Müller did not classify root-heads within barnacles, but as sister taxa, so he could not start his scenario with an existing lineage of barnacles. However, a reasonable, and one of the commonest evolutionary intuitions, is that parasitic animals are derived with respect to free-living relatives, even when these are sessile. Müller therefore started his scenario with an ancestor that looked like a barnacle, which gradually evolved into a rhizocephalan via a seamless series of plausible intermediate forms. He further proposed that the increasingly sophisticated mode of parasitic feeding and the physical protection afforded by the host were the selective advantages that drove this transformation of body plans. Müller did not identify distinct ancestors along this evolving lineage, but he used *Anelasma squalicola* to evoke a vivid picture of what an intermediate ancestor might have looked like about halfway along this phylogenetic trajectory.

Müller sketched his scenario in less than five short pages, but it nicely exemplifies the diversity of ingredients that can be integrated into phylogenetic narratives. He wove together observations and speculations about morphology, ontogeny, behavior, ecology, functional morphology, physiology, selection pressures, and extinct forms into a picture rich enough for the mind's eye to trace the transformation of body plans. Given the range of ingredients that can be synthesized into scenarios, it is unsurprising to find there is huge variation in their detail and texture. Some are cartoonish and devoid of detail (Figure 5.1), while others are book-length narratives of almost cinematographic richness. At the core of all this variety there is one constant: chains of hypothetical ancestors make up the epistemic backbone of every phylogenetic scenario.

Why, you may wonder, do I call the backbones of hypothetical ancestors in scenarios epistemic? Isn't it true that these ancestors are hypotheses about real, once-living organisms? Aren't they historical inferences? That's true of course. But it wasn't always so. In modern research, hypothetical ancestors are the products of phylogenetic inference. They are equipped with traits observed in their descendants, which are given a soft focus as taxon-specific details are filtered out by ancestral state reconstruction. Before the rise of cladistics and numerical phylogenetics, hypothetical ancestors were used as tools in the narrative reconstruction of lineage evolution. As we will see, they were often specifically

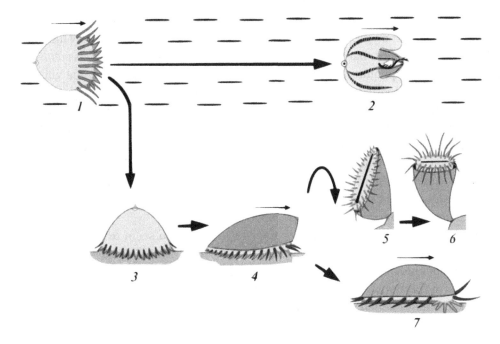

Figure 5.1 Scenario for the evolution of eumetazoan body plans
(after figure 6 in Malakhov et al, 2019, reprinted by permission from Springer, Copyright, 2019).

equipped with precursor structures from which descendant traits could have been derived. They were tools as much as they were evolutionary inferences. And to allow them to do their epistemic work, phylogeneticists made a number of assumptions.

5.3 No *De Novo* Origins

Few evolutionary morphologists have cast a more creative eye on animal evolution than Anton Dohrn (1840-1909). In 1875 he published his famous study *Der Ursprung der Wirbelthiere und das Princip des Functionswechsel. Genealogische Skizzen.* It is available in English translation as *The Origin of Vertebrates and the Principle of Succession of Functions. Genealogical Sketches*, prepared by Michael Ghiselin (1994). Dohrn presents a detailed scenario for the origin of vertebrates from annelid ancestors that is powered by his principle of the succession of functions, which he had conceived to help explain the evolutionary origin of complex new structures (see Section 5.4). Dohrn dedicated his paper to the aging Karl Ernst von Baer (1792-1876), whose scientific pedestal Dohrn considered second in height only to Darwin's. In the correspondence that followed, von Baer gracefully accepted the dedication, but he admitted that in reading Dohrn's paper "there has been, as you assumed, no lack of head-shaking" (von Baer in Groeben and Oppenheimer, 1993: 76). Von Baer did not reject the idea of evolutionary transmutation out of hand, but he considered the evidence that could support it as a general theory to be as yet unconvincing (von Baer,

1876). Even though von Baer was willing to entertain the theoretical possibility that annelid ancestors had turned over on the road to becoming vertebrates, he remained deeply uncomfortable with transformations that crossed the boundaries of the animal *embranchements*.

Much to Dohrn's disappointment, von Baer bristled at the suggestion that the gills of worms could have transformed into the limbs of vertebrates. Yet, this wasn't because von Baer rejected Dohrn's principle of the succession of functions. Von Baer assured Dohrn that he had "nothing against shift of function" (von Baer in Groeben and Oppenheimer, 1993: 79). In fact, von Baer accepted that certain characters in snails, fish, and crustaceans had gone through a succession of functions, and he was actually impressed that Dohrn applied his principle to show that the root-like feeding organs of the parasitic root-heads had likely developed from attachment organs (von Baer, 1876: 479). What really irked von Baer was the most fundamental principle of any phylogenetic investigation.

Von Baer wrote to Dohrn that "we differ completely in that you seem to assume, as do most Darwinists, that an animal can have no part that has not been inherited" (von Baer in Groeben and Oppenheimer, 1993: 79). He asked, "[a]re animals then supposed not to be able any longer to form anything from their own needs, but only to modulate what they inherited?" (von Baer in Groeben and Oppenheimer, 1993: 76). "I do not feel any need to derive the vertebrates from another larger group" (von Baer in Groeben and Oppenheimer, 1993: 76). Strict historical continuity of characters was as unpalatable to von Baer as natural selection was as an agent of change. That evolution could only fashion novelty from inherited parts contradicted von Baer's deeply rooted conviction that a transcendental and teleological harmony pervaded nature. "I cannot break away from my old view that nature reasons and strives towards goals," he wrote to Dohrn (von Baer in Groeben and Oppenheimer, 1993: 80). He suspected that there "were and are countless life forms that exist without inheritance, but have developed from primitive primordia, while they are destined for a kind of existence on Earth and accordingly therefore develop the organization necessary for this kind of existence" (von Baer, 1876: 480). Ironically, while von Baer was perfectly happy with his own teleological arguments, he ridiculed transmutationist ideas with a similar flavor, as we saw in Chapter 3.

It is interesting to note that von Baer's rejection of the idea that apparent novelties are always rooted in inherited traits is not paralleled by the same argument in embryological time. He inserted in the margin of his discussion of the appearance of novel traits during development this nutshell summary: "Nirgends ist Neubildung, sondern nur Umbildung" ("Nowhere is novelty, instead just transformation") (von Baer, 1828: 156). If a particular way of life required a certain trait, von Baer thought that nature would simply provide it. For Dohrn this argument was no better than the pseudo-explanation of spontaneous generation. Dohrn thought that evolutionists could commit no greater sin than to invoke "that Deus ex machina 'new formation', which is neither better nor worse than *Generatio aequivoca* [the accidental origin of a thing from a fundamentally dissimilar starting point]" (Dohrn in Ghiselin, 1994: 41). The insect morphologist and phylogeneticist Guy Chester Crampton (1881–1951) went so far as to link some hypotheses for *de novo* origins of organs to creationist leanings. He thought that two entomologists who had proposed hypotheses for the *de novo* origins of insect wings in the early nineteenth century "were doubtless influenced in their belief by the then prevalent idea of 'special creation'" (Crampton, 1916: 2). Genealogical continuity is the ontological bedrock of the Darwinian worldview, and tracing characters back to their evolutionary precursors is the only explanatory tool phylogeneticists can wield.

If extinction would not erase phylogenetic history, biologists should be able to trace even the most complex features back along seamless chains of ancestors, at least in principle. In the *Origin* Darwin twice mentioned French zoologist Henri Milne-Edwards' dictum that "nature is prodigal in variety, but niggard in innovation" (Darwin, 1859: 223, 445). Natural selection can only braid together novelties from inherited variation. The apparently sudden origin of complex features is an artifact resulting from a lack of evidence. Darwin's mechanism of evolutionary change demands that nature doesn't leap to complex novelties in a single bound. The only place where Darwin's theory allows for the origin of novelty from nothing is on the much smaller scale of new variations. Once natural selection gets purchase on new variation, novelties of great complexity can gradually evolve. In theory, this allows one to trace the gradual emergence of novelties along evolutionary lineages. But to be able to fulfill their professional calling and jumpstart their scenarios, phylogeneticists preferred something more visible than the barely detectable pinpoints of incipient phenotypic variations. After all, most early phylogeneticists were morphologists, and they could only work with what they could see.

In his famous monograph on the horseshoe crab *Limulus*, evolutionary morphologist E. Ray Lankester stipulated as "one of the fundamental principles of phylogeny... that new organs do not arise *de novo* as new parts, but by the modification of pre-existing parts" (Lankester, 1881: 646). This was the core principle that underpinned all phylogenetic morphology (Huxley, 1858: 382; Lankester, 1875: 480; 1876: 54; 1877: 434; Hubrecht, 1887: 644; MacBride, 1895: 342; Meyrick, 1895: 10; Patten, 1912: 253; Crampton, 1916: 2; Bock, 1959: 210; Raw, 1960: 500; Rensch, 1960a: 275; Ghiselin, 1969: 114; 1991: 292; 1994: 11; Willmer, 1974: 327; Rieppel, 2001b: 70; Nyhart, 2003: 165; Cracraft, 2005: 354; Gudo, 2005: 194; Kluge, 2007: 217, 224; Arthur, 2014: 232; Brunet et al., 2015: 836; Havstad et al., 2015; Minelli, 2016: 42; West-Eberhard, 2019: 359). It is the sole foundation of the power of phylogenetic hypotheses to explain the evolution of form.

Complexity might emerge from nothing in a world ruled by intelligent design, or through the formative power of ethereal archetypes, but in a material world where organisms are ensnared by the concrete bonds of descent, nothing comes from nothing. In his paper where he derived vertebrates from ribbon worms (nemerteans), zoologist Ambrosius Hubrecht (1853–1915) explicitly linked the principle of no *de novo* origins to Darwin's insights: "At the base of all the speculations contained in this chapter lies the conviction, so strongly insisted upon by Darwin, that new combinations or organs do not appear by the action of natural selection unless others have preceded, from which they are gradually derived by a slow change and differentiation" (Hubrecht, 1887: 644). In the twentieth century, Ernst Mayr sketched the Darwinian consensus view, as he saw it, by recognizing two modes for the evolutionary origin of novelties. It either involved an intensification of function, or a change of function, but both modes required preexisting structures for natural selection to work upon (Mayr, 1971: 362–363; 1988: 408–409).

One didn't have to be a Darwinian to reject *de novo* origins, as the principle roots the explanatory power of phylogenetic hypotheses within any evolutionary theory based on unbroken lineages of ancestors and descendants. Lamarckian zoologist Ernest MacBride (1866–1940) was deeply disdainful of natural selection. He thought that it "affords no explanation of mimicry or of any other form of evolution" (MacBride, 1929a: 713). Indeed, he was "convinced that the view that 'natural selection' explains evolution is one of the greatest hindrances to biological science" (MacBride, 1929b: 225). Yet, even MacBride admitted that "new organs never arise de novo, but by the differentiation of older organs" (MacBride, 1895: 342), even though his reasons for relying on this principle appear distinctly circular when he writes about "that rule which seems to hold in all cases where we can by means of comparative anatomy show with reasonable probability that evolution has occurred" (MacBride, 1895: 342). After all, what else could comparative morphologists and embryologists do to demonstrate evolutionary change other than tracing traits to plausible precursors in hypothetical ancestors?

The phylogenetic principle of no *de novo* origins is a logical corollary of a commitment to a materialist metaphysics. In a biological world of genealogical continuity across generations, explanations of evolutionary origins must be rooted in identifiable precursors. This reveals an interesting parallel to preformationist theories of development. Preformationism was popular in the eighteenth century and provided a mechanistic explanation for the developmental origins of organismal traits in terms of preexisting material precursors in embryos (Rieppel, 1986; Pinto-Correia, 1997; Maienschein, 2000). Of course this strategy "compels us to postulate what we cannot perceive" (Gould, 1977b: 18), which situates its explanatory force and its greatest weakness in exactly the same place. Evolutionary morphologists faced a similar situation. In his *Generelle Morphologie*, Haeckel laid out the same explanatory logic for the evolutionary realm. The subtitle of his book pointed out that the science of form was "mechanically grounded in Charles Darwin's reformed theory of descent." Evolutionary morphologists tried to explain the origin of organismal traits by tracing them back to unobservable hypothetical ancestors. Whereas the mind's eye of a preformationist embryologist might see a tiny homunculus crouching inside a sperm cell, a morphologist might perceive a distant ancestor in the embryo of a sponge. But to those not committed to strict materialism or preformationist logic, such explanations were unconvincing.

The neo-Lamarckian teleologist Edward Stuart Russell, whose 1916 classic *Form and Function* chronicled the history of the science of morphology, sympathetically echoed the vitalistic sentiments of von Baer and others who thought that animal body plans could sprout novelties virtually without constraint in response to environmental stimuli. Evolutionary morphologists had to understand body plan evolution within the boundaries of homology because homology alone paves the pathways of descent. For Russell, the principle of no *de novo* origins was nothing but a restrictive "prejudice" (Russell, 1916: 306) that led evolutionary morphologists to be "singularly unwilling to admit the existence of convergence or of parallel evolution" (Russell, 1916: 305) and "unwilling to admit the creative power of life" (Russell, 1916: 307). According to Coleman (1976: 172), evolutionary morphologists were burdened with the older "type concept, [which] strictly interpreted, allowed only a permutation of the basic elements of a type among the members of that type." Whereas Darwin "had understood that descent with modification by means of natural selection demanded the production of genuine evolutionary novelties" (Coleman, 1976: 172), "typological morphologists, by contrast, discovered in the presumed novelty of any given biological form only an additional realization of a potential already possessed by its type" (Coleman, 1976: 172).

I don't think typological morphologists and Darwinian evolutionists necessarily thought differently. The biological present only contains that which was bequeathed by the past. As Russell summarized it, the followers of Geoffroy and Darwin thought that "Nature is so limited by the unity of composition that she can and does form no new organs" (Russell, 1916: 305). One discerns here the contours of the concepts of phylogenetic and historical constraints, which became topics of a sizeable literature. Some modern writers point out that evolutionary biology and phylogenetics have not just grown in preformationist soil, but that they remain essentially preformationist disciplines today, limiting the light they can shed on the origin of taxa and traits (Rieppel, 2001b; Bruce, 2016). Historical and material continuity are the sinew of phylogenetic ties, and truly novel traits with *de novo* origins are therefore chinks in the explanatory armory of phylogenetics.[3] But there is also an important epistemological dimension to the preformationist approach to understanding the origin of novelty.

Evolutionary morphologists didn't deny that there were organismic traits that really did appear to be true novelties without any ancestral trace. Such features simply fell outside their explanatory ambit. Without the principle of no *de novo* origins, comparative morphology couldn't become evolutionary morphology and systematics couldn't become phylogenetics. Without it, the present couldn't be explained in terms of the past, leaving these disciplines devoid of explanatory power. Homologies and systematic relationships between collateral relatives record the residue of ancestral resemblance, and it is the task of morphologists and phylogeneticists to reconstruct its source. Dohrn understood this more clearly than many of his contemporaries. Lankester (1875: 480), in an article in *Nature* in which he drew the attention of English-speaking readers to Dohrn's annelid theory of vertebrate origins, wrote that "most of Dr. Dohrn's readers will feel that there really is not much novelty in this proposition [that organs do not evolve *de novo*], since it is already involved in the doctrine of homologies to a very large extent." However, "Dr. Dohrn raises this into a hypothesis of *universal* application, and proposes to apply it stringently in speculations as to the genealogical relationships of organisms" [italics in original] (Lankester, 1875: 480). Lankester understood why Dohrn did this. Dohrn's evolutionary principle of the succession of functions assumed the transformation of preexisting structures.

Nikolaus Kleinenberg, who was an assistant of Dohrn's during the early years of the Naples Zoological Station, developed a theory of phylogenetic change that was strikingly at odds with the received view. Kleinenberg thought that body plans didn't just evolve by the

---

[3] Interesting literature has been emerging about the evolution of novelty that forces a fundamental rethinking of the traditional explanatory power of phylogenetic hypotheses that is rooted in the transformation of primitive into derived character states. Günter Wagner's book *Homology, Genes, and Evolutionary Innovation* and the thought-provoking article by Havstad et al. (2015) are good starting points to delve into this body of work.

slow, continuous transformation of form, but that discontinuities that resulted from the replacement of organs were also frequently involved. He called his idea "evolution by substitution of organs" (Kleinenberg, 1886: 212). He applied it to understand the phylogenetic history of the annelid nervous system. He traced the prototrochal nerve ring of polychaete trochophore larvae back to the nerve ring of a hypothetical medusa-like ancestor, but here the trail of homology went cold. Observing that the adult nervous system of the worm developed largely independently from the larval nervous system, he concluded that evolutionarily it was a new structure without homology to either the larval nervous system or the nervous system of coelenterate ancestors. Therefore, during body plan evolution, ancestral characters hadn't just transformed into those of descendants, but some had also been substituted by functionally equivalent but nonhomologous novelties. According to this view, as body plans evolve, they erase traces of their phylogenetic history. Kleinenberg understood the epistemic consequences: "It is theoretically possible that the final organization of an animal rests solely on substitutions. In this case the comparative anatomical method will not be able to unlock the natural relationships. But also under all other circumstances is developmental history the only reliable way to determine the nature of organs" (Kleinenberg, 1886: 223). In Kleinenberg's scenario, only one thin strand of homology binds the adult annelid body plan to its coelenterate ancestor: the endoderm of the archenteron (Kleinenberg, 1886: 222).

Some wielded the epistemic principle of no *de novo* origins ruthlessly. For them even the presence of an empty space inside an organism was suspicious if it could not be ascertained from whence it came. MacBride, for instance, dismissed the possibility that the gut and coelom had initially emerged in evolution as splits in solid masses of cells (a process known as schizocoely) because this "violated" (MacBride, 1895: 342) the rule of no *de novo* origins. Instead, he traced the archenteron back to a part of the surface of a blastula-like ancestor that had invaginated, while the coelom had originated as an evaginated differentiation of the archenteron. Such reasoning is by no means restricted to early phylogeneticists. Like MacBride, Ghiselin (1972: 141) considered the schizocoel theory for the origin of the coelom in bilaterians to be less plausible than the competing enterocoel and gonocoel theories because it required the coelom to evolve *de novo*.

Even the most densely painted scenarios couldn't be expected to connect all the dots. Within the complex character constellations of animal body plans, there are always traits whose ancestral origins remain obscure. Where, for example, did nephridia come from? According to the traditional view nephridia evolved at the base of Bilateria, but from what?[4] Nonbilaterians, such as sponges and cnidarians, don't have any obvious precursor structures, but as Slovenian zoologist Jovan Hadži (1884–1972) put it, "[i]t cannot be supposed that they have developed as such out of nothing" (Hadži, 1963: 137). The widely accepted hypothesis that derives animals from colonial protists provided no clues to the evolutionary

---

[4] Spectacular light was recently shed on this enduring question by a team led by Andi Hejnol (Gąsiorowski et al., 2021). They showed that conserved sets of transcription factors and structural proteins are expressed in ultrafiltration nephridia across protostomes and deuterostomes, supporting the homology of bilaterian proto- and metanephridia.

origin of nephridia. Hadži thought he could solve this puzzle by deriving protonephridia from excretory structures possessed by the unicellular ancestors of Bilateria. Hadži had developed the heterodox idea that bilaterians had evolved directly from ciliates, with cnidarians and ctenophores being degenerate bilaterian offshoots, and sponges sprouting independently from protozoan soil. As we will see in Chapter 9, his ideas gained little following, despite loud applause in some quarters of the zoological community (de Beer, 1954, 1958).

The explanatory power of phylogenetic hypotheses resides in them tracing back traits to ancestral precursors. Indeed, many hypothetical ancestors were chosen or designed to maximize their precursor potential. This explains why the authors of some of the most outrageous scenarios ever conceived were especially keen not to leave any descendant traits dangling loose. Even the origins of seemingly inconspicuous structures needed to be addressed lest they became the Achilles' heels of phylogenetic theories. Consider the vertebrate pituitary gland, for instance. Dohrn traced it back to the degenerated remnant of the esophagus of an annelid ancestor, Hubrecht located its origins in the shriveled remains of the nemertean proboscis, Gaskell found its phylogenetic beginnings in the excretory organs of ancestral arthropods, Patten derived its anterior lobe from excretory organs in ancestral arachnids and the posterior lobe from the remnants of the ancestral arthropod foregut, and Masterman placed its cradle in a small esophageal invagination found in the planktonic actinotroch larva of phoronids (Hubrecht, 1883: 351; Masterman, 1897a: 323; Gaskell, 1908: 322; Patten, 1912: 250, 260; Ghiselin, 1994: 28). The existence of such a menagerie of different vertebrate ancestors highlights another characteristic of evolutionary scenarios: they are highly perspectival.

5.4 The Precursor Potential of Hypothetical Ancestors

When comparative morphologists, embryologists, and paleontologists engaged each other in the phylogenetic arena that emerged in the second half of the nineteenth century, they probed the ability of competing hypotheses to best explain the origin of animal body plans. Which hypothesis best avoided the explanatory blind spots associated with *de novo* origins? Which one managed to connect the most, or the most important, ancestral and descendant dots? Who presented the most plausible morphological transformations? And which ideas remained most firmly tethered to empirical evidence? Together these considerations combined into what I will call the precursor potential of hypothetical ancestors, and I will argue that it functioned as an early phylogenetic optimality criterion (Jenner, 2019).

Many hypothetical ancestors did their epistemic work in the space provided by the uncertainty about the systematic relationships of the major animal groups. Today, our vastly improved understanding of metazoan relationships places strong constraints on what ancestral state reconstructions can be considered plausible. The current consensus view on higher-level deuterostome phylogeny, for example, rules out many of the older hypotheses for the origin of vertebrates that involved ancestral annelids, arthropods, phoronids, nemerteans (ribbon worms), or platyhelminths (flatworms). But during the late nineteenth and early twentieth centuries, it wasn't unreasonable to bring such taxa into these debates. Yet the great diversity of ancestors that were proposed to explain the origins of major animal groups, such as bilaterians and vertebrates, were not chosen haphazardly. Although early phylogenetic debates may appear as a riotous cacophony of disagreeing voices, with participants not infrequently willfully talking past each other, hypothetical ancestors were constructed with great care.

Hypothetical ancestors were deliberately conceived to maximize their precursor potential for the descendant traits that needed explanation. Which traits to focus on depended on the researchers' interests, as well as their professional expertise. Convinced that amphioxus and ascidian tadpole larvae were degenerate fish rather than the ancestral beacons they were held to be by Haeckel and others (Hopwood, 2015a), Anton Dohrn focused on the ancestral promise of annelids. That decision placed the segmented aspects of the vertebrate body plan at center stage. The body plan of annelids offered many routes for evolutionary modification via fusion, elaboration, concentration, reduction, and specialization of segments and structures (see Chapter 9 for more discussion of annelid ancestors). Dohrn deftly wielded these mechanisms to fashion vertebrate traits from annelid antecedents, from gill slits to the pituitary gland, and from the penis and clitoris to post-anal tail and ribs.

It was his evolutionary principle of the succession of functions that allowed Dohrn to connect the most dissimilar dots, and it catapulted his phylogenetic fancy to unprecedented heights. By deriving the vertebrate penis from annelid gills, he turned organs of respiration into organs of procreation, and by deriving the mouth of sea squirts from the nasal duct of cyclostomes, he transformed organs of olfaction into organs of ingestion. His former mentors Haeckel and Gegenbaur were quick to condemn what they saw as reckless forays into the realm of fantasy (Ghiselin, 1994, 2003; Maienschein, 1994; Caianiello, 2015). Gegenbaur (1876: 5–6) deemed Dohrn's homologizing of the ventral nerve cord of annelids and the dorsal nerve cord of chordates to be anatomical anathema, while Haeckel (1875: 88–89) ridiculed Dohrn's rather fantastic suggestion that amphioxus, urochordates, and cyclostomes are all degenerated fish. Even with his evolutionary imagination running at full speed, his dexterity in connecting phylogenetic dots floundered when confronted with the notochord, the diagnostic character of chordates. He admitted that "it is rather difficult to say anything certain about its origins" (Dohrn in Ghiselin, 1994: 44).

This weakness was seized upon by Dutch zoologist Ambrosius Hubrecht (1853–1915), a contemporary of Dohrn's who had also studied with Gegenbaur. Dismissing Dohrn's phylogenetic scenario as a "fata morgana" ("mirage") (Hubrecht, 1887: 641), he proposed his nemertean theory for the origin of vertebrates primarily to explain the evolution of the notochord and pituitary gland. Hubrecht suggested that the nemertean proboscis apparatus, which they use to catch prey, was the source of both. Hubrecht thought the "proboscidean sheath" – now generally known as the rhynchocoel, the coelom surrounding the nemertean proboscis – was the evolutionary precursor of the notochord, while the proboscis itself had given rise to the pituitary gland (Hubrecht, 1883). Because nemerteans don't show the pronounced metamery that vertebrates do, Hubrecht had to devise a hypothesis to bridge this gap in organization. He proposed that incipient metamery had evolved in the nemertean ancestors of vertebrates to afford protection against predatory attacks. An animal might more easily survive damage if its body comprised more or less repeated units, which could be the basis for regenerating smaller, but functional individuals. But other than saying that such incipient metamerism could eventually evolve into more fully differentiated segmentation, Hubrecht did not explain how and why this would have happened along the lineage leading to vertebrates.

This example shows that hypothetical ancestors modeled on annelids and nemerteans had different precursor potentials for vertebrate traits. The choice of a hypothetical ancestor shapes the explanatory texture of a phylogenetic scenario by specifying its explanatory foreground. Dohrn's annelid scenario put the origin of vertebrate segmentation at center stage, but had to relegate the notochord to the explanatory background. In contrast, Hubrecht gave the notochord explanatory priority, but he asked readers not to judge his scenario by his speculations on the origin of other characters, which he labeled "merely the sequel in a train of thoughts" (Hubrecht, 1883: 368). Indeed, committing oneself to a specific explanatory foreground may entail logical consequences that may be hard to swallow, at least for others.

The cardiac physiologist and nervous system expert Walter Gaskell (1847-1914) was a British contemporary of Hubrecht and Dohrn. He judged any theory about the evolution of vertebrates that failed to explain the origin of "that striking rudimentary organ of the brain, the pineal eye" to have "an absolutely fatal stumbling-block" (Gaskell, 1906: 538). Gaskell thought he could explain the evolutionary origin of the pineal eye by placing one homology assumption in the explanatory foreground from which everything else followed – the infundibulum (hollow pituitary stalk) in the vertebrate brain represents the old arthropod esophagus. Given this "one fixed point... all the resemblances between the two groups of animals to which I have drawn attention, naturally follow. The devil is not in my ingenuity but in Nature's facts" (Gaskell, 1910: 46). Alas, most readers took Gaskell's facts for fantasy.

If the infundibulum is the remnant of the ancestral esophagus, Gaskell reasoned, the ventricles of the vertebrate brain must represent the ancestral stomach, and the central canal of the spinal cord must be the lumen of the ancestral gut. In Gaskell's scenario, the arthropod ancestor's nervous system had become fused with the alimentary canal to form the new hollow nervous system of vertebrates (Figure 5.2). Gaskell noted that others before him, such as Anton Dohrn, had also identified the vestige of the old esophagus in the pituitary gland, but they had been prevented from drawing the logical conclusion that the arthropod alimentary canal had been incorporated *in toto* into the vertebrate nervous system because they had been under the spell of "the fetish of the inviolability of the alimentary canal" (Gaskell, 1910: 10). The homology of the gut had been the invariant grid for comparisons between animal body plans. But Gaskell did not wish to be incarcerated within the logical constraints set by Haeckel's Gastraea theory. For Gaskell the most important organ system, both physiologically and evolutionarily, was the nervous system. Therefore, any believable scenario for the origin of the brainiest of all animals had to place the nervous system at center stage.

Gaskell noted that the topological relationship between the supra-esophageal ganglia of arthropods and annelids and the esophagus that pierces them would be the same as that between the vertebrate brain and the ventricular spaces connected to the central canal of

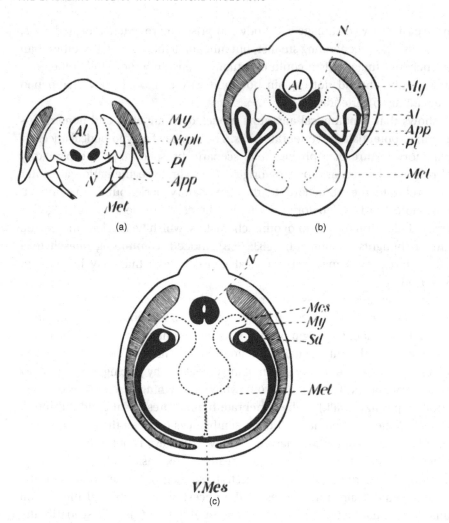

Figure 5.2 Schematic cross sections depicting how a trilobite-like arthropod (a) could evolve into a chordate (c) via a hypothetical intermediate (b). Note the expansion of the nephridial cavities ("Neph") into the main body coelom ("Met"), and the fusion of the alimentary canal ("Al") and ventral nerve cords ("N") to form the dorsal nerve cord
(after figure 160 in Gaskell, 1908).

the spinal cord (Figure 5.3). This provided him with a new framework for comparison. The fact that a pineal eye is present in the dorsal midline of the brain of some vertebrates proved for Gaskell that this side couldn't correspond to the ventral side of an arthropod ancestor because mid-ventral eyes were unheard of in annelids or arthropods (Gaskell, 1908: 15). But some arthropods do have median eyes on the front of the head. This meant that Gaskell could derive vertebrate pineal eyes from these arthropod precursors without the need for a dorsoventral flip of the body, a hypothesis that he, like others, found unacceptable (Hubrecht, 1883, 1887; Willey, 1894; Bernard, 1898).

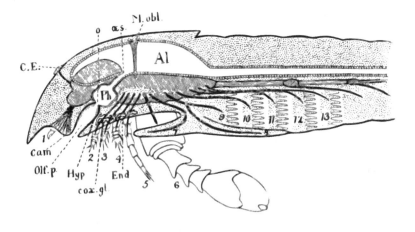

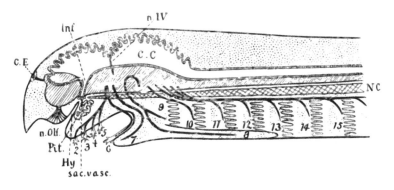

Figure 5.3 Schematic sagittal sections of a eurypterid (sea scorpion) (top) and a young ammocoetes larva of a lamprey (bottom), illustrating the topological correspondences of the arthropod alimentary canal plus nervous system and the chordate nervous system. Note that "inf" in the lower figure labels the infundibulum
(after figure 106 in Gaskell, 1908).

The morphological complexity of an ancestral arthropod provided Gaskell with many precursors for vertebrate traits. The same was true for zoologist William Patten, who likewise traced vertebrates back to arthropod ancestors, although he allowed a dorsoventral inversion (Patten, 1912). Starting with complex ancestors, Gaskell and Patten engaged in elaborate games of connect the dots that culminated in complex scenarios presented in fat books. They were able to produce far richer, if also more farfetched, scenarios for the origin of vertebrates than Hubrecht could manage with the morphologically less complex nemerteans or even Dohrn with his annelids. Hypothetical ancestors became a conspicuous component of the premodern phylogenetic literature, playing important roles as starting points and intermediate forms in scenarios. Some biologists, such as the embryologist Francis Maitland Balfour and the entomologist Guy Chester Crampton, acquired a certain notoriety for conjuring series of hypothetical ancestral groups to form the backbones of

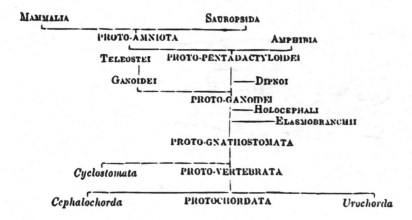

Figure 5.4 Francis Maitland Balfour's chordate phylogeny, with hypothetical ancestral groups in capitals and with the prefix "proto". The other taxa contain living representatives, with degenerate taxa in italics
(after Balfour, 1881: 271).

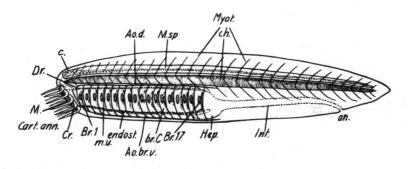

Figure 5.5 Hypothetical last common ancestor of cephalochordates and vertebrates as envisioned by Alexei Sewertzoff (1866–1936), a leading Russian evolutionary morphologist
(after figure 1 in Sewertzoff, 1929, by permission of John Wiley and Sons).

their scenarios and trees (Balfour, 1881; Crampton, 1938) (Figure 5.4). Others followed Haeckel's lead and composed detailed schematic drawings of hypothetical ancestors to show exactly what such intermediates looked like in their mind's eye (Sewertzoff, 1929) (Figure 5.5).

Although the aforementioned examples concern hypothetical ancestors constructed with building blocks from extant organisms, fossils have played the same role. A nice example is discussed in detail in Rosen et al. (1981) and summarized by Forey (2004: 163–164). It concerns the problem of the crossopterygian (lobe-finned fish) ancestry of tetrapods. Different paleontologists nominated different crossopterygian fossils as possessing the best-suited precursors for the tetrapod skull, depending on what evolutionary transformations in bone patterns they found to be most plausible. But agreement proved elusive, depending on what shifts, deformations, fusions, or fragmentations of bone were considered credible. The debate was unresolvable because irreconcilable differences of opinion existed about what evolutionary processes could best bridge the crossopterygian-tetrapod divide. Strikingly, a commitment to a certain theory of bone evolution could blind one to the relevance of observable similarities of bone patterns between taxa. Process convictions could trump pattern recognition. The spinning of phylogenetic scenarios that filtered incomplete fossil data through a process theory was a common enough paleontological strategy for George Gaylord Simpson to conclude, in a book aimed at general readers, that this practice "explains diversity of opinion among competent students and shows how all can cite evidence for their conflicting theories" (Simpson, 1956:37).

Even though hypothetical ancestors did most of their epistemic work within the confines of phylogenetic scenarios, they have also influenced how hypotheses of systematic relationships are judged. When veteran Russian invertebrate zoologist Olga Ivanova-Kazas, who died in 2015 at the age of 101, was confronted with the revolutionary results of molecular phylogenetics that placed annelids and arthropods in two different, deeply diverging clades, she was baffled. How was it possible that annelids and arthropods were only distant relatives? "[N]o morphologist would agree to separate arthropods from annelids; all peculiarities of the organization of arthropods could easily be understood as a result of transformation of characteristics that they have inherited from polychaetes; the causes of such transformation could also find some rational explanations" (Ivanova-Kazas, 2008: 400). Scanning the body plans of what molecular trees suggested were the new phylogenetic neighbors of arthropods, she asked: "What members of the group of molting acoelomate protostomes could give rise to arthropods? How and why did their organization change and become complicated so strikingly, with the sudden creation of the coelom, segmentation, and branched appendages? Let these questions be answered by the followers of the 'ecdysone hypothesis' [sic]" (Ivanova-Kazas, 2008: 401). Several years later, when she attempted to answer this question herself, she wrote: "Could Arthropoda evolve from lower Ecdysozoa? . . . It could be, but it is unlikely" (Ivanova-Kazas, 2013: 228–229). In contrast to polychaetes, she thought none of the ecdysozoan phyla offered any convincing precursor structures that could be turned into arthropod traits. She concluded that accepting the clade Ecdysozoa "contradicts simple logic" (Ivanova-Kazas, 2013: 230).

Ivanova-Kazas wasn't alone in pitting the precursor potential of hypothetical ancestors against the unpalatable implications of the Ecdysozoa hypothesis, the flagship clade of the molecular view of animal phylogeny that houses arthropods, but not annelids. When faced with the rising tide of molecular evidence that conflicted with this traditional union, Claus Nielsen's (2003) creative solution was to make Ecdysozoa in its entirety the sister group of annelids. Although no morphological or molecular dataset has ever existed to support this hypothesis, it allowed Nielsen to maintain the homology of the segmented body plans of annelids and arthropods. Around the same time, Wolfgang Wägele and his colleagues presented their case against the validity of Ecdysozoa and for the recognition of the traditional Articulata hypothesis in terms of the superior precursor potential of an annelid-like arthropod ancestor, exactly like Ivanova-Kazas did (Wägele et al., 1999; Wägele and Misof, 2001). More than a decade later, and despite what can only be described as the accumulation of absolutely overwhelming molecular phylogenetic evidence to the contrary, Wägele and Kück (2013: 289) continued to prefer an annelid-like ancestor for arthropods because "the basic anatomy is already there." Such can be the stranglehold of hypothetical ancestors.

These examples show how the precursor potential of hypothetical ancestors has been used as an optimality criterion, a narrative standard by which evolutionary scenarios were judged. Indeed, I have argued that it represents the first, more or less objective, phylogenetic optimality criterion (Jenner, 2019). Phylogenetic explanations consisted of tracing back traits to plausible precursors in hypothetical ancestors. Of course, determining what precursors were most suitable remained largely subjective, but there was general agreement that traits that could not be traced back were the Achilles' heels of scenarios.

5.5 The Unequal Eye and the Perspectival Nature of Scenarios

Historian of biology Robert Richards noted that "[n]arratives, as I have suggested, are radically perspectival and, because of that, radically incomplete" (Richards, 1992b: 29). This is as true for phylogenetic storytelling as it is for historical narratives generally. While systematists do well to base their inferences of systematic relationships on as much evidence as possible, phylogeneticists sketching evolutionary narratives have no choice but to apply the unequal eye. Stories cannot be told from all perspectives simultaneously. Nor can all events or evidence be given equal weight or attention. The ubiquity of conflicting data makes selective use of evidence virtually inescapable. The seeds that can sow disagreement abound in the phylogenetic arena.

Scenarios paint perspectival views of evolution defined by their choice of explanatory foregrounds and the nature of the hypothetical ancestors that are at center stage. The selective viewpoint of a scenario is immediately obvious when it is represented graphically. A quick glance at drawings that depict the archicoelomate and Trochaea theories of early animal evolution, for example, reveals that they have an entirely different focus. The archicoelomate theory foregrounds the morphology of adult animals evolving on the bottom of the sea to explain the origin of coelomic body cavities. The Trochaea theory has an embryological focus and foregrounds a series of hypothetical ancestors evolving in the plankton that trace back to Haeckel's Gastraea, whose modified remains are reflected by the ciliated larvae of living invertebrates (Figure 5.6). These scenarios are more or less distorted and incomplete representations of reality in much the same way as portraits drawn by caricaturists are – they are dominated by attention to a few traits, while relegating everything else to the background.

One factor that adds to the perspectival nature of scenarios is what type of evidence is consulted. Even though evolutionists know that the fossil record contains the only direct evidence of the past, its scarcity means that most phylogeneticists have starved at the buffet of paleontology. Most scenarios constructed from the nineteenth century onward were prepared with ingredients drawn from comparative embryology and morphology. Many scenarios incorporated evidence from both adults and embryos, but what to do when these conflict? This issue came to a dramatic head in the late nineteenth century in what is known as the *Kompetenzkonflikt* (Conflict of Competence) (Nyhart, 1995, 2002). It was precipitated by a debate about the evolutionary origins of tetrapod limbs, but it developed into a war of words about whether comparative anatomy or embryology had ultimate phylogenetic

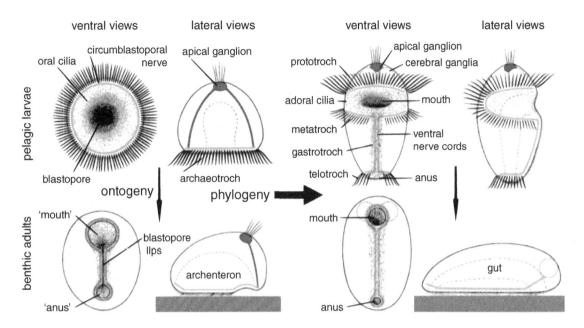

Figure 5.6 Claus Nielsen's Trochaea theory for the evolution of metazoan life cycles. The ancestral life cycle of Trochaea (left) transforms into the life cycle of the ancestral protostomian Gastroneuron (right)
(after figure 22.8 in Nielsen, 2012, by permission of Oxford University Press).

jurisdiction. Would the anatomy of lungfish fins offer the most reliable clues to the origin of vertebrate limbs, or were the embryos of sharks and sturgeons better places to look? The debate rapidly ballooned into an increasingly vitriolic polemic between embittered factions, a schism that eventually signaled the end of the first era of phylogenetic biology. Evolutionary morphologists' inability to overcome the hurdle of conflicting evidence and produce a consensus view on the origins of something so conspicuous and important as tetrapod limbs left a nasty epistemological stain on historical biology's status as a bona fide scientific discipline.

Even when debaters agreed on using the same type of evidence, consensus could easily be thwarted by the unequal eye adopting a different focus. During the same years that Carl Gegenbaur (1826–1903) developed his theory for the origin of tetrapod limbs that precipitated the *Kompetenzkonflikt*, his Jena colleague Ernst Haeckel was busy constructing his Gastraea theory, one of the most influential and enduring phylogenetic theories of all time. Based on observations of embryos from many different phyla, Haeckel constructed the image of the tiny, cup-shaped Gastraea as the ancestor of all animals (Haeckel, 1874, 1877). E. Ray Lankester and Élie Metschnikoff (1845–1916), however, developed rival theories by giving priority to observations from early branching lineages, especially cnidarians. Metschnikoff reasoned that gastrulation by invagination is not typical for cnidarians, and the animal ancestor was therefore unlikely to have looked like an invaginated gastrula. He instead envisioned a solid, gutless two-layered form as the metazoan ancestor, a creature he called Phagocytella (Metschnikoff, 1882, 1886). We will explore this debate in detail in Chapter 8 to reveal how the fires of phylogenetic debates are stoked by disagreements about what the most relevant evidence is.

Another manifestation of the unequal eye in phylogenetic debates was the common tactic of carefully choosing observations that conformed to one's hypothesis, but ignoring any conflicting data. Haeckel was frequently accused of this strategy. Turbellarian expert Otto Steinböck called this tactic "Haeckel's method" (Steinböck, 1958b: 147, 155), and he deplored that many of his own contemporaries still used it to promote their views four decades after Haeckel's death. Steinböck singled out Adolf Remane – who is best known today for his work on homology criteria – for special criticism, and he ridiculed Remane's phylogenetic method as the attempt to connect taxa via hypothetical paper animals, cemented together by "apodictic formulations" (Steinböck, 1958b: 159). Chapter 9 presents in more detail a phylogenetic debate that pitted Steinböck against Remane.

The flames of phylogenetic debates were further fanned by disagreements about the role of functional reasoning in the construction of scenarios. Although all scenarios try to reconstruct real historical events, there were striking differences in how much attention researchers paid to conceiving hypothetical ancestors as real organisms that once lived and evolved in a particular ecological context. Haeckel's program of evolutionary morphology was primarily concerned with the structural transformations that turned ancestors into descendants and generally paid much less attention to how form relates to function in the context of the environment. As Nyhart (1995: 203) has observed, evolutionary morphologists witnessed "one of the most important developments in zoology in the 1870s and 1880s: the creation of a morphology that was divorced from questions of adaptation and the living organism." From the 1870s onward, evolutionary morphologists produced a flood of unfettered phylogenetic speculations, often on the basis of little more than serial sections on microscope slides. Even though the founder of phylogenetics was no stranger to the realm of speculation, in 1880 even Haeckel worried that the "next generation of 'scientific zoologists' will only know *cross-sections* and *colored* tissues, but neither the *entire* animal nor its mode of life!" [italics in original] (Nyhart, 1995: 203).

Haeckel's rebellious former student Anton Dohrn, whose voice also echoed loudly in the embryological camp of the *Kompetenzkonflikt*, was deeply committed to treating form and function as inseparable parts of phylogenetic thinking (Ghiselin, 1994; Maienschein, 1994; Caianiello, 2015). He placed great emphasis on explicating the functional continuity and selective plausibility of evolutionary transformations in body plan. Dohrn thought, for instance, that the selective context for a crucial step in the evolution of vertebrates was enhanced locomotion abilities to facilitate more effective escape from predators and capture of prey. To evolve a more hydrodynamic fish-like body, the annelid ancestor needed to concentrate its gills in protective pockets at its front end and remove the gills from the posterior segments. Dohrn proposed that the anterior gills would gradually move into the openings of the nephridia located nearby, until they were completely internal. The terminal parts of the anterior nephridia would enlarge to form gill cavities. And while the more posterior gills were reduced and finally lost, their cartilaginous support arches would evolve into ribs, providing attachment points for the muscles required for powerful swimming.

Anton Dohrn's late-nineteenth century work was in the vanguard of attempts to enrich phylogenetic scenarios by integrating arguments about functional transitions, physiological continuity, and the adaptive context of body plan evolution (Ghiselin, 1994, 1996). Bowler (1996: 65, 254) has pointed out that it wasn't until the early decades of the twentieth century that structuralist phylogenetic scenarios, which largely ignored functional reasoning, were increasingly superseded by adaptive scenarios. As paleontological and geological clues about the ecological context of evolution were discovered, they provided information for answering why questions in addition to how questions. However, there is no sharp historical demarcation separating structuralist and adaptationist scenario building. As we saw earlier, Fritz Müller braided together conjectures about the how and why of evolutionary transformations in the 1860s, and Élie Metschnikoff and William Keith Brooks (1848–1908), among others, likewise seriously thought about the functional and adaptive aspects of body plan evolution in the later decades of the nineteenth century (Metschnikoff, 1886; Brooks, 1893a, 1893b; Benson, 1979; Ghiselin and Groeben, 1997). In the early decades of the twentieth century, Alexei Sewertzoff (1866–1936) was one of the more prominent architects of adaptive scenarios, and his student Ivan Schmalhausen (1884–1963) continued to emphasize the ecological and selective contexts of evolutionary change (Sewertzoff, 1929, 1931, 1932; Ghiselin, 1994, 1996; Levit et al., 2004, 2006). However, many modern scenarios for body plan evolution, often heavily relying on developmental genetic data, continue to be purely morphological as they pay little or no attention to the functional implications of the proposed transformations.

Ron Amundson (2002, 2005) distinguished two methods of phylogeny reconstruction in the late nineteenth and early twentieth centuries based on the extent to which functional reasoning was used in scenario building. According to Amundson (2005: 127), morphologists of a structuralist bend used the "Generative Rule of Reconstruction: Identify an ancestral ontogeny that can be modified into the ontogenies of the descendent groups," while others used the "Adaptive Rule of Reconstruction: Identify ancestral characters and selective forces such that the forces might have caused populations that possessed the characters to diverge into the descendent forms." He rightly argues that both methods view body plan evolution from a limited perspective, because neither ontogeny nor adaptation alone suffice to paint a full picture. Moreover, he argues that "[t]ypologists attended to the Generative Rule and ignored the Adaptive Rule; adaptationists did the opposite" (Amundson, 2005: 127). Although I think these positions are useful as markers for opposite ends of a spectrum of approaches to scenario building, I think this dichotomy fails as a general claim about how phylogeneticists traced evolution.

Thinking about morphological transformations purely in structural terms was certainly part of the epistemology of idealistic morphologists like Adolf Naef (see Chapter 6), but many, perhaps most, evolutionary morphologists thought about both form and function when drafting their scenarios. Amundson (2005: 117) writes that morphologists who applied the Generative Rule did not ask "how one adult form could change into another adult form, but which ancestral ontogeny could be modified to give rise to the ontogenies of descendant organisms." Yet the examples discussed earlier and in the following chapters show that many morphologists thought about the evolution of animal body plans purely in terms of adult morphology.

Even when embryological evidence was used, body plan evolution wasn't necessarily viewed as a transformation of entire life cycles. Recapitulationist reasoning was frequently used to extract hypothetical ancestors from embryological evidence, and these ancestors were then treated as starting points for scenarios in the same way as those based on adult morphology (see Chapter 8). The common thread I see running through the early literature on animal body plan evolution is that no matter whether hypothetical ancestors were dressed in the garb of embryos or adults, every scenario identified ancestral traits that could be modified into descendant traits. The anatomists and embryologists that were pitted against each other in the *Kompetenzkonflikt* disagreed vehemently about the value of ontogeny as a phylogenetic oracle, and functional arguments were used to varying degrees, but everyone traced back traits to precursors in their hypothetical ancestors.

5.6 When Imagination Falters

William Patten, who proposed the much maligned scenario of vertebrates evolving from crustacean ancestors via arachnid intermediates (Figure 5.7) called the researcher's imagination "the real seven-league boots and the helmet of invisibility with which he annihilates time, and space, and matter, and with which, in effect, he may project himself far beyond the narrow confines of his physical sanctuary into realms his physical body may not enter" (Patten, 1920: 294–295). The imagination is a powerful engine for animating scenarios, but not all seven-league boots stride in the same direction. A person's imaginative powers are a more important source of subjectivity in scenario building than any of the other factors we have identified so far. Ironically, a selective lack of imagination is a potent weapon with which to attack critics. No phylogenetic scenario is immune to claims that it is implausible. The potency of this weapon lies less in its capacity to convince opponents than the impossibility of mounting an effective defense. What one biologist thinks should be obvious to all who are familiar with the relevant facts, another considers to be "the most improbable of improbabilities" (Brooks, 1893b: 210). And what one American expert in invertebrate morphology called "fantastic nonsense" that deserves to be "dead and buried" (Hyman, 1959: 750, 752) – the evolutionary origin of coeloms from cnidarian gastric pouches – was the phylogenetic totem at the heart of a zoological tradition in German-speaking countries that endured for generations (see Chapter 9).

It is easy to use one's imagination, or lack thereof, as a tool in phylogenetic debates because one can open one's mind for one's own creations, while narrowing it sharply to squint disapprovingly at the constructs of others. But the imagination is a protean tool that is hard to wield consistently. Paleontologist Frank Raw called the fin-fold theory for the origin of vertebrate limbs that was developed by the embryologists embroiled in the debates of the *Kompetenzkonflikt* a "pure invention" (Raw, 1960: 510, 523), even though it continues to inspire research today, nearly a century and a half after it was first proposed.

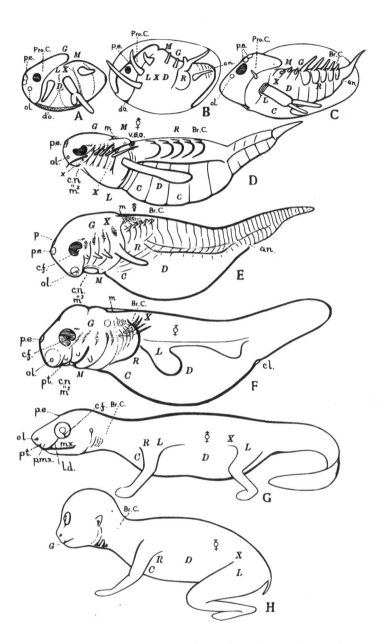

Figure 5.7 Stages in the evolution of vertebrates from arthropods according to William Patten (after figure 307 in Patten, 1912).

Raw's disapproval, however, did not stop him from claiming that any evidence for the fin-fold theory could easily be considered support for his own scenario that traced vertebrates back to arthropod and annelid ancestors. Moreover, although Raw had no difficulty imagining the most extraordinary body plan transformations – he claimed, for instance, that the arthropod ancestors of vertebrates had completely lost their gut and replaced it with a new digestive system that evolved by internalizing an external ventral food groove – he considered it "extremely improbable" (Raw, 1960: 518) that jawed vertebrates could have evolved from agnathan ancestors because agnathans have only one nostril.

Viewing evolution through the lens of functional morphology illustrates the seeming paradox that the epistemic power of hypothetical ancestors is greatest precisely where they can't be imagined. Not all imaginable character combinations are compatible with viable or functional hypothetical ancestors, which is why hypotheses of systematic relationships that imply implausible ancestors may be rejected: "To my mind, the most important test of this hypothesis [of ancestral states optimized to a node in a phylogenetic tree] is the question: *Is the postulated ancestor imaginatively credible as a living species, with a definite mode of life to which its characters are well adapted?*" [italics in original] (Crowson, 1982: 247). When such reasoning was applied to the heterodox hypothesis that mammals rather than crocodilians are the extant sister group of birds, it implied that their last common ancestor must have been a small, naked animal with a high metabolic rate, but no adaptations for achieving a high rate of respiration or rapid food processing (Kemp, 1988; Lee and Doughty, 1997). Such an animal would not be functionally well-adapted. Indeed, the sister group relationship between birds and mammals is soundly rejected by available morphological, molecular, and fossil evidence. But judging systematic hypotheses and phylogenetic scenarios strictly in terms of the plausibility of the ancestors they imply or postulate is fraught with danger. How do you know if a particular hypothetical ancestor or body plan transformation is truly implausible or whether in the context of severely limited evidence you just cannot imagine it?

Patsy McLaughlin (1932–2011) was the leading taxonomic expert on hermit crabs in the twentieth century. During the last decade and a half of her career, she and her colleagues investigated whether the crab-like king crabs could have evolved from hermit crab ancestors, a hypothesis considered plausible by many carcinologists since the 1880s. This extraordinary evolutionary transformation would have turned a lobster-like hermit crab, which concealed itself in gastropod shells to protect its soft and asymmetrical abdomen, into a compact king crab that holds its reduced abdomen folded under its thorax, thereby removing the need for shelter. After reviewing larval and adult morphology in great detail and failing to convince themselves of the selective advantage or even the possibility of the morphological changes involved in the supposed evolution of kings from hermits, McLaughlin and her colleagues concluded that this evolutionary transition was a myth (McLaughlin and Lemaitre, 1997; McLaughlin et al., 2004): "As the earth has been shown not to be flat, we have shown that the lowly hermit crab did not evolve into a mighty king" (McLaughlin et al., 2004: 196).

Instead, the authors proposed that it is more likely that the opposite had happened – king crabs evolved into hermit crabs as increasing decalcification, which happened for unknown reasons, softened the animal and forced it to retreat into the safety of mollusc shells. At the time of its publication, this heterodox hypothesis was in conflict with molecular phylogenetic evidence, and comprehensive analyses published since then leave no doubt that these crustacean experts were wrong. King crabs are deeply nested within asymmetrical hermit crabs and have therefore indeed evolved from hermit crab ancestors, although the details of this transformation remain obscure (Tsang et al., 2011; Bracken-Grissom et al., 2013; Noever and Glenner, 2018).

There is a danger of overwhelming one's imagination if one tries to imagine an evolutionary transformation in body plans in too much detail. This may be a particular peril for experts in functional morphology. Thinking about hypothetical ancestors and evolutionary transformations from the perspective that organisms are complex, functionally integrated systems certainly offers a path toward a deeper and more realistic understanding of evolution than just plotting disjointed character states on a tree. But by comparing taxa in ever greater detail, inevitable differences may come to obscure the ties of homology. Assigning the same character state to two taxa in a cladistic analysis always involves a degree of abstraction because any taxon-specific differences are disregarded, but such differences often matter greatly to the functional morphologist. Barnacle expert Donald Anderson (1982: 159) wrote that what functional morphology had done for animal phylogenetics was "mostly negative. The recognition that animals are highly functionally integrated and adapted in relation to habit and habitat has led to the further realization that they are often more different than they seem. Common ancestries then become much less plausible." Nudibranch expert Michael Ghiselin echoed this sentiment, writing that "the proper role of functional reasoning is in refuting hypotheses, not in showing that they are true" (Ghiselin, 1988: 89). That differences can more easily deter functional morphologists from connecting the dots than structuralist morphologists has been apparent at least since the debate between Cuvier and Geoffroy in 1830 (see Chapter 2). Cuvier's ability to detect morphological correspondences between disparate body plans was limited by his insistence that form and function were tightly linked, while Geoffroy's abstraction of function allowed him to see a much more encompassing unity of plan.

Don Anderson's embryological research tied in with the work of Sydnie Manton (1902–1979), the most influential arthropod functional morphologist of the twentieth century, to suggest that the body plans of the major arthropod taxa were fundamentally distinct (Anderson, 1973; Manton, 1973; Schram, 1978). When looked at in functional detail, body plan commonalities seem to be shrouded in a veil of differences. Manton considered the functional differences that she observed between the legs of different groups of arthropods so profound that she could not imagine that one could have given rise to another. She adopted a kind of neo-Cuvierian perspective according to which the main arthropod taxa – Chelicerata, Crustacea, and Uniramia (insects, myriapods and onychophorans) – represented isolated pinnacles of functional organization, separated by unbridgeable gaps. She accepted that evolution had happened, but she thought that these three groups of arthropods use their limbs in such different ways that she could not imagine an ancestor with jointed legs that could have given rise to the diversity of their limbs. She therefore concluded that the biting jaws and articulated walking legs of arthropods had evolved from soft lobopodial limbs independently on several occasions.

Geoffrey Fryer (1996: 23), a follower of Manton, considered independent evolutionary origins with a kind of epistemic relief: "[p]olyphyly is unencumbered by the need to demonstrate transitions from one basic design to another or the derivation of two such designs from a common ancestor. If trilobites, crustaceans and uniramians each had a separate origin in a different pre-arthropod then the evolution of their limbs from unsegmented precursors presents no problems." Here, the assumed convergent evolution of characters comes to the rescue of a lack of imagination.

Manton's conclusion that the mandibles of onychophorans, myriapods, and insects could not have given rise to each other because any such transition would break functional continuity may well be correct. Our current understanding of arthropod phylogeny doesn't place any of these taxa in an ancestor–descendant relationship. However, molecular phylogenetics has exposed some of her other conclusions as being due to a lack of evidence and imagination. The nesting of insects within crustaceans (Regier et al., 2010; Oakley et al., 2012; von Reumont et al., 2012; Schwentner et al., 2017; Lozano-Fernandez et al., 2019) proves beyond a doubt that both the walking legs and jaws of insects have evolved from the jointed limbs of a crustacean ancestor and that jointed legs have not evolved five times independently in the insects, as she believed (Manton, 1973). Although many details of this divergence in body plans remain to be reconstructed, Manton clearly underestimated the precursor potential of a crustacean ancestor for the evolution of insects. Dazzled by differences, she could not imagine how to connect the dots. This flawed logic of using differences as evidence for denying homology and common descent is dismayingly

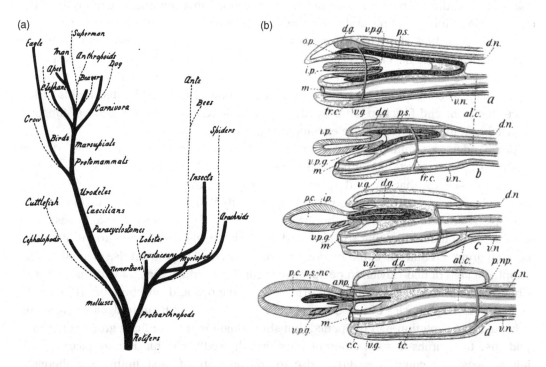

Figure 5.8 John Muirhead Macfarlane's 1918 view of animal phylogeny (a), represented by the solid lines, with rotifers rooting the tree and chordates evolving from nemertean ancestors. The stippled lines represent the degree of cognitive development of the "highest representatives" of these groups, with superman being the label he reserved for the intellectual crème de la crème among contemporary people. Macfarlane's scenario for the evolution of nemertean ancestors (top) into hemichordates (bottom) via two hypothetical intermediate states (b), with the nemertean proboscis becoming the hemichordate proboscis and the proboscis sheath evolving into the notochord (Macfarlane assumed homology of the hemichordate stomochord and chordate notochord) (after figure 27 and figure 26, respectively, in MacFarlane, 1918).

common in the literature (Jenner, 2006), and it continues to affect current debates (see Chapter 6).

This chapter introduced hypothetical ancestors as the epistemic core of narrative phylogenetics. Whether they were modeled on living species, represented by fossils, or amalgamated from disjunctly distributed traits, hypothetical ancestors have been used to root the explanatory power of scenarios since the birth of phylogenetics. The rise of explicit tree-based ancestral state reconstruction has largely replaced the deliberate design of hypothetical ancestors to maximize their precursor potential, but the traditional approach has proved tenacious. One illustration of this is the fact that some narrative phylogenetic hypotheses have endured across the centuries, such as the nemertean theory for vertebrate origins that was first promoted by Ambrosius Hubrecht in the late nineteenth century. It was subsequently taken up by John Muirhead Macfarlane (Figure 5.8), a Scottish neo-Lamarckian botanist in the early twentieth century, was further developed by Edward Neville Willmer, an English cytologist in the second half of the twentieth century, and was finally brought into the new millennium by Donald Jensen, an American comparative psychologist (Jenner, 2005a, 2005b). Together they identified, sometimes conflicting, nemertean precursors for a large number of vertebrate cell types, tissues, and organs, including the uvula and the pineal organ. Willmer and especially Jensen promoted their scenarios despite increasingly compelling evidence that placed nemerteans and chordates in different parts of the invertebrate tree of life. Still, their attraction to the explanatory strategy of narrative phylogenetics caused these workers to reject more modern approaches.

The late nemertean expert Janet Moore told me that when she confronted Willmer with genetic evidence that might refute his ideas, he responded by saying that genes were recipes for phenotypes and he wasn't interested in recipes. In the next four chapters, we will take a deeper look into narrative phylogenetics and how it has framed some of the major debates about animal body plan evolution and will show that some of its procedures and products continue to be very much with us today.

6

Intuiting Evolution

6.1 Intuiting the Arrow of Evolution

The phylogeneticist faces a steep epistemic hurdle. She finds herself in the observable present, but the story she wishes to tell is located in the empirically inaccessible realm of evolution. How can she sink beneath the synchronic surface of systematic diversity to penetrate the diachronic depths of phylogenetic history? How can she flesh out portraits of ancient ancestors with shreds of skin-deep evidence gleaned from microscopical slides, broken fossils, and pickled embryos? In short, how can her mind's eye gaze across the phylogenetic frontier? The modern tool that assists phylogeneticists in this endeavor is known as ancestral state reconstruction (Cunningham et al., 1998; Joy et al., 2016). The procedure is straightforward, at least in outline, and it applies to both living and fossil taxa. First, a phylogenetic tree is reconstructed, with taxa as tips. Characters are then mapped onto the tree using an optimality criterion, such as maximum parsimony or maximum likelihood. This procedure estimates ancestral character states for the nodes in the tree. Outgroup comparison then installs a relative time axis, and the result is a hypothesis for the pattern of character states and the succession of character state transformations along the lineages in the tree. The key is that systematic relationships between taxa are used to guide these inferences, which ensures that every ancestral state has been informed by empirical data from at least two tips of the tree. It is in this sense that ancestral states can properly be said to be inferred rather than merely imagined, irrespective of whether you think phylogenetic inference best fits the philosophical mold of induction (Ghiselin, 1997), deduction (Kluge, 2009), or abduction (Fitzhugh, 2006).

Modern phylogenetically informed ancestral state reconstruction accords epistemic priority to the inference of patterns of systematic relationships over the inference of hypothetical ancestors and evolutionary transformations. Before this epistemological advance, however, many phylogeneticists leavened their attempts to crack the hard problem of elucidating evolutionary patterns with hunches, intuitions, and speculations about the evolutionary process. Many faced the reconstruction of evolutionary history head-on, speculating boldly about the nature of distant ancestors and the morphological flux that connects disparate body plans. Truth be told, in some areas of research they had little choice because there was no tree that could act as an interpretative scaffold for evolutionary inference. This was especially true for those who wanted to understand the origin and evolution of the major animal body plans. Strikingly different conceptions about metazoan evolution emerged from the late nineteenth to the mid-twentieth century, each driven by different evolutionary intuitions and ancestral attractions, as we will see in the following chapters. Although a broad consensus on metazoan phylogeny finally seems to be emerging (Giribet and Edgecombe, 2020), controversies that are infused by different views about the evolutionary process remain a fixture of the contemporary literature.

From our modern perspective, attempts to infer or intuit ancestral states and the direction of evolutionary change without the strict guidance of a preestablished tree are shortcuts. Anointing fossils as ancestors without first systematizing them in a tree is a shortcut. So is lining them up in an ancestor–descendant lineage based solely on their stratigraphic order or according to a morphological gradient for a few of their characters. Deciding that a character state is primitive compared to another one because it is morphologically simpler is a shortcut, as is the attempt to read evolution from the sequence of morphological changes observed during embryonic development. Thinning the field of possible phylogenetic paths simply by declaring that you find certain character transformations difficult to imagine is also a shortcut.

The influential Swiss zoologist and systematist Adolf Naef classified such shortcuts as attempts to do "direct phylogenetics" (Naef, 1919: 51). They remove the inferential framework that necessarily interposes between observation and historical insight. In phylogenetics this framework is the pattern of systematic relationships between taxa. No modern phylogeneticist would countenance any of these shortcuts. Using these shortcuts fails to ensure that hypotheses about ancestral states and character transformations are informed and constrained by empirical data from relevant taxa, and they open the door to unsupported assumptions and preconceptions about the evolutionary process.

Although such shortcuts substitute speculation and preconceived ideas for inference, they have been storytelling tools since the inception of phylogenetics. They are at the heart of the narrative phylogenetics practiced by biologists and paleontologists before the conceptual and methodological advances of cladistics and numerical phylogenetics transformed comparative biology from the mid-twentieth century onward. As we will see in the next several chapters, these narrative tools provided sparks of life for a slew of hypothetical ancestors, phylogenetic hypotheses, and evolutionary theories. Because many phylogenetic shortcuts rely on subjective and often questionable intuitions about the evolutionary process, they have, with good reason, been called "authoritarian" by cladists (Cracraft, 1981: 21; Rieppel, 2010a: 487; Wheeler, 2012: 14; Brower, 2015: 575). Colin Patterson called the procedures of evolutionary systematics a "pernicious black box" (Patterson et al., 2011: 131). Yet, many evolutionists continue to unholster their ancestral hunches and intuitions when creating scenarios and judging phylogenies. Let's open this black box to discover what assumptions phylogeneticists have been feeding into their evolutionary inferences.

Finding the arrow of evolutionary time is the central challenge faced by phylogeneticists. If you want to tell an evolutionary story, you need to know where to begin. By the early nineteenth century, comparative morphologists, embryologists, and paleontologists had started to discern a biological directionality in the form of a threefold parallelism. Formulated initially in nonevolutionary terms, it emerged from parallels in the sequential changes of form seen in embryonic development; the stratigraphic sequence of fossils; and the systematic sequence of types, taxa, and traits (Russell, 1916; Coleman, 1973; Gould, 1977b; Patterson, 1981; Rieppel, 1988b; Richards, 1992a; Bryant, 1995; Jahn, 2002; Göbbel and Schultka, 2003).

This directionality meant different things to different people, and it was incorporated into different shades of idealistic and evolutionary morphology with equal ease. For Johann Friedrich Meckel and Étienne Serres in the early nineteenth century, it suggested that the same natural laws ruled the development of individuals and their position in a linear series of animal types. E. S. Russell (1916: 94) named this twofold parallelism the Meckel-Serres law – for Meckel and Serres the third parallel strand comprised malformations that arrest the development of an organ or the entire embryo at different stages of development. For Louis Agassiz in the mid-nineteenth century, the threefold parallelism reflected the static harmony of God's design incarnate in nature. And for evolutionists in the late nineteenth century, the threefold parallelism became convincing evidence for the theory of descent with modification.

Our understanding of what these parallel series of form changes mean for phylogenetics was transformed in the twentieth century (Hennig, 1966; Nixon and Carpenter, 1993; Bryant, 1995; Williams and Ebach, 2008; Rieppel et al., 2013; Grant, 2019). Although we now know that the arrow of evolution cannot be read more or less directly from any of these series, generations of biologists and paleontologists were seduced by the allure of direct phylogenetics. Phylogenetics today is a conceptually and analytically sophisticated discipline, but the intuitions that underpin narrative phylogenetics run deep and continue to bubble to the surface when modern authors explore the implications of trees. In the next chapters, we will examine several phylogenetic debates associated with each of the strands of the threefold parallelism. In Chapter 7 we will see how late-nineteenth and early-twentieth-century paleontologists plotted their stories by lining up fossils based on their stratigraphic order or arranging them along morphological gradients. In Chapter 8 we will see how intuitions about the ontogenetic leg of the threefold parallelism drove the late-nineteenth century debate about the origin of animals that was jumpstarted by Haeckel's controversial Gastraea theory. And in Chapter 9 we will see how several influential and enduring scenarios for the evolution of bilaterian body plans reveal the action of powerful ancestral attractions and evolutionary intuitions. But before getting into these specific debates, we first need to take a more detailed look at how narrative phylogeneticists plot the direction of their evolutionary stories.

6.2 Morphological Seriation: The Arrow at the Heart of the Threefold Parallelism

Two of the three strands of the threefold parallelism have an inbuilt time axis. Many biologists and paleontologists have used the stratigraphic sequence of fossils and the succession of embryonic forms as convenient shortcuts for divining ancestors and plotting the direction of evolutionary change. But what about the third strand of comparative morphology and systematics? It lacks an intrinsic time axis, so what exactly was thought to run parallel to the arrows of ontogeny and paleontology, and how could it provide a story arc?

The simple answer is the systematic hierarchy of the natural system, the nested set of subordinated taxa that are diagnosed by characters with different degrees of generality. Darwin considered the subordination of taxonomic groups to be "that great and universal feature in the affinities of all organic beings" (Darwin, 1859: 414). His theory of descent with modification explains why the natural system is structured in this way and why a branching evolutionary tree precisely corresponds to it. When interpreted evolutionarily, a relative time axis appears when you descend the systematic hierarchy, for example, from chordates to vertebrates to tetrapods to amniotes to mammals. It traces a path that parallels the evolutionary origins of these taxa as well as the sequence of their first appearance in the fossil record. The evolution of morphology emerges when you follow the transformations of homologues along this trajectory. But this simple answer ignores some relevant twists of history.

When the systematic arrow of the threefold parallelism was conceived by morphologists like Meckel the Younger and Louis Agassiz in the early nineteenth century, it was atemporal, unlike the arrows of embryology and paleontology. They saw it as a synchronic, linear morphological series of organisms arranged along a single scale according to a criterion of morphological complexity or the perfection of form (*Vollkommenheit* or *Vervollkommnung*) (Rieppel, 1985; Bryant, 1995; Jahn, 2002; Göbbel and Schultka, 2003). Unsurprisingly, this arrow pointed from simple to complex organisms, in line with the deep tradition of arranging nature's productions along a single *scala naturae* (Lovejoy, 1936; Archibald, 2014). Agassiz's thinking is particularly interesting in this regard because he formulated his ideas at the dawn of the evolutionary era, and while he never embraced evolution, his vision of the threefold parallelism nevertheless fertilized evolutionary theorizing.

In his *Essay on Classification*, originally published in 1857, Agassiz saw the systematic strand of the threefold parallelism most clearly in a linear, nonhierarchical component of the natural system. He classified species within each of the four animal types – Radiata, Mollusca, Vertebrata, and Articulata – in the traditional Linnean hierarchy of classes, orders, families, and genera. For Agassiz the strands of the parallelism stayed within the boundaries of each type, and they revealed glimpses of themselves on various levels of the hierarchy. But the systematic strand stands out most strongly on the level of orders because, according to Agassiz, each has a distinct degree of structural complication, which allows them to be ordered linearly from simple to complex (Agassiz, 1962: 105, 150, 156–159, 179). As Agassiz (1962: 156) noted, the hint is in the name "order." For instance, he arranged echinoderm orders in a series of increasing morphological complexity that runs from sea lilies to sea stars to sea urchins to sea cucumbers, paralleling the appearance of these orders in the fossil record. The reader will look in vain for Agassiz's criteria for judging crinoids "lowest" and holothurians "highest," but we are assured that this structural series is "perfectly plain" (Agassiz, 1962: 106). This lining up of living taxa on the basis of their perceived morphological complexity is the backbone of Agassiz's systematic strand of the threefold parallelism. It gets complicated when the same rationale is deployed to give the parallelism an evolutionary interpretation.

For the first Darwinian evolutionists, the threefold parallelism was prized evidence for the theory of descent with modification. Haeckel gave it a prominent place in his *Generelle Morphologie*, and despite their fundamental misgivings about Haeckelian phylogenetics, early-twentieth-century conceptual reformers of systematic biology, including Adolf Naef and Sinai Tschulok, agreed with Haeckel on this. Naef (1919: 32) called it the "touchstone" of idealistic morphology. Haeckel (1866b: 384–386) rejected Agassiz's linear ordering of orders to represent the morphological series at the heart of the systematic strand of his parallelism. This was in part because he thought that Agassiz had gone too far in identifying these progressive series. Orders containing species of comparable morphological complexity couldn't be arranged along a complexity gradient. However, Haeckel also saw no reason why, if this parallel did occur, it should be restricted to orders. Some classes within types and some families within orders could likewise be aligned into progressive series. Tracking the degree of morphological differentiation and perfection of organisms along these series rooted the systematic strand of the threefold parallelism in Haeckel's evolutionary soil (Haeckel, 1866b: 254–256, 422; 1889: 312–315).

Haeckel conceived of the systematic strand of the threefold parallelism as a chain (*Kette*) or a ladder of forms (*Stufenleiter*) (Haeckel, 1866b: 422; 1889: 312) that ran across the living tips of the tree of life. It measured the degree of morphological differentiation between living taxa (Haeckel, 1866b: 254–256). He was explicit that it did not measure the divergence from a common ancestor. Instead, he thought of it as running orthogonal to the arrow of paleontology. Although he avoided reading this synchronic systematic series as a phylogenetic *scala naturae*, he nevertheless did read it as a series of forms, or a "ladder of progress" (Haeckel, 1889: 312). I haven't found any explicit and unambiguous statement that Haeckel thought of this strand as running through, or representing, the systematic hierarchy, although his writing is consistent with this interpretation. It was in the minds of his successors, such as Adolf Naef and Sinai Tschulok, that this strand of the threefold parallelism was unequivocally identified with the systematic hierarchy of taxa, and with a hierarchy of characters rather than a chain of taxa (Rieppel, 2010b; Rieppel et al., 2013).

Who discovered the hierarchical natural system? The standard narrative tells us it was achieved by pre-Darwinian systematists from Linnaeus onward, enabling Darwin to use it as proof for his theory of descent with modification (Mayr, 1982; Brady, 1985; Amundson, 2005; Williams and Ebach, 2020). There is much truth to this. Darwin believed it. He called the subordination of taxonomic groups "the grand fact in natural history" (Darwin, 1859: 398), and he explained it as the product of the branching of evolutionary lineages. This then allowed evolutionists to use it to frame and guide their evolutionary stories. But this summary of history hides some revealing complexities. The hierarchical structure of the natural system was by no means obvious to all pre-Darwinian systematists, and it wasn't obvious everywhere. As we saw in Chapter 3, many thought of the affinity relationships between species and higher taxa in terms of complex geometries, including webs and maps because of the conflicting distributions of characters. These complex geometries can neither be neatly captured in a taxonomic hierarchy, nor do they allow one to trace the directional transformation of characters under an evolutionary interpretation. It is in this sense that I think there is an unresolved conceptual tension between Linnaeus's classificatory hierarchy and the fact that he simultaneously conceived of the relationships between taxa as a geographical map (Nelson and Platnick, 1981: 95).

The same is true for the views of another taxonomic pioneer we encountered in Chapter 3, Antoine-Laurent de Jussieu. Although he constructed hierarchical classifications (Amundson, 2005: 41), he also imagined affinity relationships as a map (Wilkins, 2009: 81). De Jussieu didn't think that nature was hierarchical (Stevens, 1994), which suggests that for him the classificatory hierarchy was just a convenient tool. I don't think one should go as far as Bryant (1995: 213) and suggest that pre-Darwinian classifications "cannot be used to support arguments that the genealogical hierarchy can be discovered without recourse to evolutionary theory." But when my friends Dave and Malte conclude that "the hierarchy of life was largely an empirical discovery" (Williams and Ebach, 2020: 86), I can't help but think that the word "largely" is significant here. I don't think that the hierarchical structure of the natural system was as obvious in all places and to as many people as Darwin's earlier quote suggests. I agree with the more carefully phrased conclusion of Nelson and Platnick (1981: 159) that there may have been some empirical basis for pre-Darwinian conceptions of the classificatory hierarchy, but "the belief that nature was best described in that fashion" as a matter of epistemic convenience played a significant role as well.

Signs of the existence of subordinated groups on higher taxonomic levels can disappear when zooming in on lower levels. Here reticulate relationships loom into view. In the same sentence where Darwin lauds the hierarchical structure of the natural system as a grand fact, he observes that this realization "from its familiarity, does not always sufficiently strike us" (Darwin, 1859: 398). Could it be that instead of familiarity blinding biologists, Darwin so easily discerned the hierarchy of nature because his evolutionary theory demanded something hierarchical to explain? As I have argued in Chapter 3, I see this as part of the explanation for why tree-like evolutionary imagery sprouted so suddenly and prodigiously from the post-Darwinian soil. If one accepts that evolution happens by descent with modification along branching lineages, then evolutionary trees and their associated hierarchical classifications are the expected outcomes of systematic research. Still, this does not solve the problem that affinity relationships looked reticulated to many systematists.

6.2.1 Disentangling Reticulate Affinity Relationships

Ukraine-born biologist Sinai Tschulok concisely declared, in his influential book *Deszendenzlehre* (*Doctrine of Descent*), that "[t]he natural system is the evidence for the theory of descent" (Tschulok, 1922: 152). He realized that the empirical evidence for the natural system consisted of reticulated affinity relationships (Tschulok, 1922: 151–152). Although one rarely encounters Tschulok in the modern literature, with Olivier Rieppel's recent writings being an exception (Rieppel, 2010b, 2016), he deserves credit for fashioning a tool that cut through this tangled web. As an idealistic morphologist, Tschulok was acutely attuned to the complex relationship between the empirical evidence upon which the natural system was based and the phylogenetic inferences that it might underpin. The key to revealing the tree in the tangles was to distinguish primitive and derived character states (Tschulok, 1922: 197). A large amount of seemingly conflicting systematic evidence can be dissolved this way and nonsensical relationships avoided.

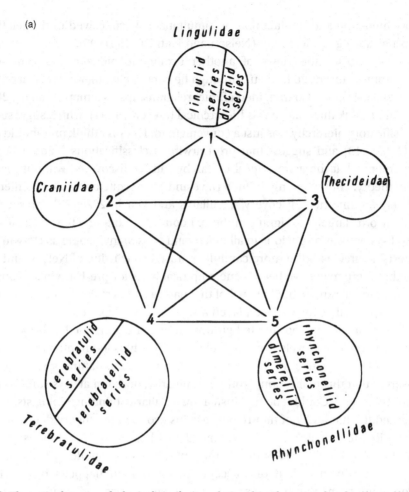

Figure 6.1 The reticulate morphological similarity relationships between five families of brachiopods (a) and their phylogenetic relationships based on distinguishing primitive and derived character states (b) (after figures 46 and 47, respectively, in Hennig, 1966. Copyright 1966, 1979 by Board of Trustees of the University of Illinois. Used with permission of the University of Illinois Press).

Tschulok took explicit aim at Lamarck and Haeckel, who he thought should have paid much closer attention to separating primitive and derived character states, and who obscured relevant details by populating their trees with higher-level taxa instead of species. He ridiculed Lamarck for deriving terrestrial from aquatic mammals with reduced hindlegs, a clearly derived trait (Tschulok, 1922: 200–205). Willi Hennig adopted precisely this strategy in his 1966 *Phylogenetic Systematics* when he addressed the seeming paradox "that phylogenetic systematics tries to order species into a hierarchic system by analyzing relations that have a reticular structure" (Hennig, 1966: 148). Illustrating the power of this approach with an example of the relationships between five brachiopod families (Figure 6.1), Hennig managed to turn a previously published pentagonal reticulum of relationships into a fully resolved dichotomous tree by distinguishing primitive from

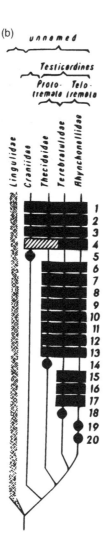

Figure 6.1 (*cont.*)

derived character states. Of course, branching evolutionary trees had been published for decades before Tschulok's insights, but as I argued in Chapter 3, one reason why many of these trees, including the grove planted by Haeckel, were drawn that way was because Darwinian evolutionary theory dictated this shape. It was with the help of the epistemological insights of twentieth-century systematists like Tschulok, Naef, and Hennig that the shape of evolutionary trees increasingly reflected the detection and separation of real hierarchical signal in systematic data from nonhierarchical noise.

If trees can be made to rise from webs of raw morphological similarity by discriminating primitive and derived character states, how are these to be distinguished? What did Tschulok do that others before him hadn't done? I will not rehearse the large literature on how to root trees and polarize character transformation series here, but Tschulok and his former student Adolf Naef made the primary principle that came to form the basis of outgroup comparison crystal clear. Primitive character states can be distinguished from derived character states based on their relative generality. Primitive character states characterize more inclusive groups than derived character states. Illustrating this with an example about the origin of modern elephants, Tschulok (1922: 191–196) lined up a series of fossils based on their stratigraphic occurrence and morphology. Characters with a wider taxonomic distribution, such as small body size, and a larger number of toes and teeth, mark out primitive states compared to their opposites. Following Othenio Abel's lead (see Chapter 7), *Stufenreihen* (character transformation series; literally "step series") could be distinguished from *Ahnenreihen* (series of ancestors) by resolving the conflicting distribution of derived character states. In this way, the phylogeneticist could increasingly home in on possible hypothetical ancestors, which in this case allowed Tschulok to trace elephant phylogeny back to the Eocene proboscidean *Moeritherium*. However, since the incomplete preservation of fossils could always hide relevant evidence, one could only hope "to approach the real ancestors ever closer, but without ever reaching them" (Tschulok, 1922: 209).

There is a catch if you want to use the principle of generality to polarize characters. Naef spelled it out in the last paragraph of his introduction to the unpublished typescript of a textbook he was writing at the time of his death in 1949: "To scientifically treat the phylogeny of existing organisms means to first determine their systematic relationships" (quoted in Rieppel et al., 2013: 502). But what if no systematic framework is available, or if your speculative urge cannot be constrained by it? The escape route has been to go back to what has been the epistemic core of comparative morphology ever since its inception: the construction of morphological series. Rieppel (2016, 2020) has argued that comparative morphologists root their analyses in what he calls the "seriability" of form. He is the only author writing in English that I have seen using this word, and dictionaries are silent on it. But it makes intuitive sense, and it indicates generating insights into homology and evolution by arranging observed or hypothetical taxa or their traits into linear series, and exploring the meaning of the transformations that can be imagined to have taken place between them. This approach extends from the heart of Goethe's pre-evolutionary speculations about archetypes in the eighteenth century, to the morphological musings of twentieth-century idealistic and evolutionary morphologists alike (e.g. Zangerl, 1948; Salvini-Plawen, 1978). Equally, it was at the epistemic core of Gegenbaur's evolutionary morphology in the nineteenth century, and it informed criteria to diagnose homology in the twentieth century, including Remane's third homology criterion (continuity through intermediate forms) and Maslin's morpho-clines [sic] (Maslin, 1952; Remane, 1956).

As we will see in the following three chapters, the creation of morphological series was a major component of paleontological, embryological, and morphological research from the second half of the nineteenth century onward. It was the attempt to directly construct evolutionary lineages by arranging forms into smoothly graded series rather than knitting them together indirectly by following the thread of reconstructed ancestral states on a phylogenetic tree. This approach assumes that certain living forms have departed very little from their presumed ancestors and can therefore function as ancestral stand-ins, or as the prolific entomologist Guy C. Crampton called them, "contemporaneous ancestors" (Crampton, 1921: 65). Bernard Rensch nicely illustrates this type of linear lineage thinking in his book *Evolution Above the Species Level*, especially when he discusses living taxa that approximate intermediate types between different body plans (Rensch, 1960a: 268–271).

Even after the cladistic revolution, some authors preferred this traditional approach over modern methods, especially when a broadly accepted phylogenetic framework was lacking. The work of invertebrate zoologist Luitfried von Salvini-Plawen (1939–2014), who was one of the most visible narrative invertebrate phylogeneticists of the last half century, provides instructive examples. For instance, in his classic 1978 paper "On the origin and evolution of the lower Metazoa," he depicted several hypothetical lineages as sequences of living forms that he thought could function as models of possible ancestors. These morphological series allow readers to mentally trace the evolution of simple bilaterians from blastula-like ancestors via planula-like intermediates (figures 14–18 in Salvini-Plawen, 1978).

The smooth gradation of steps was a core criterion for the direct construction of such morphological series, both for evolutionary morphologists interested in tracing lineages and for idealistic morphologists interested in tracing purely formal series of forms. But this approach could be complemented by functional considerations as well, in which the arrow of change followed an assumed direction of increased adaptation (Bonik et al., 1977; Gutmann, 1981). Yet this direct approach to (re)constructing lineages has well-known dangers.

The biggest danger is to commit the sin of the scala: lining up collateral relatives into a *scala naturae* and misinterpreting them as ancestors and descendants. Some of the ortho-geneticists that we will encounter in the next chapter were certainly guilty of this crime. Naef realized that the impression of a *scala naturae* could result from series of paraphyletic taxa, which he called *Stammgruppen* (stem groups). For example, although the vertebrate classes had equal rank in the Linnean hierarchy, their types could be arranged in a linear series of fish → amphibian → reptile → bird/mammal. Naef understood that while such series told a systematic lie, they spoke the phylogenetic truth (Naef, 1919: 49–51). He saw that to bring phylogeny into correspondence with the natural system, such *Stammgruppen* needed to be resolved into their component taxa until all were properly equivalent (resolved as monophyletic sister groups, to put it in modern language). Moreover, by disassembling organisms into their component characters, and inferring series of ancestral character states, the sin of the scala can be avoided.

Another danger that Naef pointed out was seeing phylogenetic change as a literal trans-formation of adult morphologies rather than as a flow of life cycles (Naef, 1919: 41): "The members of such an ancestral series [Ahnenreihe] are not stages of a process, and the development they suggest is just imagined(!)" [exclamation mark in original]. Adult organisms are just ephemeral shoots that sprout from the continuity of the germline. Worse still, Naef accused phylogeneticists of habitually overlooking the fact that the historical explanatory power of phylogenetic hypotheses is restricted to tracing conditions back to "immediately preceding, directly and truly connected" states (Naef, 1919: 38). Trying to explain observations by reference to the remote past is "a gross mistake" that is responsible for "many mental leaps in phylogenetics" (Naef, 1919: 38). Eighty years later, Henry Gee phrased the same sentiment as "Deep Time cannot sustain scenarios based on narrative" (Gee, 2000:

114). It is for these reasons that Naef considered series of hypothetical ancestors or ancestral character states only as "a kind of symbol for phylogenetic history" (Naef, 1919: 42).

Naef's penetrating insights into the logic and methodology of phylogenetics were influential, and he can certainly be excused for calling what his predecessors engaged in as "naïve phylogenetics" (Naef, 1919: 3). The traditional approach involved second-guessing the evolutionary process and the nature of ancestors, with phylogeneticists often declaring that certain evolutionary changes were "out of the question, impossible, or unlikely, based on general, so to speak, technical considerations" (Naef, 1919: 52). But these were just "unprovable opinions or prejudices" (Naef, 1919: 66). He had practiced that art himself. It was only through his theoretical research that he had found an escape route (Naef, 1919: 52). The hope that research would reveal law-like regularities of phylogenetic change that could be used as tools to generate new insights into evolution remained unfulfilled at the time of Naef's writing (Naef, 1919: 52–54). Nevertheless, evolutionary intuitions continued to be the engine driving narrative phylogenetics. Today, one can still hear them hum when researchers step back to consider what their trees actually mean.

6.3 Evolutionary Intuitions as Narrative Tools

When I was a postdoc with Matt Wills at the University of Bath in 2008, we discussed Matt's latest paper on arthropod evolution during a lab meeting. With his coauthors Sarah Adamowicz and Andy Purvis, Matt had found that parallel increases in morphological complexity marked the evolution of crustacean body plans in different lineages since the Cambrian (Adamowicz et al., 2008). It quickly emerged that the participants had not come to the discussion as blank slates. Several of us shared the intuition that in arthropods the number of body segments would evolve more readily than the degree of limb differentiation and specialization, a phenomenon known as *tagmosis*. Where did this intuition come from? None of us had studied this topic in any detail or was able, on the fly, to cite research to support our hunches. Yet, this intuition presented itself almost immediately with minimal effort when the subject came up.

Anyone with an opinion about animal evolution has a set of intuitions about the evolutionary process that goes beyond available evidence (Figure 6.2). The sources of these conceptions may be hidden from introspection, but they likely include insights gained from previous research, the scientific literature, discussions with colleagues, as well as gut feelings. Sometimes referred to as systematic tact or morphological instinct by earlier writers (Naef, 1919; Lorenz, 1941), this protean background knowledge is purely subjective and often largely unconscious, it can change radically over time and differ between people, and it can be self-contradictory. It manifests itself in researchers formulating and accepting different hypotheses, evaluating data differently, and deciding what topics to study. It is the effortless deployment of these subjective, shape-shifting intuitions that makes evolutionary storytelling both so easy and so contentious (Gould and Lewontin, 1979).

As we will see in this and the following chapters, differing evolutionary intuitions have played key roles in the debates about the evolution of animal body plans from the

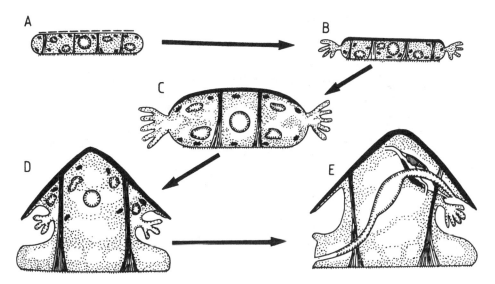

Figure 6.2 Scenario for the evolution of the molluscan body plan from a platyhelminth ancestor. Pat Willmer (1990: 261–262) presents the steps that she felt could "be predicted" as consequences of an aquatic worm evolving a dorsal shell. These include the evolution of gills and the evolution of a body cavity to aid circulation, which in turn would facilitate an increase in size. This would provide space for the evolution of pockets in which to house the gills, leading to the development of a through-gut and the condensation of the internal organs into a visceral mass
(after figure 10.6 in Willmer, 1990, by permission of Cambridge University Press).

nineteenth century to today. They press an indelible mark on evolutionary scenarios by steering authors to decisions about what evidence is relevant to their preferred hypothesis and how to fashion it into a coherent story. They are likewise used to cut down rival ideas. Although most modern phylogeneticists no longer wear their evolutionary instincts on their sleeves, our views nevertheless continue to be shaped by them. That this is so is obvious if you have ever looked at a character mapped onto a tree and asked yourself "Does this make sense?" The strong convictions of early evolutionary morphologists are remarkable given that at the time metazoan phylogenetics was in its infancy and almost none of their opinions were rooted in robust evidence. Since so many evolutionary intuitions were unmoored from empirical data, it isn't surprising that many of the scenarios they were supposed to buoy up sank without a trace. But as we will see, evolutionary intuitions themselves typically lead much longer lives.

Clashing evolutionary instincts can stimulate debate, but they can just as easily paralyze it. Malacologists Amélie Scheltema and Christoffer Schander (2006: 23) found it "difficult to imagine" that the ancestral mollusc was a vermiform creature, an idea promoted by their colleague Gerhard Haszprunar. Can such a disagreement be turned into a productive debate? The idea that living molluscs stem from a worm-like ancestor went back decades earlier and was strongly promoted by Haszprunar's erstwhile boss Luitfried von Salvini-Plawen. In a series of narrative phylogenetic papers from the 1970s onward, Salvini-Plawen and fellow aplacophoran mollusc expert Scheltema conducted a decades-long debate about

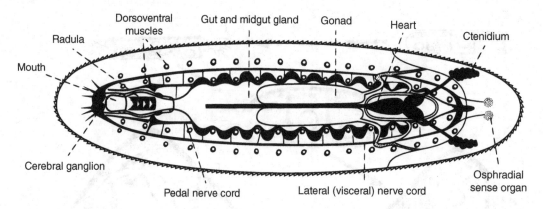

Figure 6.3 A hypothetical ancestral mollusc
(reprinted from figure 2 in Haszprunar and Wanninger, 2012, with permission from Cell Press).

whether the worm-like aplacophorans were a paraphyletic grade at the base of the molluscan crown group that reflected the ancestral molluscan body form, or whether they were a clade with a highly derived morphology. Yet, four decades of debate did nothing to change their minds. They took their disagreements to the grave when they died within eight months of each other in 2014 (Salvini-Plawen) and 2015 (Scheltema). Carrying the torch of tradition forward, in 2012 Haszprunar and his former PhD student Andreas Wanninger still drew the urmollusc as exactly the same worm-like creature that their intellectual father and grandfather had been drawing since the 1970s (Figure 6.3) (Salvini-Plawen, 1985; Haszprunar and Wanninger, 2012).

With minds left unchanged and heels dug in until death, this was not really a conversation. Such stasis of opinions is typical for many traditional phylogenetic debates. Salvini-Plawen's views on animal evolution were strongly colored by what he was able to imagine. For instance, when Scheltema (1993) adopted a more cladistic approach to molluscan phylogeny, he took issue with the evolutionary implications of the characters mapped on the tree. Salvini-Plawen and Steiner (2014: 2756) rejected the mapping of paired pharyngeal diverticula and a complex radula apparatus to the molluscan stem group because it was unclear to them how these could be "supposed to have evolved in molluscs *de novo* without preceding structures and then to be reduced again in aplacophorans." As we saw in Chapter 5, biologists often rejected hypotheses for *de novo* character origins because they lacked explanatory power. The fact that the cladistic value of characters as synapomorphies is undiminished even when their origins are unknown illustrates the enormity of the epistemological revolution brought by cladistics. The explanatory ambition of cladistics is to discover monophyletic clades, while narrative phylogenetics seeks to understand the origins of characters.

Salvini-Plawen and Steiner rejected other traits that Scheltema mapped as synapomorphies on her tree because they interpreted morphological differences as evidence for convergent origins, such as the reduced mantle cavity and ganglionated nerve cords in the two groups of aplacophorans. As we will see, this flawed strategy of regarding differences as evidence against homology and common ancestry is an enduring and pervasive component of narrative phylogenetics. This debate between molluscan experts shows how difficult it can be to dislodge entrenched evolutionary intuitions and how easy it is to talk past each other. But opinions have recently started to converge, and fresh air has been blown through dusty echo chambers as a result of new molecular phylogenetic and fossil evidence about early molluscan evolution (Vinther et al., 2017; Wanninger and Wollesen, 2019; Kocot et al., 2020). Time will tell if the experts will come to agree on what makes sense or not.

Many evolutionary intuitions were products of the search for evolutionary laws that began when phylogenetics was born. Zoologists Alexei Sewertzoff (1866-1936), Adolf Remane (1898-1976), and Bernard Rensch (1900-1990) were three influential authors on this topic. They catalogued an extensive menu of evolutionary regularities and trends observed across a wide swathe of taxa that phylogeneticists could consult to animate their scenarios (Sewertzoff, 1931; Remane, 1956; Rensch, 1960a, 1960b, 1991). Adolf Remane critically reviewed many of these phylogenetic signposts in a 125-page chapter titled "Phylogenetic laws as tools in genealogical research" in his 1956 book *The Foundations of the Natural System, Comparative Anatomy, and Phylogenetics* (*Die Grundlagen des Natürlichen Systems, der vergleichenden Anatomie und der Phylogenetik*). There was the law of perfection, the law of differentiation, the law of the decreasing number of similar organs, and the law of the increasing number of similar organs. There was the biogenetic law (Haeckel's law), the law of phyletic size increase (Cope's rule), and the law of irreversibility (Dollo's law). There were the laws of the internalization, concentration, and centralization of organs. There was the law of the increasing division of labor, the law of acceleration, Cope's law of the unspecialized, the law of specialization, and Williston's law.

Phylogeneticists could use these and a host of other rules and hunches to amplify fragmentary evidence into evolutionary stories. Add to these the countless intuitions held by individual evolutionists, ranging from broadly accepted generalities – parasitic and sessile animals generally evolve from free-living ancestors – to the most idiosyncratic of notions – hybridization between distantly related species is responsible for the evolutionary origin of larvae – and you end up with a chest full of narrative tools with which to spin evolutionary stories.

Adolf Naef classified these evolutionary rules of thumb as components of what he called "direct phylogenetics" (Naef, 1919: 51). By this he meant the traditional type of evolutionary storytelling that wasn't strictly constrained and informed by a robust framework of systematic relationships. Remane considered evolutionary rules of thumb complementary methods (*Ergänzungsmethoden*) that could assist the main method (*Hauptmethode*) of phylogenetics, namely the determination of the taxonomic distribution of homologous similarities (Remane, 1956: 148). Remane recognized that the observations underpinning these phylogenetic laws were often marred by so many exceptions that they were flimsy tools. What use was a so-called law if it applied to only 60 to 70 percent of cases (Remane, 1956: 149)? Of course, the appeal of these evolutionary intuitions never resided in their empirical support. They were attractive as narrative devices that helped authors tell their preferred evolutionary stories. Some of these stories tied in with older traditions as their guiding intuitions traced back to pre-evolutionary times. For instance, the intuitions that animal body plans generally evolve from simple to complex, that the form changes of embryos parallel the systematic arrangement of their adult stages, and that higher animals have smaller numbers of more differentiated parts than lower organisms all map to pre-evolutionary systems of thought (see further discussion in Chapter 9).

These pre-evolutionary ideas about the natural order became parts of the evolutionary *Gesetz der Vervollkommnung*, an encompassing and influential law that can be translated as the law of perfection, progress, or improvement. As we saw in Chapter 4, this law was a conspicuous component of Haeckel's evolutionary morphology as well as the thinking of several biologists both before and after him (Hossfeld and Olsson, 2003). Noteworthy in this respect was the paleontologist Heinrich Bronn, whose German translation of *The Origin of Species* was the catnip that made Haeckel crazy for Darwin's theory. Yet for Bronn, the law of progress wasn't a phylogenetic law. Bronn, whose name continues to echo in modern zoology because of his conception and inception of the zoological encyclopedia titled *Dr. H.G. Bronn's Klassen und Ordnungen des Thier-Reichs* (*Dr. H.G. Bronn's Classes und Orders of the Animal Kingdom*), rejected evolutionary change along genealogical lineages. Instead, he saw the history of nature embodied in a series of successions of spontaneously generated species (Rupke, 2005; Gliboff, 2007, 2008). Nevertheless, he noticed the same regularities, like an arrow of progressive increase in complexity, that evolutionists consecrated as phylogenetic laws.

Let's look at two examples of how biological laws could be used as tools in phylogenetic scenarios, both dealing with nervous system evolution. The first is Erich Reisinger's (1972) orthogon theory for the origin of nervous systems in spiralian protostomes, a taxon that at the time included arthropods alongside molluscs and annelids. In Reisinger's scenario (Figure 6.4), the evolution of spiralian nervous systems is a story of the increasing concentration and internalization of neurons from a primitively diffuse and peripheral network, like that found in cnidarians. Nerves condensed into ganglionated nerve cords embedded more deeply in the body, where in some taxa, like flatworms, a grid was formed of several longitudinal nerve cords connected by circular commissures, a configuration that Reisinger called an *orthogon*. In other taxa, evolution ultimately led, by the progressive reduction in the number of ganglia and nerve cords, to the concentrated nervous systems of arthropods, such as mites and ticks.

Reisinger approvingly cites Remane, who judged these "littler laws" (*kleineren Gesetzen*) (Remane, 1956: 222) of the internalization, concentration, and centralization of morphological structures to be useful tools with which to judge whether character states were more likely to be primitive or derived. Remane and Reisinger thus represent one pile of a long bridge that links these morphological instincts from the eighteenth century and Goethe, for whom differentiation and centralization were criteria of morphological perfection, through the laws of Haeckel and Bronn in the nineteenth century, to Andreas Schmidt-Rhaesa's 2007 book *The Evolution of Organ Systems*, in which the orthogon theory is still presented as a plausible explanation for the evolution of spiralian nervous systems (Haeckel, 1889: 280; Russell, 1916; Gliboff, 2007, 2008; Schmidt-Rhaesa, 2007; Levit et al., 2014). This example shows that age-old transmogrified morphological laws don't necessarily make for bad evolutionary intuitions. The problem is knowing which intuitions to trust and when.

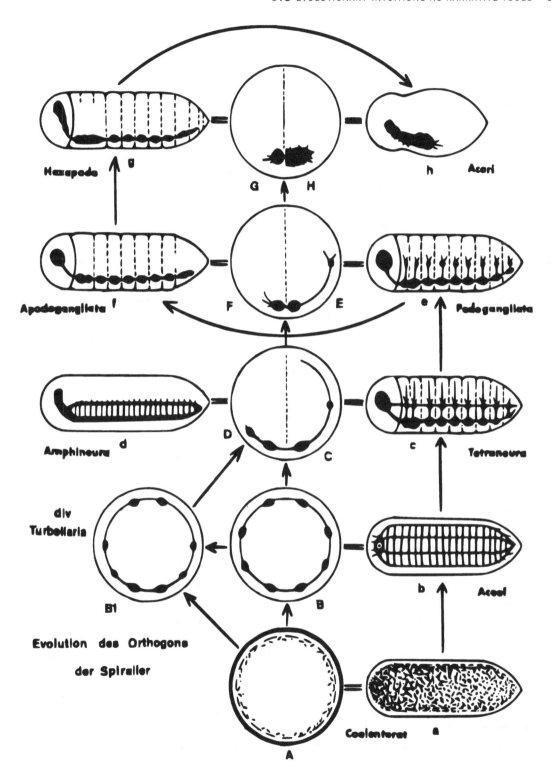

Figure 6.4 Reisinger's orthogon theory for the origin of nervous systems in spiralians. Evolving configurations of nerves, nerve cords, and brains are shown in schematic cross sections (round diagrams) and longitudinal sections
(after figure 23 in Reisinger, 1972, by permission from John Wiley and Sons).

The second example concerns the law of neurobiotaxis, a principle formulated in 1907 by Dutch neurologist and anatomist Ariëns Kappers to explain why nerve cell bodies are often concentrated near the sites from where they receive the greatest stimulation (Kappers, 1921, 1927). Kappers used this law to explain the phylogenetic shifts in the location of neurons in the nervous systems of vertebrates as well as the existence of one of the more remarkable aspects of vertebrate neuroanatomy, namely the midline crossing of nerve tracts, which is especially clear in the optic chiasm. In humans, for example, the optic nerves of the eyes converge in the optic chiasm, from where the fibers coming from the nasal halves of both retinas cross to the opposite side of the brain, while those emerging from the temporal halves remain on the same side. According to Kappers, the law of neurobiotaxis explains the meeting of the optic nerves in the midline of the brain because they are attracted there by the simultaneous stimulation of both eyes by the center of the visual field. Because the nasal fibers of one eye collaborate with the temporal fibers of the other eye in forming a single picture of peripheral areas of the visual field, the law states that the nasal fibers need to cross to the contralateral side of the brain. Although wielded with great enthusiasm by Kappers and his followers, the law of neurobiotaxis left a tiny footprint in the zoological literature, with perhaps two notable exceptions.

Half a century after Kappers (1927: 128) claimed to have "found" this law, Engelbrecht (1971) adopted it, although without mentioning Kappers, to explain the origin of the optic chiasm, but this time as part of a fanciful scenario for the origin of chordates from annelid worms. According to Engelbrecht (1967, 1971), chordates had evolved from tube-dwelling annelids that had reverted back to a free-living existence, undergone a dorsoventral axis inversion, and transformed their formerly ventral faecal groove into a new hollow, dorsal nervous system. As we will see in Chapter 9, annelids were frequently cast as hypothetical ancestors, but the idea of a dorsoventral axis inversion somewhere along the chordate lineage has long been a flashpoint in phylogenetic debates.

A more notable exception was Gavin de Beer's attempt to bolster Walter Garstang's 1894 auricularia theory for the origin of the chordate nervous system. Invoking the law of neurobiotaxis, de Beer thought that the ciliary bands and associated neurons of an echino-derm larva-like ancestor would migrate and fuse on the dorsal side because that was the "side of the body which would receive the greatest stimulation in a form swimming freely in the sea, the stimuli being the rays of light penetrating through from the surface" (de Beer, 1940: 53). Maybe so, but it requires a thick layering of presuppositions, including that a tiny, and probably transparent, larva-like creature swam with its dorsal side to the surface, that the direction of sunlight provided a sufficiently strong selective pressure to change the course of its ciliary bands, and that this change didn't disrupt their role in locomotion and feeding. It's impossible to judge such a scenario objectively. Hence, it is unsurprising that the appeal to evolutionary laws in support of scenarios has all but disappeared. Although the law of neurobiotaxis has been credited as the first explicitly causal law of brain evolution, its explanatory power is now considered extinct (Nieuwenhuys et al., 1998; Striedter, 2020).

6.4 The Flatlining of Evolutionary Intuitions

Thus far, we have discussed typical examples of how biological laws were used as narrative phylogenetic tools. During the second half of the twentieth century, this explanatory strategy came to be replaced by the parsimony, distance, likelihood, and Bayesian algorithms that drive modern phylogenetic analyses. From the late 1980s onward, this gradually transformed metazoan phylogenetics from a subjective storytelling discipline into a more objective analytical science driven by algorithms, models, and optimality criteria. This had a notable corollary, besides the appearance in papers of such novelties as Materials & Methods sections, data matrices, and statistics. Papers became boring to read. Many were no longer written to tell interesting stories. A once colorful and speculative literature filled with engaging but serious attempts to make sense of body plan evolution, now shriveled into a dull series of phylogenetic progress reports. In depth discussions of character evolution were replaced by lists of statistical measures for clade support, branch lengths, and topology tests. Although these are necessary technical components of molecular phylogenetic analyses, they provide very little that human readers can engage with or get narrative purchase on. This is important because the only way we can make sense of trees, and the only way trees have any meaning for us, is by suggesting or supporting certain evolutionary stories. But molecular phylogenetic papers typically dedicate far more space to analytical technicalities than how the trees might enhance our understanding of evolution.

The phylogenomic landmark papers on metazoan and arthropod phylogeny by Dunn et al. (2008) and Regier et al. (2010) are typical examples. The main text of Dunn et al. is dominated by a discussion of analytical details and tree topology, much of which is redundant with respect to the figures. Body plan evolution is mentioned just three times in relation to the evolution of spicules/chaetae, spiral cleavage, and general body plan complexity. Regier et al. dedicate even less space to discussing what their tree can teach us about arthropod evolution. While they mention some morphological characters, they discuss the evolution of none. Although technical details are crucial, why not relegate these to a technical supplement, and why not reduce redundancy by letting the trees speak for themselves instead of describing them in words? This would free up space to discuss what the trees might teach us about evolution. Page limitations in journals generally put a harder squeeze on the "Discussion" sections of papers than the "Results" sections, but some colleagues have confessed to me that once the phylogenomically rooted metazoan tree is in place, their work will be done. They have no interest in engaging with the challenging, messy, and subjective job of figuring out what evolutionary stories the tree can help us tell. A single tree can of course be compatible with multiple competing evolutionary narratives, but thinking through their details is surely the best type of time machine we can wish for? In any case, it is the only time machine we have.

When a new type of data is subjected to phylogenetic analysis for the first time, one needs to be open minded about the results. But in some of the first generation molecular phylogenetic studies, we find a slack-jawed reverence before the sequenced tree. The famous study of Katharine Field and her colleagues (1988) was the first attempt to gain insights into deep metazoan phylogeny with that most magic of molecules, 18S rRNA. It reported the spectacular result that cnidarians had evolved from protist ancestors separately from bilaterians. We now understand that this was due to stochastic and systematic errors, but the authors used these deeply implausible results as a catapult to launch their paper into the pages of *Science*. When they were challenged by Bode and Steele (1989), who pointed to the unique molecular, ultrastructural, and morphological traits that cnidarians share with bilaterians, Field et al. (1989: 550) unholstered the oldest weapon used in phylogenetic debates: "we should consider that an independently evolved 'animal' would have many of the same features as true metazoans." When the card of convergent evolution is played, debate ends. If the combined weight of the morphological and developmental similarities between cnidarians and bilaterians isn't considered convincing enough to argue against a molecular clade that groups cnidarians with a species of baker's yeast and a ciliate, we are witnessing the death of evolutionary intuitions.

This suspension of zoological background knowledge was not a unique event in the first wave of molecular papers on deep metazoan phylogeny. Three years after the Field et al. paper appeared, Christen et al. (1991b, a) analyzed a more broadly sampled 28S rRNA dataset and found that a clade of sponges, cnidarians, ctenophores, and placozoans was separated from a clade of bilaterians by varying sets of nonmetazoan taxa. Although the conclusion in their paper in *The EMBO Journal* was carefully phrased to suggest that they could "not exclude the possibility that triploblasts and diploblasts arose independently from different protists" (Christen et al., 1991a: 499), they pushed a more heterodox interpretation in a concurrently published conference paper. Although the common descent of all animals "could not be totally excluded" (Christen et al., 1991b: 1), they now promoted the conclusion that diploblasts and triploblasts had evolved from different protist ancestors, in each case possibly involving aggregations of more than one species of protist. The astonishing amount and precision of convergent evolution required to arrive at the eumetazoan grade of organization several times independently remained undiscussed in the papers of Christen et al and Field et al. Similarly, when Ballard et al. (1992) reported their sensational molecular phylogenetic result that onychophorans are nested deeply within arthropods, they expertly hid their excitement and surprise. In the one paragraph of their densely technical paper where the evolutionary implications of this finding are spelled out, they merely muster a single sentence: "These data imply that onychophorans are a highly specialized assemblage, neither a primitive 'missing link' nor an appropriate outgroup for analyzing arthropod phylogenetic relationships." It was up to my colleagues from the Natural History Museum, Richard Fortey and Richard Thomas, to type the words that these results "are so surprising" in a commentary on Ballard et al. (Fortey and Thomas, 1993: 206).

During the infancy of higher-level molecular metazoan phylogenetics, authors needed to keep an open mind and allow the data to speak for itself. Novel findings should not be choked by received wisdom, but in their eagerness for novelty, some authors seem to suspend their critical faculties. When Negrisolo et al. (2004) found that in their molecular tree centipedes were more closely related to chelicerates than to millipedes, they argued that feeding on liquid food could be a possible synapomorphy for this clade because this "peculiar feeding mechanism" is present in the "vast majority of chelicerates and centipedes" (Negrisolo et al., 2004: 778). But how convincing is the homology of centipedes and spiders swallowing liquid food, which centipedes chew and spiders liquefy with regurgitated digestive juices? When the tree of Nardi et al. (2003) showed that hexapods were not monophyletic because collembolans were the sister group to crustaceans and insects, they pointed out that collembolans "show a number of peculiar features" (Nardi et al., 2003: 1888). Since when are traits unique to a taxon evidence against its relationship to others? And when Giribet et al. (2006) found that monoplacophorans were nested deeply inside chitons in a molecular tree with poor clade support values, they chose to cast doubt on the consensus view that monoplacophorans are more closely related to other shell-bearing molluscs by interpreting the fact that these taxa form their shells differently as evidence for "[t]he rejection of conchiferan monophyly based on shell deposition" (Giribet et al., 2006: 7725). Again, as will be discussed further, differences are phylogenetically vacuous.

These papers appeared in *Molecular Biology and Evolution, Proceedings of the National Academy of Sciences of the United States of America*, and *Science*. Could it be that the authors suspended their disbelief, consciously or not, in the interest of publishing their work in high-impact journals? Of course, all they did was to faithfully report the results of their molecular phylogenetic analyses. But as knowledgeable zoologists, shouldn't they have scratched their heads and thought "surely this cannot be right"? I bet they did. But when novelty is the hard currency of high-impact work, there is little incentive to be too critical. Later studies refuted these hypotheses. This is how science works, and how it should work. I do not suggest for a moment that these authors should not have published these analyses. Science can only thrive on what is made public. But it does raise questions about the strategies that we all use, consciously or not, to promote or increase the novelty of our findings. These authors used weak arguments to boost the credibility of surprising molecular phylogenetic results, which smoothed their path into the pages of high-ranking journals. We will explore this topic further in the next section. On the other hand, as my friend Greg Edgecombe put it to me, at the start of the molecular era of metazoan phylogenetics, there really was a feeling that the problem of homology had been solved at the level of the nucleotide. To some this meant that evolution could be read more or less literally from molecular trees. But as molecular trees flooded the literature, clashes with evolutionary intuitions were inevitable.

6.5 Unimaginable Evolution and the Denial of Common Descent

Acoelomorphs and *Xenoturbella* are morphologically simple worms whose phylogenetic position continues to be debated (Giribet and Edgecombe, 2020). Primitively lacking nephridia, centralized nerve cords, brains, and anuses, their placement as the closest relatives to a clade of bilaterians that do have these characters neatly fits with a story of increasing complexity during bilaterian evolution. Molecular evidence for this placement is

contentious, however, with some analyses suggesting that these worms nest inside deuterostomes (Philippe et al., 2011; Kapli and Telford, 2020). Reviewing the evidence in a
2011 paper headed by Greg Edgecombe, a team of nine leading invertebrate morphologists,
paleontologists, and phylogeneticists expressed doubts about such deuterostome affinities
because these worms show no "traces of deuterostome characters, such as gill slits, that one
would expect to be present even in highly derived lineages" (Edgecombe et al., 2011: 158).
Yet, only three paragraphs later they accept that gill slits have been lost without a trace
somewhere along the lineage leading to extant echinoderms. This reveals evolutionary
intuitions to be fickle and feeble phylogenetic tools.

Yet, we cannot avoid evaluating phylogenetic hypotheses from the perspective of our
personal evolutionary background knowledge. There is no shame in that. Debates that are
currently active in the literature – on the evolution of segmentation, coeloms, nerve cells,
choanocytes, and ciliated larvae in invertebrates and on the evolution of reproductive mode
and venom in squamates, to mention just a few – are all animated by authors' differing
ideas about what could and could not have happened during evolution. What is striking is
that the participants in these debates may agree on the systematic relationships of the taxa
and the distribution of the characters involved, yet disagree vehemently about the probabilities of hidden evolutionary events and the nature of unobservable ancestors. These are
issues located in the empirically inaccessible realm of evolutionary inference and imagination. It is typically specialists most knowledgeable about the morphological minutiae of
their chosen organisms who develop the strongest evolutionary intuitions. Let's look at
several examples involving crustacean specialists and other arthropod researchers.

6.5.1 Unbelievable Arthropod Evolution

The eye-opening phylogenomic analysis of Regier et al. (2010) sent shockwaves through the
arthropod research community. Relationships between the major crustacean lineages and
insects were especially unexpected. Smithsonian Institution carcinologist Frank Ferrari
wrote one of the first papers that assessed these results from the perspective of morphology.
Ferrari wasn't happy with the new tree. Beginning his critique with a quote from comedian
Chico Marx – "Well, who you gonna believe, me or your own eyes?" – he drew the narrative
phylogeneticist's most lethal weapon: disbelief. The new tree's grouping of taxa with
dissimilar body plans was unconvincing to Ferrari because "closely related crustaceans
with disparate morphologies may require too many instances of reversals or of secondary
loss of structure to be credible" (Ferrari, 2010: 768). Moreover, the "presence of similar
morphological structures on distantly related crustaceans... is difficult to explain by convergence" (Ferrari, 2010: 768). For Ferrari evolution is apparently not "descent with
modification," but "descent with some, but not too much, modification." Too much
convergence and character evolution breached the boundaries of his evolutionary instincts.
He was particularly troubled by the fact that hexapods and their closest crustacean relatives
in the new tree have such different first trunk limbs. But how closely taxa are related to each
other is irrelevant for judging how similar they are. Hexapods and their closest living
relatives are separated by hundreds of millions of years of independent evolution, a span
of time sufficiently long to turn monkeys into humans many times over.

Ferrari was equally exercised about the fact that mystacocarids and copepods, a group that he is an expert on, have almost identical configurations of endites on the protopods of their first trunk limbs, but they are not sister groups in this molecular tree. Spelling out his intuition, Ferrari (2010: 768) wrote: "If convergence is functionally driven, as is usually assumed, then the morphology derived from convergence is not expected to be identical or even similar." This is not a reliable intuition. One resounding insight that has emerged from the growing forest of molecular trees is that independently evolved similarity is more pervasive and impressive than anyone ever expected. More recent independent phyloge-nomic analyses have confirmed the findings of Regier et al. (2010) in broad outline and cemented most of the relationships that were unacceptable to Ferrari (Schwentner et al., 2017, 2018; Lozano-Fernandez et al., 2019).

In 2009 a British carcinologist unleashed even more inscrutable evolutionary intuitions to propose a truly bizarre hypothesis. The cover of Donald Williamson's 1992 book *Larvae and Evolution: Toward a New Zoology* caught my attention as it sat on the shelves of my PhD supervisor Fred Schram at the University of Amsterdam. When I read it, I discovered that Williamson had a eureka moment in 1990 when he thought he had obtained experimental proof that sea squirt eggs could be fertilized with sea urchin sperm and that this inter-phylum fusion produced babies. Baby sea urchins to be precise. Williamson elaborated this extraordinarily unlikely result into a fully fledged theory of what he called *hybridogenesis*, or larval transfer. Although his eccentric ideas were mostly greeted with silence, they suddenly reached an august readership when he managed to publish a paper in the *Proceedings of the National Academy of Sciences of the United States of America* with the title "Caterpillars evolved from onychophorans by hybridogenesis." Williamson argued that caterpillars were inserted into the life cycle of insects when a velvet worm met an insect in the Upper Carboniferous and, instead of eating it, decided to mate with it (Williamson, 2009). The resulting hybrid baby hatched as the first holometabolous insect, with the onychophoran genome producing a caterpillar-like larval stage, but the adult insect was too different for a smooth integration of these two phases. Therefore, just as oil and vinegar will combine into a smooth emulsion after vigorous shaking, Williamson proposed this hopeful monster evolved a pupa so that larva and adult could smoothly coalesce into a single life cycle in a "pupal soup" of undifferentiated cells (Williamson, 2009: 19903).

Unsurprisingly, ridicule and condemnation of the paper and the editorial process that had allowed its publication were swift to follow. Not everyone was unsympathetic, however. Williamson's colleague Richard Hartnoll pointed out that Lynn Margulis, whose surpass-ingly soft touch had shepherded the paper through peer review, was "a fellow scientific rebel" (Hartnoll, 2016: 410). He believed that Williamson's ideas deserved to be rigorously tested. As an initial test, Williamson himself suggested this fabulous experiment: "it should be possible to attach an onychophoran spermatophore to the genital pore of a female cockroach and see if fertilized eggs are laid" (Williamson, 2009: 19904). Others quickly pointed out that our background knowledge was already sufficient for the outright rejection of his hypothesis (Giribet, 2009; Hart and Grosberg, 2009).

Williamson's hypothesis was conceived on the interface of disbelief and gullibility. He thought that the larvae and adults of many animals are too dissimilar for them to have evolved in a single lineage through descent with modification. Instead, he thought they had resulted from the fusion of adult organisms from distantly related lineages, with one of them becoming a new larval form. Seen from Williamson's perspective, the evolution of animal body plans is a grotesque reproductive bacchanal. Cephalochordates coupling with lampreys to become ammocoetes larvae. Frogs and salamanders mating with ancient agnathans to spawn tadpole larvae. Dragon flies and beetles copulating with wingless insects to acquire wingless larvae. And primitive hemichordates hybridizing with cnidarians to create pterobranchs with lophophores derived from cnidarian tentacles, an idea that was conceived by the fertile mind of Lynn Margulis (Williamson, 2012, 2013).

In Williamson's world, rhizocephalans, the parasitic barnacles we met in Chapter 5, are chimeras with three fused genomes that resulted from a crustacean parasite mating with its host, which then hybridized with a barnacle (Williamson, 2015). These examples by no means exhaust the bastardly bestiary of Williamson's zoology. In 2012 I reviewed one of his papers for the *Zoological Journal of the Linnean Society*. After offering unavoidably pointed criticisms, I concluded my review with a phrase that the late Christopher Hitchens used so effectively in a different context: "What can be asserted without evidence can also be dismissed without evidence" (Hitchens, 2007: 150). After lingering for more than two decades on the margins of zoology, and with fellow carcinologist Frank Ferrari being responsible for one of the few positive citations of Williamson's work (Ferrari et al., 2011), Williamson's theory of horizontal larval transfer petered out in two articles that appeared in the journal *Crustaceana* in the last years of his life (Williamson, 2014, 2015).

What did not perish were the evolutionary intuitions that had triggered him to propose his theory in the first place. Williamson's scenarios are sandwiched between extreme versions of two ideas that emerged during the infancy of narrative phylogenetics in the second half of the nineteenth century. The first is that observed differences between organisms are evidence against homology and monophyletic origins, in Williamson's case denying the emergence of different looking adults and larvae within a single evolving lineage. The second is that the explanatory power of phylogenetic hypotheses resides in identifying precursors to evolutionary novelties, in Williamson's case tracing larval forms back to similar looking adults from unrelated lineages. We saw in Chapter 5 that the less extreme version of this idea is the beating heart of narrative phylogenetics.

6.5.2 Differences Are Not Phylogenetic Evidence

The idea that differences are phylogenetic evidence has endured since the birth of phylogenetics, but it is logically flawed and continues to divert debates in fruitless directions. The logic of evolutionary descent and the nature of phylogenetic evidence is simple. As sister lineages diverge they become dissimilar, with retained traits becoming phylogenetic evidence for their common origin. Accumulating dissimilarities do not erase descent. Dissimilarities lack phylogenetic valence because they are compatible with hypotheses of both common descent and independent evolution. Only the tracing of traits on a phylogenetic tree can distinguish common or separate origins. Dissimilarity is nothing more than the absence of evidence for homology and monophyly, not evidence for their absence. This is why no modern phylogenetic method takes dissimilarity into account. The use of this logically flawed argument sharply separates the practice of narrative phylogenetics from modern quantitative phylogenetics. Nevertheless, it persists in the literature, but few authors have drawn explicit attention to it (Jenner, 2006; Strausfeld and Hirth, 2013).

A conspicuous example of this flawed strategy led to the widespread claim by morphologists in the late nineteenth century that arthropods were polyphyletic (Bowler, 1994, 1996). As we saw in the Chapter 5, this argument was rehearsed in the mid-twentieth century and beyond by functional morphologists, especially Sydnie Manton and her influential carcinologist converts Don Anderson and Geoffrey Fryer, as well as my PhD supervisor Fred Schram (Anderson, 1973; Schram, 1978; Fryer, 1996) (Figure 6.5). They pointed out the myriad ways in which legs, mandibles, eyes, embryonic fate maps, cleavage patterns, and more differed between the major arthropod groups. These authors couldn't or wouldn't imagine how these differences could have evolved from common ancestors. When striking similarities between groups couldn't be denied, these were defused with appeals to convergent evolution caused by evolutionary constraints, or by similarly acting selection pressures. This is a standard ploy used throughout the history of narrative phylogenetics, but the hunt for dissimilarities as evidence against homology was futile. With arthropod monophyly now firmly established and fossil evidence painting an increasingly detailed picture of early arthropod evolution, these spurious arguments for polyphyly betray a lack of imagination and a willingness to use gut feelings to extract conclusions from limited data.

The apotheosis of this kind of narrative phylogenetics is the work of another arthropod researcher, Pat Willmer. In several papers and an often maligned book, she fell prey to a form of phylogenetic defeatism in which she saw metazoan phylogeny dissolve in a sea of dissimilarities (Willmer, 1990; Willmer and Holland, 1991; Moore and Willmer, 1997). Some of her arguments seem harmless enough, reflecting the kind of subjective thinking that remains common among taxonomists. She considered onychophorans "different enough" to deserve being liberated from Manton's phylum Uniramia, which also included hexapods and myriapods, and "merit the higher rank of phylum in their own right" (Willmer, 1990: 297). This kind of taxonomic inflation is commonly produced by taxonomists creating new supraspecific taxa, but it lacks any objective phylogenetic basis and has no meaning for comparative analyses. Only sister lineages are objectively comparable. Any discrepancy in their Linnean ranks matters not. This is why it is amusing to see modern biologists argue that the Linnean rank of urochordates deserves to be elevated from subphylum to phylum because they show a "fundamental difference" (Cameron et al., 2000: 4471) compared to their sister group. Others (Satoh et al., 2014; Irie et al., 2018) went further and proposed that cephalochordates and vertebrates also deserve to be elevated to phyla because of the "profound dipleurula versus tadpole larval differences" (Satoh et al., 2014: 1) that separate chordates from echinoderms and hemichordates, two taxa that are already ranked as phyla. This type of reasoning becomes harmful when it is misused to deny common descent.

Phrases such as "fundamental differences" and "fundamentally different" are the flawed refrains of those who try to refute homology and monophyly by diagnosing dissimilarities. They hint at a threshold of morphological difference beyond which it is reasonable to reject hypotheses of common descent of taxa and traits. This hollow tune echoes throughout the history of narrative phylogenetics. There are "fundamental differences" between the body

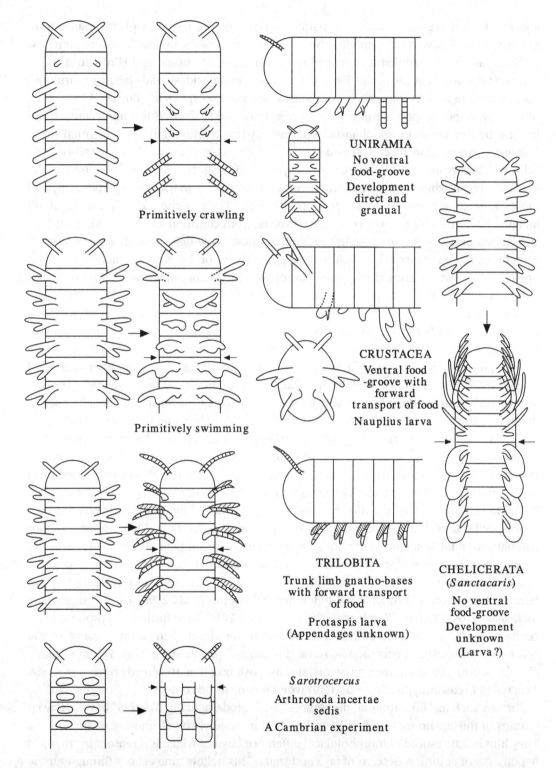

Figure 6.5 The independent evolution of arthropod body plans from segmented worm ancestors, with easy-to-imagine transformations of simple unsegmented appendages into arthropod limbs (after figure 2 in Fryer, 1996, by permission of Oxford University Press).

plans of sponges and other animals; ergo sponges and eumetazoans may have separate origins in protists (Harvey, 1961: 120), an argument that traces back to 1884 (Sollas, 1884). The origin and development of primordial germ cells in urodeles and anurans show "fundamental differences"; ergo extant amphibians evolved twice from different fish ancestors (Nieuwkoop and Sutasurya, 1976: 164). The nervous systems of trochophore larvae from different phyla show "fundamental differences"; ergo they evolved convergently (Nezlin, 2010: 381). The simultaneous appearance of serially repeated muscle cells in developing molluscs and their asynchronous appearance in developing annelids are "fundamental differences"; ergo this seriality is not homologous (Wanninger and Wollesen, 2015: 128). Again, there are apparently "fundamental differences" in the morphology and function of sponge choanocytes and choanoflagellate cells; ergo "homology cannot be taken for granted" (Mah et al., 2014: 25). Of course, what "fundamental" means is never specified. The word is simply used as an adjective to boost the believability of the author's preferred interpretation.

Darwin warned us in the *Origin* that dissimilarities cannot dissolve homology. Writing about the polymorphic zooids of bryozoans, he noted "[i]t is not easy to imagine two objects more widely different in appearance than a bristle or vibraculum, and an avicularium like the head of a bird; yet they are almost certainly homologous and have been developed from the same common source" (Darwin, 1876: 194). Indeed, our modern understanding of these maximally dissimilar structures agrees that they are homologous modifications of zooid opercula (Taylor, 2020: 77), something Darwin surmised "no one with the most vivid imagination would ever have thought" (Darwin, 1876: 194). Later in his book, Darwin is even more explicit about the phylogenetic impotence of differences. In the glossary, he defined homology as "[t]hat relation between parts which results from their development from corresponding embryonic parts" (Darwin, 1876: 434), but "dissimilarity in embryonic development does not prove discommunity of descent" (Darwin, 1876: 396). Lineages diverge by the limitless accumulation of differences that can never erase their common origin. This is why the search for similarities or correspondences between taxa is the only torch that we have to illuminate common descent. As we saw in Chapter 2, it guided Geoffroy Saint-Hilaire to the unity of composition in the early nineteenth century, and it is enshrined in Hennig's auxiliary principle, which assumes homology in the absence of evidence to the contrary. Differences cannot be evidence for phylogenetic disunity. This idea is at the epistemological heart of all modern evolutionary and comparative biology, and it should brook no dissent. Except, perhaps, if you are a creationist.

Baraminology is a creationist pseudoscience that is epistemically antithetical to phylogenetics. It strives to discover the unbridgeable gaps that separate created kinds, which it calls *baramins*. I am aware of only one baraminologist who managed to publish a paper in a peer-reviewed journal (Wood, 2011). It was written in response to a study by a scientist who used baraminological methods to undercut creation science. However, a quick internet search will reveal that baraminology is an active – I hesitate to say thriving – research field. Flawed phylogeneticists from the dark side, baraminologists attempt to quantify differences between taxa and use an arbitrary cutoff as spurious evidence against common descent. The first sentence of the conclusion of a recent baraminological paper from *Creation*

Research Society Quarterly is emblematic for the type of foregone insights that this kind of research produces: "The results of this study yield another striking confirmation of the biblical narrative" (Lightner and Cserhati, 2019: 138). The paper's title touts its prize finding: "The uniqueness of humans is clearly demonstrated by the gene-content statistical baraminology method." If you adhere to a creationist metaphysics, it makes a certain amount of sense to consider differences between taxa as evidence for their separate origins. But in an evolutionary world, this strategy is logically flawed. It is therefore surprising and somewhat disheartening to discover that evolutionists continue to employ this flimsy form of reasoning to hype the impact of their discoveries. The ongoing debate about the origin of the animal kingdom is especially infected with it.

6.5.3 The Contested Homology of Collar Cells

The origin of Metazoa and the identification of the earliest diverging animal lineage continue to be fascinating puzzles. One central question is whether sponge choanocytes and choanoflagellate cells are homologous. These collar cells possess a motile flagellum encircled by a collar of microvilli. They function in feeding, with the microvilli capturing food particles from the water flow created by the beating flagella. The proposed homology of these cells has been at the heart of phylogenetic scenarios for the origin of animals since the late nineteenth century, a topic discussed further in Chapter 8. This remains so today, with the sister group relationship between choanoflagellates and animals being robustly supported by molecular phylogenetic analyses (Brunet and King, 2017; Laundon et al., 2019). For more than a century, no serious doubt has been cast on this homology. In early theorizing, sponges were often considered to be so different from other animals that their common origin was doubted. Sollas (1884) placed sponges in a taxon Parazoa with a choanoflagellate ancestry, while the remaining metazoans were thought to originate independently from different protists (Lankester, 1900). Indeed, the separate origin of sponges from choanoflagellate-like ancestors was accepted by many zoologists well into the second half of the twentieth century (Harvey, 1961; Hadži, 1963; Willmer, 1990).

In 2014 the consensus about the homology of these collar cells changed abruptly when sponge expert Sally Leys and her colleagues published a morphological study that compared a marine choanoflagellate with the choanocytes of a freshwater sponge (Mah et al., 2014). Although they "found notable similarities in the functional morphology" (Mah et al., 2014: 35) of these cells, they also sang the flawed refrain of "fundamental differences" five times in their paper, before concluding "that homology cannot be assumed without question" (Mah et al., 2014: 35). One example of a supposedly fundamental difference is that the amplitude of the flagellar beat of the sponge cells is affected by the surrounding collar of microvilli, whereas that of the choanoflagellate cells is not. Another example, which is no doubt related, is that the microvilli of the sponge cells are joined together more tightly into a collar by surrounding glycocalyx than those of the choanoflagellate cells. To my eyes, these and the other reported differences are very minor. If these cell types are indeed homologous, they are an amazing case of evolutionary conservation between a marine protist and a freshwater sponge separated by at least 1.2 billion years of independent evolution. If they aren't homologous, their similarities reflect an astonishing degree of convergent evolution.

A year later, Leys coauthored a review paper in which the carefully stated conclusion of the previous year had hardened into an all-out rejection of the homology of sponge and choanoflagellate collar cells. Citing the 2014 paper and changing adjectives, the differences between these cells were now labeled as "key," "primary," and "significant" and were "consistent with multiple independent origins" (Dunn et al., 2015: 287). Any similarities between these cells were now "largely superficial" (Dunn et al., 2015: 285), and the hypothesis of their homology was now an "entrenched misconception" (Dunn et al., 2015: 287). What had happened? Leys' collaborators were Casey Dunn and Steven Haddock, who had been coauthors on one of the most influential molecular phylogenetic papers of the new millennium. In 2008 they published the phylogenomic tree of the Metazoa mentioned before, which surprisingly placed ctenophores as the sister group to all other animals (Dunn et al., 2008). With sponges being placed deeper in the tree, this suggested that ctenophores had either lost collar cells or that collar cells in sponges and choanoflagellates were not homologous. This unexpected result no doubt fostered their promotion of the second hypothesis.

This would be legitimate if they had performed an ancestral state reconstruction on the new topology and found that this suggested independent origins of collar cells. But they didn't do that. Instead, they used a flawed narrative phylogenetic strategy to advance a favored hypothesis, which unsurprisingly, breaks with traditional views. Debate about basal metazoan relationships continues today (King and Rokas, 2017; Simion et al., 2017; Whelan et al., 2017; Erives and Fritzsch, 2020; Kapli and Telford, 2020; Li et al., 2021; Redmond and McLysaght, 2021), but I don't think that explicit ancestral state reconstruction will be able to decide the issue even if a stable topology emerges. The alternative hypotheses differ in so few steps that maximum parsimony is unlikely to be able to decide the issue to anyone's satisfaction. The internal branches of phylogenomic trees at the base of the metazoan radiation are very short, which will limit the amount of information available for informing ancestral state reconstruction based on probabilistic methods. This debate will therefore stay in the realm of evolutionary storytelling for the foreseeable future. This makes it all the more important to stick to valid arguments.

Unfortunately, other researchers have become equally beguiled by differences. Pozdnyakov et al. (2017) sought to assess homology of choanoflagellate and sponge collar cells by studying the ultrastructure of the flagellar apparatus. They concluded that homology is not supported because, you will be unsurprised to read, the flagellar apparatus in these groups is "fundamentally different" (Pozdnyakov et al., 2017: 250). By this they mean that although there are definite similarities there are also differences. The most recent example of this flawed argument at the time of writing comes from a team of researchers led by Bernie and Sandie Degnan from the University of Queensland in Australia. They published a paper in *Nature* that compared the transcriptomes of three cell types from the sponge *Amphimedon queenslandica* with those of several nonmetazoans, including two choanoflagellates (Sogabe et al., 2019). The sponge cell types were archaeocytes, which are pluripotent stem cells, pinacocytes, which are epithelial lining cells, and choanocytes, which can transdifferentiate into archaeocytes. They found that the gene expression profile of archaeocytes was more similar to the expression profiles of the choanoflagellates than the profiles of pinacocytes and choanocytes. From this they concluded that "these analyses argue against homology of sponge choanocytes and choanoflagellates," citing in support the morphological studies of Mah et al. (2014) and Pozdnyakov et al. (2017).

Although Sogabe et al. claimed their study was a test of this homology hypothesis, it wasn't. The tree they present as the comparative framework for their study is consistent with a single origin of collar cells in the last common ancestor of choanoflagellates and animals, and their results are consistent with the hypothesis that the transcriptomic differences between these taxa are a result of descent with modification. Indeed, their conclusion that the last common ancestor of animals "had the capacity to exist in and transition between multiple cell states in a manner similar to modern transdifferentiating and stem cells" (Sogabe et al., 2019: 522) seems to suggest just this, with the collar cell type being one of such states. Instead, they were blinded by differences and contributed to hyperbolic press reports that their paper required the rewriting of textbooks (https://stories.uq.edu.au/news/article/evolutionary-discovery-rewrite-textbooks/index.html; last accessed September 10, 2021). Sadly, but unsurprisingly, at the time of writing their results are percolating into the peer-reviewed literature to cast further spurious doubt on the homology of choanocytes and choanoflagellates (Ruiz-Trillo and de Mendoza, 2020).

6.5.4 The Endless Divergence of Lineages

In 2015 I took part in a conference on the origin of the animal kingdom. I gave a talk in a session chaired by Bernie Degnan in which I laid out the arguments outlined here and cautioned against embarking on the slippery road to pseudoscience that one enters when trying to refute common descent by pointing out differences between taxa. It was to no avail. The hunt for differences as evidence against homology is seductive and widespread. You are guaranteed to find them, especially in comparative transcriptomic studies. Leonid Moroz and his colleagues are using essentially the same strategy of emphasizing differences over, admittedly limited, similarities to argue that the nervous system of ctenophores has evolved independently from that of other animals (Moroz and Kohn, 2016). But the observation of differences cannot be a test of homology. There is no logical basis for interpreting an arbitrary degree of dissimilarity as the point at which support for a hypothesis of homology weakens and disappears, let alone becomes evidence against homology. When phylogenetic analysis shows that the origin of a trait maps to a single lineage, such as collar cells mapping to the ancestral lineage of Choanozoa (choanoflagellates and animals), it is difficult to make the case that the trait is nevertheless nonhomologous based on observed differences. The homology of complex traits is highly dissociable between different levels of organization so that aspects of the developmental and genetic architectures of homologous phenotypic traits can diverge to the point of total dissimilarity (Moczek et al., 2015; DiFrisco and Jaeger, 2021). In other words, nonhomologous genes can be involved in the formation of homologous phenotypic traits, and we can expect such developmental systems drift to have especially large effects over long periods of evolutionary time.

Phylogenetic explanations are lineage explanations (Calcott, 2009), which require lineage thinking. This means that when the origin of a trait maps to a single lineage on a tree, one should at least ask this question: could the observed differences between taxa be due to descent with modification? The authors discussed in this section seem not to have asked themselves this question at all. Instead of lineage thinking, they use outdated typological thinking. The answer to this question does not need to be "yes" in every case. In a paper where they discuss how to test hypotheses of homology, Andi Hejnol and Chris Lowe (2015: 7) ask what we should conclude about the homology of wings if bats and birds turn out to be sister groups: "We would probably conclude that wings are also homologous and reconstruct a wing in their last common ancestor, including the underlying gene regulatory network." Here they echo the answer that Mário de Pinna (1991: 375) gave when he posed this question in his classic paper on homology. But should we really accept a hypothesis of homology here? I think not.

Other than being constructed on homologous forelimbs, bird and bat wings share no unique structural similarities at all that would warrant a hypothesis of wing rather than just forelimb homology. Different digits are involved, and the wing blades are formed of nonhomologous materials (feathers versus skin) that are organized differently. This doesn't mean that this dissimilarity disproves homology; it just suggests that there is no evidence to propose the homology of these wings in the first place. Not even the most unbridled lineage thinking is likely to find a reasonable route from one to the other, or from a common ancestral wing while maintaining functionality as a wing. Phylogenetic congruence can only test primary homology hypotheses for which there is observable evidence (Rieppel and Kearney, 2002). The misuses of differences in the examples discussed earlier all involve characters for which there are sufficient grounds to propose primary homology.

Narrative phylogenetics operates on the basis of the expectation that, in a world of perfectly preserved evidence, we should be able to trace evolutionary lineages along graded series of morphologies. In our world of imperfectly preserved evidence, this unattainable ideal can nevertheless be pursued by arranging observed and hypothesized character states into linear transformation series that connect hypothetical ancestors to their presumed descendants. It is important to realize that this goal is the same for all phylogeneticists, narrative and modern. In a paper on the evolution of animal guts, Hejnol and Martín-Durán (2015: 68) wrote that in the "pre-cladistic era, zoologists tried to explain evolution by assembling stages of evolutionary transformations of observed morphologies of different species... [a] misleading approach." But in a profound way we still do this today. The difference is that we now use preestablished trees to guide the inference of hypothetical ancestors and the construction of transformation series. But the unobservable evolution of lineages can still only be grasped by the mental recreation of linear transformation series rooted in observed characters.

Narrative phylogenetics uses shortcuts, which indeed become indefensible when they involve ignoring available phylogenetic frameworks. When such frameworks aren't available, which was the case for more than a century for biologists interested in reconstructing the evolution of animal body plans, the narrative approach offers a speculative way forward. It involves positing hypothetical ancestors with precursors suitable for explaining the origin of traits of interest. And it involves the use of evolutionary intuitions to propose graded morphological series that are interpreted as hypothetical lineages. In the next three chapters, we will explore how paleontologists and biologists have practiced narrative phylogenetics to trace animal evolution, especially in the late nineteenth and early twentieth centuries. We will see how they filtered fossil, embryological, and morphological evidence through their evolutionary intuitions to trace evolving body plans. And although the era of narrative phylogenetics drew to a close in the mid-twentieth century, we will see that evolutionary storytelling is as alive today as it ever was.

7

Telling Straight Stories with Fossils

7.1 Fossils and the Arrow of Stratigraphy

Ancestors live in the fossil record. Haeckel (1866a: 30) chose "scientific paleontology" as a synonym for his science of phylogenetics to reflect the privileged status of the rock record as the only direct archive of evolutionary history. The hunt for fossil ancestors offered exciting prospects. T.H. Huxley was inspired to search for fossil clues for the origin of birds after reading Haeckel's *Generelle Morphologie*. Huxley understood that although *Archaeopteryx* and the dinosaur remains known at the time were not "the animals which linked Reptiles and Birds together historically and genetically" (Huxley, 1868: 246), these fossils nonetheless "help us to form a reasonable conception of what these intermediate forms may have been" and "enable us to form a conception of the manner in which Birds may have been evolved from Reptiles" (Huxley, 1868: 246). Others mounted expeditions in the hopes of discovering fossils that would record frozen acts of evolutionary origins.

Haeckel predicted the existence of a missing link between apes and humans, an idea that inspired a young Dutchman to embark on a career-defining journey of discovery in Southeast Asia. It was on Java where Eugène Dubois discovered the remains of what is now known as *Homo erectus*. He described it in 1894 as *Pithecanthropus erectus* in honor of the hypothetical, upright walking, and speechless human ancestor that Haeckel had called *Pithecanthropus alali* (Shipman, 2001; Caspari and Wolpoff, 2012). But single fossils generally don't and, many would argue, can't shed such dramatic light on phylogenetic origins. The arrow of evolutionary time more clearly emerges from the stratigraphic arrangement of fossils.

The broad-brush progressive nature of the fossil record had already been firmly established by the early nineteenth century, and it showed that the history of life had a discernable direction (Secord, 1991; Rudwick, 2008; Jenkins, 2016). In 1851 the Austrian botanist Franz Unger published one of the first series of images that depicted successive Mesozoic and Cenozoic scenes of plant and animal life (Rudwick, 1992). The young Darwin had been greatly impressed by the geological succession of fossil vertebrates and their similarity to species now living in the same geographic areas. Marsupial fossils from Australian caves showed striking similarities to living species. Likewise for fossil birds from Brazilian caves. South American fossils of giant armadillos and sloths, although monstrously large, called to mind diminutive living counterparts. The same was true for New Zealand's gigantic extinct birds. Such discoveries stoked Darwin's transmutationist thinking during the voyage of the *Beagle*, and he came to understand them as instances of a "law of the succession of types" in his early years of theorizing (Darwin, 1859: 338; Brinkman, 2010; Eldredge, 2010). With his theory of descent with modification, these findings were "at once explained" (Darwin, 1859: 339). The fossils indicated the phylogenetic fuses of the modern biota.

Others too had noticed this link between the extinct and the living, and these insights had fed into the emergence of transformist theories and speculations before Darwin's. Three of the earliest attempts to create phylogenetic lineages linking fossils to living species focused on molluscs. Lamarck led the way in the first decade of the nineteenth century by connecting Tertiary fossils to living species, a strategy duplicated in 1814 by the Italian paleontologist Giambattista Brocchi, and in 1827 by one of Cuvier's former students, and coiner of the word paléontologie, Henri de Blainville (Appel, 1980; Rudwick, 2008; Dominici and Eldredge, 2010; Eldredge, 2010; Podgorny, 2017). In 1828 Geoffroy Saint-Hilaire proposed that a much older fossil, the Jurassic crocodile-like *Teleosaurus*, might be a link in the lineage leading to modern crocodiles. More adventurously, he connected reptiles to mammals in a rather uneven progression from a fishy ichthyosaur to the elephantine *Mastodon* via an unlikely series of intermediates: plesiosaur-pterodactyl-mosasaur-*Teleosaurus*-*Megalonyx*-*Megatherium*-*Anoplotherium*-*Palaeotherium* (Appel, 1987: 131–132; Le Guyader, 2004: 88–95, 225–226; Rudwick, 2008: 239–242). This aquatic → aerial → aquatic → terrestrial transformational series was an unconvincing sequence of alternating steps forward and backward in geological time. More convincing fossil lineages would have to track the arrow of geological time more faithfully and present a smoother gradation of morphological steps.

Identifying fossils as ancestors and interpreting stratigraphic series of fossils as evolutionary lineages constitutes a phylogenetic strategy referred to as the paleontological method, argument, or approach (Hennig, 1966: 140; Nelson and Platnick, 1981; Rosen et al., 1981; Eldredge and Novacek, 1985; Forey, 2004; Williams and Ebach, 2004). The stratigraphic ordering of fossils has been the traditional paleontological method for establishing ancestor–descendant relationships since the 1860s. It was in common use for a century before cladistic methods started to percolate through phylogenetics and substituted branching order for geological precedence as the index of evolutionary time (Schaeffer et al., 1972; Patterson, 1981; Forey, 2004). Two quotes from leading paleontologists from successive generations illustrate the basic simplicity of the paleontological method and the high hopes that were pinned on it.

In 1959 the American Philosophical Society published a celebratory issue of its house journal to commemorate the centennial of the publication of the *Origin*. It included an article titled "The nature of the fossil record," written by Norman D. Newell, one of whose claims to fame was to have trained, among others, three leading lights of the next generation of paleontologists: Niles Eldredge, Stephen Jay Gould, and Steven Stanley. In his article Newell gave a concise summary of the relationship of fossils and phylogeny: "Phylogenies in paleontology are inferred from morphological series distributed chronologically through the stratigraphic succession" (Newell, 1959: 275). Almost three decades later, British paleontologist Anthony Hallam published a chapter in the 1988 volume of the

Systematics Association Special Volume Series. In his chapter titled "The contribution of palaeontology to systematics and evolution," Hallam (1988: 131) wrote: "Notwithstanding the incompleteness of the fossil record, the relative stratigraphic order of fossil species is overwhelmingly correct and can be used as an independent test of phylogenies reconstructed on the basis of morphological comparisons," and if "there is no range overlap, for forms considered very similar, then the stratigraphic sequence coincides exactly with the phylogenetic sequence deduced from morphology" (Hallam, 1988: 131).

Although it was a widespread assumption at the heart of the paleontological method that ancestors could be recognized in the fossil record (Forey, 2004: 161), Hallam's trust in the veracity of the stratigraphic record was extreme, especially at a time that came on the heels of the cladistic revolution and the fundamental reinterpretation of the phylogenetic significance of fossils. Nevertheless, in my conversations with paleontologists over the years, especially with early baby boomers born in the 1940s and early 1950s, I have repeatedly heard the sentiment expressed that if phylogenies are in conflict with the arrow of stratigraphy, they need to be interpreted with caution. Incidentally, the architects and best-known advocates of the explicit stratigraphy-informed phylogenetic methods known as stratophenetics and stratocladistics – University of Michigan colleagues Phil Gingerich and Dan Fisher (Gingerich, 1979; Fisher, 2008) – are of this generation as well. There may be nothing here, but perhaps their efforts to highlight the continued importance of stratigraphy for phylogenetics are the result of a tension between what they had learned as undergraduates and what they encountered as graduate students during the years when the cladistic revolution was gathering pace. According to a strict interpretation of cladistic analysis, the arrow of evolutionary time tracks the branching order of taxa, irrespective of their stratigraphic addresses.

If the fossil record were complete enough, paleontologists should be able to discover finely graded series of fossils organized along local stratigraphic continua. In the first two decades after the publication of the *Origin*, a handful of studies appeared that did precisely that, with those on freshwater snails, ammonoids, and sea urchins being among the best-known cases (Russell, 1916: 359–360; Rudwick, 1985: 254–256; Reif, 1986). Early studies tracing lineages in densely sampled local stratigraphic sections were rare because such richly preserved fossil assemblages were rare. One paleontological pioneer, the Austrian paleontologist Melchior Neumayr (1845–1890), hoped to trace lineages by a literal reading of the fossil record. After having traced the gradual transition between freshwater snail species in Tertiary deposits in Slovenia, he wrote in 1889 that "for about 10 years, I have been mainly occupied with the transmutation of species and in all this time I found only a single example in which the production of series of forms was so easy that it would have been recognized by other observers without any difficulty" (cited in Tamborini, 2017: 43). Wilhelm Waagen (1841–1900), who supervised Neumayr as a student, invented both a term and a notation to capture the paleontological basis of phylogenetic lineages. He coined the term "mutation"[1] as the temporal equivalent of the geographic term "variety" (Waagen,

---

[1] It is interesting to realize that in contrast to its modern genetic meaning, the word mutation in biology originally referred to small morphological changes along phylogenetic lineages. In his 1920 Presidential address to the geological section of the British Association for the Advancement of Science, Francis Arthur

1869: 186) and defined it as the smallest morphologically distinguishable variant in a geological succession of forms (*Formenreihe*). However, in contrast to the limited systematic significance of geographic variation, the temporal variation recorded by Waagen's mutations, no matter how minute, indicated the very arrow of evolutionary descent.

Assuming a genetic relationship between forms in a *Formenreihe*, Waagen suggested a notation that simultaneously indicated the link of a mutation to its ancestor as well as their relative stratigraphic position in the form of a square root symbol with the ancestral form under the horizontal and the mutation above it. This notation allowed readers scanning the pages of a text to quickly recognize ancestor–descendant relationships among fossils. Waagen applied his phylogenetic approach to trace lineages of ammonoids using fossils collected all over Europe. To those who denied the significance of tracing minute morphological variations along series of fossil forms, Waagen responded that such work was the beating heart of stratigraphic paleontology, without which morphologically disparate forms could never be proven to be connected (Waagen, 1869: 189–190).

The number of paleontologists who enthusiastically embarked on phylogeny tracing immediately post-*Origin* was limited, and their views, especially on evolutionary mechanisms, often diverged from what could be called Darwinian (Rainger, 1985; Rudwick, 1985; Reif, 1986; Cohen, 2017). Waagen, for instance, was no fan of natural selection. His views more closely aligned with those who would later be called orthogeneticists. He thought his ammonoids evolved according to laws inherent in the organisms themselves. External circumstances could aid these intrinsic evolutionary tendencies, but more often they would "delay, prevent, or even suppress them to cause regressions. Only the law of evolution [Entwicklung], inherent in the organism, can never be destroyed, always striving, when sometimes also in a different direction, again upwards to the greater complexification of organs, to a more perfectly developed form" (Waagen, 1869: 239). The French paleontologist Jean Albert Gaudry (1827–1908), who likewise rejected natural selection, traced mammal evolution in local Miocene faunas in Greece and France and also reconstructed fossil-based lineages across much vaster geographical and time scales (Rudwick, 1985; Tassy, 2011; Cohen, 2017). However, by the time he became professor of vertebrate paleontology at the Museum National d'Histoire Naturelle in 1872, Gaudry had been barred from that chair for more than a decade because of his transformist and Darwinian leanings (Cohen, 2017). When Vladimir Kovalevsky (1842–1883) came to Paris in 1871 to study with Gaudry, he wrote to Darwin that "Darwinism is a bad smell in the Jardin des Plantes" (Cohen, 2017: 8). Like Gaudry, Kovalevsky treated fossils as ancestors that could be arranged into stratigraphically framed evolutionary lineages, the most famous of which traced the ancestry of modern horses. But unlike Gaudry, Kovalevsky interpreted evolution through the lens of natural selection, offering adaptive hypotheses that linked changes in anatomy to changing environmental conditions.

Bather complained that Hugo "De Vries has unfortunately robbed palaeontologists of the word 'mutation,' by which following Waagen, they were accustomed to denote such change" (Bather, 1920: 71).

These phylogenetic forays into the fossil record in the 1860s and 1870s earned them exuberant praise from Darwin's camp (Desmond, 1982: 165-166). In his 1870 Presidential address to the Geological Society, T. H. Huxley (1870: 47) hailed Gaudry's monograph on a Miocene mammalian fauna from near Athens "as one of the most perfect pieces of palaeontological work I have seen for a long time." Louis Dollo (1857–1931), the influential Belgian paleontologist famous for the eponymous evolutionary law, honored Kovalevsky's paleontological work by using it to symbolize the last of the three phases into which he had divided the history of paleontology and which he called the evolutionary or definitive epoch (Gould, 1998: 148).

7.2 Hilgendorf, Hyatt, and the Steinheim Snails

A notable early attempt to trace phylogenetic lineages along a densely sampled stratigraphic column dates from the early 1860s. The German paleontologist Franz Hilgendorf (1839–1904) isn't famous in Japan, but should be. He was the first visiting scientist there to lecture about Darwinian evolution (Yajima, 2007). He worked on a Miocene fauna of freshwater planorbid snails from a meteoritic impact crater lake in Steinheim am Albuch in southern Germany. Based on fieldwork conducted in 1862, he summarized his findings in an unpublished PhD thesis in 1863. Although this thesis didn't contain any figures, Hilgendorf did draw a fossil-based phylogeny on a card upon which he glued snail shells. He finally published this phylogenetic diagram in 1866, with snail specimens arranged into several diverging ancestor–descendant lineages, and with the fossils oriented from older to younger (Reif, 1983; Janz, 1999; Glaubrecht, 2012; Rasser, 2013) (Figure 7.1). Darwin cited this research in the fifth and sixth editions of the *Origin* as a rare example of a geologically traceable series of subspecies (Hilgendorf initially considered his specimens to be varieties or subspecies of *Planorbis* [now *Gyraulus*] *multiformis*, but later concluded they could be separate species; Reif, 1983).

Wolf-Ernst Reif, who rediscovered Hilgendorf's unpublished data and specimens in the 1980s, concluded that after Hilgendorf's death "the Steinheim snails were totally rejected as an example of an almost completely documented phylogeny" (Reif, 1983: 8), while Tassy (2011: 6) wrote that this work had been "totally forgotten" until Reif's rediscovery. These claims are more strongly worded than the situation merits. For instance, Germany's leading paleontologist Otto Schindewolf (1962: 62) highlighted Hilgendorf's Steinheim research among a handful of important early studies that established an evolutionary approach to paleontology. The Steinheim study was celebrated in early-twentieth century works intended for a broader readership as well, such as E. S. Russell's 1916 classic *Form and Function* and the 1914 volume on descent theory in the multivolume series *Die Kultur der Gegenwart*, edited by zoologist Richard Hertwig and botanist Richard Wettstein. No doubt Reif's 1983 paper brought Hilgendorf's fossil phylogeny back into the limelight, but the Steinheim snails had been firmly ensconced in the paleontological canon from the late nineteenth century onward (Haeckel, 1889: 397; Abel, 1914: 351; Russell, 1916; Rudwick, 1985; Bowler, 1996: 315; Haszprunar and Wanninger, 2012: R510). Indeed, leading Austrian

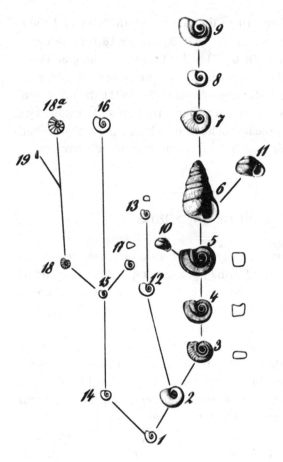

Figure 7.1 Franz Hilgendorf's phylogeny of the Steinheim planorbid snails
(after figure 4 in Rasser, 2013, by permission from John Wiley and Sons).

paleontologist Othenio Abel hailed both Hilgendorf's Steinheim study and Kovalevsky's 1874 monograph on horse evolution as curtain raisers for the paleontological approach to phylogenetics (Abel, 1914: 352; 1918: 498).

Yet, Hilgendorf's research didn't go unchallenged. His own thesis supervisor disliked phylogenetic speculation and didn't even mention the evolutionary context of the work in his written evaluation (Reif, 1983). Instead, he praised Hilgendorf's research primarily as a contribution to stratigraphy. Others reinterpreted the different snail shells as ecopheno-types rather than segments of evolutionary lineages, but recent work has reinstated the Steinheim snails as an early example of lineage evolution recorded in fossils (Reif, 1983; Glaubrecht, 2012; Rasser, 2013). One of Hilgendorf's better known contemporaries did agree that the fossils traced phylogenetic lineages, but he drew an entirely different conclusion about their significance.

When American invertebrate paleontologist Alpheus Hyatt (1838–1902) read Hilgendorf's Steinheim paper he got excited "because, if true, it was the only reliable statement of the theory of evolution, which could be considered a demonstration of the practical applicability of that doctrine to the life history of any considerable series of animal forms" (Hyatt, 1880: 3). Not content to take Hilgendorf at his word, Hyatt set out to find out for himself if it was true. In the autumn of 1872, he visited the Steinheim sand pits and collected thousands of snail fossils, carefully labeling the specimens and noting their stratigraphic origins. However, something didn't seem right: "My studies led me to think either that Dr. Hilgendorf had made the most serious mistakes with respect to the stratigraphical position of the forms, or that I had collected them without sufficient care" (Hyatt, 1880: 4). To resolve this conundrum, Hyatt revisited the site in the spring of 1873 and made further collections. It made no difference. His results continued to clash with Hilgendorf's. He recorded his discoveries in a 114-page monograph that was more than three times the length of Hilgendorf's original paper. Although he agreed with some of Hilgendorf's conclusions, he comprehensively rejected that the Steinheim fossils recorded the stratigraphically orderly evolution of several diverging lineages that were connected via common ancestors. Instead, Hyatt's phylogeny depicted four parallel series of snails without branching.[2]

Despite Hyatt's diligence in digging up snails, Hilgendorf didn't hesitate to diagnose the cause of the discrepancies in their results. He was convinced that the stratigraphic origins of some of Hyatt's specimens could not be trusted. Hyatt admitted that he had found shells away from their original locations, but Hilgendorf thought that Hyatt had failed to recognize the true extent to which shells had been displaced. He was particularly suspicious that Hyatt trusted the discovery of isolated specimens of forms that Hilgendorf had only found in younger overlaying layers. Should we really trust, Hilgendorf asked, Hyatt's discovery of *Planorbis trochiformis*, the largest and most dome-shaped of the Steinheim snails (the shell numbered 6 in Figure 7.1), in a sedimentary layer that Hilgendorf had investigated "at least a hundred times" (Hilgendorf, 1881: 96), and where he had never seen this species that "could absolutely not be overlooked" (Hilgendorf, 1881: 96)? Hilgendorf also accused Hyatt of having overlooked entire sedimentary zones, including the one where the distinctive *P. trochiformis* transitions into the very different looking *P. oxystomus*. But the vagaries of fieldwork aside, Hilgendorf and Hyatt did agree on the main reason why they obtained such different results: the adoption of completely different phylogenetic methods.

Hilgendorf's talisman of truth had been the stratigraphic order along which he had traced the transitions between forms. Although Hyatt did not ignore stratigraphy – he had carefully labeled his samples and used the stratigraphic location of *P. levis* to cement the status of this snail as the ancestor for his phylogenetic series – it wasn't his final arbiter. After praising Hilgendorf's "reputation and the thoroughness of his explorations," Hyatt (1880: 7) declared that "I require more evidence than is found in the Pit Deposits to prove the genesis of the *Planorbis trochiformis* out of *Pl. Steinheimensis/levis*, as far as the succession

---

[2] Hyatt depicted this phylogenetic scheme in line drawings in plate nine of his monograph. Stephen Jay Gould, who occupied Hyatt's office at Harvard's Museum of Comparative Zoology, found Hyatt's original of the figure in the form of a glass plate with actual snail specimens. He published a photograph of it as figure 5.5 in Gould, 2002: 375.

of the forms in time is concerned." By "more evidence" Hyatt did not mean more data, but rather the application of a distinctive evolutionary theory that arranged the data along a predetermined narrative arc that could deviate dramatically from the geological succession of fossils. At the time of Hyatt's writing, his theory had not yet received a name, but it came to be known as orthogenesis.

Orthogenesis unites a cluster of theories around the idea that the direction of evolutionary change within lineages is determined by factors or forces intrinsic to organisms, rather than external forces, such as natural selection (Bowler, 1992a; Gould, 2002b; Levit and Olsson, 2006). Most orthogenetic theories stipulated that these internal factors constrain organismal variation in specific directions, which causes internally directed evolution. Importantly, irrespective of what they believed the engine of orthogenetic change to be, orthogeneticists were non-Darwinian to the extent that they considered natural selection to be at best a handmaiden of internally generated evolutionary forces. They conceded that external factors could influence evolution, but these couldn't knock orthogenesis from its internally propelled linear course. The intrinsic factors that drive orthogenesis produced unidirectional character change, even if the traits would overshoot their adaptive value. Indeed, a hallmark of orthogenetic evolutionary theories was to bring nonadaptive traits under their explanatory umbrella. Orthogenesis could even drive lineages into maladaptive evolutionary dead ends and cause their extinction, which is something that natural selection could never accomplish. German zoologist Wilhelm Haacke named orthogenesis in 1893, and orthogenetic theories were influential, especially in paleontology, until the middle of the twentieth century. Although Hyatt was writing well before Haacke coined the term, the idea of orthogenesis was nevertheless a career-defining thread in his evolutionary studies. Gould (2002b: 374) rightly concluded that Hyatt's Steinheim study "epitomizes hardline orthogenesis."

Invertebrate zoologist William Keith Brooks (1908: 319) wrote in a posthumous biographical sketch of Hyatt that his former boss's research was driven "from beginning to end, by a single motive," which "I must admit that I do not understand." Continuing, Brooks (1908: 321) writes "I have studied his more recent writings upon the subject with all the diligence that my great respect and admiration for him demanded, I have listened attentively when he has discussed his views in public, and I have had many private talks with him about them, but I do not understand them." The bare bones of Hyatt's orthogenetic theory aren't too difficult to understand. What was difficult to understand for an ardent Darwinian like Brooks – as it was for Darwin himself (Gould, 1977b: 424) – was how Hyatt's profoundly non-Darwinian theory had any merit beyond being compelling to its inventor. The story that Hyatt had his fossils tell was one of lineages progressing through several evolutionary phases that paralleled the stages that individual organisms go through during their lives. Phylogenetic momentum was provided by "an unknown power of force" (Hyatt, 1880: 18) that drove lineages through an evolutionary life cycle from young to mature to old age to senility to death. According to Hyatt, lineages progressively added characters until they reached a zenith of phylogenetic and morphological development, after which their evolutionary force ebbed away, causing forms to revert back to stages that were superficially similar to young forms, but that were in reality senile forms shorn of their earlier traits.

The appearance of such senile or "geratologous characteristics" (Hyatt, 1880: 21) augured the lineage's imminent extinction and the end of the phyletic life cycle.

Hyatt's old-age theory, as he liked to call it, was rooted in the recapitulationist logic he had imbibed from his Harvard mentor Louis Agassiz. According to Agassiz's threefold parallelism, the form changes observed during the development of embryos recapitulated those frozen in the sequence of fossils in the geological record as well as those crystallized in the systematic hierarchy of taxonomic groups. Hyatt credits Haeckel with giving this "law of Biogenesis" (Hyatt, 1880: 27) its evolutionary interpretation and name, and Hyatt was convinced both of its universality and its infallibility as a phylogenetic guide: "In every series of animals which I have studied the same fact appears, namely, that in a given number of generations inherited characteristics of every kind tend to appear in the descendants at earlier stages than that at which they first occurred in the ancestral forms" (Hyatt, 1880: 27). According to Hyatt's recapitulationist engine of evolution, the same fundamental laws of growth and development drive organisms from egg to end, and they push lineages through their evolutionary life cycles. During their development, descendants unfailingly recapitulate adult ancestral traits, and these traits are pushed back into earlier and earlier ontogenetic stages according to what Hyatt called the "law of accelerated development" (Hyatt, 1880: 27). If evolution is a conveyor belt of progressively added characters that continuously shift to earlier developmental stages, then all phylogeneticists have to do to reveal evolutionary lineages is to arrange specimens into regular, finely graded, unidirectional series by focusing on a few traits. That was precisely what Hyatt did.

Hyatt reconstructed four parallel lineages of Steinheim snails based on variations in the size, shape, and ornamentation of the shells, each starting with a different but closely similar variety of the ancestral species *Planorbis levis*. Three of these lineages were progressive and tracked arrows of increasing size and morphological complexity, transforming flat shells into larger, more ornamented and dome-shaped shells, while one lineage, which was split into three sublineages, was predominantly retrogressive and characterized by decreasing size, the lack of addition of new characters, and the uncoiling of the shells. The phylogenetic flourishing of *Planorbis* reflected the action of an internal growth force rather than natural selection:

> [The ancestral species] *Planorbis levis* is an immature or low form, the field into which it entered was free, and it developed all its latent growth force, in order to fill it with species... the series had room to expand, or to grow and reproduce to the fullest extent in this field; that they did so in precise accordance with the laws of growth, and the succession of characteristics in the individual... Thus it may be said that the struggle for existence, and the survival of the fittest, is a secondary law grafted upon laws of growth, and governed by them in all its manifestations. (Hyatt, 1880: 19–20)

Although the three progressive series possessed a phylogenetic vitality that had inured them against detrimental environmental conditions, the retrogressive series had lost this vigor, causing the body plan to deteriorate. The specific morphologies of Hyatt's retrogressive or "geratologous" characters weren't programmed into the phylogenetic life cycle, however. Instead, a diminished growth force left lineages in a weakened position for coping with adverse environmental conditions, and their detrimental impact was passed on to future generations.

However, some fossil lineages didn't seem to go through a phyletic life cycle that accompanied the leaching of their growth force. This was the case for what T. H. Huxley had called "persistent types," forms that had endured more or less unchanged for vast periods of geological time. These persistent types, or living fossils as they came to be commonly called, caused the budding evolutionist a headache as Huxley tried to square these indications of a lack of progressive change in the fossil record with what he expected to see if Darwin's theory was true (Desmond, 1982; Lyons, 1993). Why would living forms like the nautilus and the inarticulate brachiopod *Lingula* seem to have a fossil record going back deep into the Paleozoic while changing very little? Huxley eventually concluded that such persistent types were incompatible with theories that demanded progressive change, especially natural theology, but they were compatible with a nonprogressionist Darwinian interpretation of evolution. For Hyatt persistent types meant something entirely different.

Orthogenesis was inherently directional and progressive, so why did persistent types exist at all? It was because persistent types were relics from the phylogenetic youth of their groups. These "types could only be persistent when they sprang from a point of origin near to the source of the whole group to which they belong. The instances of persistent types are all of this character, as far as I know them" (Hyatt, 1880: 22-23). By interpreting the examples he listed – *Nautilus, Lingula,* the lungfish *Ceratodus,* and the marsupial mammal *Myrmecobius* (numbat) – as early branching forms full of phylogenetic vigor, Hyatt made sense of their geological endurance. Persistent types wore their phylogenetic address on their sleeves because there was a correlation between their "life power... and the point at which they sprang from the ancestral tree" (Hyatt, 1880: 22). Persistent types were phylogenetically vibrant forever.

Hilgendorf was not impressed by Hyatt's reinterpretation of the Steinheim snails. By arranging the fossils solely on the basis of morphological similarities, while ignoring their stratigraphic context, Hyatt had given up "the main advantage of the Steinheim planorbids," namely that "the temporally constrained succession prevented theoretical arbitrariness" (Hilgendorf, 1881: 99). Truth be told, Hyatt did try to take stratigraphic information into account, but after two bouts of fieldwork, he felt unable to bring his and Hilgendorf's conclusions into agreement. In the end, it didn't matter to him because he was convinced that using his method, irrespective of "whether the forms or species occur mixed on the same level, or on different levels, there is but one natural arrangement, which has been illustrated on Plate 9" (Hyatt, 1880: 8). This is why Hyatt placed the large, dome-shaped, and heavily sculpted *P. trochiformis* at the apex of his most progressive series, rather than in the middle of a lineage, as Hilgendorf had done. As Hilgendorf concluded, for Hyatt "the ruling evolutionary law is the development of the domed shape, the presence of grooves and ridges, and increasing size. If the *oxystomus* form evolved from the *trochiformis* form [as Hilgendorf's phylogeny suggested], this evolutionary path would be completely reversed for all three characters" (Hilgendorf, 1881: 100). This would have been unacceptable to Hyatt.

7.3 Orthogenesis: A Straight Story from Beginning to End

Hyatt wasn't the only paleontologist who saw orthogenesis wherever he looked. Henry Fairfield Osborn (1857–1935) wrote in a review that "[i]t has taken me thirty-three years of uninterrupted observation in many groups of mammals and reptiles to reach the conclusion that the origin of new characters is invariably orthogenetic" (Osborn, 1922: 135). By this he meant that linear evolutionary trends started with a new character that "always arises gradually, continuously, and adaptively from its minute shadowy beginnings" (Osborn, 1922: 137). The last influential flag bearer for orthogenesis, the German paleontologist Otto Schindewolf (1896–1971), was even more categorical in his support when he wrote in his 1950 book *Grundfragen der Paläontologie* that

> Many attempts have been made to disavow orthogenesis, but it is a fact, and there is no way around it. Once any principle whatsoever is introduced into a particular development, we see it follow its course as if from inner compulsion, automatically, like clockwork, no matter whether it leads to ascendancy or decline. It continues on its way until an end is reached, a point that cannot be transcended; this usually also means the end of the lineage in question. (Schindewolf, 1993: 268–269)

But orthogenetic trends weren't just discovered in the fossil record. It was when orthogenetic speculation was unshackled from stratigraphy that it truly blossomed.

Orthogenetic trends readily emerged when extant species were lined up according to the degree to which they had developed characters of interest. In a breathless review, botanist John Schaffner detected orthogenesis in plants and animals wherever he looked (Schaffner, 1937). Beetle horns could be arranged in a series culminating in the enormous protuberances of male scarabs. Beetle antennae could likewise be aligned along a spectrum that ended in the long feelers of longhorn beetles (Cerambycidae). It was child's play for Schaffner to see orthogenetic series rooted in the relative size of traits, whether they were monkey tails, lemur tails, ground squirrel tails, moth probosces, or wasp waists. And if the increasing sizes of these traits did not immediately suggest any increased adaptive value, so much the better. Because natural selection can't guide the evolution of inadaptive features, these had to result from the operation of "an internal creative principle" (Schaffner, 1937: 286). Arthur Cronquist, one of the most influential botanists of the twentieth century, used the perceived inadaptiveness of characters as a tool to conjure orthogenetic series in angiosperms. In a 1956 review he wrote:

> I must confess myself unable to see the survival value of parietal or free-central placentation over axile placentation, or perigyny over hypogyny, or of epigyny over both perigyny and hypogyny, or of sympetaly over polypetaly, or of definite whorls of stamens over an indefinite spiral, or of one whorl of stamens over two, etc. These differences are in my opinion the result of mere orthogenetic trends, and characters of this sort are nearly all that we have to distinguish most families and orders of angiosperms. (Cronquist, 1956: 98)

Orthogenetic evolution therefore ruled over the higher taxonomic categories of angio-sperms, while "Darwinian evolution is prominent at all taxonomic levels among the higher vertebrate animals" because those taxa are "readily distinguished by structural features showing adaptation to environment" (Cronquist, 1956: 97). Neither Cronquist nor Schaffner worried about whether the species lined up in their orthogenetic series were actually part of ancestor–descendant lineages because that was "after all of secondary importance" (Schaffner, 1937: 267). The assumption that extant morphologies paralleled ancestral states in orthogenetic series with sufficient fidelity to trace an actual evolutionary path was considered to be so obvious as to not even be worth mentioning. What was important to mention was that orthogenetic series had "no relation either to utilitarian advantage or to any ecological conditions either physical or biological" (Schaffner, 1937: 267). Conveniently, here "science comes to its limit. The next step in causative reasoning, as to what is the ultimate cause of this orthogenetic, progressive, perfective principle, is a question of theistic philosophy and belongs to the domain of the philosopher rather than the scientist" (Schaffner, 1937: 286–287). It is hard to imagine an evolutionary approach more devoid of epistemic robustness and explanatory ambition than this.

Ichthyologist and ancient armor expert Bashford Dean, who during the first two decades of the twentieth century was simultaneously Curator of Fishes at the American Museum of Natural History and Curator of Arms and Armor at the Metropolitan Museum in New York, published a paper that includes a drawing that nicely illustrates how an orthogenetic series could be constructed (Dean, 1912). In the paper Dean described the dart-shaped egg capsule of an unknown species of chimaera, a group of cartilaginous fishes sometimes also referred to as ghost sharks. When comparing it to the egg capsules of other chimaeras, he found that "they arrange themselves in a series" (Dean, 1912: 37) of increasingly slender and dart-shaped egg capsules (Figure 7.2). Although the phylogenetic relationships of the species were unknown, Dean was convinced that the "character of the changes which here are indicated, are certainly definite in their direction; and it is significant that this serial arrangement in the capsules accompanies a similar serial arrangement in at least the external characters of the adult Chimaeras. It is thus an orthogenetic series narrowly defined. It cannot be explained on grounds of natural selection" (Dean, 1912: 40). Not only was it obvious to Dean that adult characters would fall in line with the series based on egg capsule morphology, the regularity of the series also allowed him to imagine both an unobserved beginning and an end of the series, in the form of a blade-shaped ancestral egg capsule, and a maximally spiniform egg capsule of an "undetermined terminal species" (Dean, 1912: 39). He was confident both "will eventually be discovered" (Dean, 1912: 38). Orthogenetic fantasizing such as this produced a sizeable literature (Jepsen, 1949), which driven by influential voices like Schindewolf's and Cronquist's, extended into the second half of the twentieth century. But by then orthogenesis was running out of steam.

7.3.1 The Epistemic Weaknesses of Orthogenesis

By the middle of the twentieth century, any explanatory potency that could be claimed for Hyatt's "unknown power or force" (Hyatt, 1880: 18) was rhetorical at best. Writing more than half a century after Hyatt, John Schaffner, who believed that the cause of orthogenesis

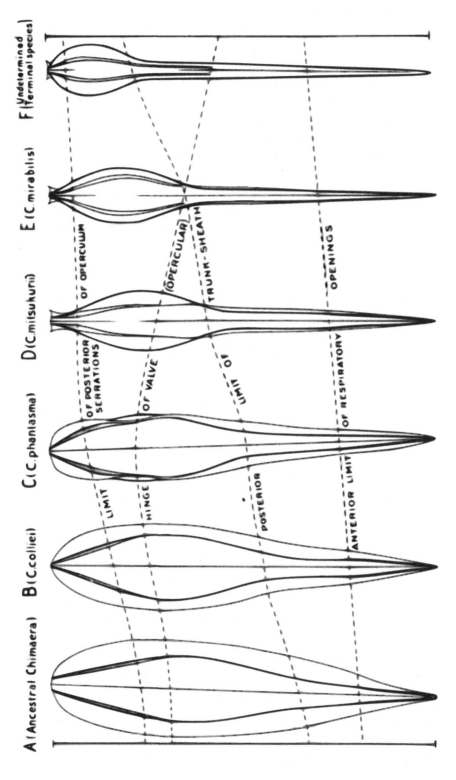

Figure 7.2 Bashford Dean's orthogenetic series of chimaeroid egg capsules, with figures A and F representing a hypothetical ancestral and descendant form, respectively (after figure 2 in Dean, 1912).

lay beyond the purview of science, nevertheless felt the need to offer this puff of air: "recognition of the orthogenetic series and of the perfective series as a result of the perfective principle in the organic kingdom does require us to recognize an internal creative principle which is just as evident and unescapable as the fact of the general atomic series and the complicated molecule systems of the chemist" (Schaffner, 1937: 286). The explanatory umbrella of the Modern Synthesis offered orthogenesis no shelter, and its architects offered no sympathy. When Arthur Cronquist submitted a manuscript on ortho-genesis to *Evolution* in 1949, Ernst Mayr not only rejected it, but also thought that the word should be banned from the journal (Ruse, 1996: 447). And Theodosius Dobzhansky, who had done his doctoral research under the guidance of the orthogeneticist Yuri Filipchenko and written a foreword to an English translation of Leo Berg's orthogenetic book *Nomogenesis*, wrote in a 1965 paper commemorating the work of Gregor Mendel that "orthogenesis was not simply a word describing the fact of directionality, but a now defunct hypothesis pretending to explain the causation of this directionality" (Dobzhansky, 1965: 214). The Synthesis architect who dedicated the most thought and pages to debunking the data claimed in support of orthogenesis was George Gaylord Simpson (1902–1984), the leading American paleontologist of his time.

Simpson critically discussed orthogenesis in his writings for both professionals and the broader public. The topic was important enough for Simpson to dedicate a 30-page section to it in his classic *Tempo and Mode in Evolution* (1944) and a 21-page section in the 1956 edition of his popular *The Meaning of Evolution*. The first of these books was Simpson's main contribution to the Modern Synthesis. In it he declared that although "[t]here is no possible doubt but that some degree of rectilinearity is common in evolution... [i]t is doubtful whether an undeviating or even a relatively straight structural line can be traced from an archetypal protozoan to any real metazoan, an ancestral fish to any real tetrapod, a protolemur to any existing primate, and so forth" (Simpson, 1944: 152). He concluded that "a tendency toward rectilinearity is not characteristic of evolution as a whole, but only of certain levels of change under certain common but far from universal conditions" (Simpson, 1944: 153). However, in most of these cases Simpson felt that linear evolution wasn't internally directed but the result of orthoselection (i.e. caused by enduring selective pressure in the same direction). In the end, Simpson found that "a dispassionate survey of many of the phenomena of orthogenesis, so called, strongly suggests that much of the rectilinearity of evolution is a product rather of the tendency of the minds of scientists to move in straight lines than of a tendency for nature to do so" (Simpson, 1944: 164).

In his popular science book *The Meaning of Evolution*, Simpson used even more pointed language to dismiss orthogenesis. One reason for this was the publication in 1950 of Otto Schindewolf's magnum opus *Grundfragen der Paläontologie* (*Basic Questions of Palentology*), a work steeped in orthogenesis. Using the cruel method of dismissal in a footnote, Simpson accused Schindewolf of being guilty "of subjective arrangement of evidence to favor this type of theory" (Simpson, 1956: 36). In a footnote two pages later he wrote:

> In a session on orthogenesis at a recent international conference on evolutionary problems held in Paris, an eminent student startled his colleagues by proclaiming that evolution consists of nothing but millions of orthogenetic lines. This is the most extreme case of

orthomania, or straight lines before the eyes, known to me, but milder cases of this affliction are fairly common. (Simpson, 1956: 38)

It is not inconceivable that Schindewolf was the target of this remark.

In his books Simpson discussed textbook examples of orthogenetic evolution ending in maladaptive disaster, such as the coiled bivalve *Gryphaea*, the antlers of the Irish elk, the horns of titanotheres, and the canines of sabertooth cats, and concluded that in his view all these trends were "strictly adaptive" (Simpson, 1956: 48). In *The Major Features of Evolution* (1953), the expanded sequel to *Tempo and Mode in Evolution*, Simpson delivered his strongest verdict on both orthogenesis and its most influential promoter. Figure 3.120 in the 1993 English translation of Schindewolf's book presents in four images of skulls the "[o]rthogenetic enlarging and overspecialization of the upper canine in the family Felidae" (Schindewolf, 1993: 284) (this figure was based on a figure from Alfred Sherwood Romer: see Figure 7.3). Simpson's one-page dismissal of this scenario of sabertooth evolution is instructive as it shows how he indicted Schindewolf's thinking on three points: stratigraphy, systematic relationships, and adaptation.

First, following Schindewolf's morphological series from small to large canines involves an erratic stratigraphic dance, with the fossils taking two steps back from the Pliocene to the middle Oligocene, followed by one step forward to the Pleistocene. Second, the four skulls are "almost certainly [from] three sharply different lines of descent" (Simpson, 1953: 269). Schindewolf explicitly acknowledged these points: "[t]he forms illustrated are not a genetic lineage but represent evolutionary stages that have only a loose phylogenetic relationship with one another" (Schindewolf, 1993: 284). Yet this admission cannot deflect Simpson's criticism because in Schindewolf's mind's eye sabertooth evolution did look like this. It illustrates that for orthogeneticists like Schindewolf stratigraphy and systematic relationships were only important if they aligned with the orthogenetic arrow they divined from morphology. Third, Schindewolf thought that the enlarged canines became less useful for food intake and as weapons, while the "original biting function was... rendered impossible" (Schindewolf, 1993: 283). Simpson's response was withering: if "the large sabers made it very difficult to eat, the animals took 40 million years to starve to death. The fact is that this famous example of 'orthogenesis,' presented as gospel over and over again in the literature does not exist. It is pure fiction" (Simpson, 1953: 269). Simpson concluded that canines became larger and smaller in different lineages. In his next short paragraph, Simpson closed his case:

> It is true that there have been many and still are some paleontologists who believe in the existence of sustained, rigidly undeviating, nonadaptive trends. Nevertheless there are more paleontologists, an increasing number in recent years, who point out that the supposed evidence for trends of this sort is worthless or misinterpreted and who conclude that such trends are entirely absent in the known fossil record. (Simpson, 1953: 269–270)

The contrast with Schindewolf's words, published just three years earlier, is stark: "all groups of invertebrates as well as vertebrates have provided countless examples of orthogenetic evolution" (Schindewolf, 1993: 272).

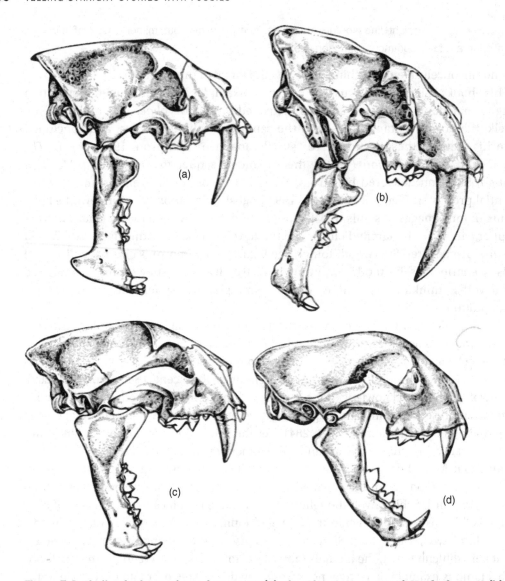

Figure 7.3 Skulls of sabertooths and true cats: (a), the Oligocene nimravid *Hoplophoneus*; (b), the Pleistocene felid *Smilodon*; (c), the Oligocene nimravid *Dinictis*; and (d), the Pliocene felid *Metailurus*.
(after figure 279 in Romer, 1964, republished with permission of the University of Chicago Press; permission conveyed through Copyright Clearance Center, Inc.)

Simpson's takedown of Germany's most influential contemporary paleontologist[3] lays bare the weakest points of orthogenesis. For committed orthogeneticists like Schindewolf,

---

[3] Ochoa (2021) provides an interesting perspective on Simpson's critique of orthogenesis and argues that Simpson's conception of parallelisms was actually similar to the conception of orthogenetic trends proposed by several orthogeneticists.

Hyatt, and Osborn, detecting orthogenetic evolution had become a self-fulfilling prophecy. The law-like unspooling of evolution through the narrow channel of directed variation created predictable patterns that could be recreated with ease. Stratigraphy could be a helpful guide, but it could be jettisoned without guilt if it didn't match a linear procession of fossils that obeyed a preferred arrow of morphological change. Indeed, fossils were unnecessary frills for orthogeneticists working on living species. As we saw for Bashford Dean and his chimaeran egg capsules, a handful of living species was enough to draw an orthogenetic arrow that pointed with equal accuracy to the ancestral start of the series and its projected end. A really competent orthogeneticist required no more than a single data point to detect an orthogenetic progression.

James Duerden was a British zoologist with eclectic interests and positions. Among the professional titles he successively held were Honorary Curator of Coelenterates at the American Museum of Natural History in New York, Professor of Zoology at Rhodes University in South Africa, and Director of Wool Research at the Grootfontein College of Agriculture in South Africa (Wyse Jackson and Maderson, 2014). Although he is celebrated for his work on cnidarians and bryozoans, the topic to which he devoted the greatest number of papers was ostriches. Based on his studies of the wings, feathers, and legs of ostriches he foresaw a bleak orthogenetic future. Since "the degenerative forces are so relentless… these is much to indicate that… we may look forward in the dim future to the sad spectacle of a wingless, legless and featherless ostrich, if extinction does not supervene" (Duerden, 1919: 192). Alas, a century later we still await the evolution of the ostrich nugget.

The idea of lineage evolution ending in an inadaptive march over the cliff edge of extinction was another weak point of orthogenetic speculations. It is easy to claim that seemingly extreme forms, such as sabertooth cats or the Irish elk, were bound to go extinct because they are in fact extinct. But so are legions of more mundane lineages. As Peter Bowler (1992a: 161) rightly points out, it is easy to dismiss seemingly bizarre traits of extinct animals as inadaptive ballast when their lifestyles are utterly unknown. In a witty and trenchant 1920 critique of orthogenesis, Francis Bather, then Assistant Keeper of Geology at the museum where I work, wrote that "as long as my knowledge of palaeontology was derived mainly from books I accepted this premiss [sic] as well founded… [but] more intensive study generally shows that characters at first regarded as indifferent or detrimental may have been adapted to some factor in the environment or some peculiar mode of life" (Bather, 1920: 77).

Here Bather puts his finger on a characteristic epistemic feature of orthogenetic thinking. Since Darwinian evolutionary theory cannot explain lineages evolving increasingly maladaptive traits (it doesn't deny that this might happen, just that natural and sexual selection cannot drive the evolution of increasingly unfit phenotypes), orthogeneticists merely had to deny the adaptive value of a trend to be able to label it as orthogenetic. In one example, Bather pokes gentle fun at John M. Clarke (1857–1925), one of America's foremost paleontologists of the late-nineteenth and early-twentieth centuries and a follower of Hyatt's recapitulationist and orthogenetic approach to the fossil record (Schuchert, 1926; Rainger, 1981). Clarke had seen the parallel evolution of ornately spinose trilobites as a sure "sign of orthogenesis in the most mystical meaning of the term" (Bather, 1920: 79).

Bather pointed out that the evolution of conspicuous spines could also have provided the animals with protection against predators, as well as stability and buoyancy during swimming, an interpretation under which "even the most extravagant spines lose their mystery and appear as consequences of natural selection" (Bather, 1920: 80).

One could argue that orthogeneticist and Darwinian interpretations were just two different ways of seeing the data. After all, a reasonable argument commonly made by orthogeneticists is that it is impossible to know if and how the many, often trivial differences between related species are adaptive and how they would key to specific environmental variables (Schaffner, 1937; Cronquist, 1956). The adaptive storytelling that was mandated by Darwinian theory provided no proof of the truth either, despite the supreme confidence of some of its devotees "that if we can see any advantage whatever in a small variation (and sometimes even if we cannot), selection sees more" (Simpson, 1953: 271). However, the weakness of the orthogenetic approach is that a failure to imagine the adaptive value of evolutionary changes or, worse, an unwillingness to even think about them could, and was, used as a tool to bolster orthogenesis. This turned a lack of intellectual ambition into the sharp edge of a tool that many orthogeneticists wielded with abandon.

Commenting on the varying lengths of the horns of male scarab beetles, Schaffner (1937: 282) writes that these "give no advantage to the individual possessing the one type or another. The notion of special adaptation is not to be entertained rationally... The sex dimorphism in these species was not originated because the female delights in a mate with long horns." Schaffner was clearly disagreeing with Darwin here, who had tried to do exactly that when discussing beetle horns in the context of his theory of sexual selection (Darwin, 1901). Likewise, regarding the varying lengths of the slender petiole, or waist, of wasps, Schaffner thought that to "see any special advantage in such a condition is to proclaim one's self [sic] a firm believer in witch stories and fairy tales" (Schaffner, 1937: 283). Yet, recent research suggests that petiole length is related to wing shape, with variations evolving in a trade-off between flight speed and maneuverability (Perrard, 2020). Classic examples of orthogenetic maladaptation have likewise been reinterpreted. For instance, the large canines of sabertooths are now thought to be specializations for specific killing strategies (Andersson et al., 2011; Figueirido et al., 2018), while the giant antlers of the Irish elk find their explanation in allometric growth and sexual selection (Gould, 1974).

My point here is not to judge orthogenetic thought by later insights, but to point out that orthogenetic storytelling adopted even less critical standards of argument than the adaptationist storytelling that was later so famously maligned (Gould and Lewontin, 1979). Like the advocates of arthropod polyphyly that we encountered in Chapter 6, orthogeneticists had fashioned a phylogenetic tool out of a lack of imagination. For orthogeneticists such as Schaffner, Cronquist, and Duerden, observing a seemingly extreme or inadaptive phenotype in a single living species was enough to declare the existence of an orthogenetic trend. Orthogenetic hypotheses are thus inured against criticism if they can be erected in spite of conflicting information from stratigraphy and systematic relationships. If they can be consolidated by a failure to think about the possible adaptive value of the evolutionary changes involved, and if they are thought to result from the operation of inescapable but unknown internal "laws" of growth, development, and heredity, orthogenetic hypotheses become virtually irrefutable. Alas, despite occasional attempted revivals (e.g. Grehan and Ainsworth, 1985), orthogenetic thinking went functionally extinct in the second half of the twentieth century. Nevertheless, a component of its epistemic core endured in phylogenetics.

7.4 Nonorthogenetic Linear Thinking with Fossils

When speaking of the "collapse of orthogenesis in the face of the modern synthesis" (Bowler, 1992a: 178), we must ask: what exactly collapsed? The theory's roots certainly withered in the inhospitable soil of the synthesis, but the fossil phenomenology consistent with orthogenesis didn't all evaporate. As Simpson conceded in *The Major Features of Evolution*, "[i]t is nevertheless a fact of observation that short trends and segments of long trends are, as a rule, essentially rectilinear" (Simpson, 1953: 258). This impression found agreement with fellow systematist Ernst Mayr, who wrote that orthogeneticists "were right in insisting that much of evolution is, at least superficially, 'rectilinear'" (Mayr, 1982: 530). Such "[e]ssentially undeviating trends on the order of 10^6 years also seem to be common," and "most of the known cases of moderately sustained, essentially rectilinear evolution come from what seem to have been abundant populations evolving at moderate rates in rather stable environmental (and adaptive) conditions" (Simpson, 1953: 259). Such trends were no threat to neo-Darwinists, although they no doubt gladdened the hearts of obdurate orthogeneticists as well.

As we have seen, the mechanisms of evolution and the length and frequency of unidirectional trends weren't Simpson's only bone of contention with the orthogeneticists. He utterly rejected their phylogenetic epistemology. Although Simpson accepted that lineal ancestors may be found as fossils, fossils could not simply be lined up in a morphological series if the steps went against the stratigraphic grain or if collateral relatives were misinterpreted as ancestors and descendants. Neontologists deserved special scorn for adopting "this fallacious procedure" (Simpson, 1953: 269) of lining up living species, confirming the verdict of his Princeton colleague Glenn Jepsen who concluded that the "tyranny of diagrams has probably been a real factor in the uncritical part of the popularity of orthogenesis" (Jepsen, 1949: 493). What Jepsen referred to here was the ingrained habit of connecting collateral relatives by straight lines for diagrammatic and didactic purposes. In a phylogenetic world, this amounts to what I call the "sin of the scala." The danger is that "*any* group of objects can be arranged in graded series according to some criterion. Unless the criterion is correctly and solely phylogenetic, the result has no relevance whatever as to how the characteristics of the objects arose" (italics in original; Simpson, 1953: 269).

By discarding stratigraphic and systematic evidence, Schindewolf constructed his orthogenetic series like a pure morphologist. But was that so unreasonable? Deliberately ignoring evidence that contradicts a favored hypothesis is of course inexcusable, but for most phylogenetic problems at the time, stratigraphic data and a proper understanding of relevant systematic relationships were unavailable luxuries. If lineages could not be observed along densely sampled stratigraphic columns, or inferred on the basis of an established phylogeny, ancestral stages could only be imagined. The phylogenetic literature is filled with a vast number of such imaginings. It was certainly reasonable to expect, including for Darwinians, that extreme morphologies evolved from less extremely developed precursors. Large structures were generally thought to have evolved from small ones, complex ones from simple ones, long ones from short ones, ornate ones from plain ones. Schindewolf chose fossils to illustrate possible steps along the way to the large canines of *Smilodon* because the steps made morphological sense.

Simpson's criticisms were appropriate because Schindewolf's chosen species could not have been lineal ancestors of *Smilodon*, but *Smilodon* must have evolved its impressive dentition somehow, and the lack of such enormous teeth in most cats left little doubt that they were a derived trait. Positing plausible morphological precursors was often all that lineage tracers could do. It is therefore worth noting that Schindewolf's sabertooth figure was copied from Romer's 1933 *Vertebrate Paleontology* (figure 279 in Romer, 1964) (Figure 7.3). Romer was no orthogeneticist, but his suggestive figure indicates that sabertooth evolution looked much the same to a Darwinian: the gradual evolution of increasingly large canines.

Bowler's (1992a: 148) summary verdict on orthogeneticists has been consensus for decades: "their 'laws' were really nothing more than trends created by arranging limited evidence into artificially regular patterns. They mistook the products of their own imagination for an order built into the structure of nature." As far as the existence of massively parallel, and often inadaptive, morphological trends is concerned, everyone agrees. On the other hand, all phylogeneticists arrange limited evidence into patterns that they hope capture real aspects of nature. Were orthogeneticists uniquely deluded by a fallacious phylogenetic method? What Bowler's dismissal misses is a tenacious fiber at the epistemological core of orthogenesis that both preceded it and endured in phylogenetics long after its demise: reconstructing lineages by arranging morphological evidence into regular, smoothly graded series. It is easy for modern authors to ridicule Alpheus Hyatt for claiming that his Steinheim snail series were "the result of no preconceived plan of arrangement as far as the author could judge" (Hyatt, 1880: 8). Gould (2002b: 379) concluded that Hyatt did not establish his "scheme by any principle of ordering specimens in a manner that could be called 'objective' or even bound by rules independent of his phyletic preferences." But how accurate is this assessment?

Having been unable to confirm Hilgendorf's stratigraphic distribution of fossils, Hyatt resorted to arranging fossils into morphological series based on their similarity. He claimed no originality for this method, "nor can, so far as I know, any one else. It has grown with the science of Natural History, and nearly every naturalist uses it more or less, whether he recognizes the ultimate meaning of gradations in their serial arrangements, or ignores them" (Hyatt, 1880: 9). He cited his Darwinian colleague Melchior Neumayr, who in his studies of ammonoids had "followed the ordinary practical method of tracing the series by the graded resemblances of the adult forms" (Hyatt, 1880: 9). Hyatt felt he had done the same. The creation of linear series of fossils to capture phylogenetic lineages was a logically justifiable approach in a Darwinian world of descent with modification, and as we saw, it was used by paleontologists of different stripes during the late nineteenth century (Tamborini, 2017).

Neumayr's mentor Waagen concluded that the only necessary assumption for this procedure was of "a genetic connection between the individual mutations of the same morphological series [*Formenreihe*], and it is really difficult to deny that such closely related forms should not have emerged from each other" (Waagen, 1869: 186). Hyatt thought that this was the only assumption that his phylogenetic method required (Hyatt, 1880: 22). We may not want to attach the label "objective" to this approach, and it involves a strong assumption, but it doesn't necessarily represent a straitjacket of preconceived preferences either. Francis Bather, who was scornful of orthogenesis, nevertheless considered the gradual evolution of form along fossil series to "afford convincing proof of descent" (Bather, 1920: 69). If a finely graded series of fossils climbed up a local sedimentary column, Bather sanctioned a genealogical interpretation: "If there are gaps in the series as known to us, we can safely predict their discovery; and we can prolong the curve backwards or forwards, so as to reveal the nature of ancestors or descendants" (Bather, 1920: 73).

Paleontologists hunting for phylogenetic lineages strung fragmentary evidence into morphological series that were as smooth and continuous as possible. Stratigraphy could guide these efforts, but unless the lineages were reconstructed in densely sampled localities, there was a risk that collateral relatives were mistaken for ancestors. It was this that undid Vladimir Kovalevsky's horse phylogeny when he slotted several side branches into the trunk of his tree (Simpson, 1953: 364–366; Gould, 1998: 150–155). But before we discuss this in more detail, we need to ask whether constructing such fossil *Formenreihen* in the absence of stratigraphic evidence simply reproduced preconceived notions about phylogeny.

Hyatt was a recapitulationist, as were many other early evolutionary paleontologists, from Neumayr to Bather. Hyatt explicitly stated that he applied recapitulationist logic in the construction of his morphological series, whereby "an animal found to repeat the stages of another animal of a closely allied species in the young, with the addition of new characteristics in the adult, may be considered to be either a lineal descendant of that species, or of some form common to both" (Hyatt, 1880: 8). Applying a phylogenetic rule like this would produce a morphological series of adults that traces the increasing development of selected focal characters as new ones were added to ontogeny and existing characters were elaborated more fully. Hyatt's three progressive series of Steinheim snails showed exactly that. But a nonrecapitulationist approach would produce an identical progressive morphological series if adult fossils were arranged according to the degree to which selected characters had developed. It would be impossible to know whether such a progressive series had been constructed by an orthogeneticist, neo-Lamarckian, neo-Darwinist, or even idealistic morphologist because such a series would fit within the epistemological and ontological parameters of each of these guiding theories.

It is interesting to note in this context that although Hyatt's commitment to orthogenesis made him reconstruct four parallel series rather than a branching tree like Hilgendorf, there are nevertheless substantial parallels between their lineages. Hilgendorf's longest lineage, which runs through the distinctive dome-shaped species *P. trochiformis*, corresponds to two of Hyatt's progressive series (the third and fourth) stacked on top of each other, with only a single step missing. Species transitions in Hyatt's other series parallel those reconstructed by Hilgendorf as well, although less accurately. Hyatt's purely morphological method clearly managed to trace some of the same phylogenetic pathways as Hilgendorf's stratigraphically informed strategy. However, he failed to track the most discontinuous morphological transition of all, from the large and dome-shaped *P. trochiformis* to the smaller and flat *P. oxystomus* (see the transition between the shells numbered 6 and 7 in Figure 7.1). Instead *P. trochiformis* capped Hyatt's most progressive series, while *P. oxystomus* started another one.

As we saw earlier, Hilgendorf blamed Hyatt's failure to see these two series as parts of a single lineage on him having overlooked the sedimentary zone where this transition occurred and on the incompatibility of such an abrupt reversal in the size, shape, and ornamentation of the shells with Hyatt's law-like view of evolution (Hilgendorf, 1881: 99–100). This may make theoretical sense, but I would argue that few morphologists would have reconstructed this transition with the evidence at hand because there are virtually no similarities by which these two very different looking species can be connected. When graded similarity is the only thread that guides the course of a morphological series, irregular transitions will be hard or impossible to detect. Hilgendorf was pleased that he had discovered such as surprising morphological transformation (Reif, 1983), but history delivered a different judgment. Recent research suggests that this transition never happened. *P. oxystomus* evolved from the morphologically much more similar *P. sulcatus*, a species also ancestral to *P. trochiformis* (Rasser, 2014). In other words, Hilgendorf had reconstructed a morphological transformation that hadn't happened.

Hyatt's Steinheim work clearly shows the footprint of orthogenesis in his creation of parallel progressive series, but Hilgendorf's results should not be portrayed as a neutral alternative. No one escapes the clutches of theory. Hilgendorf made a discovery that was even more surprising than the morphological jump from *P. trochiformis* to *P. oxystomus*. His thesis research uncovered a possible instance of two lineages fusing. But this didn't fit the "beautiful picture that Darwin showed us of the many-branched connections between species; the branches of a tree never grow together again" (Hilgendorf in Glaubrecht, 2012: 252–253). Hilgendorf therefore didn't dare publish this result (Reif, 1983; Janz, 1999). Recent research shows that he was right not to do so (Rasser, 2014), but his decision shows that theory can blind as well as illuminate.

Discussions of orthogenesis in the modern literature emphasize its incompatibility with neo-Darwinian theory and depict its advocates as misguided architects of air castles. I hope that my discussion helps nuance this interpretation. Although the ontology of orthogenesis was incompatible with Darwinism, its epistemic strategy of reconstructing morphological series and interpreting them as evolutionary lineages is part of a continuum of phylogenetic methods established in the nineteenth century and stretching into the next. Ideally, paleontologists hoped to be able to read phylogenies from the fossils by tracing morphological series along the stratigraphic column. But the circumstances that facilitated this were rare. In the absence of stratigraphic guidance, researchers resorted to inferring or imagining evolutionary lineages based on morphological and embryological evidence, and they plastered over gaps in evidence with evolutionary intuitions. What these intuitions were and how and when they were used is the golden thread running through Chapters 6 through 9 of this book.

Stephen Jay Gould accused Charles Doolittle Walcott (1850–1927), the central character in his book *Wonderful Life* (Gould, 1989), of committing the sin of the scala when he lined up several Burgess Shale fossils from the same stratigraphic layer into tentative lineages of ancestors and descendants (Figure 7.4). But when you read Walcott, it is clear he was a conventional lineage thinker who wanted to understand if and how the Burgess Shale arthropods could illuminate the lines of descent of major arthropod taxa. He therefore arranged fossils into series based on their perceived morphological distance from presumed annelid ancestors (Walcott, 1912: 161–164).

Simpson (1953: 269) chided neontologists and paleontologists alike for uncritically constructing morphological series by "some criterion" and then misinterpreting them as

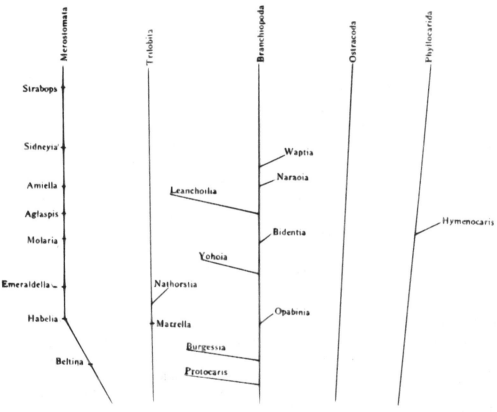

THEORETICAL LINES OF DESCENT OF CAMBRIAN CRUSTACEA

Figure 7.4 Charles Doolittle Walcott's attempt to trace lines of descent by arranging arthropod fossils into series based on their increasing morphological dissimilarity from assumed annelid ancestors (after the figure on page 161 of Walcott, 1912).

phylogenetic lineages. His call for a criterion that "is correctly and solely phylogenetic" was certainly appropriate, but Simpson's phylogenetic criteria were found wanting by the next generation of evolutionists, who were frustrated that "one is never told precisely how these criteria are used to construct a tree... subjective 'artistic' interpretations are considered a normal aspect of taxonomic procedure" (Eldredge and Cracraft, 1980: 189). The views of Willi Hennig, who also accepted orthogenesis (Rieppel, 2012a), were to transform phylogenetics, but it would take decades for the philosophy and methods of phylogenetics to develop fully and spread through the international research community. Indeed, before the cladistic revolution, the boundary between what it meant to be inferring or imagining phylogenies remained rather blurry. The sequence of cladogram → phylogenetic tree → scenario (Tattersall and Eldredge, 1976), which represents the transition from atemporal character hierarchy → tree with a time axis → causal hypothesis about evolutionary changes, was not yet enshrined, and lineage thinking held sway over tree thinking. Lining up selected evidence to create lineages was one strand of an evolutionary discipline that created countless impressionistic phylogenetic paintings. Irrespective of whether the motivating theory was Darwinism, orthogenesis, or something else, the brushstroke of fantasy often obscured a meagre backdrop of facts.

Peter Bowler concluded that "[b]y the very nature of the evidence they used, paleontologists were inclined to see evolution as a linear process" (Bowler, 1992b: 448). Indeed, the temporal arrangement of fossils reinforced the conviction that evolution is a linear process. As it turns out, this is a correct conviction, not the flawed view that Bowler thinks it is (see Chapter 11 for more discussion). I hope that the reader will agree that the spread of the linear view of evolution didn't require evidence from the fossil record. The acceptance and spread of descent theory in the nineteenth century recorded a conceptual revolution, not an empirical one. It convinced evolutionists that lineages can be understood and visualized as linear series of forms related to each other as ancestors and descendants. The challenge was how to document such series empirically. Attempts to observe, infer, or imagine such series fell along a spectrum of how concrete and convincing available evidence could be considered to be.

The early evolutionary paleontologists discussed in this chapter thought that the fossil evidence for ancestors was so rock solid in certain stratigraphic settings that lineages could simply be observed. On the opposite end of the spectrum were many of the efforts of orthogeneticists who imagined lineages based on nothing more than arranging a series of living forms along some morphological gradient, sometimes on the basis of only a single character, and irrespective of how these forms were related. Somewhere in the middle were biologists and paleontologists who considered the systematic relationships of taxa, living and extinct, to make sure their observations were properly refracted back from the realm of collateral relatives onto the ancestral lineages that had produced them. This approach developed into the core of modern phylogenetics, but it developed slowly. One reason was that phylogenetic speculations could often be published without first tackling the systematics of relevant taxa. Another was that some systematic riddles were so recalcitrant that researchers didn't have the patience to wait for them to be solved before they could start their phylogenetic speculations. This is, for example, why most scenarios for the evolution of animal body plans predate the development of a consensus about metazoan relationships, which is developing still (Giribet and Edgecombe, 2020).

I suspect that another important reason for the slow incorporation of rigorous systematic thinking into phylogenetic research is that it isn't intuitive. Linear evolutionary thinking comes natural to people, tree thinking doesn't. To think of evolution as a forward flow of transforming ancestors is easier than seeing it as a branched hierarchy of characters and taxa. A lot of conceptual change preceded the development and adoption of explicit ancestral state reconstruction methods that infer the nature of hypothetical ancestors on the basis of systematically structured comparisons of traits observed in collateral relatives. This is why the cladistic revolution, and its core precept that systematic pattern deserves epistemic priority over phylogenetic process, didn't happen until the second half of the twentieth century. Before this time, we can see a diversity of attitudes toward the epistemic merits of taking systematics explicitly into account when thinking about the evolutionary process.

Darwin knew that in our world of imperfect evidence one could only ski a straight phylogenetic course by doing systematic slalom. Lacking direct evidence for ancestors, one could visualize possible phylogenetic paths to characters of interest only with the help of observations in "the collateral descendants from the same original parent-form, in order to see what gradations are possible, and for the chance of some gradations having been transmitted from the earlier stages of descent, in an unaltered or little altered condition" (Darwin, 1859: 218). He continued, in a section that he modified and expanded in later editions of the *Origin*, to envisage the evolution of complex eyes in arthropods and vertebrates via hypothetical steps illustrated by the eyes of other animals (Darwin, 1876: 143–146). Darwin outlined the important principle that the roadmap to ancestors runs through the characters of descendants.

But how would one know whether descendants were but little altered since their divergence from a common ancestor? One common strategy was simply to assume it. George Albert Boulenger (1858–1937) was an orthogenesis enthusiast and a zoologist at The British Museum (Natural History) who described more than 2,500 species of vertebrates over the course of an astonishingly prolific career (Watson, 1940; Romero, 2001). Although he had never set foot in Africa, Boulenger was considered to be the world leading authority on African freshwater fishes. In a revision of African clariid catfishes, Boulenger aligned five extant genera into an orthogenetic series, which he thought recorded the inescapable thrust toward increasingly eel-like forms. He wrote:

> I do not believe for one moment that the more generalised forms here described represent the actual ancestors of the terminal type, as it is not likely that they should coexist at the present day; but I regard the apparent links of the chain as side branches of a continuous stem, as the close allies of these extinct forms, and for the purpose of the study of the lines of derivation they are just as good examples as if they were the actual ancestors, because they must be so very similar to them. (Boulenger, 1907: 1062)

They must be, right? Who needs fossils when extant forms freeze into neat phylogenetic series after they diverge from their relatives.

This strategy of seeing living species as reliable reflections of ancient ancestors was especially common amongst orthogeneticists, but it wasn't universally respected and was frequently ridiculed (Jepsen, 1949). One risk, as Konrad Lorenz (1941: 288) warned, was that this shoddy strategy played into the hands of evolution deniers (also Haeckel, 1891: 466; Abel, 1912: 637). But a bigger problem was that organisms are evolving mosaics. As we saw in Chapter 3, nineteenth-century morphologists and systematists had already come to realize that taxa could not simply be lined up without causing character conflict. The implication was that the linearity of descent could only emerge when organisms were mentally deconstructed.

7.5 Othenio Abel: A Straight-Thinking Orthogeneticist

Ironically, it was a straight-thinking orthogeneticist who helped the scientific community to see that many phylogenetic highways are merely byways in disguise. Austrian paleontologist Othenio Abel (1875–1946) was one of the most prominent orthogeneticists of the generation before Schindewolf. This "forgotten Darwin of the earth sciences" (Kutschera, 2007: 173) developed a distinctive evolutionary theory that tried to explain the linear course of evolution by the action of a mechanism of evolutionary inertia, in conscious analogy to inertia in physical systems (Abel, 1929; Levit and Olsson, 2006). Once evolution in a lineage had started to move in a certain direction, it would become increasingly entrenched in a unidirectional, rectilinear trend, unless knocked off course by major environmental changes. Abel believed that his law of inertia (*Trägheitsgezets*) encompassed and explained three lower-level phylogenetic laws (*Grundgesetzen*): (1) orthogenesis; (2) Dollo's law of the irreversibility of evolution, so named by Abel in recognition of his friend Louis Dollo's evolutionary insights; and (3) Rosa's law of the progressive reduction of variation in evolving lineages, which Abel held to be equivalent to Cope's law of the unspecialized (Abel, 1929: 11–13). But Abel did not construct orthogenetic series by meekly following fossils along a stratigraphic gradient. Stratigraphy could inform the reconstruction of phylogenetic lineages, but if morphological comparisons contradicted these, they took precedence (Abel, 1920: 20). Of course, as we saw earlier, this was scarcely a unique attitude, as Hyatt before him and Schindewolf after him were among many who likewise produced orthogenetic series without stratigraphic input. Where Abel did distinguish himself was the care with which he formulated his *Formenreihen*.

Abel borrowed and promoted a concept that he translated as *Spezialisationskreuzungen*, after his friend Dollo had published it in 1895 as *chevauchement des spécialisations*, or the crossing of specializations (Rieppel, 2012b). The crossing of specializations results from committing the sin of the scala: arranging collateral relatives as ancestors and descendants. Because collateral relatives have different mosaics of primitive and derived character states, arranging them as ancestors and descendants causes conflicts in character state polarity when different characters are traced. Because ancestors cannot have derived character states compared to descendants, the crossing of specializations diagnoses cases where presumed ancestors and descendants are actually collateral relatives stemming from a hypothetical common ancestor. To discover such branching of lineages requires the careful tracing of *Formenreihen* of multiple characters. It was crucial to distinguish what Abel called *Anpassungsreihen* (series of adaptations), *Stufenreihen* (step series), and *Ahnenreihen* (series of ancestors) (Abel, 1912: 632–640; 1914: 353–357; 1920: 21–22). According to Abel, *Anpassungsreihen* did not trace homologies, while the other two referred to lineages of characters and taxa, respectively.

Abel argued that tracing lineages by following single characters had frequently lured unwary phylogeneticists into mistaking *Stufenreihen* for *Ahnenreihen*. Only by tracing thick bundles of characters changing in the same direction can phylogenetic trunks be distinguished from diverging twigs. By distinguishing these different morphological series, Abel showed how branching evidence can reveal linear lineages. The evolution of characters at different levels in phylogenies was later called *heterobathmy* by the botanist Armen Takhtajan, and this ugly term became part of cladistic vocabulary, with Hennig recognizing that "Heterobathmy of characters is therefore a precondition for the establishment of the phylogenetic relationship of species and hence a phylogenetic system" (Hennig, 1965: 107). Despite Abel's contributions to transforming the impressionistic nature of phylogeny tracing, he is scarcely a household name in systematic biology today. Olivier Rieppel deserves credit for highlighting his contributions (Rieppel, 2012b, 2013b, 2016; Rieppel et al., 2013).

Abel's conceptual scalpel cut through many ancestral series. Earlier in this chapter, we saw that he heralded Hilgendorf's and Kovalevsky's work on fossil lineages as harbingers of the new paleontological approach to phylogenetics, but even these classic cases weren't perfect. Recent research suggests that Hilgendorf's longest lineage of snails lined up several collateral relatives (Rasser, 2014), and the Old World trunk of Kovalevsky's celebrated horse phylogeny turned out to be a spurious concatenation of New World migrants (Abel, 1920: 22; Simpson, 1953: 364–366; Gould, 1998: 153–155). But these revisions are not complete phylogenetic defeats. As Abel pointed out (1920: 22), *Stufenreihen* "have as phylogenetic series no less significance as *Ahnenreihen;* their recognition as *Stufenreihen* makes capturing the existing genetic connections easier and forms, so to speak, the skeleton of a phylogenetic lineage, which through continued research and the discovery of new forms can be fleshed out more and more."

Rieppel (2016: 75) considers this "an enormous conceptual step away from the Haeckelian search for ancestors and descendants, and toward the search for common ancestry instead." Rieppel (2012b: 328) regards this "replacement" or "abandonment of ancestor-descendant sequences in favour of a search for relative degrees of phylogenetic relationships" (Rieppel, 2012b: 333) as a decisive step toward Hennig's phylogenetic systematics, a view echoed by other recent commentators on Abel's work (Willmann, 2003; Nelson, 2004). I agree up to a point. We shouldn't fail to notice that in Abel's own research there is precious little evidence for such a shift in focus. His pioneering epistemological insights are embedded within the traditional ambitions of the ancestor-hunting paleontologist.

The goal of Abel's phylogenetic research was "the determination of the ancestral forms of living animals and plants, but also the tracing of the history of descent of the *Formenreihen* that went extinct during earlier times in earth history" (Abel, 1920: 20). A sentence like this could have been written by the founder of phylogenetics himself. It is not the only one. The study of comparative morphology

> can sketch in a hypothetical way the image of an ancestral form of a living type, but only paleozoology is able to acquaint us with the historical documents of the history of descent and provide the evidence on which the history of descent of the animal phyla is built. The goal of paleozoology is therefore to provide proof of such ancestral forms. (Abel, 1920: 20)

Abel appreciated that the crossing of specializations allowed the discovery of side twigs (*Seitenzweigen*) branching from the main stem (*Hauptstamm*) (Abel, 1914: 354–355). As soon as a presumed fossil ancestor was rediagnosed as a collateral relative, a new phylogenetic lineage beckoned to be fleshed out. Abel's 1920 textbook *Lehrbuch der Paläozoologie* (*Textbook of Paleozoology*), for instance, is laced with speculations about ancestors and ancestral groups. He had no qualms about labeling fossils as ancestors, like the fossil insect *Eugereon boekingi* (Abel, 1920: 127), or designating fossil ancestral groups (*Ahnengruppen*), like the Cystocidaroidea as ancestors of modern sea urchins (Abel, 1920: 297), or viewing embryos through a recapitulationist lens, for example, deriving molluscs from trochophore larva-like ancestors (Abel, 1920: 129). This shows that Abel fell comfortably within the phylogenetic tradition founded by Haeckel, even though his conceptual contribution of distinguishing *Anpassungsreihen*, *Stufenreihen*, and *Ahnenreihen* was a deliberate corrective of Haeckel's more impressionistic methods.

Abel formalized the widespread commonsensical view that any given fossil was unlikely to be a lineal ancestor of a known taxon. As we saw in Chapter 4, Haeckel certainly knew this, and in his 1870 Presidential Address to the Geological Society, T. H. Huxley had already coined the concepts of linear and intercalary types for fossils that were either directly ancestral or collaterally related to a taxon (Huxley, 1870: xlvi). Although intercalary types, such as *Archaeopteryx*, were perched on side branches of major transitions in animal body plans, they could still provide valuable clues about the possible nature of real intermediate forms. This is both the curse and promise of phylogenetics: most roads to ancestors run through collateral relatives.

7.6 Linear Thinking with Branching Evidence

An inability to determine whether a fossil was on the main line or a side branch didn't deter biologists and paleontologists from drawing conclusions. Two years after his former boss Hyatt published his orthogenetic Steinheim snail research, zoologist William Keith Brooks wrote an essay on the value of fossils for phylogenetics. Pointing out that it is difficult to determine whether a fossil belongs in the direct line of descent of a living taxon, he stressed that if the fossil has a combination of characters that is compatible with that hypothesis, one can be confident that it has "a general resemblance [to the actual ancestor] ... This, after all, is the essential thing, the gist of the whole matter" (Brooks, 1882b: 203). Although you may disagree with Brooks' conclusion that knowing "the precise line of descent has no more scientific interest than the exact pedigree of each person would have" to an anthropologist (Brooks, 1882b: 203), this nevertheless paints the epistemological outlines within which phylogeneticists still work today. We observe the tips of trees to infer what may have happened along the branches that connect them.

It is interesting to see how this tension between the linear and branching aspects of evolution played out in an early study of a segment of our favorite lineage. Anatomist Gustav Schwalbe (1844–1916) is famous in paleoanthropology for being the first to propose that *Homo erectus* and *Homo neanderthalensis* are direct ancestors of modern humans (Schwalbe, 1899, 1906). Although Neanderthals are no longer seen as our direct ancestors, recent work suggests that *H. erectus* may continue to repose on its ancestral throne (Parins-Fukuchi et al., 2019). Schwalbe's fame for recognizing this was based on scooping Eugène Dubois in the most cruel and selfish way. In 1891 and 1892 Dubois had discovered the then oldest known hominin fossils in Java, and named them *Pithecanthropus erectus* (now *Homo erectus*) in a 1894 monograph (Theunissen, 1989; Shipman, 2001; Shipman and Storm, 2002). This name honored Ernst Haeckel, who had decades earlier proposed the genus *Pithecanthropus* as a placeholder for the yet to be discovered bipedal and speechless ancestors of modern humans. Dubois concluded that his fossil skullcap, femur, and molar represented this missing link between apes and modern humans. But the monograph wasn't the final word. Dubois hadn't fully prepared and cleaned his fossils, and he had not attempted detailed comparisons with the fossils of Neanderthals.

When he returned to the Netherlands in 1895, he continued his work, now aided by access to better equipment, museums, and libraries. Interest in his discovery was great, even though criticisms rang as loudly as praise. The unkindest cut of all came from Gustav Schwalbe, a supposed ally. Schwalbe had contacted Dubois in 1896 and sent him the deformed femur from a deceased actor that showed similarities to the femur of *P. erectus*. Dubois reciprocated in kind, sending Schwalbe a plaster cast of the inside of *P. erectus'* skullcap and allowing him access to the fossils. In turn, Schwalbe supported Dubois' interpretation of the fossils against dissenters in his lectures. But when Dubois wrote to Schwalbe, at the end of 1897, that he planned to conduct a more detailed study of his fossils, Schwalbe's reply was ominous: "I shall be glad to leave the femur to you for some time before publishing. Certainly we are competitors. . . I have almost finished a manuscript on the skull and another one on the femur" (Schwalbe in Shipman, 2001: 353). This was dreadfully unwelcome news.

In 1899 Schwalbe repaid Dubois' kindness of providing him access to his fossils by publishing a 225-page monograph on the skullcap of *P. erectus* in the inaugural issue of a journal that Schwalbe edited, the *Zeitschrift für Morphologie und Anthropologie*. Schwalbe had been editor of the journal's predecessor *Morphologische Arbeiten*, and he used his study of *Pithecanthropus* to refocus the journal on human origins. Schwalbe's betrayal provided him with a springboard for a comprehensive reconstruction of human ancestry, but it probably contributed to Dubois locking his fossils away from friend and foe for the next several decades (Theunissen, 1989: 121). Based on a detailed comparison of skull characters, Schwalbe concluded that apes, humans, and Old World monkeys were more closely related to each other than either of them was to New World monkeys (Schwalbe,

1899: 211). He rejected the possibility that humans and apes had evolved from Old World monkeys in part because of the distribution of characters – they lacked a host of characters present in Old World monkeys – and in part because it would imply what Schwalbe considered an implausible reversal of an evolutionary trend in the width of the face and the prominence of the nose.

Schwalbe calculated an index for the interorbital width of primate skulls for a broad selection of species, and he accorded the progressive evolution of this character great importance as a phylogenetic guide. If apes and humans had evolved from primitive Old World monkeys with close-set eyes, this would imply a decrease in this index from a prosimian-like ancestor to Old World monkeys, followed by an increase in the index en route to apes and humans. Schwalbe, who thought about evolution in rather simple progressive terms, considered such a reversal in the width of the face as well as the size of the nose to be most unlikely (Schwalbe, 1899: 206–210). He therefore preferred rooting the human lineage in prosimians, relegating Old World monkeys with their close-set eyes to a side branch. Likewise barring living apes from the direct line of human ancestry, Schwalbe rooted our deepest origins in a series of hypothetical primate ancestors. But as we shall see, fossils made two of the later steps to humankind concrete.

Schwalbe was pleased to find that Dubois' *Pithecanthropus* skullcap contained enough information to calculate its interorbital index. It clearly fell outside the range for narrow-faced monkeys, but comfortably within that for hominids. He calculated that Neanderthals had even wider faces with an index near the top of the range for modern humans. Based on an assessment of many characters and measurements, Schwalbe concluded that, morphologically speaking, the *Pithecanthropus* skull fell between monkeys and Neanderthals, while Neanderthals bridged to some degree the gap between *Pithecanthropus* and modern humans. He did, however, not commit himself to a definite phylogenetic placement of either fossil species. *Pithecantropus*, he thought, could stand alongside apes and Old World and New World monkeys in a yet to be determined position (Schwalbe, 1899: 227), but this didn't dim the light the fossils could shed on human ancestry. Writing in a later overview of human evolution, Schwalbe explained that he didn't care whether *Pithecanthropus* and Neanderthals were part of the direct lineage to modern humans or represented side branches because "also in the latter case must the ancestors have looked similar to the preserved remains [of these fossils]" (Schwalbe, 1906: 14).

Schwalbe adopted the same attitude towards the phylogenetic significance of fossils that Brooks had articulated a few decades earlier. It didn't matter if fossils that seemed to be morphologically intermediate between taxa were part of the same lineage or were merely side branches. What was important was that they looked like intermediate forms. Demanding more precision was paleontological pedantry. Unsurprisingly, Haeckel agreed (Haeckel, 1908: 42–43). Although a direct line of ancestry could perhaps not be drawn through the three species, Schwalbe drew it through their genera instead, from *Pithecanthropus* to *Homo* (Schwalbe, 1906: 14–15).

Mentioning Schwalbe's contributions to understanding human prehistory, Ian Tattersall concluded that slotting these fossil hominins into the lineage of modern humans showed that early paleoanthropologists had been "anxious… to fit the little they knew within a logical and intelligible picture" (Tattersall, 2015: 40). I don't get the impression of an anxious mind at work when reading Schwalbe. He considered both options, realized that no firm decision could be made, and concluded that it didn't matter if the fossils were ancestors or close relatives. His crucial insight was that many of their characters, such as their brain volumes, made sense as morphological intermediates between modern humans and nonhuman primates. For Schwalbe these characters traced threads of homology, even though he could not decide whether these threads only traced series of characters or entire organisms. In his 1920 textbook, Abel considered *Pithecanthropus erectus* and Neanderthals to be part of our *Stufenreihen*, while conceding that the human *Ahnenreihen*, for the time being, remained hidden in the mists of time (Abel, 1920: 456–457).

Several recent authors draw what they regard as an important epistemological line between seeing evolution as a transforming series of whole organisms and interpreting it as a flowing mosaic of independently evolving characters (Willmann, 2003; Nelson, 2004; Rieppel, 2012b, 2016, 2020). This line is said to separate Haeckelian phylogenetics from the epistemically sharpened version advocated by practitioners of the following generation, with Dollo's and Abel's insights into the crossing of specializations marking the start of a shift from trying to discover real ancestors to reconstituting hypothetical ancestors as character blends drawn from disassembled descendants. Of course, I won't deny that an arrow of epistemological improvement runs through the history of phylogenetics, but I feel that recent histories, such as those offered by Williams and Ebach (2008) and Rieppel (2016), fail to highlight the continuity in the goals and methods of pre-Hennigian phylogenetics, while exaggerating the methodological novelty introduced in particular by Othenio Abel.

Rieppel (2016: 77) concludes that "Abel's approach to phylogeny reconstruction thus differed in fundamental ways from Haeckel's" in recognizing the crossing of specializations, in expecting that ancestors will only rarely be discovered in the fossil record, in replacing discoverable ancestors with hypothetical ancestors, in not divining ancestors from embryos, and in reconstructing lineages as evolving character mosaics rather than whole organisms (Rieppel, 2020: 219). I see no fundamental differences, only ones of degree, especially when one consults Haeckel's later writings. The biggest difference was that Abel was more careful in constructing his morphological series, and he was more diligent in diagnosing the character conflict that indicated lineage splitting. But Haeckel split lineages as well when character conflict demanded it. For example, in his *Systematische Phylogenie* he removed lobe-finned fishes (crossopterygians) from the ancestry of lungfishes (Haeckel, 1895: 261) – a hypothesis he had endorsed in earlier works such as *Anthropogenie* and *Natürliche Schöpfungs-Geschichte*. He did this because he recognized a crossing of specializations, with lungfishes being primitive compared to lobe-finned fishes in possessing a cloaca, but derived in having an autostylic jaw suspension (the upper jaws are fixed firmly to the skull). He therefore placed them as two diverging branches in his phylogeny of fishes.

Rieppel's remaining criteria similarly fail to separate Haeckel from Abel. As we saw in Chapter 4, Haeckel frequently created hypothetical ancestors. Peter Bowler must have been reading a different Haeckel than I to be able to claim that Haeckel would invoke hypothetical ancestors "[o]nly when absolutely necessary" (Bowler, 1996: 60). Although Haeckel loved to welcome fossils as ancestors, as we saw in Chapter 4, he knew that specialized characters often barred fossils from functioning in ancestral roles, just like living taxa. Finally, Abel unhesitatingly used recapitulation as a phylogenetic tool, even referencing the biogenetic law in doing so (Abel, 1920: 205).

What impresses me most when I survey the phylogenetic research of the workers discussed in this and the adjoining chapters is that we can see the outlines of a tradition taking shape. In this book I have proposed to call this tradition narrative phylogenetics (see especially Chapter 6). It originated during the last four decades of the nineteenth century with the embryological and morphological work of Ernst Haeckel and Carl Gegenbaur and their students, as well as evolutionary paleontologists like Wilhelm Waagen and Franz Hilgendorf. Although this tradition emerged most prominently in Germany, narrative phylogenetic research was done elsewhere in Europe, Russia, and the United States as well. I don't mean to imply that these workers explicitly identified themselves as part of a tradition, but their goals and methods share an epistemological core. Their goal was to reconstruct the morphological transformations involved in the evolution of traits and taxa by arranging evidence or hypothetical ancestors into continuous series. Although the cladistic revolution made much of the methodological core of this approach untenable, it still survives today, for instance, in the form of scenario-driven attempts to reconstruct the evolution of animal body plans, as we will see in the next two chapters.

The central challenge of lineage thinking is how to translate static, and mostly synchronic, evidence into dynamic and diachronic evolutionary narratives. The first step that lineage thinkers take is to arrange characters into a smooth continuous series based on a gradient of similarity. This strategy traces back to pre-evolutionary roots (Rieppel, 2020), and it blossomed in the hands of evolutionary morphologists from the second half of the nineteenth century onward. The same procedure is at the core of establishing transformation series of ordered homologous character states, and it also represents Remane's third principal homology criterion (connecting dissimilar homologues via intermediate forms) (Maslin, 1952; Lipscomb, 1992; Williams and Ebach, 2008). Surprisingly, Williams and Ebach (2008: 181) argue that considering characters as linear series of states lacks a "reasonable empirical basis." This method arranges observational evidence into an ordered series by following the purely morphological criterion of similarity. What can be more empirical than that?

Williams and Ebach's (2008: 141) more specific criticism is that Remane's third homology criterion, and by implication all attempts to arrange homologous traits into linear series, involves "dynamic rather than static comparisons – both invoking transformations." I disagree. Transformations, be it evolutionary or formal, can indeed be imagined to connect the steps in such series, but they need not necessarily be invoked. Idealistic morphologists, like Rainer Zangerl, used precisely this produce of the ordering of morphological character states based on the similarity of adjacent states, while simultaneously maintaining the epistemological independence of comparative morphology from any phylogenetic speculations that might be built upon it (Zangerl, 1948). Such orderly and ordered series form the epistemic backbone of the purely morphological conclusions of the idealistic morphologist as well as the phylogenetic inferences of evolutionists. It is only when phylogeneticists attempt to determine the polarity of the morphological series that a dynamic interpretation is introduced. It is at this point that purely observational evidence becomes animated by process speculations, and it is here that evolutionists working with living taxa mentally transport synchronic evidence back in time to furnish the steps of phylogenetic scenarios.

As the only evidence that comes with a temporal tag, it is unsurprising that fossils have special status in lineage thinking. Tracing lineages directly in the rock record requires the assumption that succeeding forms represent ancestor–descendant lineages, an assumption made explicit by early workers, like Wilhelm Waagen and Alpheus Hyatt. Even when the sedimentary record was dense enough to guide the construction of morphological series, fossils were never aligned blindly along a stratigraphic gradient. Reconstructions were kept on track by following smoothly graded transitions in morphology. Discontinuities were generally thought to betray the arrival of immigrants or the splitting of lineages, rather than a sudden change in the direction of morphological evolution along a single lineage. This is why Hyatt interpreted the two parts of Hilgendorf's longest snail lineage, which were separated by a reversal in the direction of several traits, as a distinct lineage, a correct inference as it turned out. This type of transformational reasoning is a key feature of lineage thinking.

As we saw, Gustav Schwalbe excluded Old World monkeys from the direct ancestry of humans because this would imply what he considered an unlikely reversal in the direction of the width of the face and the prominence of the nose, and Otto Schindewolf detected a lineage split in ammonoids by a sudden reversal in the shape of the shell sutures (Schindewolf, 1927: 144). This criterion of graduated morphological similarity has also been used to split transformation series and to reject them if adjacent character states are dissimilar (Lipscomb, 1992). The methodological improvements that pre-Hennigian biologists and paleontologists like Othenio Abel, Adolf Naef, and Sinai Tschulok introduced into phylogenetics (Willmann, 2003; Williams and Ebach, 2008; Rieppel, 2016) further helped with the recognition of the dangers of strictly linear lineage thinking.

The seeds of some of these reforms had already sprouted in the minds of their predecessors. Both Haeckel (1866b: 255–256) and Gegenbaur (1870: 74–75), for instance, had already warned their readers against committing the sin of the scala. When we study living taxa "even where we can identify a single feature as a transitional form, will many others in the strictest sense forbid us to interpret the entire organism in this way" (Gegenbaur, 1870: 75). In other words, don't think too linearly about lineages. Don't mistake *Stufenreihen* for *Ahnenreihen*. This message sank in gradually, and sometimes not at all. Haeckel certainly didn't always heed his own advice. Many orthogeneticists willfully ignored the significance of the crossing of specializations. By arranging taxa or hypothetical ancestors into lineages on the basis of a single or a small number of characters, they and other lineage thinkers applied a principle we might call *character hitchhiking*. The late Dick Jefferies provided a nice example in a 1988 paper where he concluded that spiral cleavage is more primitive than radial cleavage because platyhelminths, which have spiral cleavage, don't have an anus, an assumed primitive character (Jefferies, 1988: 6). Jefferies considered the lack of a through-gut in platyhelminths to be primitive, with the implication that their other characters were likely to be primitive as well.

Such reasoning was often left implicit, but W. B. Crow (1926: 95) made it explicit in his review on the relationship between phylogeny and the natural system, writing that it had been "definitely established in a large number of cases that. . . if a number of species may be arranged in a series in respect to the degree of development of one character they are also *ipso facto* arranged in a series in respect to other characters." The crossing of specializations put paid to this argument, as did parallel evolution, as Crow realized. Yet, the principle of character hitchhiking can be discerned at the heart of much lineage thinking, and this is not surprising. It was a narrative shortcut that allowed lineage thinkers to exclude conflicting evidence that would split the scenario into multiple directions and so maintain one lineage as the central subject of a single storyline. As discussed in Chapter 5, scenario builders must apply the unequal eye to evidence, and the assumption of a tight correlation between evolving characters helped keep storylines straight, even when they weren't. The use of character hitchhiking as a narrative device to streamline stories was still common enough in 1941 that Konrad Lorenz felt he needed to warn against it in one of the only italicized sentences in his 100-page monograph on duck behavior (Lorenz, 1941: 289).

Olson (2012) discusses a striking example of applying an unequal eye to evidence in the service of telling linear evolutionary stories. Botanist Irving Widmer Bailey (1884–1967) developed a comprehensive and influential view of angiosperm evolution based on what he perceived were trends in the evolution of xylem vessels. He viewed angiosperm evolution through the lens of a scenario that focused on how one character system had evolved. His linear phylogenetic perspective paid close attention to ontogeny, implicitly relied on character hitchhiking, assumed that the morphology of extant species retains "comprehensive and reliable evidence regarding successive stages of the origin and specializations of vessels" (Bailey, 1944: 424), and posited that morphological trends in vessel specialization were irreversible. The happy consequences of this focus and these assumptions were that one didn't have to rely on fossils or have to confront "serious uncertainties regarding the interpretation of hypothetical phylogenetic sequences" (Bailey, 1944: 424). If one arranged angiosperms according to graded series of vessel specializations, one could simply scan the pathways of evolution from the barcodes of modern morphology.

Bailey's is a good example of the kind of scenario-driven lineage thinking that emerged with the birth of phylogenetics. Although he didn't appear to have been an orthogeneticist, he shared their epistemic approach to phylogeny reconstruction. Olson (2012), however, fails to recognize the place of Bailey's thinking in this tradition. He faults the Baileyian scheme for linear thinking, for using just one character system, for making strong and unproven assumptions about the evolutionary process, for using homoplasies, and for using recapitulationist reasoning. Olson is right that modern systematists will not be convinced by Bailey's approach, although I was surprised to read that Olson thinks "that homoplasies are useless for the inference of relationships" (Olson, 2012: 161). Olson's criticism that the Baileyian approach uses too few characters to be able to infer a tree, and that it is unclear "how a linear primitive–advanced scheme can relate to a branched phylogeny" (Olson, 2012: 155), show that he is misjudging lineage thinking from a taxic or tree-thinking perspective. Bailey didn't try to group taxa into nested sets on the basis of a broad sample of characters. He understood that angiosperm classification ultimately required the consilience of different character systems (Bailey, 1944: 425–426), but he believed that his vessel-based phylogeny provided an essential framework for filling in these details. By accepting that the linear view of evolution is incompatible with evolution's branching chronicle, Olson falls into the same conceptual trap as other contemporary writers (see Chapter 11).

We see an increasing awareness of the complex relationship between the linear and branching aspects of phylogenies in works from the late nineteenth and early twentieth centuries. Gustav Schwalbe resolved the tension between lineage and tree thinking by admitting that *Pithecanthropus erectus* and Neanderthals were possibly sidelines of our lineage, but he preserved the linearity of descent by lining up their genera. Othenio Abel did the same (e.g. Abel, 1919: 157). Abel's contemporary, the biologist Sinai Tschulok (1875–1945), considered trees that lined up supraspecific taxa to be useful for teaching, but he pointed out that they hid details that could emerge only when zooming in on lower taxonomic levels, even though one could never zoom in sufficiently to see real ancestors (Tschulok, 1922: 209). Sadly, many modern commentators, like Olson (2012) and Härlin (1998), fail to understand how different the goals, perspectives, and approaches of their predecessors were when accusing them of committing the sin of the scala because they lined up paraphyletic higher taxa into lineages. Paraphyletic taxa may be bad taxonomy, but as theoretically astute thinkers realized (Naef, 1919: 49–51), it reveals an underlying phylogenetic truth. While on the surface many earlier workers may appear to be locked in the straitjacket of the scala, a closer look often reveals a more sophisticated understanding of the relationship between the branching realm of systematics, and the linear realm of evolutionary lineages. As we will see in Chapter 11, a cladistic blindfold is hampering the theoretical acuity of a surprising number of modern researchers, educators, and science popularizers as they fail to see the phylogenetic trunks through thickets of taxonomic twigs.

Bishop, occultist, and biologist William Bernard Crow (1895–1976) raised the principle that hypotheses of the direct descent of supraspecific taxa could not be disproven by character variation on lower taxonomic levels to a phylogenetic law. He called it Cope's Law (not to be confused with Cope's law of the unspecialized or Cope's rule of phyletic size increase), to honor Edward Drinker Cope's elaboration of this argument (Cope, 1896: 84–85). Illustrating it with the same equine example that Cope had used, Crow (1926: 144) wrote that "although *Hippotherium mediterraneum* cannot be the ancestor of *Equus caballus* this is no proof that the genus *Hippotherium* is not ancestral to the genus *Equus*." Although there was little chance of observing ancestors, and mosaic evolution often fractured ancestral portraits, the characters of near relatives could still give an impression of the likely flow of character transformation when arranged into smooth morphological gradients. Ancestral state reconstruction through the careful triangulation of traits was an approach that only achieved methodological hegemony after Hennig, but these early-twentieth-century workers already understood that tracing the evolution of morphology required one to consider the systematic relationships of taxa, if these were known, and include fossil taxa.

If stratigraphy, morphological similarity, and systematic relationships clashed, linear thinking often prevailed. Abel's "phylogenetic development of the horse hand" (Abel, 1919: 863) presents the familiar sequence of the progressive development of the middle toe and the regression of the remaining digits (Figure 7.5). Abel ordered seven fossils in a linear series according to stratigraphic age, except for the fifth fossil, which is an offshoot of the third fossil, and is actually the youngest fossil of all. That Abel nevertheless presents this as a phylogenetic series (albeit a *Stufenreihe* rather than an *Ahnenreihe*) shows that

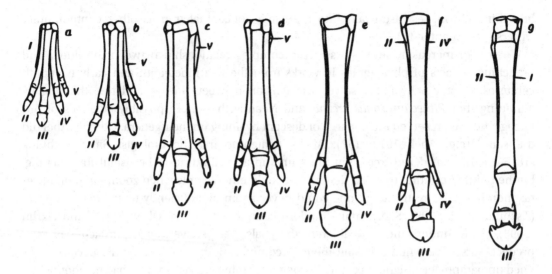

Figure 7.5 Othenio Abel's phylogenetic development of horse limbs. See text for explanation (after figure 657 in Abel, 1919).

morphological intermediacy was a more important criterion for arranging evidence than either stratigraphy or systematic branching order when the objective was to form an impression of evolutionary lineages. Most contemporary researchers probably agreed with Abel that stratigraphic evidence could buttress phylogenetic hypotheses, but not dictate them (Tschulok, 1922: 191–192; Crow, 1926: 99). Abel was just as susceptible to the beguiling power of linear lineage thinking as his contemporaries when he slotted a young fossil from a side branch into the role of an older ancestor.

Although the main focus of this chapter was orthogenesis, I have argued that the epistemic core of reconstructing fossil lineages of ancestors and descendants in the late nineteenth and early twentieth centuries was accepted by evolutionists of all stripes, and it traces back to pre-evolutionary idealistic morphology. Orthogeneticists and Darwinian evolutionists disagreed fundamentally about the evolutionary mechanisms responsible for the patterns they reconstructed, but commentators (e.g. Bowler, 1992a; Levit and Olsson, 2006; Ulett, 2014) have often failed to highlight this shared epistemic core. Observable evidence was arranged into finely graded linear series according to morphological criteria. Where possible, fossils were aligned with the arrow of stratigraphy, but many scenarios were put forward even when they went against the sedimentary grain. Because many orthogeneticists and non-orthogeneticists alike used recapitulation as a phylogenetic guide, the morphological steps in many of these lineages were thought to be suggested by the form changes observable in embryos (Smith, 1900; Rainger, 1981; Bowler, 1992a; Allmon, 2020). Alpheus Hyatt's 1866 paper on fossil nautiloids and their relatives was an important starting shot for this recapitulationist approach to fossil phylogenetics (Hyatt, 1866). As historical fate would have it, Haeckel independently laid the foundations of his own version of recapitulation theory in the same year as Hyatt. In the next chapter, we will see how he wielded this tool to understand the origin of animals.

Seeing Animal Ancestors in Embryos

The last common ancestor of animals was popular quarry in the hunt for ancient ancestors. Understanding the origin of Metazoa was one of the first evolutionary questions that occasioned an intense period of competitive storytelling in the second half of the nineteenth century. Leading biologists from Britain, Germany, and Russia drew up imaginative scenarios for the emergence of animals from protist ancestors. The evidence gathered by the study of animal embryos was their shared foundation, but it was patchy and open to conflicting interpretations. Guided by their personal evolutionary intuitions, the participants in this debate forged diverging narrative paths to the origin of animals. In this chapter we retrace the main tracks in these debates to discover which one leads to a hypothesis of maximal explanatory power.

8.1 Haeckel's Gastraea Theory

A tiny cup-shaped animal ancestor floats at the heart of one of the most famous, fertile, and controversial phylogenetic theories ever proposed. Haeckel named it Gastraea (Haeckel, 1872, 1874, 1877, 1885) (Figure 8.1). Gastraea's birth certificate was a monograph on calcareous sponges that Haeckel published in 1872. Gastraea appeared as the fifth ancestor in a lineup that culminated in Olynthus, the last of nine hypothetical ancestors in the lineage of calcareous sponges. It escaped from these specialist confines when Haeckel presented Gastraea to the wider zoological community a year later and to English-speaking readers the year after as the last common ancestor of all animals (Haeckel, 1874). Despite Gastraea's simplicity, the details of its morphology were key to its explanatory power. Its inner and outer cell layers represent the phylogenetic origin of the primary germ layers, while its blind-ending gut was the first animal organ ever to evolve. Haeckel believed that Gastraea's simple structure could unite the phylogenetic diversity of the entire animal kingdom. Previously, while traveling aboard *H.M.S. Rattlesnake* in 1848, a young T. H. Huxley had his first major scientific breakthrough when he realized that the two-layered structure of polyps and jellyfish didn't just indicate their close affinities as cnidarians, but also suggested much more distant ties of homology. Following a paragraph where he explains what criteria he used to discover homologous structures, Huxley writes that "it is

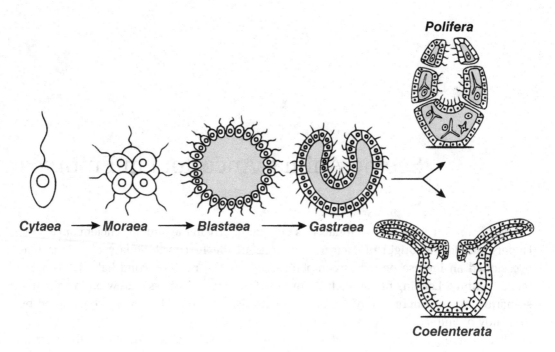

Figure 8.1 Schematic of Haeckel's Gastraea theory
(after figure 1 in Malakhov et al., 2019, reprinted by permission from Springer, Copyright, 2019).

curious to remark, that throughout, the outer and inner membranes [of cnidarians] appear to bear the same physiological relation to one another as do the serous and mucous layers of the germ" (Huxley, 1849: 426). Earlier in the century Heinrich Pander and Karl Ernst von Baer had documented the layered structure of vertebrate embryos, and Pander had used the terms "serous" and "mucous" to refer to the ecto- and endodermal layers. But for Huxley these homologies weren't evolutionary. It was Haeckel who, two decades later, interpreted these correspondences as the evolutionary glue that united the entire animal kingdom.

Haeckel founded his Gastraea theory based on observations collected from all corners of the animal kingdom. He noted that two-layered stages were present during the development of all animals and that an invaginated gastrula stage, "in a more or less modified form" (Haeckel, 1874: 156), could be observed in the development of some sponges, cnidarians, ctenophores, platyhelminths, nematodes, chaetognaths, bryozoans, tunicates, amphioxus, phoronids, annelids, arthropods, echinoderms, molluscs, and gephyreans[1] (Haeckel, 1874: 155–156). In a great irony, Haeckel got gastrulation wrong in sponges. Although sponge embryologists disagree about whether sponges have a form of gastrulation (Leys and Eerkes-Medrano, 2005; Ereskovsky and Dondua, 2006), Haeckel mistook the metamorphosis of calcarean sponges for gastrulation. Sponges lack clear homologues of the

---

[1] A now obsolete taxon comprising priapulids, echiurans, and sipunculids.

ecto- and endodermal germ layers of eumetazoans. This, however, hasn't diminished the staying power of the Gastraea theory.

Haeckel drew gratefully and explicitly upon the embryological work of the great Russian embryologist Alexander Kovalevsky, who had described these invaginated developmental stages across a wide range of animal phyla. Haeckel knew that there was variation in how a two-layered organization was achieved during the development of different taxa, but he considered invagination to be the primitive mode of gastrulation and the cup-shaped stage to reflect the nature of an ancient animal ancestor. He thought that embryos with more yolk had changed their mode of gastrulation to a greater or lesser extent, and seemingly completely different gastrulation modes, such as the transverse division of blastula cells to form an inner and outer cell layer in a process called delamination, would have evolved from invagination to shorten development (Haeckel, 1874: 157–160; 1877: 113–114, 149–150).

Haeckel didn't just propose a widespread developmental homology. His biogenetic law stipulated that the last common ancestor of animals didn't just have a gastrula, but *was* a gastrula (Haeckel, 1874: 157). For him the existence of Gastraea killed the widely held type theories of Cuvier and von Baer (Haeckel, 1874: 239–243). These zoologists had independently recognized the existence of four basic animal types – Radiata, Vertebrata, Mollusca, and Articulata – that represented distinct and incompatible functional morphological and developmental plans. If Gastraea was the last common ancestor of all animals, these types didn't stand in eternal isolation. Instead, they would be higher-level taxa that housed the endlessly modified descendants of Gastraea. They would signify nothing special, either morphologically or developmentally. They would just be like lower taxonomic ranks write large (Haeckel, 1874: 243).

The existence of Gastraea as an ancient but real organism was the beating heart of Haeckel's theory, so he had to at least try to answer the question of why and how Gastraea had evolved. The starting point was a hollow spherical protist ancestor, which he variously called Blastaea or Planaea. It was composed of a single cell layer made up of similar cells. Haeckel envisioned the evolutionary emergence of a division of labor between cells driven by the adaptive benefit of dedicating different specialized cells to providing locomotion and catching and digesting food. He considered the division of labor to be "the most important organizational law" of all (Haeckel, 1866b: 250). The animal hemisphere of this ancestral animal differentiated into locomotory cells that eventually became the ectoderm, while cells of the vegetal hemisphere specialized into nutritive cells (Haeckel, 1877: 157). He thought that food uptake and extracellular digestion would be enhanced if the vegetal pole would become increasingly concave, which would allow food to remain in contact with the nutritive cells for longer. He even proposed a hypothetical ancestor he called Depaea that was approximately halfway along this adaptive trajectory from spherical to fully invaginated form (Haeckel, 1889, 1895). It had a small invagination at the vegetal pole, which would attain its full extent in its descendant Gastraea.

Summarizing, Haeckel proposed his Gastraea theory to explain the phylogenetic origin of the primary germ layers and the gut of animals. The theory was based on developmental observations from a broad range of phyla, but the theory's real engine was the biogenetic law. Recapitulationist reasoning produced detailed ancestral portraits from a limited but widely distributed set of observations. Finally, the theory relied on two key assumptions: all modes of gastrulation other than invagination are derived, and the evolution of Gastraea was driven by the adaptive benefits of a division of labor between locomotory and nutritive cells. All these observations and evolutionary intuitions came under attack.

8.2 Lankester's Planula Theory

In the year that Haeckel published the first outline of his Gastraea theory, a young disciple of T. H. Huxley sketched his own scenario for the evolutionary origin of the primary germ layers and their relevance to animal phylogeny. As a schoolboy, Edwin Ray Lankester (1847–1929) had played truant to hear Huxley lecture (Lester and Bowler, 1995), and as a 25 year old he presented his thoughts on early animal evolution to students in a course on animal classification at the University Museum of Oxford in 1872 (Lankester, 1873). Lankester had studied with Haeckel in 1871, but he conceived his own ideas about early animal evolution before Haeckel's 1872 paper had appeared (Goodrich, 1931). Like Haeckel, Lankester had noticed that a two-layered stage was common in the development of a wide range of animals, and like Haeckel he used recapitulationist logic to turn this developmental homology into a hypothetical ancestor. Instead of using Haeckel's term gastrula, however, Lankester described this two-layered form as a planula, but otherwise the original formulation of his Planula theory closely agrees with Haeckel's Gastraea theory. Like Haeckel, Lankester observed that the blastulae of many animals invaginated to become gastrulae, but he noted that delamination was the more common mode in cnidarians (Lankester, 1873: 328). Yet Lankester agreed with Haeckel that blastulae primitively gas-trulated by invagination, with other modes evolving to accommodate increased amounts of yolk or to abbreviate development (Lankester, 1873: 328). Despite these agreements, Lankester's thinking soon diverged from Haeckel's.

In 1876 Lankester responded to Haeckel's 1874 paper that introduced Gastraea to English-speaking readers. He reviewed the phylogenetic distribution of the different modes of gastrulation and agreed with Haeckel that invagination could indeed be primitive, but that more comparative data were needed to settle the question (Lankester, 1876). A year later Lankester rejected Haeckel's conclusions. He acknowledged that invagination may be the more common mode of gastrulation, but he now decided that delamination was primitive. However, he didn't change his mind because new and convincing data had tipped the balance. He could only point to "one *apparently* well-observed case of the formation of an endoderm by delamination" [italics in original] (Lankester, 1877: 402). In fact, he claimed that even if this one piece of data in support of his views didn't actually exist, his Planula theory would still be preferable over the Gastraea theory. What caused Lankester to change his mind?

Lankester's greatest problem with the Gastraea theory was an issue that Haeckel had tried to dodge. In his 1876 paper Lankester wrote that on the "important question of the persistence of the blastopore Haeckel maintains a discreet though disappointing silence" (Lankester, 1876: 60). In truth, Haeckel did write a few words on body openings:

> [t]he original oral opening of the gastrula, the rudimentary mouth or the Prostoma, seems only to have descended to the Zoophyta [sponges and cnidarians], and, perhaps, to a part of the Vermes. It, nevertheless, seems to reappear in the Rusconian anus of the Vertebrata. On the other hand, the oral openings of the Vertebrata, the Arthropoda, and the Echinodermata, are peculiar new formations, and certainly not homologous with the rudimentary mouth. (Haeckel, 1874: 235)

What was the phylogenetic significance of Gastraea's Urmund if it wasn't homologous to either the mouth or the anus in a host of bilaterians? More problematic still, what to make of the variations in blastopore fate known to occur within individual phyla? If the openings that do develop from the blastopore are indeed homologous "we are driven to the conclusion that the mouth of a whelk is the homologue of the anus of a water-snail" (Lankester, 1877: 415). For Lankester, such reasoning was "a *reductio ad absurdum*" and "a fatal objection" to Haeckel's theory (Lankester, 1877: 415). Our current, more sophisticated understanding doesn't demand a homology of snails' mouths and anuses because we now know that the development of homologous structures, including the blastopore (Martín-Durán et al., 2017), can diverge dramatically even between closely related species. But Lankester wanted to get rid of this interpretive quagmire, and arguing that the last common ancestor of animals didn't have a gut opening at all allowed him to do just that.

One apparently well-observed case of gastrulation by delamination in a cnidarian doesn't impress as a strong factual foundation for a new theory about animal origins. But Lankester felt that even without this evidence he could keep his theory afloat solely by the buoyant power of an adaptive scenario. The scenario at the heart of the Gastraea theory didn't make sense to Lankester. Why would a round and hollow blastula-like creature gradually evolve an invagination? What adaptive benefit would drive such a transformation? Primitively, the blastula-like ancestor must have phagocytized its food over its entire surface, as protists do. How would the animal benefit if some cells started to invaginate and "to cease the habit of seizing and ingesting solid particles, and took to the outpouring of digestive juices and the passive function of absorption? By what influences are we to suppose that the depression was sufficiently deepened and its margin sufficiently narrowed to retain a digestive fluid?" (Lankester, 1877: 406). What would the adaptive benefit be of such an incipient specialization of cells?

In Lankester's scenario, a fully functional digestive cavity emerged more or less at once when a round, hollow blastula-like animal transversely divided its single cell layer to form a double-layered sac composed of an inner endoderm and an outer ectoderm. This instantaneously turned the blastocoel into a primitive gut. The ectoderm cells could continue to take up food particles and secrete these into the gut where they could be digested. This instantaneous division of labor was the reason that, for example, Ambrosius Hubrecht preferred Lankester's theory over Haeckel's (Hubrecht, 1905: 404). In this scenario, food ingestion would eventually become restricted to one area: "[t]he development of cilia on the general surface, and of locomotion, would account for this localization. A rupture of the sac at this point and the establishment of an open way into the already actively secreting and absorbing digestive cavity, would constitute the mouth" (Lankester, 1877: 405). Once this level of organization was achieved, development of the primitive gut by invagination could evolve by the precocious segregation of ecto- and endodermal cell fates during early cleavage, leading to the formation of a compact blastula lacking a blastocoel that could invaginate to form a gut with a blastopore (Lankester, 1877: 411–412, 415). The evolution of the blastopore was therefore tied to the origin of gastrulation by invagination, which had evolved independently in different species. According to Lankester's Planula theory, mouth and anus evolved before the blastopore did, in sharp contrast to the Gastraea theory. The relation between the blastopore and a persisting adult mouth or anus were derived conditions without phylogenetic significance (Lankester, 1877: 413).

Lankester's and Haeckel's theories had very different implications for the homology of developmental processes and structures. Haeckel thought that gastrulation by delamination had evolved repeatedly in different groups as an abbreviation of the development of the primary germ layers. In contrast, Lankester thought gastrulation by invagination had evolved from delamination as a form of developmental retardation. The homology between the blastocoel and the primitive gut was the phylogenetic totem of Lankester's scenario, but it was rejected by Haeckel. Conversely, the homology of Gastraea's blastopore and the mouths of at least nonbilaterians was lost on Lankester. In the years after their seminal papers had been published, Haeckel gladly acknowledged Lankester's views, but only selectively. While he sometimes cited Lankester's 1877 paper, his explicit comments just relate to Lankester's 1873 paper for its agreement with the central tenet of the Gastraea theory: gastrulation by invagination is primitive for animals (Haeckel, 1885: 210–211; 1889: 494). Lankester wasn't the only biologist who couldn't swallow Haeckel's theory for the origin of the digestive system.

8.3 Balfour's Amphiblastula Theory

Lankester's dear friend Francis Maitland Balfour (1851–1882) was as passionately interested in the evolution of animal body plans as Lankester was. But whereas Lankester preferred to illuminate animal phylogeny with the light shed by comparative morphology, the 26 papers that Balfour penned and published in the 11 short years of his career were predominantly embryological. His publications ranged widely across the zoological spectrum, from the development and morphology of chickens, fishes, reptiles, and mammals to sponges, onychophorans, spiders, and amphioxus. But his enduring fame probably rests most solidly on the two volumes of his *A Treatise on Comparative Embryology*. In more than 1,200 pages, Balfour told readers everything known about the embryological and evolutionary significance of animal development. The books appeared in 1880 and 1881, the years in which Balfour turned 29 and 30 years old. He never turned 31. In the summer of 1882, Balfour went mountain climbing in the Swiss Alps. Before his departure, Huxley's wife Nettie had admonished him to please be careful. Lankester also thought that climbing mountains was too dangerous a pursuit (Lester and Bowler, 1995; Desmond, 1997: 141). Alas, their warnings were to no avail. Five days after Balfour and his guide set off to ascend Mont Blanc, they were found dead from a fall. With Balfour fell one of the brightest stars from the zoological firmament.

The 22-year-old Balfour sat the Trinity College Fellowship exam in September 1874. T. H. Huxley had been appointed as his examiner, but was told not to even bother coming up to Cambridge for Balfour's *viva* (oral exam) (Desmond, 1997: 108). Balfour had flunked his written exams. The Fellowship would have been lost to him if these exam results were all that counted. Luckily, they weren't. Balfour was a brilliant researcher. Before he sat for his Fellowship exam, he had already coauthored a geological paper and published three papers on chick development, and he was shortly to publish the initial results that had come out of his research on the development of elasmobranch fishes at Anton Dohrn's zoological station in Naples. These original contributions were of such a caliber "that the examiners did not hesitate for a moment to neglect altogether the formal written answers... and unanimously recommended him for election" (Foster and Sedgwick, 1885: 10). For eight more years, Balfour pursued ground-breaking empirical research and conceptual synthesis on the development of animals.

In 1880 he published two papers (Balfour, 1880a, 1880b) that departed from his previous densely empirical studies. They were republished together as Chapter 13 in the second volume of his comparative embryology in 1881, and they stand out because of their broad comparative sweep and their evolutionary focus. The first one was dedicated to the evolutionary enigma of the germ layers, and the second one to the problem of larvae. In the first paper, Balfour discussed Haeckel's and Lankester's Gastraea and Planula theories. He wasn't convinced by either of them. Both theories hinged upon the belief that two-layered developmental forms were fair representations of ancient ancestors. Balfour had absolutely no problem with recapitulationist reasoning, but his survey of gastrulation modes in the animal kingdom failed to nominate either invagination or delamination as primitive (Balfour, 1880b: 253–257). Because essentially two-layered animals exist today, Balfour reasoned that a two-layered ancestor must have existed as well. But what would it have looked like? Like Lankester he saw the unexplained variation in the relation between the blastopore and the adult mouth as the greatest obstacle to Haeckel's Gastraea theory. But he also faulted Lankester's delaminate ancestor for its tiny empirical base – a type of development only seen in some cnidarians. However, his own animal ancestor scarcely sat on a broader empirical foundation.

Balfour proposed that an ancestor like the amphiblastula larva of calcareous sponges could connect animals to protists (Balfour, 1880b: 259–260), which was why Metschnikoff (1886: 141) called it the Amphiblastula theory. This creature would be divided into two hemispheres, one composed of ciliated locomotory cells and another comprising nonciliated amoeboid feeding cells. Then, recapitulating the course of development observed in the sponge *Sycon raphanus*, the ciliated cells would invaginate into the hemisphere of amoeboid cells, creating a cup-shaped, two-layered animal (Figure 8.2). Et voilà, recapitulationist logic had cranked out another ancestor. But of what exactly? Balfour wasn't sure himself. The problem was that this scenario produced an ancestor with a body plan that was inverted relative to that of most animals, except sponges. Balfour found it impossible to

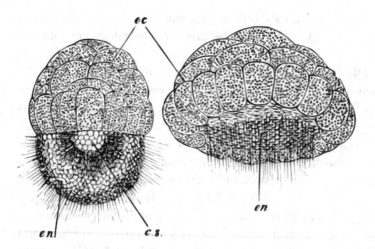

Figure 8.2 The amphiblastula stage (left) and the stage following it of the sponge *Sycon raphanus* (after figure 65 in Balfour, 1880c).

imagine how an ancestor with its nutritive cells on the outside, and its ciliated respiratory cells on the inside could have given rise to animals with the reverse orientation. Maybe amphiblastula was just the ancestor of sponges. Maybe sponges deserved their own "special division" (Balfour, 1885: 150) in the animal kingdom, cordoned off from the ancestral lineage of all other animals. In the previous year, William Sollas had proposed the name Parazoa for sponges to indicate their descent from choanoflagellates independently from other animals (Sollas, 1884), a view that was destined to live a long life in zoological handbooks. The tragedy of amphiblastula was that no sooner had it entered the phylogenetic arena as an ancient animal ancestor than its creator dispelled its phylogenetic magic.

That the calcarean sponge *Sycon raphanus* had an amphiblastula stage that underwent a process of invagination was never going to be convincing evidence for the existence of a similar ancestor, either for sponges or animals. Its taxonomic footprint was simply too small. But when it was seen as just one of many species with invagination gastrulas found throughout the animal kingdom, it could contribute its own little facet to the ancestral mosaic that other biologists were trying to construct. For Haeckel, it contributed a welcome piece to complete his portrait of Gastraea. When he first formulated the Gastraea theory in 1872, gastrulation by invagination was unknown for sponges. This didn't stop Haeckel from labeling the gastrula-like larvae of calcareous sponges as gastrulae, but he didn't know how they developed. So when in 1875 the German zoologist Franz Eilhard Schulze described invagination of the amphiblastula larva of *Sycon raphanus*, Haeckel couldn't praise his discovery enough (Haeckel, 1877: 158–160). For Haeckel this showed that sponges truly belonged in the metazoan pantheon. They hadn't risen independently from protistan ancestors. They were the phylogenetic offspring of Gastraea, the common ancestor of all animals.

There was a problem, however. Schulze had observed the invagination of the nonciliated amoeboid cells into the ciliated cells, the opposite of the observations that Balfour had placed at the heart of his scenario. However, both sets of observations were Schulze's. The ones that Haeckel relied on were published in 1875, while the ones used by Balfour appeared in 1878. Schulze's 1875 observations were never repeated by any researcher. His 1878 observations were confirmed in 1908 by one of his students, and then by no one else, until 2005. It took sponge biologist Sally Leys and her colleague Dafne Eerkes-Medrano four years and the study of thousands of *Sycon raphanus* embryos to confirm that, indeed, the anterior ciliated cells of the amphiblastula larva invaginate into the posterior nonciliated cells. They made this observation on four specimens of settled larvae that they had to pry off the substrate to reveal the small invagination. Such can be the immense effort required to confirm observations of perceived phylogenetic importance.

According to our modern understanding of early animal embryology, the development of a two-layered organization and the development of the gut are two separate processes, with only the first being gastrulation. Leys and Eerkes-Medrano interpret the invagination process observed in *Sycon raphanus* not as gastrulation but as the metamorphosis of settling larvae. According to their interpretation, the amphiblastula is already a gastrula with different cell layers that started forming much earlier during embryogenesis. Other embryologists, however, deny that sponges have gastrulation and germ layers and instead hypothesize that these evolved along the eumetazoan stem (Ereskovsky and Dondua, 2006; Nakanishi et al., 2014). In any case, the invagination that forms at metamorphosis in *Sycon raphanus* is not homologous to the archenteron of other animals. The central stipulation of Haeckel's Gastraea theory that the germ layers are formed simultaneously with the gut is wrong. A gut doesn't develop in sponges, and in many cnidarians two-layered gastrulae develop well before the gut appears. Leys and Eerkes-Medrano speculate that either ingression or delamination of cells into the blastula is the primitive mode of gastrulation. In this they agree with one of Haeckel's strongest critics, whose own scenario for metazoan origins became the most important alternative to the Gastraea theory.

8.4 Metschnikoff's Parenchymella/Phagocytella Theory

Members of the public are unlikely to recognize the name of the Russian zoologist Elie Metschnikoff (1845–1916),[2] but his fame endures in more specialized pockets of humanity. Those interested in prolonging life and mitigating the effects of aging might know Metschnikoff as the yogurt-gobbling father of probiotics. Immunologists certainly know Metschnikoff as the father of the theory of cellular immunity, work for which he was awarded the Nobel prize. Invertebrate zoologists probably know that today's phylogenetic sisterhood of echinoderms and hemichordates is honored with the name Ambulacraria, which Metschnikoff first gave to this grouping. And historians of biology know that Metschnikoff constructed an influential theory for the origin of Metazoa that was designed to dethrone Haeckel's Gastraea. These achievements, and more, mark a career of

---

[2] In English Elie Metschnikoff is also spelled as Ilya Mechnikoff or Mechnikov.

extraordinary accomplishment, and no small measure of personal tragedy (Metchnikoff, 1921; Vikhanski, 2016).

In the autumn of 1882 Elie Metschnikoff moved to Messina, a seaside city in southern Italy, with his young wife Olga and five of her siblings. Before then Metschnikoff had been teaching and doing research at the university in Odessa, a port city on the Black Sea in what is now Ukraine. However, after his mental and physical health declined when Olga fell gravely ill with typhoid fever in 1880, he attempted suicide for the second time in seven years. In the posthumously published biography of her husband, Olga tells us that he chose a method that she found both quite considerate and understandable for a scientist: "In order to spare his family the sorrow of an obvious suicide, he inoculated himself with relapsing fever, choosing this disease in order to ascertain at the same time whether it could be inoculated through the blood. The answer was in the affirmative: he became very seriously ill" (Metchnikoff, 1921: 104). As Metschnikoff recovered, the Russian emperor Alexander II was assassinated in March 1881, and the ensuing political turmoil "made normal teaching and scientific work impossible" (Metchnikoff, 1921: 101). Using an inheritance left by Olga's parents, they decided to move to Italy so Metschnikoff could find the quiet needed to pursue his research.

The family took up residence in Messina in a small flat a stone's throw from the sea. In an article in the intriguingly named *Invertebrate Survival Journal*, Cammarata and Pagliara (2018: 235) write that Metschnikoff "started up a private research laboratory" there. The reality was rather less grandiose than these words suggest. The family of seven was crammed into a small flat, which meant that Metschnikoff had to set up his laboratory in the drawing room. But cramped quarters were no deterrent to doing science. One day in the winter of 1882, his family went to the circus to see "some extraordinary performing apes" (Metchnikoff, 1921: 117). Metschnikoff stayed at home by his microscope because he wanted to see a different and much smaller kind of spectacle. He was curious to see what would happen when he stabbed rose thorns into the transparent larvae of starfish. As Metschnikoff recalled later in life, the result of that experiment was "the great event of my scientific life... A zoologist until then, I suddenly became a pathologist" (Metchnikoff, 1921: 115). He was too excited to sleep that night. Early the next morning he saw his expectations confirmed. Clumps of mobile cells had gathered around the thorns in a manner similar to white blood cells swarming around a splinter in a man's finger. The thorns had caused an inflammation in the tiny larvae, and an army of mobile phagocytic cells had acted to defend against the intrusive object. Other pointed experiments followed. The same cells mobilized when human blood, goat's milk, or bacteria were injected into the larvae, with the cells attacking and devouring the intrusive particles (Metschnikoff, 1883).

Metschnikoff biographer Luba Vikhanski (2016: 38) thinks the timing of this discovery supports "the only widely respected 'law' of creativity in music, art, and science," which is known as the 10-year rule, or the 10,000 hours rule, which Malcolm Gladwell popularized to nonuniversal acclaim in his book *Outliers*. It states that creative individuals must immerse themselves in their chosen crafts for at least this long before the sweat of their toil can kindle a spark of genius. Metschnikoff's characterization of his thorny experiments as a kind of eureka moment or epiphany 28 years or some 245,280 hours after he had done

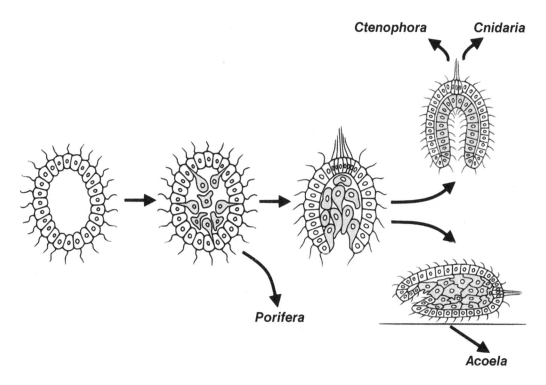

Figure 8.3 Schematic of Metschnikoff's Phagocytella theory
(after figure 2 in Malakhov et al., 2019, reprinted by permission from Springer, Copyright, 2019).

them probably contains at least a kernel of deliberate myth-making. That Metschnikoff had discovered phagocytosis is definitely a myth (Cavaillon and Legout, 2016; Gordon, 2016). What is beyond a doubt, however, is that it galvanized him to start a research program that established the central role played by phagocytes in the body's defense against pathogens and foreign substances, work for which he received the Nobel Prize in Physiology or Medicine in 1908. But his many observations on mobile cells were not just absorbed into his phagocyte theory of cellular immunity.

Both before and after the Messina experiments, they also fed into a scenario that Metschnikoff was developing to explain the origin of animals. After first being called the Parenchymella theory, it appeared in its final form as the Phagocytella theory (Figure 8.3). By the time he finalized it in 1886, Metschnikoff had done firsthand studies on an impressive bestiary of animals, including polychaetes, flatworms, insects, rotifers, molluscs, crustaceans, hemichordates, echinoderms, nemerteans, scorpions, pseudoscorpions, ascidians, millipedes, centipedes, sponges, cnidarians, mesozoans, ctenophores, and vertebrates. These mostly embryological studies, supplemented by published literature, convinced Metschnikoff that the scenarios that Haeckel, Lankester, and Balfour had proposed for the origin of animals were wrong.

I agree with Ghiselin and Groeben (1997) that the Gastraea theory was Metschnikoff's principal target, but I don't find his hostility towards Haeckel "astonishing" (Levit, 2007:

134; Olsson et al., 2017: 21). These authors point out that Metschnikoff was friends with Alexander Kovalevsky, and Haeckel had gratefully incorporated Kovalevsky's embryological observations into the foundation of his Gastraea theory. So why would Metschnikoff be so hostile to Haeckel's theory? For Metschnikoff it was a simple matter. He considered Haeckel an unforgivably sloppy researcher and his Gastraea theory an indefensible air castle. When Haeckel published his monograph on sponges in 1872, the founding document of his Gastraea theory, Metschnikoff felt compelled to publish his own unfinished embryological observations on sponges because he considered Haeckel's work to be so inadequate ("so mangelhaft") (Metschnikoff, 1874: 1). He accused Haeckel of getting facts wrong, making unwarranted extrapolations, and promoting unobserved phenomena as facts. And he had no kinder words to say about Haeckel's work on cnidarians.

His indictment of Haeckel culminated in an 1876 essay in which he accused Haeckel of dilettantism, of abandoning the rigors of the scientific method for the arm-waving rhetoric of the popularizer (Metschnikoff in Gourko et al., 2000: 89). As for the Gastraea theory, "[e]verything really valuable and scientifically proven in this theory belongs to others, mostly to Kovalevsky" (Metschnikoff in Gourko et al., 2000: 90). Not averse to polemics himself, Haeckel gave as good as he got: "Unfortunately, most statements made by this diligent but uncritical observer are unreliable because of his superficiality and ignorance about the most important basic concepts of morphology, especially histology" (Haeckel, 1877: 110). Sadly, apart from such ad hominem attacks, there is no evidence of any constructive scientific dialogue between Haeckel and Metschnikoff. There is no known correspondence between them (Levit, 2007), and you will look in vain for any discussion of Metschnikoff's Phagocytella theory in Haeckel's writings. Metschnikoff did, however, offer a substantive critique of the Gastraea theory.

Metschnikoff assassinated competing scenarios for animal origins in Chapter 7 of his book *Embryologische Studien an Medusen* (*Embryological Studies of Jellyfishes*) (Metschnikoff, 1886). No guesses necessary about his main target. In the first sentence of the book's foreword, he cites his rejection of the Gastraea theory as the key motivation for his embryological and physiological research for the past 10 years. He demolished the Gastraea theory with three arguments. First, although invagination gastrulation may be widespread in the animal kingdom, it is not typical for lower animals ("niederen Thieren"), and it is these – sponges, cnidarians, ctenophores – that should hold the most promising clues to the origin of animals.

Second, Metschnikoff considered the greatest weakness of Haeckel's theory that it lacked explanatory power. It didn't, and couldn't, according to Metschnikoff, explain how invagination gastrulation could have given rise to the various modes of gastrulation seen in cnidarians and sponges. He therefore constructed an adaptive scenario for the evolution of gastrulation modes that was fundamentally at odds with the Gastraea theory.

Third, Metschnikoff couldn't see any adaptive reason why invagination would have been the first mechanism to evolve to produce a two-layered organization. Haeckel had tied the evolutionary origin of invagination gastrulation to the emergence of a division of labor between surface ectoderm cells dedicated to locomotion and internalized endoderm cells dedicated to taking up and digesting food. Metschnikoff agreed that the emergence of a division of labor was an important component of early animal evolution, but he couldn't understand what adaptive pressure could act to restrict the uptake and digestion of food to just one area of the body. The same ammunition sufficed to slay Lankester's and Balfour's theories. Lankester didn't explain how delamination, his preferred ancestral mode of gastrulation, could give rise to invagination gastrulation, and the invaginating ancestor of Balfour was just as unsuitable as Gastraea as a starting point to account for the evolution of the other modes of gastrulation.

The empirical foundation of Metschnikoff's theory had two pillars. First, the diversity of gastrulation modes in nonbilaterians and, second, the ubiquity of intracellular digestion in nonbilaterians, something they had in common with protists. The theory's main ambition was to arrange these observations into a unifying adaptive scenario that best explained transitions between different gastrulation modes and the shift from intracellular to extracellular digestion. The criterion of epistemological success was the theory's agreement with the largest number of facts (Metschnikoff, 1886: 132). That's where Metschnikoff thought the Gastraea theory fell short (Metschnikoff, 1886: 137). Both theories started with a premetazoan ancestor that was a hollow, swimming colony of flagellates that was recapitulated as the blastula stage in animal development, but from that point on they diverged. The landscape of facts had changed considerably since Haeckel had devised his theory. In the intervening years, much had been learned about the developmental diversity and the prevalence of intracellular digestion in nonbilaterians, in no small measure as a result of Metschnikoff's own research.

In 1880 the description of the freshwater choanoflagellate *Proterospongia (Protospongia) haeckeli* (Figure 8.4) provided another crucial piece of information that Metschnikoff could built into his scenario. Before this time, he had sought the origin of animals in a hypothetical ancestral colonial flagellate with sexual reproduction, intracellular digestion, and with both longitudinal and transverse cell divisions (Metschnikoff, 1886: 133–135). *Proterospongia haeckeli* was a welcome model for this hypothetical ancestor (Metschnikoff, 1886: 145-147).

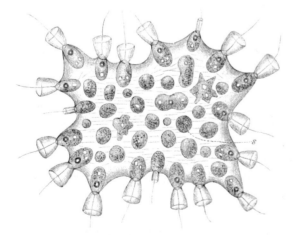

Figure 8.4 *Proterospongia haeckeli*
(after figure 2 in Metschnikoff, 1886).

It had both flagellated and amoeboid cells that could move into the colony's interior, where they could divide and multiply, and its overall organization was reminiscent of a sponge larva. Although it wasn't clear how *Proterospongia* fed, Metschnikoff hypothesized that its differentiation of motile amoeboid cells mirrored a first step toward the evolution of true endoderm.

In his Phagocytella scenario (Metschnikoff, 1886: 145–156), a *Proterospongia*-like ancestor produced internal amoeboid cells both by immigration of surface cells and by their transverse division. Such a mixed mode of internalizing cells was retained in the development of some cnidarians, and it was this mode of gastrulation, which he called *mixed delamination*, that he considered ancestral for animals. Importantly, he didn't consider it primitive because it was common in cnidarians, but rather because he thought it was how the choanoflagellate ancestors of animals had internalized cells, and it provided a convenient starting point from which to derive the other modes of gastrulation found in nonbilaterians. He agreed with Haeckel that a division of labor between locomotory and nutritive cells underpinned the evolution of the first two-layered animal organization, but in his theory, this was not associated with a restriction of the area where food was taken up. Whereas Gastraea had ingested food with its mouth and digested it extracellularly in its gut, Phagocytella took up food all over its body and digested it intracellularly. Small food particles were initially taken up by the flagellated surface cells, but also by the mobile amoeboid phagocytic cells through small pores between the surface cells. At first the functional and morphological demarcation between surface and internal cells wasn't sharp, as mobile cells ferried between them, but the layers gradually differentiated. Food uptake increasingly shifted toward the mobile phagocytic cells as they specialized in taking up increasingly large food particles at the surface pores, which Metschnikoff thought corresponded to the pores distributed over the surface of sponges. He reasoned that it made functional sense that heavy food-laden cells would be kept in the interior of the colony so as not to hamper swimming. These nutrient-laden cells would divide fast and eventually form an internal mass of amoeboid cells. Phagocytella was born when this process of functional and morphological differentiation established two distinct layers: an external ectodermal cell layer that Metschnikoff called kynoblast and a solid internal mass of phagocytic cells that he called phagocytoblast.

Phagocytella was a solid ancestor for the origin of animals. At its heart was the phagocytoblast, which was retained in sponges and which formed the evolutionary precursor to endoderm and mesoderm in other animals. These germ layers had not originated as epithelial folds, as the Gastraea theory and its derivatives claimed (see later discussion), but as confederations of amoeboid, phagocytic cells. Among the distant witnesses to this phylogenetic path, Metschnikoff could point to the nonepithelial digestive parenchyma of acoel flatworms, the wandering phagocytes of echinoderm larvae, and the solid two-layered larvae of sponges and cnidarians. And somewhere along this route, one could even encounter an ancestor very similar to Gastraea.

Starting with an ancestor that developed two layers by a mixture of immigrating and transversely dividing cells (mixed delamination), Metschnikoff plotted three evolutionary routes to other gastrulation modes (Metschnikoff, 1886: 147–151). One led to what he called *primary delamination*, in which the two-layered organization developed exclusively through the transverse division of cells. Another led to what he called *secondary delamination*, in which the two layers developed from a solid mass of blastomeres by the gradual differentiation of surface and internal cells. Gastrulation by invagination was the end station of the third route to a two-layered organization. Mixed delamination gave rise to multipolar ingression, in which surface cells ingressed individually into the blastocoel, without help from transversely dividing cells. Multipolar ingression in turn evolved into unipolar ingression, especially in species with very mobile blastulae, in which ingressing cells were restricted to the posterior end of the embryo. And here, finally, was the point in Metschnikoff's adaptive scenario where gastrulation by invagination could evolve. He noted that in some cnidarians with unipolar ingression, the ingressed amoeboid cells would arrange themselves around an epithelialized gut. If these prospective gut cells would invaginate all together, rather than ingress individually, a hollow gut would develop more quickly. Invagination gastrulation thus evolved as a developmental shortcut and with it the mouth, which evolved to satisfy the dietary demands of an increasingly active ancestral animal.

Metschnikoff's Phagocytella scenario departed from Haeckel's, Lankester's, and Balfour's by starting with a different ancestral mode of gastrulation and by denying that the evolutionary origin of two-layered body plans was associated with the origin of extracellular digestion. Haeckel maintained a deafening silence in response to Metschnikoff's theory. Balfour (1880b: 258) accepted that intracellular digestion was primitive for animals, and he explicitly cited some of Metschnikoff's work in support of this conclusion. But when Balfour discussed the early Parenchymella version of the Phagocytella theory, he merely noted that he found it "improbable in itself" even though it "fits in very well with the ontogeny of the lower Hydrozoa" (Balfour, 1880b: 259). In 1884 a German zoologist dismissed all these scenarios and their supporting evidence and arguments. He was unconvinced by these adaptationist stories, and he didn't think that embryos were ancestral beacons. Instead he thought that a tiny and mysterious invertebrate could lift the phylogenetic veil on animal origins. His name was Otto Bütschli, and its name was *Trichoplax adhaerens*.

8.5 Bütschli's Plakula Theory

One day after Otto Bütschli's (1848–1920) death, the iconoclast geneticist Richard Goldschmidt wrote an obituary that painted a detailed and tenderly sympathetic picture of Bütschli's accomplishments. He hailed Bütschli as one of the "großen Führer der Biologie" ("great leaders of biology") (Goldschmidt, 1920: 543) during the dawning decades of the Darwinian era, but he noted that Bütschli's contemporaries often didn't accord him the scientific recognition that he craved and deserved. Microscopy was Bütschli's tool of choice and a microscope could scarcely serve a keener set of eyes. Bütschli's fame rests safely on his observations on cell and nuclear division, fertilization, conjugation in protists, the structure of protoplasm, and his recognition that unicellular eukaryotes are equivalent to the cells of multicellular creatures (Churchill, 1989; Jacobs, 1989; Fokin, 2013). His massive three-volume monograph on protists, which appeared between 1880 and 1889 in the famous book series *Dr. H. G. Bronn's Klassen und Ordnungen des Thier-Reichs* was a discipline-founding document, and it remains an important reference work for researchers today. But these achievements are not the reason why I first encountered Bütschli as an undergraduate student or why his name rings familiar to invertebrate zoologists today. Bütschli earned his place in invertebrate zoology textbooks because he constructed a scenario for the origin of animals, which he named for a hypothetical ancestor that he called Plakula (incorrectly spelled Placula by many authors).

The title of the 1884 paper in which he proposed his Plakula theory, *Bemerkungen zur Gastraeatheorie*, (*Comments on the Gastraea Theory*) reveals what Bütschli thought was the most commonly accepted theory for metazoan origins at the time. Like Haeckel, Bütschli aimed to trace the evolutionary origin of the primary germ layers, but he didn't think that the clues of embryology would get him there. He had already tried to take the wind out of the sails of the biogenetic law four years after Haeckel had formulated it (Bütschli, 1876), and now it was time to topple the theory that it had built. Bütschli thought that it was impossible to decide whether available embryological evidence best supported the Gastraea, Planula, or Phagocytella theory, and he considered none of them to be a convincing adaptive scenario (Bütschli, 1884: 416–419). Like Metschnikoff, Bütschli couldn't understand why ecto- and endoderm would evolve as separate cell layers, as proposed by the Gastraea theory, rather than as cell types that initially remained distributed throughout the organism. Like Metschnikoff, he wondered what the adaptive benefit would be of restricting the uptake of food to only one area of the animal, an Achilles heel of the Gastraea theory that continues to be laid bare by modern authors (Cavalier-Smith, 2017). And like Lankester, he couldn't see the adaptive benefit of evolving an incipient invagination either. But neither was he convinced by Lankester's idea that the two-layered organization evolved suddenly by the splitting of a single-layered blastula, because this would disadvantage the ectoderm by robbing it of access to food digested in the newly evolved gut. Wouldn't it be adaptively advantageous to simultaneously evolve a mouth to provide a means for the direct uptake of food, rather than having to rely on the shunting of food from the ectoderm through the endoderm into the blind-ending gut? He criticized Metschnikoff's Phagocytella theory for the same reason. What was the adaptive benefit of individual cells transporting food from the surface of the animal to the interior, eventually leading to the evolution of a central digestive mass of cells, without the evolution of a mouth giving direct access to the digestive tissue?

A good scenario, according to Bütschli, portrayed the evolutionary transformations of body plans as "easy to understand, and gradual, not saltational, as well as really advantageous" (Bütschli, 1884: 416). To show that steps in a phylogenetic scenario were plausible and that hypothetical ancestors could, at least in principle, have existed, he thought it was important to find living or extinct organisms that exemplified ancestral body plans (Bütschli, 1884: 416). He was of course not the only or the first researcher to think like this. As we saw in Chapter 4, although Haeckel didn't believe that extant species could simply be interpreted as perfectly preserved ancestral portraits, he did like to highlight living taxa that represented ancestral body plans to a greater or lesser extent (e.g. Haeckel, 1891: 616; Haeckel, 1895: 41). Bütschli rooted his scenario of animal origins in an ancestor modelled on volvocine algae, a group of flagellates that continues to be studied today for clues about the evolutionary origins of multicellularity. Haeckel had done that before him, according special status to a species called *Magosphaera planula*, which he saw as a phylogenetic mirror for his ancestor Blastaea. But in his protist monograph, Bütschli wrote that if Haeckel's description of *Magosphaera* was accurate, it couldn't be a flagellate (Reynolds and Hülsmann, 2008). Yet, despite citing Bütschli's monograph in his *Anthropogenie*, Haeckel nevertheless retained *Magosphaera* as a model for an early animal ancestor (Haeckel, 1891).

Bütschli modelled his own ancestor on a group of colonial flagellates that are constructed as flat discs. A flat, two-layered disc would be formed if the cells divided perpendicular to the surface. Proposing that such an evolutionary transformation had indeed occurred, Bütschli called the resulting two-layered ancestor Plakula. But this evolutionary step immediately presented an adaptive conundrum. What was the adaptive advantage for a colonial flagellate of evolving a two-layered structure? Bütschli admitted he didn't know. But as he happily noted, he also couldn't see any disadvantage of evolving it (Bütschli, 1884: 419–420). Once this two-layered structure was in place, Bütschli could easily imagine the evolution of increasing differentiation between the layers, with the upper layer becoming dedicated to locomotion and the lower layer specializing in the uptake of food. Further evolution of these ectoderm and endoderm precursors would then be driven by the advantages that increased bending of the layers would offer to feeding. Bending would allow the animal's endoderm cells to collaborate in grabbing and digesting food, especially larger food items. As the creature became increasingly bent, a gut-like cavity emerged with an increasingly narrow opening. Bütschli thought that any disadvantage that would arise from narrowing the opening would be compensated for by the animal's increased ability to keep food in its gut. Eventually, this evolutionary process of curving would result in a gastrula-like animal with a blind gut and a blastopore, the common ancestor of all complex animals. Here his scenario converged on Haeckel's Gastraea theory.

It was easy enough for Bütschli to imagine this scenario, but he initially didn't publish it because its backbone was composed of purely imaginary creatures with body plans that had never been observed. He started with an ancestor that resembled some living colonial flagellates, and his endpoint was recapitulated in the embryos of invertebrates, but Plakula and its hunchbacked descendants were mere figments of his imagination. This situation changed in 1883 when Franz Schulze – the same German zoologist whose observations on sponge development had nourished Haeckel's and Balfour's scenarios for metazoan origins – described a tiny flat invertebrate that lived in the marine aquarium of the University of Graz in Austria. The animal looked like nothing that had ever been described before. Scarcely more than a millimeter in diameter and one fiftieth of a millimeter thin, it was covered with cilia and adhered tenaciously to the aquarium wall. Schulze named this hairy, flat, and sticky creature *Trichoplax adhaerens*. *Trichoplax* was constructed as a two-layered pancake, with two epithelia enclosing a connective tissue-like space. Its body plan wasn't like that of any sponge, cnidarian, worm, or mesozoan,[3] and Schulze therefore placed it in "isolation on the lowest rung of Metazoa" (Schulze, 1883: 97).

---

[3] First named by Belgian zoologist Edouard van Beneden in 1876, mesozoans are tiny parasites of invertebrates first found in the kidneys of squid. They were long considered to be an early offshoot of the animal kingdom with a primitive two-layered body plan. They are now understood to be parasitic and morphologically simplified protostomes placed in the large clade Lophotrochozoa (Schiffer et al., 2018).

Bütschli, however, couldn't have dreamt up a better modern analogue for one of the intermediate ancestors in his scenario, and Schulze's paper triggered him to publish it. For Bütschli, *Trichoplax* exemplified a phylogenetic stage just a little beyond Plakula, after the beginning of a third layer had been added to the body plan in the form of the connective tissue that was sandwiched between *Trichoplax*'s two epithelial layers. This made *Trichoplax* a perfect "transitional form to the higher animals that are constructed on the plan of the gastrula" (Bütschli, 1884: 425). The beginning, middle, and end of Bütschli's scenario were now pegged to empirical data that reflected, however dimly, the nature of distant animal ancestors. *Trichoplax*'s identity became so tightly linked to the Plakula theory that, when it finally received its own phylum in 1971, it was called Placozoa. That, however, didn't sit well with everyone.

Soon after the German zoologist and protistologist Karl Grell had erected Placozoa to honor the phylogenetic theory of his fellow countryman Bütschli, Russian zoologist Artemii Ivanov took exception. Ivanov was intensely interested in animal origins and in 1968 he had published a monograph that elaborated and reinforced Metschnikoff's Phagocytella theory. Appropriately enough, the book earned him the I. I. Mechnikov (same man, different spelling) Gold Medal in 1975 (Kotikova and Mamkaev, 2006). Ivanov also undertook a study of *Trichoplax*, published in 1973, in which he tried to recruit it to support the Phagocytella theory because he thought that it was "a living model of Phagocytella" (Ivanov in Grell, 1979: 288). He therefore proposed to house *Trichoplax* in a new taxon called Phagocytellozoa, a proposal Grell disagreed with because although Phagocytella was thought to have fed via phagocytosis, *Trichoplax* (and Plakula) seemingly didn't. Metschnikoff himself had tried to feed *Trichoplax* when he studied it firsthand in 1883, but he was disappointed that the little creature didn't want to eat anything at all (Metschnikoff, 1886: 144). However, Wenderoth (1986) discovered that *Trichoplax* is in fact able to feed via phagocytosis, which removed the objection to seeing it as a more or less distorted reflection of Phagocytella (Ivanov, 1988). Therefore, in some corners of the zoological literature, *Trichoplax* lives in a telescoped Linnean hierarchy, inside phylum Placozoa, inside superphylum Phagocytellozoa, with these names reminding us of the battle between two hypothetical ancestors inspired by this simplest of all animals.

8.6 Judging a Phylogenetic Face-Off

The scenarios discussed here were at the center of the first post-Darwinian generation of phylogenetic debates. At the time of their construction, their architects were, or were to become, influential scientists. Haeckel's Gastraea theory was the ideal tinderbox for setting off an explosion of interest and polemics about the origin of animal body plans. It had immense ambitions – to explain the origin of the animal kingdom and establish its mono-phyly, to show the homology of the primary germ layers, and to demonstrate the phylogenetic power of embryological data refracted through the biogenetic law. But it was a delicate intellectual construct. It was unabashedly speculative and built upon a thin and debatable empirical foundation. It could be tripped up easily by denying that gastrulation by invagination was primitive, which was precisely the strategy adopted by Haeckel's opponents. But despite factual inaccuracies, debatable interpretations, a frequently dogmatic and polemical style, questions about the veracity of his drawings,[4] and claims that he had given insufficient credit to other authors, Haeckel presented and defended the Gastraea theory with unwavering confidence to the end of his career. Unsurprisingly, it generated intense debate and provoked the rival theories of Lankester, Balfour, Metschnikoff, and Bütschli. Perhaps more surprisingly, it is still very much with us today.

These biologists had the same goal: explain the origin of diploblastic animals and the primary germ layers by reconstructing the last common ancestor of animals. A comparison of their competing scenarios in terms of three variables highlights their different perspectives on the issue: phylogenetic framework, explanatory structure, and evolutionary intuitions. First, the phylogenetic framework determined which taxa and facts were considered most relevant. At the time, there was no consensus about basal metazoan relationships. The phylogenetic position of sponges was especially uncertain. Haeckel had been inspired to formulate the Gastraea theory by his research on sponges, and he incorporated embryological observations on sponges into its empirical foundation. The importance of sponges for understanding metazoan origins was almost the only point that Haeckel and Metschnikoff agreed upon when constructing their scenarios. But Balfour and Bütschli both doubted that sponges shared a common ancestry with other animals (Balfour, 1880b: 260; Bütschli, 1884: 424). This made Balfour unsure about whether his Amphiblastula theory could actually explain animal origins, and Bütschli used the separate ancestry of sponges to undermine Metschnikoff's Phagocytella theory. Indeed, doubts that sponges and other animals had a monophyletic origin in a shared protist ancestor survived well into the twentieth century. Accordingly, sponges were placed in the taxon Parazoa (Sollas, 1884), outside of Metazoa, the clade of all other animals.

Aside from the question of the monophyly of the animal kingdom, participants in these debates disagreed about the phylogenetic distribution of the most relevant evidence. Haeckel cast a wide phylogenetic net to capture facts supporting his theory. He was confident that Gastraea was real because its cup-shaped reflection could be observed during the development of representatives of virtually all animal phyla. Developmental variation didn't thwart him because the light of the biogenetic law led him straight from gastrula to Gastraea. Strikingly, some modern authors still use the exact same roadmap to Gastraea (Nielsen, 2012: 35; Arendt et al., 2015: 4). Metschnikoff adopted a different focus. Instead of taking a broadly comparative perspective, he thought that the most promising

---

[4] It is well known that Haeckel was accused by several contemporaries of deliberate inaccuracies in some of his embryo illustrations included in his popular science books *Natürliche Schöpfungsgeschichte* and *Anthropogenie* (Gould, 2002a; Richards, 2008, 2013; Hopwood, 2015b). But his professional publications weren't exempt from criticism either. Just as Louis Agassiz was disgusted by the inaccuracies of Haeckel's embryo drawings in *Natürliche Schöpfungsgeschichte*, his son Alexander expressed disappointment about some of the embryo drawings that Haeckel included in a professional paper supporting the Gastraea theory: "It is a great pity that such a skillfull draughtsman should give such untrustworthy figures to illustrate so fundamental a theory, and quietly fall back upon the righteousness of his cause" (Agassiz, 1876).

clues to animal ancestry would occur in the lower animals, especially the early branching nonbilaterians. Consequently, observations on the development and physiology of cnidarians and sponges are at the heart of his Phagocytella theory.

Contemporaries, like the Canadian zoologist Playfair McMurrich, who presented the competing theories to the broader public during an evening lecture at the Marine Biological Laboratory in Woods Hole in 1890, agreed that ancestral clues should be sought especially in the lower animals. Criticizing the Gastraea theory along the same lines as Metschnikoff, McMurrich pointed out that gastrulation by invagination is not typical of lower animals: "It is hardly logical to take phenomena occurring in comparatively few cases to be the most typical, and consequently we must assume that the formation of a solid planula and a subsequent hollowing out of the central mass is typical in the Porifera and Cnidaria" (McMurrich, 1891: 91). This is a fair criticism if you think that early diverging lineages are most likely to retain ancestral traits. Libbie Hyman preferred Metschnikoff's theory over Haeckel's for the same reason (Hyman, 1940: 252). As discussed earlier, this type of linear evolutionary thinking, according to which side branches were interpreted as more or less frozen snapshots of a main line of evolution, was common in the nineteenth and twentieth centuries, and it persists even today.

For example, Alfred Tauber, a Metschnikoff expert, wrote:

> Metchnikoff called his hypothetical urmetazoan parenchymella and, because he modelled it on more primitive animals than Haeckel's gastrea, the Russian could claim the phylogenetic priority of introgression as a more ancient mechanism of gastrulation. Simply, in the competition to describe the earliest metazoan, Metchnikoff upstaged Haeckel on claims that the older ancestry showed a more basic developmental process. (Tauber, 2003: 898)

Alas, as a logical argument this is flawed. Cnidarians and sponges don't have an older ancestry than bilaterians. Early branching taxa may indeed retain more primitive characters than taxa that are more deeply nested in a phylogeny, but whether or not this is true in any given case is a matter of contingent historical fact. We should of course not convict nineteenth-century thinkers of thought crimes that remain common in the modern literature, as we will see in Chapter 10. The important point is that Haeckel and Metschnikoff cast unequal eyes upon the available evidence, and Haeckel's broad phylogenetic view clashed with Metschnikoff's deep focus on early branching taxa. Neither was a perfect strategy, and both differ fundamentally from how we infer ancestral character states today. In comparison, Lankester, Balfour and Bütschli adopted neither a broad and shallow, nor a deep and narrow phylogenetic perspective, but precariously balanced their scenarios on just a few observations about cnidarians, sponges, and *Trichoplax*, respectively.

Our storytellers did agree that their scenarios could obtain support, or at least a degree of plausibility, from the existence of living organisms that resembled their hypothetical ancestors. Metschnikoff used the choanoflagellate *Proterospongia* to illustrate the starting point of his scenario, Bütschli only dared to publish his scenario when *Trichoplax* was described, and it was an important aspect of Haeckel's phylogenetics to list the least modified descendants of his hypothetical ancestors, as he did, for instance, for all 30 steps in his sequence of human ancestors in the little book he aptly titled *Unsere Ahnenreihe* (*Our Ancestral Lineage*) (Haeckel, 1908).

The second factor that helps us understand why these biologists proposed such different theories in the context of pretty much the same available data is the explanatory structure of their scenarios. All of them reconstructed the evolutionary origin of the primary germ layers, but not all of them also focused on the origin of the gut and its opening. The Gastraea theory did. Haeckel saw the gut and mouth of Gastraea faithfully reflected in the archenteron and blastopore, respectively, of the embryos of living species. It was for this reason that he preferred to call them Urdarm and Urmund, rather than archenteron and blastopore, the more neutral terms that Lankester had given them (Haeckel, 1877: 259). By using these phylogenetically tinted terms for the morphology of living embryos, Haeckel highlighted they were hitched to a lineage of ancient ancestors. What better way to promote his theory that phylogeny explains ontogeny than by baking it into the very language used to describe embryos?

The evolutionary origin of the primary germ layers and the gut were parts of the explanatory structure of the Gastraea theory, but explaining the evolution of different modes of gastrulation observed in nonbilaterians wasn't. That was, however, a central feature of Metschnikoff's theory. In an article on the impact of Haeckel in Russia, Kolchinksy and Levit (2019: 84) claim that "there was a deeper methodological discrepancy between Haeckel and Metschnikoff" with Metschnikoff "trying to be closer to empirical data" than Haeckel. They don't explain why they think this, but as a general statement it misses the mark. Haeckel drew on a much broader phylogenetic range of taxa to inform his conception of Gastraea than Metschnikoff did to conceive Phagocytella. But the Phagocytella theory had several explanatory ambitions that the Gastraea theory lacked. It tried to explain the evolution of the variation in the early development of nonbilaterians, and it tried to trace the primitive mode of feeding in animals. Haeckel hadn't tried to reconstruct how invagination gastrulation had evolved into the other modes, and his scenario for the origin of extracellular digestion in a hollow gut didn't make sense to Metschnikoff. Haeckel never addressed these issues in any detail because he had placed them outside the explanatory scope of the Gastraea theory. I therefore disagree with Levit (2007: 135), who concludes that Haeckel's theory "presupposed a rigid ontogeny mechanically reflecting a phylogenetic history, [while] Metschnikoff's hypothesis allowed adaptive changes at every stage of embryogenesis."

The third factor that helps us understand why such different scenarios were proposed is the evolutionary intuitions of the authors. As argued earlier, evolutionary intuitions are a mix of preconceived ideas and hunches about what kinds of evidence are reliable phylogenetic guides and what is and isn't likely to happen during evolution. Some intuitions were widely shared – for example, that evolution generally proceeds from simple to complex morphologies. Others were deeply influential but not universally held ideas about what kinds of phylogenetic evidence were most relevant – for instance, recapitulation, the idea that ontogeny is a condensed summary of phylogeny. But many evolutionary intuitions were just purely personal views about what is plausible during evolution, and such differences of opinion pervaded evolutionary reasoning. Because the phylogenetic folklore accepted by different authors could differ dramatically, they often built very different scenarios from the same factual substrate. As we have seen, scenarios for early animal evolution provide many examples.

For instance, all authors accepted that an increasing division of labor between locomotive and nutritive cells was the central driver of the evolution of a two-layered body plan, but they envisioned it in different ways. There were still other perspectives on this issue. In his public discussion of these scenarios, McMurrich (1891) expressed doubts that the first division of labor was one between locomotive and digestive cells. He drew attention to the fact that protists such as *Proterospongia* and *Volvox* internalized their reproductive cells and that the same could be observed in the supposedly primitive mesozoans. He therefore thought that the prime driver for the evolution of two-layered animals was the internalization of reproductive rather than digestive cells, and he offered this as an amendment to Metschnikoff's Phagocytella theory, which he otherwise endorsed.

Even accepting the same rule of thumb – for instance, that embryos recapitulate (parts of) adult ancestral body plans – did not guarantee a convergence of opinions. To use the biogenetic law as a phylogenetic tool, one had to distinguish between what Haeckel had called palingenetic and cenogenetic traits. Only the former were reliable guides to evolutionary descent, as they preserved relatively unmodified traits in the order in which they had evolved, while the latter represented evolutionary novelties or adaptive modifications that disrupted the parallel order in which traits emerged in ontogeny and phylogeny. But how to tell them apart was very much in the eye of the beholder. Haeckel's palingenetic beacon in the chaos of metazoan ontogeny – the gastrula with its invaginated Urdarm – was for Metschnikoff a modification of a modification of a modification. Like Lankester, Metschnikoff saw it as a cenogenetic diversion from the palingenetic path, a convergently evolved shortcut to speed up development. Profound disagreements are inevitable when the balance between empirical evidence and unbridled imagination leans so heavily toward the latter. Scenarios for the evolution of animal body plans are irreducibly perspectival, and that is why it is so astonishing that these scenarios are still with us today.

Active research into early animal evolution waned during the first half of the twentieth century, before picking up again in the second half. These scenarios found sanctuary in zoology textbooks and that is where we can find them today (Brusca and Brusca, 2003). Remarkably, with the exception of Balfour's Amphiblastula theory, they still continue to frame ongoing research into the early events of animal evolution, alongside several newer scenarios (Grasshoff and Gudo, 2002; Rieger, 2003; Manuel, 2009; Martindale and Hejnol, 2009; Mikhailov et al., 2009; Schierwater et al., 2009b; Kutschera, 2011; Nielsen, 2012; Osigus et al., 2013; Arendt et al., 2015; Budd and Jensen, 2015; Ivanova-Kazas, 2015; Brunet and King, 2017; Cavalier-Smith, 2017; Martín-Durán et al., 2017; Malakhov et al., 2019; Ros-Rocher et al., 2021). The great variation in the early development of sponges and cnidarians still stubbornly refuses to reveal what is primitive, and continuing uncertainty about basal metazoan metazoan relationships (King and Rokas, 2017; Simion et al., 2017; Whelan et al., 2017; Laumer et al., 2018; Giribet and Edgecombe, 2020; Kapli and Telford, 2020; Pandey and Braun, 2020; Li et al., 2021) prevents the unambiguous reconstruction of the nature of early animal ancestors. As a result, the facts and fantasies claimed in support of these nineteenth-century scenarios remain with us today, and opinion remains as divided as ever.

8.7 Modern Perspectives on Early Animal Evolution

Early in the twenty-first century, Bütschli's Plakula theory underwent a Renaissance in the mind of German zoologist Bernd Schierwater. It had been a long time coming. On a Tuesday evening in 1979, when Schierwater was a zoology student in Braunschweig, he decided that he wanted to study *Trichoplax* and test the Plakula theory (Schierwater, 2005). And so he did. Over the years, he and his collaborators have authored a string of interesting papers on placozoans, and one of Schierwater's goals has been to search for traces of the urmetazoan in *Trichoplax*. He thinks they are there. According to Schierwater and his colleagues, the Plakula theory "is an improved version of Haeckel's 'gastraea hypothesis'" (Schierwater et al., 2011: 290), and although theories about metazoan origins are difficult to test, "the placula hypothesis is the only one that can claim even a tiny grain of empirical support" (Schierwater et al., 2009a: 371). And tiny it is indeed. In their molecular phylogeny, Schierwater and his colleagues found that *Trichoplax* was the sister taxon to a clade comprising sponges, cnidarians, and ctenophores, which they interpreted as "evidence for a basal position of Placozoa relative to other diploblasts" (Schierwater et al., 2009b: 0039). They considered this excellent news because it "identifies Placozoa as the most basal diploblast group and thus a living fossil genome that nicely demonstrates, not only that complex genetic tool kits arise before morphological complexity, but also that these kits may form similar morphological structures in parallel [in cnidarians and bilaterians]" (Schierwater et al., 2009b: 0037).

Except, of course, for the following problem. Sister taxa are equivalent (the problematic inference that basal taxa can be considered living fossils will be discussed in Chapter 10). Neither of two sister taxa is more basal than the other. No arrow of time points from one to the other. They are at the tips of two diverging lineages rooted in a shared common ancestor. According to this sadly common misinterpretation of phylogenetic relationships, Bilateria is even more basal in the tree of Schierwater and his colleagues because it is the sister group to the clade including all these nonbilaterians. Does this mean that bilaterian genomes are ancestral to those of nonbilaterians? Of course not. Based on a tree like this, should *Trichoplax* "spark a modernized 'urmetazoan' hypothesis," as Schierwater et al. enthusiastically claim? Perhaps not. A stronger case for the ancestral status of the placozoan body plan could be made if placozoans were the sister group to all other animals, but in their tree Bilateria unhelpfully occupies that coveted position. The latest phylogenomic analyses show that placozoans are no longer in the race to be the earliest-diverging animal lineage. Ctenophores and sponges are the only remaining contestants (Simion et al., 2017; Whelan et al., 2017; Kapli and Telford, 2020; Li et al., 2021).

So what can we conclude when we assess Schierwater's modified Plakula theory in terms of the three factors that we used to evaluate the competing scenarios for metazoan origins in the nineteenth century? First, our current understanding of the phylogenetic position of placozoans indicates that they may well retain ancestral animal traits, such as the lack of a basal lamina, but they are not the sister group to the remaining metazoans. The latest trees place them either as sister group to a clade of cnidarians and bilaterians, or just cnidarians. When a consensus about basal animal relationships eventually emerges, if it ever does, only rigorous ancestral state reconstruction will be able to tell us which parts of the placozoan body plan are probably primitive and which derived. Second, Schierwater and his colleagues did expand the explanatory ambit of the Plakula theory by proposing a scenario that derived cnidarians and bilaterians from a placozoan-like body plan, but their tree doesn't unambiguously support such a reconstruction of events. Its topology makes Placozoa just as informative about the ancestral state of bilaterians as humans are for understanding the evolution of gorillas. Nevertheless, they propose a seductive scenario for the origin of bilaterians:

> the placula could also be transformed into a Bilateria bauplan, i.e., a bilaterally symmetric bauplan with an anterior–posterior body axis. One of the easiest models for adopting a bilateral symmetry suggests that the "urbilaterian" kept the benthic lifestyle of the placula but adopted directional movement. The latter almost automatically leads to an anterior – posterior and ventral – dorsal differentiation. The pole moving forward develops a head and becomes anterior, the body side facing the ground carries the mouth and thus by definition becomes ventral. (Schierwater et al., 2009b: 0041)

An easy model indeed, but one for which empirical and phylogenetic support is very much in the eye of the beholder. Which brings us to the third factor, evolutionary intuitions. Schierwater and his colleagues share Bütschli's reasoning that an initially flat, two-layered animal ancestor gradually evolved into a bullet-shaped creature, driven by the increased bending and differentiation of the lower surface for the uptake and digestion of food. These events occurred on the benthos rather than the plankton, and they are easy to imagine. But did they happen? Paleontologists Graham Budd and Sören Jensen (2015) speculate that they might have. They conclude that some of the large Ediacaran organisms could have been stem-animals that evolved from a Plakula-like ancestor in a benthic world of microbial mats. Such a scenario is indeed consistent with a modified Plakula theory, but it does not lend real support to it. In my view, current evidence provides as little support for the Plakula theory today as it did in Bütschli's time.

How are the other nineteenth-century scenarios for animal origins doing today? Balfour's Amphiblastula theory is no longer around. Molecular phylogenies indicated the paraphyly of sponges for several years around the turn of the twentieth century (Sperling et al., 2007, 2009). This motivated invertebrate zoologist Claus Nielsen to propose a scenario for the origin of Eumetazoa (all nonsponge animals) from sponge larvae, albeit not from an amphiblastula-like larva (Nielsen, 2008). Because poriferan paraphyly is no longer supported by phylogenomic analyses, this hypothesis went extinct. Lankester's Planula theory is likewise clinically dead today. Although it still sometimes pops up in reviews of the origin of Metazoa (Rieger, 2003; Mikhailov et al., 2009), no one has breathed life into it for a long time.

In striking contrast, the Gastraea and Phagocytella theories have been doing very well. Unlike Balfour's and Lankester's theories they are not impaled upon tiny pinpoints of empirical data, but they are based on broad but conflicting syntheses of animal development and physiology. How have they been holding up? Our current understanding of animal phylogeny doesn't unambiguously support either theory. A single True Topology hasn't emerged yet, and what the hypothetical ancestral animal will eventually look like will depend very much on how the early-diverging animal lineages turn out to be related. Maybe we will never be able to distill a convincing image of Urmetazoa from the fleeting faces of developing embryos, given the profound developmental differences in these lineages. The stark disparity of the adult body plans of these taxa is just as unlikely to yield an unequivocal ancestral portrait. But these uncertainties haven't discouraged supporters of the Gastraea and Phagocytella theories. They became part of influential zoological traditions during the twentieth century, and they are still with us today.

As we will see in the next chapter, the reason why they have persisted so long, other than providing scenarios for the origin of animals, is their incorporation into different hypotheses for the origin of Bilateria. Gastraea and Phagocytella provide different starting points for tracing the origin of complex bilaterian body plans, and authors have been attracted to them for different reasons. Gastraea's appeal is its precursor potential for explaining the origin of a host of body plan features, especially coeloms, a through gut with mouth and anus, a central nervous system, and larvae. In contrast, Phagocytella's allure is the ease with which it can be transformed into a simple, acoel-like bilaterian that lacks a through gut, coeloms, and larvae. These hypothetical ancestors are therefore at the heart of different kinds of scenarios with different explanatory ambitions, and these flourished in different parts of the world. The Phagocytella theory was part of the twentieth-century zoological tradition in Russia, but it was much less known outside of it (Ivanov, 1988; Ghiselin and Groeben, 1997; Kotikova and Mamkaev, 2006; Levit, 2007; Ivanova-Kazas, 2015; Kolchinsky and Levit, 2019). However, it is less widely appreciated that the Phagocytella theory was also incorporated into the American zoological tradition through the somewhat reluctant, yet massively disseminated phylogenetic views of the twentieth-century doyen of invertebrate morphology, Libbie Hyman. In contrast, the Gastraea theory was particularly prominent in German-speaking parts of Europe, chiefly as a result of being forcefully promoted by leading zoologists, like Adolf Remane. How their views clashed is part of the story of the next chapter.

9

Ancestral Attractions and Phylogenetic Folklore

Enduring phylogenetic traditions sprouted from the Gastraea and Phagocytella scenarios for the origin of animals. Being incorporated into scenarios for the evolution of bilaterian body plans, they became strands of phylogenetic folklore stretching from the late nineteenth century to today. Seemingly refractory to accumulating evidence and changing views about phylogenetic relationships, these scenarios seem almost self-sustaining as they continue to be advocated for their perceived explanatory power. This situation is reminiscent of what has been observed about scenarios for the evolutionary emergence of humans created by paleoanthropologists from the early nineteenth century onward. Many such so-called hominization scenarios follow a template of storytelling conventions imported from hero myths and folktales, and they are fueled by the deployment of narrative elements commonly found in fictional literature (Landau, 1984, 1991; Caporael, 1994). Hominization scenarios, like folk tales, see central characters face life-changing challenges, which they have to overcome with the help of newly found abilities. For instance, food scarcity resulting from climate change was overcome by the adoption of hunting, increased cooperation, and the use of tools. This required a more intense use of brain power, which set the evolving lineage on an inexorable march toward modernity. In his book *Explaining Human Origins*, archeologist and ethnologist Wiktor Stoczkowski (2002) calls this approach to scenario building common-sense or naive anthropology. It produced an imaginative literature filled with stories that were judged primarily by their plausibility, with the tide of empirical evidence only lapping feebly at their base. This recalls the naive or direct phylogenetics (Naef, 1919: 3, 51) we encountered in Chapter 6, and which Adolf Naef and others sought to replace with a more rigorous and less subjective approach.

We will immerse ourselves one last time in traditional evolutionary storytelling. A massive amount of literature is dedicated to understanding the origin and evolution of animal body plans. It sprouted and blossomed during the second half of the nineteenth century, before entering a relatively dormant state around the end of the century. After reawakening in the mid-twentieth century and experiencing a resurgence of interest largely as a result of new morphological insights produced by electron microscopical work, the field exploded in the 1990s on the back of molecular metazoan phylogenetics and the coming of age of evolutionary developmental biology. At the time of writing, in early 2022, studies of animal body plan evolution continue to appear in high-impact journals, and the historical literature has once again become primary literature. Hypothetical ancestors that have long attracted phylogeneticists, including annelids, anemones, and Gastraea, once again take their place in scenarios of body plan evolution, where they shine with a veneer of novelty lent by the expression patterns of developmental genes. In the sections that follow, I will dip into this sprawling body of work to retrieve a series of vignettes that will illustrate some of the most influential ancestral attractions and enduring phylogenetic folklore in invertebrate zoology.

9.1 Amphistomy, Anemones, and the Origin of Bilateria

The origin of Bilateria has fascinated zoologists since the late nineteenth century. The origin of the name Bilateria is usually credited to the Austrian zoologist Berthold Hatschek (Cavalier-Smith, 1998; Manuel, 2009; Minelli, 2009; Nielsen, 2012; Giribet and Edgecombe, 2020), who is said to have defined it in his 1888 textbook (Hatschek, 1888). This is a common misconception. It was Ernst Haeckel who in 1873 first grouped "all the primitively bilaterally symmetrical descendants of Gastraea... in a natural main division [natürliche Hauptabteilung]" that he named Bilateria (Haeckel, 1877: 35). Haeckel preferred it over E. Ray Lankester's term Triploblastica (Lankester, 1873) because he thought that many cnidarians and ctenophores were triploblastic rather than diploblastic, as Lankester thought. When Haeckel presented these views in 1873, he saw Gastraea giving rise to two lineages. One led via a sessile ancestor called Protascus to sponges and coelenterates, and the other led via a freely moving ancestor called Prothelmis to the bilaterian phyla. Prothelmis was a bilaterally symmetrical acoelomate worm that lacked a through-gut. In Haeckel's later writings, Prothelmis disappeared as he changed the lineup of his early bilaterian ancestors, but throughout his career he derived coelomate bilaterians with a through-gut from coelenterate-grade ancestors via flatworm-like intermediates. As we will see, others preferred a different route to the origin of Bilateria, a route that was marked by a contentious character.

Disagreement about the homology of characters is standard fuel for phylogenetic debates. What one normally doesn't expect to see is disagreement over the very existence of a character. Amphistomy is a such a character. Amphistomy received its name in 1997 (Arendt and Nübler-Jung, 1997: 14), but it had already become the beating heart of a family of influential scenarios for the origin of bilaterians more than a century earlier. Yet the phylogenetic value of amphistomy has been contested ever since. Some zoologists deem it typical for protostomes (Garey, 2002: 612; Telford, 2006: R983). The popular invertebrate zoology textbook of Ruppert et al. (2004: 219) considers amphistomy primitive for bilaterians. Yet, in 2009 two invertebrate developmental biologists wrote that it "has never been shown to occur in any real organism" (Martindale and Hejnol, 2009: 168), an opinion echoed a few years later: "no description of a bona fide amphistomy exists for any animal embryo" (Hejnol and Martín-Durán, 2015: 69). But in 2018 a team of invertebrate morphologists and embryologists concluded that available data suggests that the last common bilaterian ancestor probably had amphistomy (Nielsen et al., 2018). What is going on? This is the story of one of the most influential and contested ancestral attractions in the history of metazoan phylogenetics.

Amphistomy is a type of gastrulation in which the lateral blastopore lips start closing from the middle, creating a slit-like blastopore before leaving an opening at either end. The anterior opening becomes the mouth and the posterior opening the anus. It was already appreciated in the nineteenth century that a slit-like blastopore occurs in several different groups of invertebrates (Caldwell, 1885: 21). What is contentious, however, is whether the two openings left after closure of the lateral blastopore lips develop into the larval and/or adult mouth and anus. Before 1883 there was no evidence that portions of the blastopore persisted into later stages. Before he died in 1882, Francis Balfour had been working on a monograph about a South African species of onychophoran, *Peripatus* (now *Peripatopsis*) *capensis*. After his death, his friend and former pupil Adam Sedgwick (1854–1913) teamed up with Henry Moseley, who had monographed this species when he was biologist on the 1872–1876 *Challenger* expedition and had provided Balfour with specimens. Together, they completed his monograph with the help of Balfour's sister. Before it appeared in 1883, they published a preliminary note to highlight some "embryological results [that] are of especial interest and of general morphological importance" (Moseley and Sedgwick, 1882: 390).

One notable result was that the onychophoran embryo had a slit-like blastopore. Not only that, but the "embryonic mouth and anus are derived from the respective ends of the original blastopore, the middle part of the blastopore closing up" (Balfour, 1883: 256). With these openings becoming the adult mouth and anus – a discovery confirmed by Sedgwick (1885) – the character that came to be called amphistomy was born. Sedgwick laid out the full significance of these embryological observations in an influential paper.

9.1.1 Adam Sedgwick's Enterocoel Theory

Published in 1884 in the *Quarterly Journal of Microscopical Science* and titled "On the origin of metameric segmentation and some other morphological questions," Sedgwick's paper presents one of the most influential phylogenetic scenarios for the evolution of animal body plans ever conceived. Sedgwick (1884: 67) wanted to follow Haeckel's Gastraea theory "to its logical conclusion" and to extend Balfour's theory for the origin of bilaterian nervous systems from a coelenterate ancestor (Balfour, 1880a) to other major organ systems, including gut, coelomic body cavities, and excretory organs. The spark that lit the fuse of Sedgwick's scenario was an observation that only a comparative morphologist intent on tracing deep ancestries could have made: "the great resemblance between the embryo of P. Capensis with its elongated blastopore and somites, and an adult Actinozooid polyp" (Sedgwick, 1884: 45). What, exactly, is the great resemblance between a tiny onychophoran embryo and a sea anemone? To Sedgwick it was the similar shape of the slit-like blastopore of the embryo and the elongated mouth of anemones.

Sedgwick drew on the work of the brothers Oscar and Richard Hertwig, a star team of Jena-trained morphologists, who had published a monograph on sea anemones a few years earlier. Sedgwick was impressed by their depiction of the dumbbell-shaped appearance of

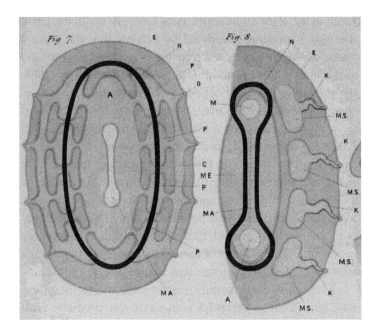

Figure 9.1 Adam Sedgwick's diagram for the sea anemone-like "ideal ancestor of segmented animals" (left) and its early bilaterian descendant (right), both viewed from the ventral side. The ancestral slit-like opening into the gut gave rise to the bilaterian mouth and anus, the gastrovascular pouches gave rise to coeloms (M.S) and associated excretory organs (K), and the nervous system is indicated with N
(from plate III in Sedgwick, 1884).

the closed mouth of anemones, with openings left on either side that lead into ciliated grooves (siphonoglyphs) opening into the gastrovascular cavity. For Sedgwick (1884: 53) these similarities were "very strong support" for a homology tying coelenterates to bilaterians. He reasoned that if the mouth of Gastraea is homologous to the mouth of cnidarians, and if the blastopore recapitulates this ancestral opening during development, then the two openings framing the partially closed blastopore of the onychophoran embryo recapitulated the siphonoglyph-framed mouth of a sea anemone-like bilaterian ancestor. In the hands of Sedgwick, this hypothetical ancestor had great phylogenetic fecundity.

With a plane of symmetry running through the long axis of the mouth, Sedgwick's ancestor had started on the path to bilateral symmetry (Figure 9.1). Permanent closure of the midline of its mouth would give rise to a through-gut with a separate mouth and anus. The mesenteries that divided its gut into several pouches could develop further and evolve into a series of closed coelomic cavities arranged on either side of the gut. This tied the bilaterian ancestor directly to segmented coelomate descendants such as annelids, arthropods, and vertebrates. T. H. Huxley had introduced the term enterocoel several years earlier for gut-derived coeloms, and Sedgwick therefore considered bilaterians without coeloms "degenerate Enterocoela" (Sedgwick, 1884: 44). The nerve ring that had circled the ancestral mouth would elongate as the animal stretched its antero-posterior axis and would differentiate a supraesophageal brain in the front and a suprarectal commissure in the back. Sedgwick's scenario accepted the new hypothesis endorsed by Balfour that the ventral surface of bilaterians is homologous to the oral surface of coelenterates, rather than the surface running from the oral to the aboral side[1] (Balfour, 1880a: 395–397; Wilson, 1891: 72–74). But this didn't exhaust the precursor potential of Sedgwick's ancestor.

Whence nephridia? They arose as perforations from the gut pouches to the outside, and they are homologous to the pores in the column of sea anemones. Whence the tracheae of terrestrial arthropods? They evolved from ectodermal pits that the aquatic ancestor used for respiration, and they are homologous to the subgenital pits found in certain jellyfish that serve the same purpose. The canal of the central nervous system in vertebrates? Like arthropod tracheae, it represents an elaboration of an ancestral respiratory invagination. Locomotory appendages, like polychaete parapodia and arthropod legs? These are body wall outgrowths and homologous to the tentacles of cnidarians. The triumph of Sedgwick's scenario was "that all the most important organ systems of these [coelomate] Triploblastica are found in a rudimentary condition in the Coelenterata; and that all the Triploblastica referred to must be traced back to a common diploblastic ancestor common to them and the Coelenterata" (Sedgwick, 1884: 68). Because Sedgwick's ancestor was pregnant with phylogenetic precursors, it had great explanatory power. Others saw this too.

In a lecture intended for a broad audience and delivered at the Woods Hole Marine Biological Laboratory, Edmund Beecher Wilson (1856–1939) summarized the outlines of a scenario for the origin of Bilateria that rested on the hypothesis that the amphistomy of *Peripatus* recapitulated the elongated mouth of a coelenterate ancestor, although he cited neither Balfour nor Sedgwick. Ernest MacBride (1866–1940) studied under Sedgwick in Cambridge and succeeded his mentor and friend as Professor of Zoology at Imperial College. MacBride enthusiastically endorsed the principle of recapitulation and the Gastraea theory throughout his career. He supported the central tenets of Sedgwick's scenario, including amphistomy and enterocoely as primitive characters for Bilateria (MacBride, 1895: 341; 1909: 335; 1914: 661).

9.1.2 The Emergence of Archicoelomate Theories

Another Cambridge student saw things differently. William H. Caldwell (1859–1941) had attended Balfour's lectures, and he had worked with Sedgwick on testing a new type of microtome that had been designed for the ribbon method of cutting paraffin samples, a technique that Caldwell had accidentally discovered in 1882 while studying phoronids as an undergraduate (Bidder, 1941; Hall, 1999, 2007). Caldwell received the first Balfour studentship (still in existence) after Balfour's death, which allowed him to travel to Australia in 1883 to study lungfish and platypus reproduction.[2] While en route to Australia, Caldwell pondered the developmental and evolutionary fates of blastopores. He thought that his

---

[1] This riddle still remains to be solved, with the most recent research suggesting that the oral-to-aboral axis of cnidarians corresponds to the posterior-to-anterior axis of bilaterians (Lebedeva et al., 2021).

[2] One major question he wanted to answer was whether platypuses laid eggs. His discovery that they did led to the famous cable that arrived on 2 September 1884 at the annual meeting of the British Association for the

observations on the development of phoronids fit a different scenario than his friend Sedgwick had proposed. He started a manuscript that he finished while encamped on the banks of the Burnett river in Northern Queensland, almost exactly one month before dispatching his discovery about platypus eggs that made him famous. Caldwell concluded that the development of phoronids showed even more primitive traits than that of *Peripatus*.

Like Sedgwick, he considered amphistomy to be a primitive bilaterian character, but unlike Sedgwick, who derived fully segmented bilaterian body plans directly from a coelenterate-grade ancestor, Caldwell inserted a coelomate, phoronid-like intermediate form. Caldwell described his conclusions briefly and not very clearly. It took me some effort to puzzle out exactly what he was and wasn't saying. I was amused to discover that Thomas Hunt Morgan had had an identical experience: "[s]o brief, indeed, is the presentation, and so obscurely worded, that I am not certain that I have entirely understood the meaning of the author" (Morgan, 1895: 466). I am therefore careful not to read things into his words that aren't there. What is there is Caldwell's observation that the mesoderm in *Phoronis* originates in three different ways in the anterior, middle, and posterior part of the embryo. What is also clear is that Caldwell (1885: 26) thought that the evolutionary implications of this are that "Phoronis is the first step towards a complete division of the blastopore... [which] caused the division of mesoderm," and which ultimately resulted in metameric segmentation. What isn't there is an explicit statement that says that metameric segmentation in bilaterians evolved from a tripartite coelomate body plan, although this is consistent with what he wrote. Twelve years later, another British zoologist made this thought explicit. He came, like Caldwell, to his conclusions based on the study of phoronid development.

Arthur Masterman (1869–1941) supported the basic features of Sedgwick's scenario, including the primitiveness of amphistomy and enterocoely. Based on his study of phoronid development, he introduced several modifications that mark the emergence of a family of enterocoel theories for the origin of Bilateria (Masterman, 1897b, 1897c, 1898). Masterman started with a pelagic rather than a benthic coelenterate ancestor, possessing four gastric pouches. These pouches evolved into coeloms that Masterman called archimeres. As this radially symmetrical jellyfish-like animal changed its preferred direction of movement from the vertical to the horizontal, it evolved an elongated mouth and it reorganized its coeloms in a bilaterally symmetrical fashion, with one anterior, one posterior, and two lateral coeloms. This tripartite, archimeric ancestor fathered a group of phyla that Masterman called Archicoelomata (Masterman, 1897b, 1898), replacing the name Trimetamera that he had experimented with in a preliminary report (Masterman, 1897c).

Archicoelomata housed phoronids, brachiopods, ectoprocts, hemichordates, echinoderms, chaetognaths, and sipunculids. Masterman provisionally included entoprocts and rotifers as well because he thought they had probably lost their archimeric organization

---

Advancement of Science that was being held in Montreal: "Monotremes oviparous, ovum meroblastic" (Bidder, 1941: 558).

(Masterman, 1898). Truly segmented, metameric body plans had evolved independently from archimeric ancestors in the lineage leading to annelids and arthropods and in the vertebrate lineage. The assumption that the first bilaterians had an archimeric body plan was one of the significant differences between Masterman's and Sedgwick's scenarios. As we will see, in the twentieth century archimeric ancestors became the defining hallmark of scenarios for the origin of Bilateria in the German zoological tradition.

For Sedgwick *Peripatus* embryos had an ancestral sparkle. When he wrote that the "structure and distribution of Peripatus all point to its being an extremely primitive type" (Sedgwick, 1884: 56), he echoed Henry Moseley's monograph. Their Gondwanan distribution suggested onychophorans were a group of great age, and Moseley painted *Peripatus* as the last remnant of an ancient taxon that Haeckel had called Protracheata (Moseley, 1874). With a body plan morphologically intermediate between annelids and arthropods, *Peripatus* showed what the last common ancestor of trachaea-breathing insects, myriapods, and arachnids might have looked like.[3] Sedgwick attempted to buttress the ancestral nature of *Peripatus* by using the common, but flawed narrative tactic that I called character hitchhiking in Chapter 7. Because the morphology and distribution of *Peripatus* seemed primitive, Sedgwick argued that "[w]e should, therefore, a priori, expect to find that its development showed primitive features" (Sedgwick, 1884: 56). Not everyone agreed.

The Dutch zoologist H. C. Delsman made the reasonable objection that, if the amphistomy of *Peripatus* was really primitive, we would expect to see it more widely distributed among animals (Delsman, 1922: 167–168, 205–206). Others rejected amphistomy as an ancestral beacon because it was associated with the Gastraea theory. Elie Metschnikoff saw the recapitulationist interpretation of amphistomy as an unconvincing maneuver to rescue the Gastraea theory. It had the unlikely implication that yolk-rich embryos with slit-like blastopores, such as those of *Peripatus*, were more primitive than the radially symmetrical, yolk-poorer gastrulas of other taxa (Metschnikoff, 1886: 157). Metschnikoff's friend and close colleague Nikolaus Kleinenberg (1842–1897) increased the vitriol in a paper published the same year. He considered Sedgwick's scenario as so much mental origami:

> He puts in the place of the nervous system of jellyfish a ring around the mouth opening of actinians, which is apparently thought to consist of rubber: when it is tugged in this way and that, if a piece is cut off here, a scrap attached there, the most beautiful nervous systems of all existing and non-existing animals arise instantly before the eyes of the happily amazed spectator – including the cerebral hemispheres of man. Because facts can only become uncomfortable during the practice of such salon art, Sedgwick therefore completely abolishes developmental history [Entwicklungsgeschichte], in as far as it is based on observations. (Kleinenberg, 1886: 185)

In the first few pages of his monograph, Kleinenberg assassinates both Haeckel's scientific reputation and the Gastraea theory in even more caustic prose. As a student,

---

[3] In 1874 Lankester had not yet published his 1881 *Limulus* monograph that united the gill-breathing horseshoe crabs with terrestrial arachnids and posited a common aquatic origin for all chelicerates (Lankester, 1881).

Kleinenberg had been Haeckel's unpaid assistant, just like his then friend Anton Dohrn. But, as for Dohrn, inspiration had eventually turned to bile. Kleinenberg was certainly not averse to phylogenetic speculation. What he rejected was the substitution of featureless schematic paper animals, like Gastraea, for hypothetical ancestors that reflected in more detail the body plans of real animals. However, to a modern reader the ambitions of Kleinenberg and Sedgwick appear similar. Both tried to trace back bilaterian characters to precursors in cnidarian-grade ancestors. Both started with an ancestor with a central nerve ring, which was sea anemone-like for Sedgwick and medusa-like for Kleinenberg. Indeed, Kleinenberg had already placed coelenterates on "the throne" (Oppenheimer, 1940: 10; Müller, 1973: 145) as the most primitive animal type in the 1872 monograph on *Hydra* that he had dedicated to Haeckel. But their methodological strategies differed.

Whereas Sedgwick worked within the strictures of the principle of no *de novo* origins, according to which continuous transformations connect the characters of ancestors and descendants, Kleinenberg allowed frequent discontinuous origins of novel traits. This reduced the observable bond of homology between coelenterate and adult annelid to one bare thread: "The thing is simply this: the annelid evolves from a coelenterate, but to the extent that the organs of annelids appear as the coelenterate organs are cancelled and destroyed, the annelid organism is not a transformation but a substitution of the coelenterate organism. It keeps from the initial form... nothing but the endoderm of the archenteron" (Kleinenberg, 1886: 222). As we saw in Chapter 5, Kleinenberg's distinctive view of evolution by substitution of organs set him apart from many of his contemporaries. Most narrative phylogeneticists preferred to create hypothetical ancestors equipped with suitable precursors so they could track evolution along more tangible paths of homology.

9.1.3 A Tradition Grows

A through-gut and coelomic body cavities were the two dominant traits in the explanatory foreground of scenarios for the origin of Bilateria from the late nineteenth century onward. Consequently, cnidarian-like and Gastraea-like hypothetical ancestors equipped with a slit-like mouth and gut pouches have been a major theme in the literature ever since. North American and European authors around the turn of the nineteenth century discussed Sedgwick's scenario as a serious hypothesis (Andrews, 1885; McMurrich, 1891; Heider, 1914). But as the twentieth century progressed, these literatures diverged. One could still see sympathetic discussions of the Gastraea theory and the scenarios it seeded in early-twentieth century textbooks in North America (e.g. Van Cleave, 1931), but these started to disappear under the influence of the phylogenetic views of Libbie Hyman (e.g. Barnes, 1974). After the Gastraea theory was initially incorporated into mainstream Russian zoology, it and its offspring were eventually outcompeted by the Phagocytella theory, in large part, so Kolchinsky and Levit (2019) argue, because leading Russian biologists such as Metschnikoff disdained Haeckel's arm-waving phylogenetics and instead preferred "to be closer to empirical data" (Kolchinsky and Levit, 2019: 84; Kolchinsky, 2019). Although I don't doubt that Metschnikoff fancied himself to be more empirically rooted than his German colleague, his use of embryological evidence to conjure ancient animal ancestors was just as speculative as Haeckel's. As I argued in Chapter 8, whereas Haeckel used a wide and shallow approach to sampling biodiversity, Metschnikoff chose a narrow and deep one. There are two fundamentally different tactics, neither of which can claim to be more empirical.

Gastraea and its various modified descendants came to form the core of the German zoological tradition (Remane, 1967; Salvini-Plawen, 1998a). The Gastraea theory was still a standard ingredient of biology textbooks in the former German Democratic Republic during the second half of the twentieth century (Hossfeld et al., 2019). Although the beginnings of the concept of enterocoely can be traced back to works by T. H. Huxley and Rudolf Leuckart, and various smaller tributaries have fed into the family of enterocoel scenarios for the origin of Bilateria, I consider the main founding documents to be those of the Hertwig brothers (Hertwig and Hertwig, 1882), Sedgwick (1884), and Masterman (1897b, 1898). They established the features that characterize most members of this family of scenarios: (1) the bilaterian mouth and anus both evolved from a slit-like mouth; (2) coeloms evolved from gut pouches; (3) the bilaterian ancestor had a tripartite organization, variously referred to as archimeric, trimeric, or oligomeric, with coeloms defining three distinct body regions from anterior to posterior (Figure 9.2). As this selection of references shows, the endurance of this evolutionary story is impressive: Willey (1896), Hubrecht (1908), Bütschli (1921), Naef (1928), Remane (1950, 1963), Jägersten (1955), Marcus (1958), Siewing (1969, 1972, 1976, 1980), Ulrich (1972), Malakhov (2013), Arendt et al. (2015), Arendt (2018), Malakhov et al. (2019), Steinmetz (2019).

The cores of these scenarios are strikingly similar, although there is some variation. For example, most authors posited coelenterate-like ancestors, but Jägersten (1955) preferred a bilaterally symmetrical Gastraea-like creature that gave rise to both coelenterates and bilaterians. Werner Ulrich (1972) claimed to have developed his archicoelomate theory independently to that of Masterman, but he did not derive bilaterians from coelenterates, and he remained deliberately agnostic about the nature of the archicoelomate ancestor:

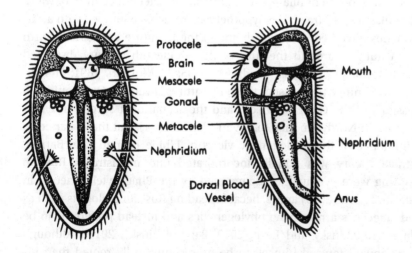

Figure 9.2 Adolf Remane's hypothetical coelomate ancestor
(after figure 5.2 in Hartman, 1963).

"No one knows what these ancestors looked like" (Ulrich, 1972: 306). And like Sedgwick, but unlike most others, Hubrecht (1908) and Arendt et al. (2015) made the bilaterian ancestor metameric instead of trimeric. If Sedgwick had been able to see the scenarios of Arendt et al. (2015) or Malakhov et al. (2019), he would have been amazed by how faithfully his central ideas had been transmitted into the new millennium. Malakhov and his colleagues even derive bilaterian appendages from coelenterate tentacles, just as Sedgwick did.

Although the enterocoel theory has been the dominant scenario for the origin of coeloms when judged by its endurance, other evolutionary precursors have been proposed as well, with nephridia and gonads being the leading contenders (Hyman, 1951; Clark, 1964; Schmidt-Rhaesa, 2007). The attraction to an ancestral slit-like mouth to explain the evolutionary origins of the bilaterian mouth and anus has been even greater, and it extends well beyond the family of enterocoel theories. The reason for this is that almost all scenarios that start with a Gastraea-like or a coelenterate-like ancestor align the emerging main body axis with the long axis of the ancestral slit-like mouth. This ancestral trait therefore crops up in a wide range of scenarios with different explanatory foregrounds, including central nervous systems, larvae, and of course the digestive system itself (Snodgrass, 1938; Beklemishev, 1969a; Salvini-Plawen, 1980; Nielsen, 1985, 2012, 2013, 2017; Nielsen and Nørrevang, 1985; Rieger, 1994; Lacalli, 1996, 2010; Holland, 2003; Baguñà and Riutort, 2004; Martindale and Hejnol, 2009; Arenas-Mena, 2010; Martín-Durán et al., 2017).

The inclusion of Salvini-Plawen and Beklemishev in this list shows that even those who rejected the enterocoel theory were attracted to amphistomy and an ancestral slit-like mouth. These two authors supported the planuloid-acoeloid theory of bilaterian origins, which was most famously promoted by Libbie Hyman (see later discussion). According to Beklemishev (1969a: 414), bilaterians evolved at least three times independently from coelenterates. Claus Nielsen's Trochaea theory deserves special mention as well. It is a scenario for the evolution of invertebrate body plans that builds on the Gastraea theory, and Nielsen has actively promoted it since 1985. It posits a radially symmetrical pelagic Gastraea-like ancestor, called Trochaea, and focuses primarily on explaining the origin of the bilaterian through-gut, ciliated larvae, and central nervous systems. The Trochaea theory proposes the through-gut emerged in a creeping benthic form by the lateral closure of the single gut opening, an event Nielsen claims is recapitulated in the amphistomic gastrulation of living bilaterians. Which brings us back to the conundrum posed at the start of this section: what evidence can the Trochaea theory, and others scenarios like it, claim in support of an ancestor with a slit-like gut opening?

9.1.4 The Death and Rebirth of Amphistomy

Neither the slit-like mouth of sea anemones, which may be a synapomorphy for Anthozoa (Daly et al., 2007), nor the rare occurrence of amphistomic gastrulation would map convincingly to the ancestral node of Bilateria on a modern phylogeny. Biologists have equipped their hypothetical ancestors with a slit-like mouth for more than 135 years because of its precursor value. Descriptions of amphistomic gastrulation form a tiny trail in the literature, from Balfour's posthumous paper on *Peripatus* in 1883 and Woltereck's

1904 paper on the polychaete *Polygordius*, to the more recent papers that continue to cite these studies as primary literature (Nielsen, 1985, 2012; Arendt and Nübler-Jung, 1997; Van den Biggelaar et al., 2002). Only a handful of studies from the twentieth century have reported amphistomy in isolated species of nematodes and flies. A thorough review of the literature might yield more potential cases – Kingsley (1885) claimed that *Limulus* embryos show amphistomy, for example, and Siewing (1976: 71) listed several others, but without providing sources. However, although a slit-like blastopore was observed in these cases, confirmation is lacking that the adult mouth and anus are indeed continuous with the openings left on either side of the laterally closing blastopore lips.

Furthermore, recent studies have dramatically contracted the phylogenetic spread of amphistomy. Andi Hejnol and Mark Martindale (2009) reexamined the original descriptions for the two oldest and most celebrated examples of amphistomy and concluded that there is no real evidence for it in onychophorans and *Polygordius*. Moreover, gene expression studies of onychophoran embryos similarly failed to yield evidence for amphistomy (Janssen et al., 2015; Janssen and Budd, 2017). Even admitting that onychophoran developmental patterns vary and that some real cases of amphistomy may exist elsewhere in the animal kingdom, the rarity and extremely disjunct distribution of this character dulls its ancestral gleam. That is, until Claus Nielsen and Detlev Arendt revived it by redefining it.

Nielsen and Arendt teamed up with Arendt's former postdoc Thibaut Brunet to conduct a massive review of the developmental literature. Citing hundreds of papers and mapping embryonic fate maps and developmental gene expression patterns onto a bilaterian phylogeny, they concluded that "Sedgwick's proposal of evo-amphistomy [their term to distinguish evolutionary from developmental amphistomy] appears the most likely course of mouth and anus evolution in the bilaterian stem, supported by axes relationships, blastoporal tissue fates, by the presence of oral and anal nerve cord loops, and by the circumoral and circumanal ciliary bands" (Nielsen et al., 2018: 1367). They admitted that developmental amphistomy is rare. But because "the fate of the blastoporal tissue is indeed less prone to evolutionary change than the fate of the actual blastoporal opening" (Nielsen et al., 2018: 1362), they redefined amphistomy based on embryonic fate maps as a form of gastrulation in which blastoporal tissue contributes to both the mouth and anus as well as the midline of the body. This redefinition allowed them to revive Sedgwick's old hypothesis in modern garb and trace the origins of the bilaterian gut and nerve cords to a Gastraea-like ancestor.

Their 2018 paper is the culmination of a debate that Nielsen and Arendt had been conducting in the literature for several years with other zoologists, notably Andi Hejnol and his collaborators. The disagreements between these camps about the evolution of major body plan characters has been fueled by fundamental epistemological differences. Hejnol and his colleagues use the modern comparative approach to diagnose and test homologies in the context of comprehensively sampled phylogenies, and they attribute to hypothetical ancestors only those characters that have passed this test (Hejnol and Lowe, 2015). In contrast, Nielsen and Arendt are prominent narrative phylogeneticists who often propose homologies between distantly related taxa, but may not test them in the context of a well-sampled tree. Moreover, they use hypothetical ancestors as epistemic tools by deliberately equipping them with traits that can function as evolutionary precursors. Both Nielsen and Arendt had long been under the spell of the ancestral attraction of amphistomy, as had generations of zoologists before them.

Thurston Lacalli (2010: 7) phrased it well when he considered the value of amphistomy, more or less reluctantly, for understanding the origin of chordates: "despite objections based on precise sites of mouth and anus formation in modern taxa, the idea has considerable utility as a conceptual tool for thinking about body plan evolution." This echoes Remane's praise, more than 40 years earlier, for the continuing explanatory value ("Erklärungswert") of the Gastraea theory in his review of animal evolution that was published in a volume written to commemorate the publication of Haeckel's *Generelle Morphologie* (Remane, 1967: 594). This type of rationale is the beating heart of narrative phylogenetics, but today it is rejected by most biologists. As Gerhard Scholtz and I noted about the Articulata versus Ecdysozoa debate, it is inevitable that authors will talk past each other if they adopt such divergent explanatory strategies (Jenner and Scholtz, 2005). By using explicit ancestral state reconstruction, Nielsen et al. (2018) have now joined the debate with modern methods and resolved it to their own satisfaction. Whether the other camp will agree remains to be seen. At the time of writing, in early 2022, no response has yet appeared.

In his otherwise wonderful book about Haeckel, *The Tragic Sense of Life*, Robert Richards gave Gastraea little more than a cameo role. It floats through a few pages of the book before Richards concludes that it wasn't Haeckel's greatest triumph: "Haeckel's projection of an ancient ancestor to all metazoans drew immediate fire from the enemies of evolution and has since been regarded as one of his typical flights of fancy" (Richards, 2008). As this and Chapters 4 and 8 have hopefully made clear, the Gastraea theory and the scenarios it inspired were unmatched for their influence and endurance in European invertebrate zoology. Their fate in America, however, was starkly different.

9.2 A Dissenting View from America

By the mid-twentieth century, the zoological tradition that had formed around the Gastraea and enterocoel theories had become centered in Germany. It was promoted there by influential zoologists like Adolf Remane and Rolf Siewing, but the Germanic view of animal phylogeny received support from prominent zoologists in other countries as well, notably Gösta Jägersten in Sweden and Ernst Marcus in Brazil (Jägersten, 1955; Marcus, 1958; Olsson, 2007). In America, however, a respectful critique of the core ideas of this tradition turned to derision in the writings of one the most influential invertebrate zoologists of the twentieth century. In the first volume of *The Invertebrates*, Libbie Hyman (1888–1969) noted the widespread acceptance of the Gastraea theory, but for the same reason as Elie Metschnikoff, she didn't think that gastrulation by invagination was primitive. Haeckel had raked the developmental diversity of animals to find enough hollow gastrulas to support an ancestral inference, but Hyman agreed with Metschnikoff that not all ancestral pointers were equal. The development of coelenterates was most informative because it was "the very group nearest the hypothetical gastraea" (Hyman, 1940: 252). She accepted Metschnikoff's hypothesis that the earliest metazoans were solid, two-layered planula-like creatures.

In the second volume of her treatise, Hyman discussed the origin of Bilateria and noted that although the "enterocoel theory has met with much favor... cogent arguments can be raised against it" (Hyman, 1951: 25). Eight years later, when she composed the last chapter of the fifth volume of *The Invertebrates* and titled it "Retrospect," her previously civil tone turned caustic. "The author hoped that the enterocoel theory was dead and buried, as it deserves, but it is being kept alive by a group of zoologists, notably Remane..., Jägersten..., and Marcus... The author regards the enterocoel theory as fantastic nonsense, for which there does not exist a single scrap of genuine evidence. The theory is a pure fabrication" (Hyman, 1959: 750). Although Werner Ulrich, who had developed his concept of archicoelomates a decade earlier, granted that Hyman had some valid points, he thought "she put herself in the wrong with her mode of expression" (Ulrich, 1972: 310). Jägersten's attempt to avoid some of the pitfalls of the enterocoel theory by creating his Bilaterogastraea ancestor was met with ridicule: "This is a nice little figment of the imagination, a sort of junior dipleurula" (Hyman, 1959: 752). In the last sentences of her book, Hyman concluded that such speculations are just a waste of time: "The author regards such phylogenetic questions as the origin of the Metazoa from the Protozoa or the origin of the Bilateria from the Radiata as insoluble on present information. Also insoluble are such questions as to whether entoderm, mesoderm, and coelom have or have not some original mode of formation from which other modes are derived. Anything said on these questions lies in the realm of fantasy" (Hyman, 1959: 754). She said no more on them before her death.

Why was Hyman so vehement in her phylogenetic criticisms? She had speculated on these problems herself in the earlier volumes of *The Invertebrates*, and she had drawn a metazoan phylogeny in the first volume that became an icon in American textbooks (Schechter, 1959; Meglitsch, 1967; Barnes, 1968, 1974), albeit a frequently distorted one (Jenner, 2000, 2004a). She may not have been too pleased that her phylogeny became iconic because she had warned readers that it had "to be regarded merely as suggestive" (Hyman, 1940: 39). However, her planuloid-acoeloid scenario for the origin of Bilateria became a major alternative to the enterocoel theories (Salvini-Plawen, 1978; Rieger et al., 1991). Hyman had rooted her scenario in Metschnikoff's view of metazoan origins, and she included a drawing of a Phagocytella-like ancestor in the first volume of *The Invertebrates* (Hyman, 1940: 251–252). Mid-twentieth century Russian zoologists such as Artemii Ivanov and Vladimir Beklemishev (Beklemishev, 1969b, a) also preferred this route into Bilateria, forming a joint American and Russian counterpoint to the continental European zoological tradition.

I suspect that Hyman's vitriol was the result of frustration. She didn't know what to believe anymore. A few paragraphs before she dismissed phylogenetic speculations altogether, she defended her own planuloid-acoeloid scenario for the origin of Bilateria from 1951, in which she envisaged a planula-like ancestor giving rise to an acoel-like early bilaterian. A page later her convictions had dissipated. Hyman liked to dismiss rival theories by claiming that they lacked support. But were her own speculations any more rooted in evidence? Her planuloid-acoeloid scenario relied on embryological evidence, just like rival theories, and it was supplemented by morphological comparisons and her personal evolutionary intuitions. In 1951 she wrote that it "is evident that there is no difficulty in passing from a planuloid type of ancestor to an acoeloid form... As no intermediate forms are known, the steps by which these changes occurred may be inferred from the embryology of the lower Turbellaria" (Hyman, 1951: 9). One can argue that the morphological transformations of the enterocoel theories were more fanciful than those of Hyman's scenario, but all of them are typical products of narrative phylogenetics. Gaps between body plans were bridged by what the authors thought were plausible morphological intermediates.

What was plausible was of course entirely in the eye of the beholder. Disagreements hinged more on imagination than on evidence, as this series of Hyman's judgements from four consecutive pages in her "Retrospect" chapter shows: "it becomes very difficult to account... We are asked to believe... appears to the author to present an insuperable difficulty... It is very difficult to conceive... It is stretching matters rather far to suppose... This is absurd... It is inconceivable... this view is totally unacceptable to the author... We are asked to believe... appears impossible to account... it is inconceivable... it is very difficult to conceive" (Hyman, 1959: 750–753). This is the typical locution of a heated narrative phylogenetic debate. It records the clash of evolutionary intuitions, not disagreement about observable facts. But intuitions can have a long reach. Tom Cavalier-Smith blamed Hyman's skewering of enterocoel theories for his inability to get his own views on the origin of Bilateria published even decades later, until he finally succeeded in 2017 (Cavalier-Smith, 1998, 2017).

Hyman's evolutionary intuitions that clashed with the main assumptions of enterocoel theories were that (1) the anthozoan body plan is too complex to be primitive for cnidarians, (2) anthozoan gastric pouches and enterocoels aren't homologous because they develop differently (outgrowth of septa versus evagination from the gut), (3) platyhelminths are not degenerate coelomates, and (4) archicoelomates cannot be primitive bilaterians because they are sedentary or sessile. In a response, Remane (1963: 84) rebuffed Hyman's first intuition, writing that "[t]hese criticisms of Hyman's arise from certain principles for reconstructing phylogenetic lines – but principles that are not shared by many other scientists." He pointed out that simple cannot be equated with primitive, although it was a common shortcut in phylogenetic reasoning. To deny homology because of observed differences is another common strategy of narrative phylogenetics, as we saw in Chapter 6, but not a convincing one, as Remane (1963: 86) noted in somewhat muted terms. On the third point, Remane had no choice but to disagree with Hyman as all enterocoel theories posit a coelomate bilaterian ancestor. Still, Remane dodged Hyman's fourth objection. Although he could imagine a sessile bilaterian ancestor, he thought it had probably been motile.

Whether sedentary or sessile lineages are evolutionary dead ends or not has been an important point of contention throughout the history of studies on animal body plan evolution. Proponents of enterocoel theories for the origin of Bilateria, from Sedgwick

(1884) to Cavalier-Smith (2017), made sessile or semi-sessile taxa the backbones of their phylogenies. So did authors who traced bilaterian body plans back to colonial ancestors, and so did those who endorsed the once widespread scenario that derived chordates from more or less sessile, filter-feeding forms similar to lophophorates, pterobranchs, or ascidians (Salvini-Plawen, 1982, 1989, 1998b; Williams, 1996; Holland, 2011; Peterson and Eernisse, 2016). Others, however, have considered sessile ancestors a "stumbling block" in scenarios (Baguñà et al., 2008: 1489). Paedomorphosis was thought to offer one evolutionary escape route from the strictures of sessile adult body plans, and larval forms were therefore imagined as ancestors that founded new motile lineages, most famously the vertebrates (Romer, 1972; Gould, 1977b).

The widespread assumption that sessile animals are morphologically degenerate is an evolutionary intuition that traces back to nineteenth-century narrative phylogenetics. Haeckel (1877: 35) considered the evolution of a sessile lifestyle to be the "mechanical cause" of the origin of radial symmetry in the ancestors of sponges and coelenterates. In his book *Degeneration. A Chapter in Darwinism*, Lankester (1880: 39) introduced the general reading public to the idea that sessile bilaterians, such as sea squirts, were "the degenerate descendants of very much higher and more elaborate ancestors." And MacBride (1914: 116) informed the students of his embryology textbook that "[t]he history of the Metazoa, so far as we have yet traced it, is that of a main pelagic group increasing in complexity of structure as time goes on, and at each level throwing off creeping and sessile stocks which are more or less degenerate in structure." MacBride's criterion for diagnosing what he judged as degenerate body plans had a low threshold, but the less extreme view that the sessile descendants of free-living ancestors often have reduced morphological complexity and/or substituted radial for bilateral symmetry to varying degrees was widespread. The influential Russian evolutionary morphologist Alexei Sewertzoff (1866–1936) incorporated degeneration into his theory of morphological evolution, and he considered the adoption of sessile habits and parasitism as its two causes (Sewertzoff, 1929; Levit et al., 2004). Modern authors have continued to use this evolutionary intuition to help polarize character state transformations (Ax, 1989: 230; Northcutt, 2012: 10631).

Few zoologists took the assumption that sessile habits lead to evolutionary dead ends as seriously as Slovenian biologist Jovan Hadži (1884–1972). He developed a heterodox view of animal phylogeny that saw bilaterians as direct descendants of ciliate ancestors and cnidarians as degenerate descendants of bilaterians. Endorsement by some high-profile zoologists, notably Gavin De Beer, made Hadži's ideas briefly popular enough for Libbie Hyman to reward them with a response in her "Retrospect" chapter: "Hadzi has written many pages in three or four languages... From all this verbiage it is possible to extract two arguments of some worth" (Hyman, 1959: 753). According to Hyman, one of these ideas was the weakening of bilateral symmetry and the appearance of radial symmetry in sessile taxa. But Hyman rejected a bilaterian ancestry for cnidarians. So did almost everyone else, except for two biologists, Otto Steinböck and Earl Hanson, who independently developed virtually identical views to Hadži's around the same time in the mid-twentieth century. In the next section, we will explore the reasons behind this remarkable convergence of views.

9.3 The Ciliate Ancestry of Animals

That early in the evolution of Metazoa there had been a transition from radial to bilateral symmetry, irrespective of any later reversals, has been near universal consensus since phylogenetic speculations began. This symmetry transition can be followed as round eggs and embryos develop into bilaterally symmetrical forms, and it tracks the arrow of morphological complexity that points from diploblasts to triploblasts. It was an evolutionary transition that even those who disagreed about everything else agreed upon (Lankester, 1877; Balfour, 1880a: 396; Kleinenberg, 1886: 178; Wilson, 1891: 72). And it is the primary reason why coelenterates, especially cnidarians, were cast in central roles in so many scenarios for bilaterian origins. Jovan Hadži's suggestion that the radial symmetry of cnidarians betrays a phylogenetically infertile body plan was unacceptable to Hyman. She grudgingly granted Hadži "that a sessile mode of life leads to alteration in the radial direction" (Hyman, 1959: 753), but she cited her intellectual adversary Remane to argue that even then bilateral symmetry wouldn't completely disappear. She felt there were two reasons that made it "impossible to bypass the Cnidaria in the evolutionary story" (Hyman, 1959: 753).

First, she considered cnidarians to be the first animals that had differentiated ectoderm and endoderm, a conclusion clearly dictated by the traditional evolutionary intuition that simpler body plans precede more complex ones. Second, she felt she couldn't explain instances of tetraradiate symmetry in bilaterians, like spiral cleavage, other than by interpreting them as "an ancestral reminiscence" (Hyman, 1959: 753) from a cnidarian past. For these reasons, cnidarians "cannot be ejected from their long-established position as direct ancestors of the remaining Eumetazoa" (Hyman, 1959: 754). Here she contradicts her earlier writings where she explicitly denied that cnidarians were in the direct ancestry of bilaterians and therefore drew them as an early-diverging branch on her animal phylogeny (Hyman, 1940: 38). I suspect it is because she sensed these contradictions in her thinking that she ended the book a few sentences later with her frustrated remarks that such problems were intractable.

Jovan Hadži's studies of cnidarians fueled his remarkable views on animal phylogeny (Hadži, 1953, 1958, 1963). Yet, almost identical views were developed independently by Austrian turbellarian researcher Otto Steinböck (1893–1969) and American protist expert Earl Hanson (1927–1993) based on studies of their own favorite organisms (Hanson, 1958, 1963, 1972, 1977, 1981; Steinböck, 1958a, 1958b, 1963). Although Steinböck developed his ideas prior to World War II, it was Hadži who published his views first and in greatest detail. The ideas of these three authors are not identical, but Hadži's views, as I will outline them here, are representative for their thinking about animal origins. In 1944, Hadži published a book with a long title: *The Turbellarian Theory of the Cnidaria. The Phylogeny of the Cnidaria and Their Position in the Animal System.* Since it was published during World War II and was written in Slovene, it lacked impact. This situation improved after Hadži published an English précis in *Systematic Zoology* in 1953, although he was sad that his presentation at the Fourteenth International Congress of Zoologists in Copenhagen that same year generated "no visible reaction" from the audience (Hadži, 1963: 24). Hadži's views received a major boost in 1954 with the appearance of the essay collection *Evolution as a Process* (Huxley et al., 1958), in which the then Director of The British Museum (Natural History) acted as Hadži's cheerleader. Sir Gavin de Beer dedicated his entire chapter to promoting Hadži's phylogenetic views:

> Few of the most optimistic (or should it read pessimistic?) zoologists could hardly have thought it possible that, in the middle of the twentieth century, all text-books of zoology should require the drastic revision of the current theory of the origin and evolution of the Metazoa. Yet that and nothing less is the situation resulting from the researches which for half a century Jovan Hadzi has made into the structure and development of the lower animals. (de Beer, 1954: 34)

In his introductory essay, Julian Huxley adopted a more measured tone, laced with a hint of sarcasm, when he wrote of de Beer's "useful summary of Hadzi's somewhat revolutionary theory of the origin of the Metazoa. . . we should be grateful to Hadzi for reminding us that phylogenetic morphology is by no means a dead subject" (Huxley, 1958: 31). De Beer also put Hadži's views on prominent display in his book *Embryos and Ancestors* (de Beer, 1958), but his 1954 essay is remarkable because he wrote it before Hadži had published his views in English. De Beer had to rely solely on Hadži's works written in German, Slovene, and Serbian.

Hadži presented his views most fully, and in English, in his 1963 book *The Evolution of the Metazoa*, where he boosted his narrative with references to the papers of Steinböck and Hanson. If you are looking for 500 pages of scintillating prose about animal body plan evolution, you should look elsewhere:

> It was in 1903, 58 years ago, that I, then a young man who had just left the classical grammar school at Zagreb, went to Vienna to study Natural Sciences and above all my beloved Zoology at Vienna University. For this study I was well prepared. Whilst at school, I had made a large collection of zoological objects. I had learnt the richness of forms and the ways of life of the animal world through my diligent study of Brehm's work on the Lives of Animals as well as through my field studies. (Hadži, 1963: x)

These opening sentences set the pulse and pace of a book that comprises four fat chapters containing an exhaustive and exhausting discussion of cnidarian morphology and evolution and their significance for understanding animal phylogeny. It is a typical example of narrative phylogenetics. Hadži follows his evolutionary intuitions and guides readers through a morass of data toward an ancestral attractor that roots the explanatory power of his scenario. Two overarching ideas coordinated Hadži's thinking. First, he rejected recapitulation as a phylogenetic principle, as well as any theories based on it. He didn't just think that the biogenetic law had important exceptions; he rejected it outright (Hadži, 1963: 34). Ontogenetic data had strictly limited phylogenetic value for Hadži. Instead, ancestral cues should be sought primarily in the anatomy of adult organisms (Hadži, 1963: 43, 274–275). Second, although Hadži accepted the traditional notion that evolution was on the whole progressive, he thought this was true only in free-living lineages

(Hadži, 1963: 97). The adoption of a sessile lifestyle would lead inexorably to regressive evolution.

The emergence of radial symmetry was one of the hallmarks of this process (Hadži, 1963: 56), and it marked cnidarians as a regressive lineage. Evolution within cnidarians had therefore not progressed from simple to more complex body plans, as Hyman had maintained and many accepted since (Schuchert, 1993), but had gone in the reverse direction. In a section titled "The right sequence of the Cnidaria groups," Hadži argued that the arrow of evolution read Anthozoa → Scyphozoa → Hydrozoa. For Hadži the signs of bilateral symmetry in anthozoans – an elongated mouth with siphonoglyphs and the arrangement of the gastrovascular septa and associated muscles – were traces of a bilaterian ancestry, not incipient stages in the evolution of Bilateria, as the proponents of enterocoel theories supposed. And if cnidarians were primitively sessile, they couldn't be the cradle of Bilateria. Where else, then, could their origins lie?

The morphological gap between complex anthozoans and a Gastraea-like creature was too great to bridge (Hadži, 1963: 248). If cnidarians evolved from free-living bilaterian ancestors, what type of creature foreshadowed their sessile body plan? One group of bilaterians had long been thought to pack ancestral clout. As Hadži (1963: 254) wrote,

> In view of the fact that there is now hardly a single zoologist who could doubt that the Turbellaria represent the starting point in the evolution of many other animal types, it may be possible to accept at least as a possibility that Cnidaria, too, had evolved from some turbellarian ancestors, provided that the anthopolyps are the most primitive among the free and solitary polyps of Cnidaria.

Hadži noticed that one group of turbellarians was truly polypoid. Temnocephalans are epizoic turbellarians that can attach themselves to other animals with a terminal sucker, and they have a series of tentacle-like projections near the mouth that assist in feeding. Hadži speculated that an ancient group of temnocephalan-like worms could be the cradle of Cnidaria, and they themselves could have evolved from morphologically simpler acoels.[4] On page 276 of his book, Hadži reveals his ancestral attraction: "Some zoologists have already called our attention to the extraordinary similarity that exists between the structure of a holotrichous infusorian and of a turbellarian... at the same time they had no intention of making phylogenetic conclusions on this basis." Actually, some had.

In 1877 Hermann von Ihering dedicated a few sentences of a massive monograph on molluscan nervous systems to suggest that animals had evolved from multinucleated infusorians and that specific cell organelles had given rise to the mouth and nephridia of turbellarians. Like Hadži, he rejected the biogenetic law, downplayed the importance of embryological data, and argued that "comparative anatomy offers the possibility to trace the same organ system through a series of types" (Von Ihering, 1877: 21) from protists to animals. Unsurprisingly, Haeckel (1877: 241) rejected this "artificial attempt" at

---

[4] Acoels are no longer considered platyhelminths, and today they feature in debates about the origin of Bilateria in their own right.

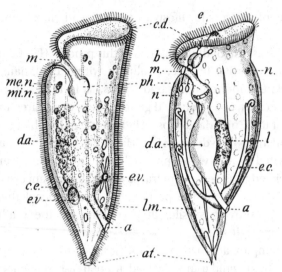

FIG. 14.—*a*, ciliate infusor; *b*, rotifer; *c.d.*, ciliated disc; *m*, mouth; *me.n.*, meganuclei; *mi.n.*, micronuclei; *n.*, nuclei; *ph.*, pharynx; *d.a.*, digestive area; *l.*, liver; *e.v.*, excretory vacuoles; *e.c.*, excretory canals; *a*, anus; *l.m.*, longitudinal muscles; *b*, brain; *e*, eyes.

Figure 9.3 Suggested homologous structures in the body plans of a unicellular ciliate (left) and a rotifer (right)
(after figure 14 in Macfarlane, 1918).

homologizing. Gegenbaur (1870: 156) had likewise speculated that turbellarians could have evolved from ciliate ancestors. The Scottish botanist John Muirhead Macfarlane attempted the same trick in his massive 1918 book *The Causes and Course of Organic Evolution*. His ciliate ancestors gave rise to rotifers instead of flatworms (Figure 9.3).

Hadži, as well as Steinböck and Hanson, suggested a diphyletic origin of Metazoa, with sponges originating separately from flagellate ancestors. The separate ancestry of sponges and eumetazoans was still a commonly held notion in the mid-twentieth century, and it only petered out in the last two decades of the century. Hadži proposed that multinucleate ciliate ancestors gave rise to acoel-like worms through a process of cellularization (Figure 9.4). This transformed organelles and differentiated parts of the ancestral multinucleate cell into syncytial and multicellular organs and structures (Hadži, 1958: 174; 1963: 138–140, 277–278). Cytostome became mouth. Cytopharynx became pharynx. Cytopyge became anus. Endoplasma and food vacuole became gut. Contractile vacuoles became protonephridia. Regarding complex new structures such as these, Hadži reaffirmed the phylogenetic credo of no *de novo* origins and noted that it "cannot be supposed that they have developed as such out of nothing" (Hadži, 1963: 137).

The morphologically complex cells of ciliates provided a far richer substrate of potential precursor structures than a Gastraea or a morphologically simple cnidarian. Ciliate ancestors therefore provided superior phylogenetic explanatory power. The advocates of the ciliate scenario considered the presence of multinucleate syncytial tissue in acoels (digestive syncytium) as a convincing trace of a multinucleate protist ancestor. But superior

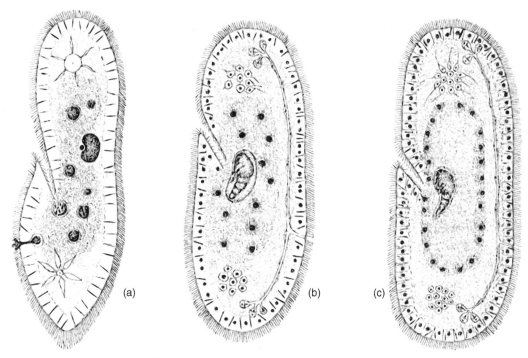

Fig. 48. Hypothetical transformation of a ciliate into primitive turbellarian (the nervous system and muscular system are omitted.) (After Hadži.)

Figure 9.4 The transformation of a ciliate (a), via a hypothetical intermediate (b), into an acoel (c) (after figure 48 in Hadži, 1963, Copyright, 1963).

precursor potential was not the only advantage that ciliates had over alternative hypothetical ancestors. Hadži (1963: 322–323) and Steinböck (1963: 49) also pointed out that their theory was preferable because there are no convincing living models for hypothetical ancestors like Blastaea and Gastraea. They thought that the strategy adopted by proponents of the Gastraea and Plakula theories of using volvocine algae, like the famous *Volvox*, as living models for Blastaea (Reynolds and Hülsmann, 2008) was unconvincing because they were too distantly related and specialized. Although the ciliate ancestors of eumetazoans were extinct, living ciliates at least approximated this ancestral body plan closely enough to give the mind's eye some purchase.

The ciliate theory for bilaterian/eumetazoan origins did receive some wider support (de Beer, 1954, 1958; Harvey, 1961; Cloud Jr., 1968). Stephen Jay Gould discussed Hanson's views sympathetically for general readers as late as 1980 (Gould, 1980). But although this scenario was included in textbooks, notably those written by Hanson himself (Hanson, 1972, 1981; Barnes, 1974; Brusca and Brusca, 1990), it did not have the revolutionary impact presaged by De Beer. Many zoologists continued to support the traditional belief "that the Cnidaria occupy the position that the phylogenetic principle, *simple before complex*, dictates; that is, they are pre- not post-turbellarian in origin" [italics in original] (Hand, 1959: 202). The accumulation of ultrastructural and molecular evidence eventually established beyond any doubt that sponges were real animals and led to the widespread acceptance of the monophyletic descent of metazoans from colonial protist ancestors. Nielsen (2012: 15) is therefore certainly right when he writes that the ciliate theory "is now only of historical interest." However, the lessons that the rise and fall of the ciliate theory can teach us are perhaps not the ones that Nielsen had in mind.

First, Hadži, Steinböck, and Hanson were attracted to this theory for the same reason: the superior precursor potential of hypothetical ciliate ancestors for deriving bilaterians. They rejected recapitulationist reasoning and sought ancestral clues in the anatomy of adult organisms rather than ontogeny. This allowed more dots to be connected and provided a broader bridge of homology between ancestors and descendants and, hence, produced greater explanatory power. Nielsen has adopted fundamentally the same narrative phylogenetic strategy in promoting the Trochaea theory for more than 35 years. Nielsen's scenario traces back the origin of central nervous systems, larval ciliary bands, and through-guts with mouth and anus to identifiable precursor structures in hypothetical ancestors.

Second, the Trochaea theory illustrates the enduring attraction of the biogenetic law to summon hypothetical ancestors, including Gastraea or what Kleinenberg (1886: 2) called Haeckel's "skinny animal ghost." Nielsen's scenario is rooted in recapitulationist reasoning, a phylogenetic strategy that is just as much part of the tradition of narrative phylogenetics as deliberately equipping hypothetical ancestors with useful ancestral traits. Neither strategy rigorously infers ancestral states by considering the distribution of relevant character states in the context of an established tree. Both are phylogenetic shortcuts designed to maximize a scenario's explanatory power.

Third, both the Trochaea and ciliate theories cast an inescapably unequal eye on available evidence. In their preference for adult anatomy, Hadži, Steinböck, and Hanson aligned themselves with the anatomists of the late nineteenth century who battled embryologists in the phylogenetic *Kompetenzkonflikt* (see Chapter 5). But by rejecting ontogenetic evidence, the ciliate scenario lacked explanatory power relevant to embryological observations. This had already been pinpointed as the major flaw of von Ihering's proposal for deriving animals from ciliates in the late nineteenth century (Haeckel, 1877: 241; Metschnikoff, 1886: 132–133; McMurrich, 1891: 80). But the opposite is equally true. The morphological simplicity of Gastraea offers few clues to the origin of evolutionary novelties other than the primary germ layers, mouth, and gut. This helps explain why zoologists who have used the biogenetic law or the hypothesis of paedomorphosis to fashion hypothetical ancestors have been so attracted to more complex larval forms. Larvae like the crustacean nauplius, the echinoderm auricularia, and the spiralian trochophore larva possess enough characters to be suitable starting points for the origin of more complex body plans (Müller, 1869; Hatschek, 1888; Garstang, 1928). Haeckel enthusiast Karl von Heider in effect pimped his hypothetical bilaterian ancestor, called Sphenula, by tacking traits taken from trochophore and tornaria larvae onto a ctenophoran frame (Heider, 1914: 476) (Figure 9.5). Indeed, by upgrading Gastraea with a blastoporal ring of compound cilia, an apical ganglion, and the precursors of nerve cords, Nielsen adopted precisely this strategy to create his Trochaea.

The ciliate theory for animal origins rose and fell before phylogenetics was modernized by the spread of cladistics and the advent of molecular phylogenetics. Earl Hanson was not

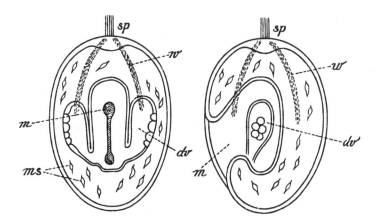

Figure 9.5 Heider's hypothetical bilaterian ancestor Sphenula, with a slit-like single opening into the gut and gut pouches that are the precursors to coeloms
(after figure 16 in Heider, 1914).

a traditional narrative phylogeneticist, unlike Hadži and Steinböck. He was keenly interested in refining phylogenetic practice and dedicated a substantial part of his 1977 book, *The Origin and Early Evolution of Animals*, to a discussion of phylogenetic methodology. Alas, his ideas gained little traction, although the ever open-minded Lynn Margulis saw fit to use them (Margulis et al., 2006). It wasn't until the 1980s and 1990s that narrative phylogenetic approaches started to be replaced by modern methods to understand animal body plan evolution. This transition is nicely illustrated in the beloved invertebrate zoology textbook of Brusca and Brusca (1990). It contains the first morphological computerized cladistic analysis of the animal phyla, which is used to organize the discussion of animal evolution. On the page preceding their cladogram, the Brusca brothers indulge in traditional narrative speculations when discussing the origin of Metazoa. They write that one of the "very compelling lines of reasoning" that animals evolved from a flagellated Blastaea-like form is that "early stages in the origin of the metazoan condition must have included the formation of layered tissues, an event akin to the embryonic process of gastrulation. A blastula-like ancestor would have set the stage, so to speak, for such a gastrulation-like phenomenon" (Brusca and Brusca, 1990: 881). Readers can judge for themselves if this argument deserves to be called very compelling, but using the imagined precursor potential of a hypothetical ancestor as a springboard for a scenario is a strategy that is as old as phylogenetic speculation itself.

9.4 Annelids as Ultimate Ancestors

This section will close our discussion of the narrative reconstruction of animal phylogeny by highlighting a bundle of related evolutionary intuitions that has dominated traditional thinking and that continues to exert its influence today.

Annelids are special. They attract praise that is generally withheld from other phyla. According to Brusca and Brusca (2003: 295), their "body plan is the classic example of the metameric, triploblastic, coelomate Bilateria, and provides a good model for comparison with other protostomes." Jim Valentine (2004: 511) echoes this sentiment and calls them "classic spiralians with trochophores." Detlev Arendt and colleagues amped up the adjectival flourishes when promoting the nereidid polychaete *Platynereis dumerilii* as a model organism for comparative molecular developmental biology. We learn that this species was chosen for its "ancestrality" (Tessmar-Raible and Arendt, 2003: 335), specifically its "ancestral development, body plan and genes, making it ideally suited for comparative developmental and evolutionary studies" (Tessmar-Raible and Arendt, 2003: 338). In addition "the polychaete rope-ladder-like central nervous system exhibits prototype invertebrate design" (Tessmar-Raible and Arendt, 2003: 335). Winchell et al. (2010: 1) agree that nereidid polychaetes have a "generalized morphology [that] is traditionally considered archetypal within Annelida." On his website (www.embl.de/research/units/dev_biology/arendt/; last accessed September 24, 2021), Detlev Arendt awards *Platynereis* the ultimate evolutionary epithet: "living fossil."

These are old sentiments. Annelids have been considered standards for comparative studies and attractive ancestral models for more than a century and a half. Why is this? A short and simple answer may be familiarity. It is safe to assume that no biologist, and probably no high school student either, has ever graduated without having dissected an annelid. Anyone who has taken an interest in invertebrate diversity will be able to tell you that annelids have traits also found in other conspicuous invertebrate groups, like segments, spiral cleavage, and trochophore larvae. They are therefore more or less representative for larger groups of animals, including arthropods and molluscs. But a deeper and more interesting answer emerges if we do a little historical digging.

The idealistic morphologists of the late eighteenth and early nineteenth centuries developed the ideas of the unity of type and the archetype in order to understand the nature and origin of complex morphologies. Many of the most conspicuous and complex body plans are composed of repeated parts. It was these, especially vertebrates and arthropods, that became the focus of morphologists' efforts (Russell, 1916; Ospovat, 1981; Van der Hammen, 1981; Schmitt, 2004, 2017; Richards, 2008). Their strategy was to conceive of these complex forms as combinations of more or less modified parts that were transformations of elements belonging to a shared type. For example, in the early nineteenth century, the French zoologists Jean Victor Audouin, Jules César Savigny, and Pierre André Latreille tried to understand the body plans of arthropods as modifications of a type with a specific number of segments and appendages. Around the same time in Germany, Johann Wolfgang von Goethe and Lorenz Oken independently developed the vertebral theory of the skull, explaining its unique structure as being a composite of variously modified vertebrae, and Carl Gustav Carus saw the entirety of the vertebrate skeleton sprouting from an endlessly modified ideal vertebra. Somewhat later in the century Richard Owen developed the most influential British product of this movement, his vertebrate archetype (see Chapter 2).

It is not surprising that archetypal thinking swirled mostly around two of Cuvier's four *embranchements*, the vertebrates and articulates (arthropods and annelids). It was intuitively obvious to try and understand their composite, modular body plans in terms of the deployment of a number of variously modified elemental parts. Indeed, Goethe used the same strategy to understand the diversity of plant morphologies in terms of a metamorphosing primordial leaf (see Chapter 2). However, the hodgepodge of taxa grouped in each of Cuvier's two other *embranchements* defied any neat archetypal reduction. What was the common denominator of Radiata, a mixture of echinoderms, parasitic worms, and cnidarians? What common plan underpinned Mollusca, a wastebasket taxon including molluscs, brachiopods, barnacles, and ascidians? Even when archetypes were eventually invented for sanitized parts of these *embranchements*, they lacked the morphological fecundity possessed by modular archetypes. In a paper he was very proud of, the 26-year-old T. H. Huxley drew a diagram of "the archetypal form of the Mollusca" (Huxley, 1853: 65), a vaguely pig-like creature equipped with the major characters of molluscs (Figure 9.6). Underneath it he drew seven modifications of the archetype that captured the body plans of different gastropods and cephalopods. But the archetype's creative potential was limited. The modifications that Huxley drew were essentially deformations – the gut would curve this way or that, the dorsal side would bulge upward or backward, the main body axis would shorten or bend. Huxley's molluscan archetype produced strictly molluscan offspring. Its creative potential was that of clay, not Legos.

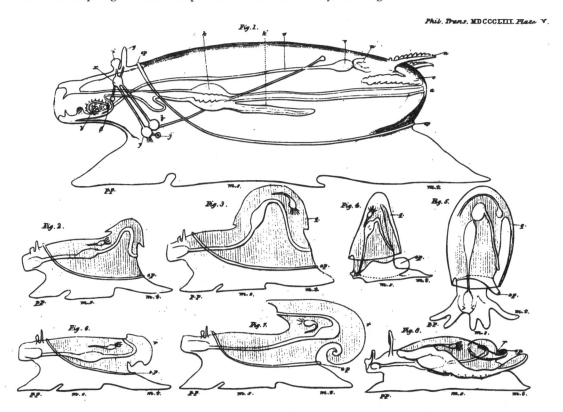

Figure 9.6 T. H. Huxley's archetypal mollusc and the modified body plans that he derived from it (after plate V in Huxley, 1853).

Like Legos, modular archetypes are much more versatile. Units can be multiplied, divided, fused, modified, and rearranged. Start with a homonomous series of parts, or just a single one like an ideal vertebra, and you can build almost anything. The less specialized the parts, the easier it is to mold them into the most diverse forms. The vertebrate archetypes of Carus and Owen from the 1820s and 1840s are nearly homonomous strings of vertebrae and ribs. In the 1820s, Audouin similarly envisioned his arthropod archetype as a string of identical segments and legs, with centipedes approaching this ideal in the flesh (Schmitt, 2004: 160). As we saw in Chapter 2, Owen's omnipresent polarizing force produced repetitive structures in nature, including the serially repeated parts of animals. Homonomous segmentation was baked into the core of his concept of serial homology (Owen, 1848: 171–172). Morphological differentiation of structures was due to what Owen called the adaptive force, working in opposition to the polarizing force. The more the former suppressed the latter, the more the resulting forms diverged from the archetype, and the more highly they would be ranked along the scale of being. The degree of morphological differentiation of serially homologous parts was Owen's index of organizational grade. He wasn't the only one who thought along these lines.

9.4.1 Milne-Edwards and the Division of Physiological Labor

An influential zoologist who had thoughts similar to Owen's was the Frenchman Henri Milne-Edwards (1800–1885) (Ospovat, 1981: 216–218; Richards, 1992a: 135–136; Ghiselin, 1995; Ruse, 1996: 242–243; Schmitt, 2004: 221–226; Gliboff, 2007; D'Hombres, 2012, 2016). He developed the concept of the division of physiological labor in 1827 to explain why higher organisms are more productive and efficient than lower ones. According to this economic principle, greater perfection is achieved through the increasing localization of particular functions to particular parts of the body. Greater functional and morphological differentiation allows organisms to perform more efficiently. This division of labor simultaneously provided Milne-Edwards with an index of an organism's rank along the scale of being. Darwin, who dedicated his second monograph on living barnacles to Milne-Edwards, incorporated this idea of the division of labor into his thinking in two ways: to understand the advantages for the functioning of individual organisms and to understand its effect on the divergence between species in ecosystems.

Richards (2013) and D'Hombres (2016) have recently argued, convincingly I think, that Darwin's thinking on the second point was muddled. D'Hombres (2012) has also argued, again convincingly, that the spread of Darwinian evolutionary theory removed the theoretical justification for using the principle of the division of physiological labor and its correlation with morphological complexity as a criterion for ranking organisms along a scale of perfection. But I disagree with his conclusion that it is remarkable "how obsolete this notion became in just a few decades" (D'Hombres, 2012: 25) after the spread of Darwin's theory. D'Hombres argues that Milne-Edwards' principle disappeared from morphological and embryological manuals and treatises before the mid-twentieth century. Although he may well be right that explicit references to the principle have declined over time, its influence certainly did not. The principle of the division of labor was fused with other ideas to form one of the most powerful and enduring evolutionary intuitions.

By using the principle of the division of labor for ranking species along the animal scale, Milne-Edwards reinforced the oldest biological ordering principle of all: simple to complex. Attempts to arrange nature's productions along this scale go back to the thinking of the ancient Greeks, especially Aristotle and his principles of plenitude and continuity (see Chapter 3). Centuries later, they found a clear expression in the development of the idea of the *scala naturae* (Lovejoy, 1936; Ruse, 1996; Kutschera, 2011; Archibald, 2014). Natural objects and organisms were ranked based on their perceived perfection or complexity, typically with humans as the ultimate goal and standard. But these criteria were imprecise and subjective. How does one rank the complexity of animal body plans along a single scale? Should molluscs outrank annelids, as Cuvier and Lamarck thought, or should worms lord it over molluscs, as the successor to Lamarck's chair, Henri de Blainville, proposed (Appel, 1980; Burkhardt, 1995)? The distinction between simple and complex morphologies is relative, blurry, and hard to define. Yet, it came to occupy a central place in the canon of pre-evolutionary comparative morphology. Milne-Edwards's principle of the division of physiological labor helped create clarity by supplying a more or less objective criterion and an associated functional rationale for measuring the degree of organismal differentiation. Carl Gegenbaur highlighted it in his *Grundzüge der vergleichenden Anatomie* (*Principles of Comparative Anatomy*):

> The differentiation of organs that can first be seen in the localization of functions, determines by the degree of difference we observe between individual organisms whether we designate them as lower or higher. The first are those in which, relative to others, the differentiation and thus the function of organs shows, by a lower complexity of parts, a lower level. The concept of higher organisms emerges from the opposite behavior. (Gegenbaur, 1859: 3)

Differentiation was especially easy to see in repeated parts and modular body plans. Richard Owen used it to judge how far species had departed from the serially similar morphology of his vertebrate archetype. And Goethe judged creatures with more dissimilar parts to be more perfect (Goethe in Ebach, 2005: 269). Evolutionists transformed these morphological ordering principles into powerful evolutionary intuitions.

Darwin's Edinburgh mentor Robert Grant provides an early example. Grant considered freshwater sponges as ancestral to marine sponges because they have a "greater simplicity" and "a more imperfect structure" (Grant, 1826: 283), including less differentiated and more loosely attached spicules. After escaping their unchallenging ancestral environment, the marine "descendants have greatly improved their organization, during the many changes that have taken place in the composition of the ocean" (Grant, 1826: 283). Morphological simplicity became an ancestral marker, while increasing complexity and differentiation revealed evolutionary progress through functional improvement. Referring back to the morphological insights of Goethe, Bronn, and Milne-Edwards, Haeckel (1866b: 250) praised the "exceedingly important and great law of the division of labor," noting that "it has recently been recognized generally and without contradiction by all thinking zoologists and botanists as the most important organizational law."

Pre-evolutionary morphologists had used notions such as increased morphological differentiation, decreased numbers of similar parts, and internalization of organs as criteria for placing organisms lower or higher along the scale of being. Evolutionists inherited these criteria and bundled them together into a principle or law of *Vervollkommnung* (perfection or improvement) that measured the degree of evolutionary progress of organisms. It came to guide phylogenetic thinking for more than a century (Crow, 1926: 96, 125; Arber, 1950: 64; Remane, 1956; Kusnezov, 1959; Simpson, 1959: 296; Ospovat, 1981: 216; McShea, 1991). Citations from two Ed Wilsons, a century apart, illustrate the endurance of this way of thinking. Discussing the origin of vertebrates, Edmund B. Wilson wrote: "under the evolutionary interpretation of nature, every higher and more complex form has arisen from a lower and simpler one" (Wilson, 1891: 55). Discussing the social evolution of insects a century later, Edward O. Wilson (1990: 22) wrote: "Biological diversity is structured by many transformational series that pass from simple to complex, wherein the simple states appeared first and the more complex later. Many reversals may have occurred along the way, but the overall average has moved from simpler to more complex rather than in the reverse direction."

Returning to D'Hombres' argument (2012), did the rise of Darwinian evolutionary theory quickly cause the extinction of the concept of the division of labor in the work of evolutionary morphologists? As I have shown, this wasn't the case. It was amalgamated with others into a biological law of organismic improvement that was smoothly integrated into emerging evolutionary perspectives, whether of internally driven change, like orthogenesis, or of evolution by natural selection. Virtually all scenarios discussed in this and the previous chapters explicitly or implicitly drew on one or more aspects of this law to determine the direction of evolutionary change. Irrespective of the extent to which an increase in complexity is a prevailing trend in evolution, which remains an active area of research (e.g. McShea, 1991, 2016), the assumption that evolution is in fact progressive in this sense is built into the very methodology of traditional narrative phylogenetics. As we saw in Chapter 7, orthogeneticists frequently pinned their evolutionary lineages on a backbone of one or a few characters oriented along a gradient of increasing complexity, but this strategy wasn't uniquely theirs. Committed to tracing complex characters back to simpler precursors, narrative phylogeneticists created scenarios in which relatively simple hypothetical ancestors give rise to more complex descendants via series of morphological intermediates.

The age-old ancestral attraction to acoels as the most primitive bilaterians illustrates this. Their relatively simple body plan – lacking coeloms, nephridia, circulatory system, and anus and often possessing just a poorly centralized nervous system – was easy to connect to simple hypothetical ancestors, be they like a Gastraea, Phagocytella, planula, *Trichoplax*, or ciliate (Balfour, 1885; Metschnikoff, 1886; von Graff, 1891; Haeckel, 1896; Hyman, 1951; Hadži, 1963; Salvini-Plawen, 1978). This view interprets platyhelminths, to which acoels were generally thought to belong before the late 1990s, as an early-diverging bilaterian phylum, and presents bilaterian body plan evolution as a gradual climb of morphological complexity to a coelomate apex (Adoutte et al., 1999; Adoutte et al., 2000; Jenner, 2000; Halanych, 2016). The point here is not to deny that animal body plans have evolved toward greater complexity, but to emphasize that progressive scenarios are an almost inevitable product of working with narrative phylogenetic tools. As it happens, the phylogenetic position of acoels and their close relatives has been hotly debated in recent years, with the last word at the time of writing suggesting that they are closely related to echinoderms and hemichordates, and therefore morphologically simplified rather than primitively simple (Kapli and Telford, 2020).

9.4.2 The Key Is Homonomy

The ancestral attraction of annelids derives from the same set of evolutionary intuitions. With its homonomous series of segments, a generalized annelid is the archetypal invertebrate, Legos made flesh. By division of labor and morphological differentiation, biologists mentally transformed it into arthropods, vertebrates, molluscs, and more. The key is its homonomy, a lack of differentiation that gives it the creative versatility of Owen's vertebrate archetype, aided by morphological simplicity. The more generalized a hypothetical ancestor is, the less its creative potential will be hemmed in by details. Gegenbaur (1878: 380) expressed this logic when discussing the evolution of molluscan reproductive organs: "These organs are fairly distinct in the different classes of the Mollusca, so that it is only possible to derive them from a ground-form common to the whole group, by looking for this form at a very low stage of differentiation."

For a century and a half, this type of rationale has guided the narrative reconstruction of generalized hypothetical ancestors. For instance, in their search for arthropod ancestors, morphologists focused on tiny forms "for only small limbs may be generalized enough to have the potentiality of giving rise to the great array of present-day crustacean appendages. Increase in size alone leads to complexity of structure" (Tiegs and Manton, 1958: 317). The evolutionary intuition that small creatures are on average simpler and therefore more primitive has been applied widely. For example, Bernard Rensch, who was an avid student of evolutionary laws, considered Cope's rule of phyletic size increase to be one of evolution's commonest trends (Rensch, 1960a: 206–218). His advice was that if you wanted to find the most primitive member of a clade, it was best to focus on the smallest ones (Rensch, 1960a: 271).

When Howard Sanders discovered cephalocarids, small homonomously segmented crustaceans, it revealed a near-perfect picture of what the Urcrustacean was expected to have looked like (Sanders, 1955, 1957) (Figure 9.7). Cephalocarids were judged to have "almost ideally primitive flattened biramous legs" (Tiegs and Manton, 1958: 292), with a near homonomously jointed endopod and a flap-like exopod of two parts (Figure 9.8). This quickly became textbook dogma: "cephalocarids are very primitive crustaceans, largely because of their relatively homonomous body form, undifferentiated maxillae, and generalized appendage structure" (Brusca and Brusca, 2003: 519). Discussing early arthropod fossils, Bergström and Hou (2005) homed in on the similar but even simpler limbs of the Cambrian arthropod *Fuxianhuia*, in which the more numerous articles of the endopod were even less differentiated, and with the flap-like exopod consisting of just one element. Echoing Sanders, Bergström and Hou (2005: 76) considered such an appendage a good proxy for the ancestral arthropod limb because it "is simple enough to allow the rise of every possible limb specialization we see in modern arthropods."

Narrative phylogeneticists gave their hypothetical ancestors generalized morphologies to aid in the construction of progressive scenarios. But this purely epistemic principle was also

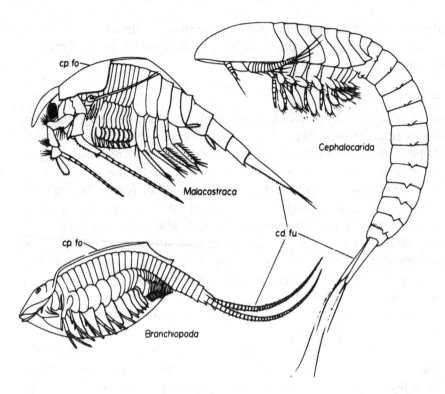

Figure 9.7 Three "primitive crustaceans" with more or less homonomously segmented bodies and flap-like limbs
(after figure 3 in Hessler and Newman, 1975, with permission).

lent weight by the idea that specialized forms were phylogenetically infertile, which was especially widespread among orthogeneticists. Toward the end of the nineteenth century, the American paleontologist Edward Drinker Cope packed this logic into his "law of the unspecialized" (Cope, 1896; Raia and Fortelius, 2013). The engine of evolutionary novelty hummed in generalized, small-bodied forms, but stalled in their specialized, larger-bodied descendants. Evolutionary potential was concentrated in "the 'golden mean' of character" (Cope, 1896: 174) or, as Huxley (1870: l) had put it, in "the average form." Orthogeneticists like Cope, his student Henry Fairfield Osborn, and Otto Schindewolf saw parallel evolutionary lineages rise from generalized ancestors and go to extinction in specialized senility. But like archetypes, generalized hypothetical ancestors are diagrams devoid of details. They are not faithful representations of real ancestral organisms, but abstractions stripped of taxon-specific adaptations. As Alfred Sherwood Romer (1946: 41–42) wrote,

> [i]n animal evolution we constantly assume the derivation of 'specialized' and complexly built forms from simple 'generalized' ancestors. But no one has ever found a diagrammatically generalized ancestor lodged in other family tree-trunks... An animal cannot, after all, spend its time being a diagrammatic ancestor. It must make a living, eat, escape enemies, reproduce. To live it must be adapted to its environment. Such adaptations are in themselves specializations. The evolution of an animal group is not in general from simplicity to complexity, but from one complexly specialized condition to another.

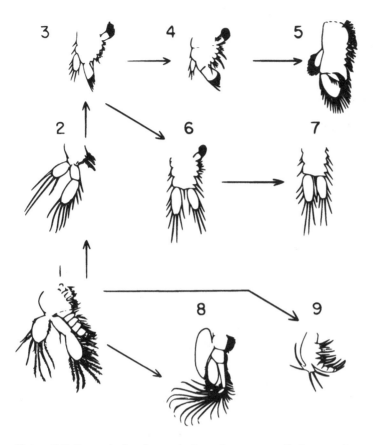

Figure 9.8 Scenario for the evolution of crustacean limb types from an ancestral cephalocarid-type thoracic limb
(after figure 6 in Sanders, 1957, by permission of Oxford University Press).

Romer therefore sought for ancestors in the fossil record instead. Still, generalized hypothetical ancestors exerted a large and enduring influence on evolutionary interpretations in some fields.

William Thomas Calman (1871–1952) was one of the foremost carcinologists of the twentieth century and Keeper of Zoology in the museum where I work. He created what he called the "caridoid facies," a schematic of the hypothetical common ancestor of malacostracan crustaceans (Calman, 1904, 1909). Its generalized shrimp-like body plan with undifferentiated thorax appendages provided an ideal starting point for imagining the evolution of more specialized living forms (Figure 9.9). Although Calman devised this scheme when three of the 15 known malacostracan orders (Mictacea, Spelaeogriphacea, and Thermosbaenacea) hadn't been discovered, and another one hadn't been recognized (Bathynellacea), researchers have used the caridoid facies as a tool for understanding crustacean evolution for more than a century (Schram, 1982; Hessler, 1983; Newman, 2005; Poore, 2005; Jones et al., 2016). Some dissenters have argued that the caridoid facies isn't primitive for malacostracans or eumalacostracans (Tiegs and Manton, 1958; Dahl,

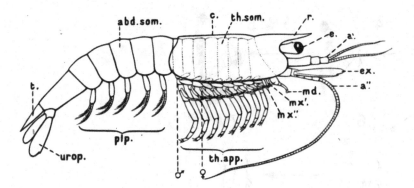

Figure 9.9 W. T. Calman's caridoid facies as a generalized malacostracan (after figure 85 in Calman, 1909).

1983; Watling et al., 2000), and they are absolutely right. If you look at malacostracan diversity with squinting eyes, you can see how the caridoid facies represents some sort of average, undifferentiated form. Like the generalized archetypes used by idealistic morphologists (e.g. Zangerl, 1948), such a scheme can be useful for thinking about body plan transformations in a purely formal way. But phylogenetic transformations need to be rooted in accurately reconstructed ancestral traits.

Although our understanding of malacostracan phylogenetic relationships continues to evolve (Schwentner et al., 2018), we can reject the caridoid facies as a well-supported ancestral groundplan. Calman gave it a ventrally bend pleon and completely undifferentiated biramous thorax appendages with a slender endopod and flagelliform exopod. Neither of these can be supported as malacostracan or eumalacostracan plesiomorphies. The two earliest diverging malacostracan lineages, leptostracans and stomatopods, lack a ventrally bend pleon. And while leptostracans do have undifferentiated thorax appendages, these are foliaceous. Virtually all other malacostracans have at least one pair of differentiated maxillipedes. But the pull of the caridoid facies is strong. As my former PhD supervisor Fred Schram put it, ""[w]e have our 'sacred cows' in carcinology, and like the holy beasts of India, they are allowed to roam through our papers and discussions at will" (Schram, 1982: 133). Schram's own papers show the tracks of this wandering beast, while sometimes accepting (Jones et al., 2016) and sometimes rejecting (Watling et al., 2000) its ancestral status.

Staying with ancestral shellfish for a moment, malacologists David Lindberg and Michael Ghiselin (2003) published an amusing paper that traces the trails left in the literature by the descendants of Huxley's archetypal mollusc. Since T. H. Huxley sketched its outline in 1853, numerous variations have appeared in the pages of papers and textbooks. Some were direct modifications of Huxley's HAM, or Hypothetical Ancestral Mollusc, as the creature is affectionately called by malacologists. Others were novel hypotheses based upon new syntheses of evidence. These HAMs served different purposes. Huxley used his HAM to understand the morphological (not evolutionary) transformations that could connect the body plans of extant molluscs. Textbook HAMs are generally pedagogical tools that capture what an ancestral, or average, mollusc might look like in soft focus. In contrast, the HAM of malacologists like Luitfried von Salvini-Plawen was a best estimate of the last common ancestor of molluscs and an integral part of a scenario for molluscan evolution. Lindberg and Ghiselin (2003: 663) conclude that "HAM has not aided evolutionary biologists or paleontologists in solving problems, but it has often had the opposite effect, by requiring that theories be treated within its framework."

One example they discuss is how the HAM has hindered the acceptance that "molluscs are modified annelids" (Lindberg and Ghiselin, 2003: 676). They argue that the unsegmented body plan of a HAM has biased researchers into accepting the rival theory that molluscs are derived from flatworm-like ancestors. When Lemche and Wingstrand (1959) described the morphology of *Neopilina galathaea*, the first discovered living monoplacophoran, they highlighted its serially repeated structures, including muscles, kidneys, gills, and nerve commissures. They argued that this remarkable repetition is homologous to the segmentation of annelids and arthropods and that the three phyla had descended from a common segmented ancestor (Lemche and Wingstrand, 1959: 67). Lindberg and Ghiselin argue that the failure of more recent HAMs to incorporate these findings shows the deadweight of tradition. Uncertainties about deep molluscan relationships, the challenges of interpreting stem group molluscs, and debates about the sister group of molluscs have long hindered unambiguous ancestral state reconstruction at the base of the phylum, although encouraging progress is emerging (Vinther et al., 2017; Wanninger and Wollesen, 2019; Kocot et al., 2020). However, it is clear to me that Lindberg and Ghiselin's conclusion that molluscs evolved from a segmented annelid-like ancestor also bears the unmistakable signs of historical burden.

At the time their paper appeared, the traditional union of annelids and arthropods as Articulata – first as a Cuvierian *embranchement* and then as a clade – had dissolved irreversibly. To maintain homology of annelid and arthropod segmentation, one had to argue that the Urprotostome or Urbilateria had been similarly segmented. Some authors did (Balavoine and Adoutte, 2003), and Lindberg and Ghiselin duly cited them. At the time, the hypothesis of a segmented Urbilateria was not the result of ancestral state reconstruction on a robustly resolved tree. A lack of phylogenetic resolution prevented this. Instead, "striking similarities in mesodermal patterns" (Balavoine and Adoutte, 2003: 137) were simply assumed to be homologous, and a segmented Urbilateria was used as the a priori starting point of a scenario that treated unsegmented phyla as simplified descendants. In principle, there is nothing wrong with such a scenario, but although Balavoine and Adoutte included two trees in their paper, theirs was a classic exercise in narrative phylogenetics, echoing a similar attempt in the same journal a few years earlier (Balavoine, 1998).

Lindberg and Ghiselin's conclusion that molluscs stem from annelid ancestors, which is historically one of the two major hypotheses for the origin of molluscs (Vagvolgyi, 1967; Haszprunar, 1996) and one that Ghiselin (1988) had endorsed before when the Articulata hypothesis still had some life in it, was likewise the product of evolutionary storytelling. It is here that the ancestral attraction of annelids comes into sharp focus. According to the Articulata hypothesis, annelids and arthropods are sister taxa. With molluscs being more or less closely related to this clade, their serially repeated structures did not have to be interpreted as the remnants of once more fully developed annelid-like segments.

Interpreting them as less fully developed homologues or as independently evolved seriations were obvious alternatives. The fact that Lindberg and Ghiselin nevertheless interpreted the repeated structures in molluscs as remnants of an annelidan past illustrates the hold of annelids as hypothetical ancestors.

Richard Owen's concept of serial homology, and the associated notion that the homonomous repetition of parts is an original condition, provides a bridge between pre-evolutionary and evolutionary morphology. As Darwin (1859: 418) said,

> An indefinite repetition of the same part or organ is the common characteristic (as Owen has observed) of all low or little-modified forms; therefore we may readily believe that the unknown progenitor of the vertebrata possessed many vertebrae; the unknown progenitor of the articulata, many segments; and the unknown progenitor of flowering plants, many spiral whorls of leaves.

What made logical sense as a morphological principle also made sense as an evolutionary intuition. Increasing specialization through division of labor transforms homonomy into heteronomy, turns long-bodied forms into shorter-bodied forms, and increases the differentiation of repeated elements in appendages. It increases efficiency and morphological complexity and transforms the index of morphological perfection sought by pre-evolutionists into a tangible criterion of evolutionary progress. I cannot think of any narrative phylogenetic principle that has had a greater and longer lasting effect on our thinking about the evolution of animal body plans than this (Gegenbaur, 1859: 135, 201; Hatschek, 1888: 403, 406, 411; Haeckel, 1896: 615, 619; Lankester, 1904; Handlirsch, 1908; Heider, 1914: 496–497; Crow, 1926: 123–124; Tiegs and Manton, 1958; Beklemishev, 1969a: 243, 263, 343; Van der Hammen, 1981: 39–40; Schram, 1982; Averof and Akam, 1995; Gould, 2002b: 84, 1104, 1143–1144, 1163, 1169; Bergström and Hou, 2003: 328-329; Bitsch and Bitsch, 2004: 521, 534; Schmitt, 2004, 2017; Fusco and Minelli, 2013: 198; Siomava et al., 2020).

These intuitions were made flesh in annelids. For a century and a half, homonomously segmented generalized annelids have been starting points in scenarios for the evolution of arthropods, molluscs, and vertebrates, as well as larger clades of bilaterians, and of course annelids themselves. They were thought to provide perfectly pliable precursors for the origin of more differentiated body plans and near infallible guides to the direction of evolutionary change. Once a certain degree of segmental specialization, which Lankester (1904: 533) called "tagmosis," has developed, a reversal back to homonomy would be highly unlikely (Calman, 1936: 202). These assumptions have reverberated through the literature.

Before the first broadly sampled morphological cladistic and molecular phylogenetic analyses of annelid relationships were published in the 1990s, "an untested assumption behind all schemes was that evolution appeared to have tracked morphological differentiation from homonomous to heteronomous segmentation, perhaps along several lines, but certainly going from 'simple' to 'complex'" (Rouse and Fauchald, 1997: 72). (Note this idea is also supported in other works: Fauchald, 1974; Westheide, 1997; Westheide et al., 1999; Bartolomaeus et al., 2005.). One of the zoologists most responsible for pushing annelids into an ancestral mold was Berthold Hatschek (1854–1941), an Austrian annelid expert who is best remembered today for his Trochophore theory, according to which the trochophore larvae of annelids and related phyla recapitulate a rotifer-like hypothetical bilaterian ancestor (Hatschek, 1878, 1888). He considered *Polygordius*, a genus of superficially nematode-like polychaetes, as one of the most primitive living annelids. These are the same worms that were long claimed to have amphistomic gastrulation, even though Woltereck himself explicitly denied it (Woltereck, 1904: 391). Hatschek used them to create a taxon he called Archiannelida, which was characterized by "complete (outer and inner) homonomy of metameres" (Hatschek, 1888: 411).

It was not just the adult morphology of *Polygordius* that Hatschek saw as ancestral. Being an enthusiast of the Gastraea theory, he saw in the trochophore larva of *Polygordius* an ancient protostome ancestor he called Trochozoon, a creature reflected in its least modified adult form in certain rotifers (Hatschek, 1888). From archiannelid beginnings the other annelid lineages diverged along different pathways of specialization, a scheme accepted by many zoologists for the better part of a century before archiannelids were reinterpreted as a polyphyletic group of morphologically simplified forms (Fauchald, 1974; Rouse and Pleijel, 2001; Andrade et al., 2015; Worsaae et al., 2021). Although several different systematic systems were devised for annelids, they all shared the assumption of a homonomously segmented ancestor, and this carried over into virtually all scenarios for the origin and evolution of arthropods proposed in the nineteenth and twentieth centuries.

Before the blade of molecular phylogenetics cleaved them apart, there was no greater agreement in animal phylogenetics than that arthropods had evolved from annelids or annelid-like ancestors. Such is the power of the ancestral attraction of annelids that some believe it still (Wägele and Kück, 2013). When phylogenetic speculation began in the second half of the nineteenth century, consensus about the annelid ancestry of arthropods was broad but not universal. As Peter Bowler (1994, 1996) has shown, several polyphyletic schemes for the origins of arthropods were proposed. Some authors derived crustaceans, or even all arthropods, from three-segmented nauplius-larva-like forms, which had descended from trochophore-larva-like ancestors. This implied that the segmentation of annelids and at least some arthropods wasn't homologous. Yet by the early twentieth century, many zoologists believed that the tracheates, comprising the insects, myriapods, and onychophorans, hailed from annelid ancestors.

As the twentieth century progressed, the pendulum swung toward the general acceptance of the monophyly of arthropods. The popularity of scenarios for the origin of arthropods from nauplius-like ancestors largely waned among zoologists, although paleontologists retained their enthusiasm for recapitulation longer. By the time Sydnie Manton started the Renaissance of the polyphyletic view of arthropod origins in the 1950s, arthropods were generally seen as the monophyletic descendants of homonomously segmented annelid-like ancestors. Some saw these worms as equipped with parapodia, like modern polychaetes, while others preferred the smoother contours of a more clitellate-like creature without parapodia. The scenarios painted by Bernard (1892), Haeckel (1896), Korschelt and Heider (1899), Lankester (1904), Crampton (1921), and Raw (1953), for example, derived onychophoran and arthropod limbs from ancestral parapodia (Figure 9.10). The famous scenario drawn by Robert Snodgrass (Snodgrass, 1935, 1938) (Figure 9.11), on the other hand, is the best-known example of a scheme that derives

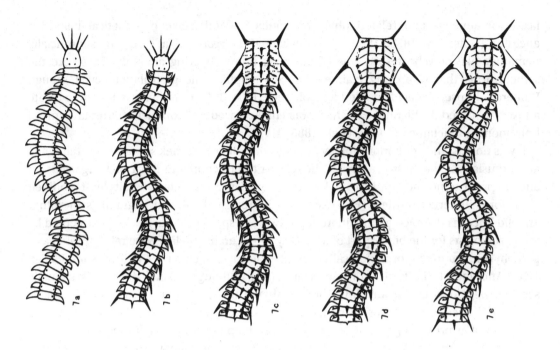

Figure 9.10 The evolution of a polychaete (left) into a hypothetical arthropod ancestor
(after figure 7 in Raw, 1953, republished with permission of the Society of Economic Paleontologists and
Mineralogists; permission conveyed through Copyright Clearance Center, Inc.).

onychophorans and arthropods from clitellate-like generalized annelids. According to
Snodgrass, polychaete parapodia and the legs of onychophorans and arthropods had
evolved independently. Salvini-Plawen (2000) proposed an identical scenario without,
however, mentioning Snodgrass.

All these schemes saw the homonomy of the ancestral annelid carried over into the
ancestors of the major lineages of arthropods. Snodgrass, for example, envisioned a
centipede-like ancestor giving rise to prototrilobites, which started the chelicerate lineage,
and protomandibulates, which started the lineage containing crustaceans, insects, and
myriapods. For Snodgrass centipedes "are the conservatives among the arthropods"
(Snodgrass, 1938: 147). Similar views of homonomously segmented ancestors penetrating
deeply into the phylogeny of arthropods have been the consensus for a century and a half.
Peter Ax's cladistic scheme of metazoan relationships provides one example (Ax, 1999).
Evolutionary transformations were generally thought to track arrows of increasing tagmosis,
and these narrative phylogenetic assumptions infused cladistic analyses as well. For
instance, Ax (1999) placed the long-bodied and homonomously segmented geophilomorph
centipedes as the sister group to a clade he called Heteroterga, which grouped the other
four living orders based on their heteronomous segmentation (alternating long and short
tergites, and spiracles associated with the long tergites only). Although this hypothesis
clashes with one that positions geophilomorphs deeply in the centipede tree, which is

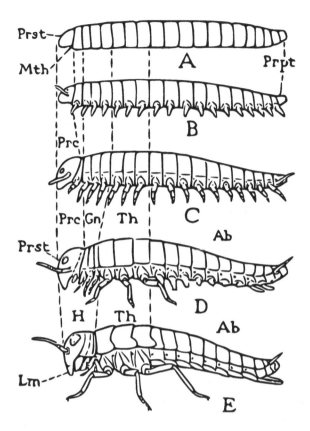

Figure 9.11 Robert Snodgrass' scenario for the evolution of arthropods from annelid ancestors (after figure 24 in Snodgrass, 1935). (This figure is taken from *Principles of Insect Morphology*, by R. E. Snodgrass, a Comstock book published by Cornell University Press. Copyright © 1993 by Ellen Burden & Ruth Roach. Included by permission of the publisher).

supported by molecular data and accepted by most centipede systematists today (Fernández et al., 2014), it echoes previous precladistic views on centipede systematics and evolution that considered geophilomorphs as the most primitive living centipedes (Beklemishev, 1969a; Lewis, 1981). Likewise, Schileyko and Pavlinov (1997) rooted the first broadly sampled cladistic analysis of scolopendromorph centipedes with a homonomously segmented hypothetical ancestor because they thought that centipedes ultimately stemmed from annelid-like ancestors.

The area of arthropod evolutionary biology where the assumption of long-bodied, homonomously segmented ancestors has been especially prominent is carcinology (Tiegs and Manton, 1958; Hessler and Newman, 1975; Schram, 1982; Fryer, 1992; Schram and Hof, 1998; Schram and Koenemann, 2004; Newman, 2005). Attempts to identify the most primitive living crustaceans and to glimpse in them the contours of the Urcrustacean has been a popular carcinological pastime since the late nineteenth century (Figure 9.12). Before the description of cephalocarids in 1955 (Sanders, 1955, 1957), branchiopods were

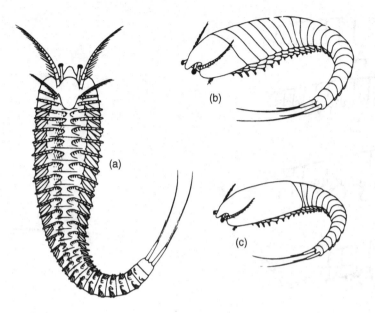

Figure 9.12 Hypothetical Urcrustacean (a), with (b) and without (c) a carapace (after figure 6 in Hessler and Newman, 1975, with permission).

generally considered to be the most primitive living crustaceans (Bernard, 1892; Crampton, 1921), a view maintained until quite recently (Martin and Davis, 2001). The many-segmented bodies and poorly differentiated limbs of anostracans (fairy shrimp) and notos-tracans (tadpole shrimp) were thought to betray their lowly descent. The same was true for cephalocarids, but their biramous limbs were thought to be even more primitive than the foliaceous limbs of branchiopods (Sanders, 1955, 1957; Hessler and Newman, 1975). After remipedes were described in 1981 (Yager, 1981), their completely homonomously seg-mented trunk equipped with biramous limbs suggested to Fred Schram that they instead should be seen as the most primitive living crustaceans (Schram, 1983, 1986). This ancestral assumption culminated in Peter Ax (1999) placing them as the sister group to all other crustaceans, and Matthew Wills using them to root his cladistic analyses of crustacean phylogeny (Wills, 1998).

The phylogenetic fertility of annelid ancestors penetrated deeply into other areas of metazoan diversity as well. Zoologists have imagined many different ways of turning segmented worms into chordates (Dohrn, 1875; Semper, 1875; Eisig, 1878; Kleinenberg, 1886; Beard, 1889; Minot, 1897; Bernard, 1898; Raw, 1960; Gutmann, 1966, 1981; Engelbrecht, 1967) and made annelid-like ancestors into cradles for larger clades of animals (Cope, 1896: 80–81; Christoffersen and Araújo-de-Almeida, 1994; Almeida et al., 2003; Gudo and Syed, 2008). The annelid body plan can itself be seen as a special case of a modular construction in which units are arranged linearly. Indeed, several pre-evolutionists in the early nineteenth century conceived of segmented body plans as colonies of lower-level units (Schmitt, 2004, 2017). Evolutionists, such as the French zoologist Edmond Perrier, elaborated on these ideas later in the century and interpreted segmented body plans as

Figure 9.13 A hypothetical, pelagic, ancestral chordate, modeled on the body plan of salps, with zooids budding along a posterior stolon. The stolon would eventually evolve into a locomotory trunk, with a series of somites and associated structures
(after figure 6 in Lacalli, 1999, by permission of John Wiley and Sons).

linear colonies that had resulted from budding or fission in free-moving ancestors, while sessile ancestors would give rise to branching colonies. Recent authors, some of whom were inspired by the ideas of Perrier, have continued to be attracted to the versatility of modular body plans to explain the origins of bilaterians, deuterostomes, vertebrates, and indeed annelids (Rieger, 1986, 1994; Lacalli, 1997, 1999; Dewel, 2000; Dewel et al., 2001; Martynov, 2012; Starunov et al., 2015) (Figure 9.13).

9.4.3 *Platynereis dumerilii*: The Ultimate Annelid?

As we have seen, a substantial part of the literature on animal body plan evolution over the past century and a half assumes that homonomously segmented ancestors existed on various phylogenetic levels. The power of this ancestral attraction can be traced back to pre-evolutionary thinking about archetypes and its power can still be felt today. What is the evidence supporting it, given our current understanding of metazoan phylogeny? The previous lack of phylogenetic resolution provided space for annelid-like ancestors to be promoted based on their precursor potential, and the endurance of the clade Articulata until about two decades ago continued to root scenarios for arthropod evolution. But with the collapse of Articulata and the separation of annelids and arthropods into different metazoan superclades, this pervasive ancestral appeal now lies exposed as a colossal mental straitjacket. I mean this epistemologically, of course. Homonomously segmented ancestors may well have existed, but the reasons for thinking this have been a priori for most of the history of phylogenetic thought. Fossils suggest that stem annelids may well have been homonomously segmented (Parry et al., 2014), but there is no evidence that this is relevant for understanding evolution outside annelids. The annelid sister group remains elusive (Bleidorn, 2019; Laumer et al., 2019; Marlétaz et al., 2019), but annelids are nested deeply within Lophotrochozoa, which limits the light they can shed on ancestral traits shared with other phyla.

Yet, there does seem to be a remarkable degree of fortuitous congruence between new insights and previous expectations based on evolutionary intuitions that had originated as morphological principles formulated by pre-evolutionists. The fossil record does suggest that homonomous segmentation is probably primitive for arthropods (Jockusch, 2017; Chipman and Edgecombe, 2019), but this does not mean that it is a primitive state that is retained at the origin of all major crown group arthropod lineages. The discovery of new fossils and increased phylogenetic resolution across major taxa shows that evolution of segment and limb differentiation in arthropods is a complex developing story. With the demise of Articulata and Tracheata, the foundation for the traditional Snodgrassian view of arthropod body plan evolution has disappeared. Arthropods did not evolve from annelids, centipedes and other myriapods are no longer acceptable ancestral models for the origin of hexapods, and the crustaceans that have long been seen as proxies for the Urcrustacean (branchiopods, cephalocarids, remipedes) are now deeply nested inside Pancrustacea, which suggests that their long, many-segmented bodies may have evolved from short-bodied ancestors (Regier et al., 2010; Lozano-Fernandez et al., 2019). Nevertheless, traditional ancestral attractions continue to influence the thinking of some zoologists (Wägele and Kück, 2013), in particular those who have promoted annelids as model organisms.

To close this chapter we return to the rhetorical flourishes with which this section started. After being a focus for studies in reproductive biology and embryology, the polychaete *Platynereis dumerilii* was developed into an influential model organism for molecular and comparative developmental biology during the past two decades. No one has done more to promote *Platynereis*, especially its supposed ancestral nature, than Detlev Arendt, a prolific researcher from the European Molecular Biological Laboratory in Heidelberg, Germany. But Arendt did not convince himself and others of the ancestral nature of this annelid on the basis of broadly and deeply sampled comparative analyses. Instead his promotional campaign adopted a traditional narrative phylogenetic approach, rooted in age-old evolutionary intuitions. Three examples spanning almost two decades illustrate this.

An example from 2003 reads "According to fossil records, the earliest bilaterians found were marine worms of considerable size, with a morphology somewhat in between the body plans of today's polychaete annelids, molluscs, and brachiopods... Comparative morphological arguments point in a similar direction, that urbilaterians may have resembled annelids" (Tessmar-Raible and Arendt, 2003: 334–335). An example from 2009 reads "the high level of synteny between humans and *Platynereis* is widespread beyond the ParaHox loci, [and] would provide a further example of the prototypical nature of *Platynereis* biology and its huge potential for revealing the nature of the PDA [protostome-deuterostome ancestor]" (Hui et al., 2009: 10). And finally, an example from Arendt's website in 2021 reads "As a 'living fossil', *Platynereis* represents an ideal connecting link between vertebrates and the fast-evolving protostome models *Drosophila* and *Caenorhabditis*" (www.embl.de/research/units/dev_biology/arendt/; last accessed September 24, 2021). These allusions to the ancestral nature of *Platynereis* are the result of rolling a loaded dice. They depend heavily on phylogenetic shortcuts that consider similarity as nearly synonymous with homology, that make comparisons within the context of severely pruned trees, and that plug gaps in the evidence with rhetoric.

In their paper on emerging model organisms, Tessmar-Raible and Arendt (2003: 335) close a section suggestively titled "Living Urbilateria?" with this sentence: "the nereid polychaete *Platynereis dumerilii* – and its close relative *Nereis virens* – have been chosen primarily for their ancestrality: 'From the standpoint of comparative neurology there could hardly have been a better choice of a polychaete as the common classroom type than

Nereis.'" This may sound reassuring, but when you pause to ponder this for a moment, things start to look different. The authority they cite is Bullock and Horridge's 1965 classic text *Structure and Function in the Nervous Systems of Invertebrates.* The sentence following the cited one reads: "Well provided with sense organs, its brain and hence the rest of the nervous system is well differentiated and exhibits virtually all the features of advanced polychaetes" (Bullock and Horridge, 1965: 735). So, advanced rather than primitive then? What about the standard by which *Platynereis* is judged to have ancestral genes? David Ferrier usefully summarized the case in a paper published nine years later: "Many comparative genomics studies involving annelids have established this group of animals as a relatively conservative lineage that is less derived from the ancestral bilaterian state than many other, more established, model systems" (Ferrier, 2012: 2647). Looking at its genome, *Platynereis* is less divergent from Urbilateria than the nematode *Caenorhabditis elegans* and the fly *Drosophila melanogaster.* But would Urbilateria recognize its reflection in the genome of *Platynereis*? Ferrier provides a helpful picture in which the branch lengths of a molecular tree illustrate these genetic distances. The insect and nematode branches are indeed substantially longer than *Platynereis*'s. The length of its branch, however, is practically identical to the human branch. Maybe we are ancestral too.

In reality, *Platynereis* is a nereidid polychaete that is nested deeply within annelids (Weigert and Bleidorn, 2016), which are nested deeply within lophotrochozoans, which is a protostome clade nested within Bilateria. Any inferences based on the biology of *Platynereis* to inform the reconstruction of ancient animal ancestors have to clear a number of nodes on any well-resolved metazoan phylogeny. This doesn't mean that they can't, just that it depends on the traits in question. Like any living taxon, *Platynereis* is a mosaic of more and less derived characters, some of which an ancient ancestor might see itself reflected in, and some of which it won't. As we will see in the next chapter, researchers commonly pimp their models to give them an ancestral gloss. This approach is understandable, if not excusable. It is understandable that researchers will try to sell the value of their organisms for comparative studies by suggesting that they provide unique glimpses into the evolutionary past. It is understandable that researchers might think that their model organisms will reveal ancestral traits if their comparative analyses are limited to only a handful of other species that have been studied in equal detail. But they should know better.

Narrowly sampled comparisons lack the power to properly test homology and will underestimate the real evolutionary dynamics of traits. These are inevitable consequences of doing comparative research on a small set of deeply studied model organisms. But these inescapable limitations should not be misconstrued as benefits and used to promote particular model organisms. Any biological insights gleaned from the deep study of model organisms are treasures, but their evolutionary significance does not inhere in any qualities intrinsic to the animals. The value of any organism for generating evolutionary insights resides entirely in the comparative context in which it is studied. Without impugning any of the many fascinating and sometimes brilliant studies that Arendt and his colleagues have published, their promotion of *Platynereis* as a uniquely valuable ancestral beacon is just a modern incarnation of the ancestral attractions and phylogenetic folklore that have been the beating heart of narrative phylogenetics for the past century and a half.

10

Narrative Shortcuts and Phylogenetic Faux Pas

10.1 Narrative Ghosts in the Cladistic Machine

One cannot exaggerate how deeply the adoption of cladistics and quantitative methods has transformed phylogenetics over the last 60 years or so. Modern phylogenetics covers a wide range of approaches and opinions about how best to reconstruct phylogenies, and finding strongly worded disagreements is easy. For our purpose of understanding the history of evolutionary storytelling, the key change was the replacement of the directly narrative approach to reconstructing evolutionary lineages with approaches that draw inferences strictly on the basis of preestablished systematic patterns. To ensure that evolutionary inferences take all relevant, but only relevant, relationships into account, phylogenetic analyses need to be comprehensively sampled. Today, the availability of a wide range of data sources makes it possible to construct phylogenetic frameworks with evidence that is independent from the traits that we want to study. This is a major boon for shedding light on the evolution of characters that by themselves lack sufficient information for establishing a comparative framework, and it helps avoid circular reasoning. In addition, rigorous ancestral state reconstruction methods have replaced often fanciful speculations about hypothetical ancestors and evolutionary events.

A paper that appeared when I was four years old captures the essence of this epistemological transformation. Ian Tattersall and Niles Eldredge from the American Museum of Natural History wrote their article titled "Fact, theory, and fantasy in human paleontology" for the *American Scientist* (Tattersall and Eldredge, 1976). They suggested that phylogenetic hypotheses can be formulated on three different levels, each more removed from observable data. On the most basic level, cladograms depict the hierarchical branching of taxa based on the distribution of shared derived character states (synapomorphies). On the second level, cladograms are interpreted as phylogenetic trees that depict the relative recency of common ancestry of taxa and their traits. Trees have a time axis, their internal branches depict lineages, although Tattersall and Eldredge don't use this word, and nodes in trees can be interpreted as hypothetical common ancestors. Tattersall and Eldredge point out that each cladogram is consistent with several different phylogenies. This one-to-many pattern of correspondence is commonly used to illustrate that cladograms and trees are not necessarily isomorphic and therefore that deciding on a given tree involves speculation and moving away from observable data (Patterson, 1981: 197; Forey, 1982: 122–123; Williams and Ebach, 2020: 163–165). However, it is important to realize that a cladogram will correspond to multiple trees only if one or more of the included taxa are not monophyletic. If all taxa are monophyletic, the cladogram and phylogeny will be isomorphic, and each terminal taxon will represent the tip of a distinct lineage.

On the third level we find scenarios, several of which may be consistent with a given phylogeny. Tattersall and Eldredge (1976: 204) defined a scenario as "essentially a phylogenetic tree fleshed out with further types of information, most commonly having to do with adaptation and ecology." They point out that as one moves from cladogram to tree to scenario, the ratio of data to speculation decreases, as does the testability of propositions. The systematic relationships summarized in the cladogram provide an interpretative framework for adding further layers of speculation about evolutionary history. According to this epistemological revision, the scenario-first strategy of narrative phylogenetics is a storytelling shortcut that fosters speculation without a safety net. Systematic patterns come first, phylogenetic speculations may follow, if one is so inclined.

The theoretical and methodological advances that were among the main drivers of modern phylogenetics, especially Hennig's phylogenetic systematics, the development of quantitative phylogenetic methods, and the development of substitution models, all trace back to the 1950s and 1960s. But the transition from narrative to modern phylogenetics took time. In the field of metazoan phylogenetics, scenario-based approaches were replaced by morphological cladistic and molecular phylogenetic analyses during the 1980s and 1990s, but narrative phylogenetics didn't disappear. Claus Nielsen, for instance, has been an active and influential representative of the narrative phylogenetic tradition for four decades. In his 1985 paper "Animal phylogeny in the light of the Trochaea theory," he sketched a picture of animal phylogeny that was organized around a backbone of hypothetical ancestors, including Blastaea, Gastraea, and Trochaea (Nielsen, 1985). He used the Trochaea theory to structure his subsequent review of metazoan ciliary bands (Nielsen, 1987), and it remained the coordinating idea for his widely read 1995 book, *Animal Evolution. Interrelationships of the Living Phyla*, now in its third edition, while bolstering it with a wide-ranging survey of invertebrate morphology and development. I vividly remember browsing Nielsen's book as an undergraduate at Utrecht University and being amazed how someone could master such a massive amount of disparate details and then herd them into such a neat narrative for body plan evolution. There seemed to have been no cilium in the sea with which Claus wasn't intimately familiar.

Nielsen's 1985 paper was a major inspiration for Fred Schram to conduct one of the first computer-assisted morphological cladistic analyses of metazoan phylogeny. Like the Brusca brothers (Brusca and Brusca, 1990), Schram published his cladistic analysis in a textbook (Meglitsch and Schram, 1991) and in a conference proceedings (Schram, 1991), two unlikely outlets for conceptual innovations. These efforts marked the beginning of a decade and a half during which morphological cladistic analyses were an important driver of research into higher-level animal phylogeny (e.g. Eernisse et al., 1992; Nielsen et al.,

1996; Zrzavý et al., 1998; Giribet et al., 2000; Sørensen et al., 2000; Peterson and Eernisse, 2001; Jenner and Scholtz, 2005). But by the second decade of the new millennium, this approach was effectively dead. Gonzalo Giribet, Greg Edgecombe, and their collaborators were among those who have been trying to make morphological cladistic analyses more objective by substituting exemplar species scored for actual observations for the traditional higher-level terminal taxa with ground patterns of assumed primitive character states. They attempted the construction of a new morphological monster matrix as part of the National Science Foundation-funded project *Assembling the Protostome Tree of Life* (Giribet, 2008), but this attempt to resolve metazoan phylogeny with morphological data foundered. Gonzalo and Greg pinpoint my own research as one of the nails in this cladistic coffin (Giribet and Edgecombe, 2020: 17).

During my PhD research in Fred Schram's lab in the late 1990s and early 2000s, I studied metazoan phylogenetics in great detail to understand in what way morphological cladistic analyses were an improvement over narrative attempts to understand animal evolution (Jenner, 1999; Jenner and Schram, 1999; Jenner, 2000, 2001, 2002, 2003, 2004b, 2004c, 2004d). When I say "great detail," I mean that I read hundreds of articles and books to discover to what extent the many thousands of entries in cladistic data matrices were reliable reflections of observed character variation, or just speculations and assumptions. One particularly large morphological matrix contained 15,732 data entries, which invited close scrutiny because the authors reported that their paper resulted from an undergraduate phylogenetics course (Zrzavý et al., 1998). I disgorged hundreds of pages into the literature myself, with one article clocking in at 161 printed pages and filling an entire issue of *Contributions to Zoology* (Jenner, 2004c). I first submitted this massive manuscript to *Biological Reviews*. The editorial office was just down the road from the University Museum of Zoology in Cambridge where I was a postdoc with Max Telford, so I printed out the required number of double-spaced copies of the manuscript and put them into the now empty box of printer paper. I carried it to the editorial office, where a staff member looked at me, and before I could speak, said "Just put the box over there," gesturing to the printer. "No, I'm sorry, I'm here to submit this manuscript," I said. "Ohh!," was the surprised reply. I left the box there, but a message quickly followed, apologizing that the paper was too long for them to consider publishing it. I imagine Martin Sørensen shared this sentiment when he accepted the unenviable task of reviewing it for *Contributions to Zoology*, and likewise editor Ronald Vonk, who had to reformat the hundreds of references by hand because I hadn't used any reference management software.

One of the conclusions that emerged from this ocean of ink was that narrative ghosts continued to dwell in the cladistic machine. Cladistic datasets make explicit the data used to infer trees. This greatly enhances the repeatability of phylogenetic analyses, but it also lays bare biases in data selection and flaws in character coding and scoring. I discovered that many cladistic analyses of deep metazoan phylogeny were marred by uncritical data selection, which compromised their power to test competing hypotheses. They often produced results that were just as subjective, biased, and constrained by a priori assumptions as the traditional narratives they were meant to replace. My analysis of morphological cladistic studies of the phylogenetic position of myzostomids around the turn of the twenty-first century powerfully illustrates that such supposedly objective analyses may barely transcend subjective storytelling (Jenner, 2003).

At the time myzostomids were generally considered to be highly modified parasitic polychaetes, until molecular phylogenetic analyses started to indicate possible affinities with different phyla. Four morphological cladistic analyses tested these competing hypotheses (Haszprunar, 1996; Rouse and Fauchald, 1997; Zrzavý et al., 1998; Zrzavý et al., 2001), with varying results. Strikingly, none of the analyses included all relevant characters for evaluating the competing hypotheses, and only Rouse and Fauchald's study found that myzostomids were nested within polychaetes, a finding in line with our current understanding of annelid relationships (Weigert and Bleidorn, 2016). None of the other studies had subdivided polychaetes into its constituent taxa, and they could therefore never test this relationship. Whereas the other studies focused on resolving inter-phylum relationships, Zrzavý et al. (2001) focused specifically on the question of the phylogenetic position of myzostomids. They confidently concluded in the title of their paper that "Myzostomida are not annelids," a conclusion that was "contradicted by no single analysis" (Zrzavý et al., 2001: 186). Yet, with the exception of a character coding for parapodia, their morphological dataset included not a single character that could indicate the polychaetan affinities of myzostomids, crippling its power to test competing hypotheses. Consciously or not, Zrzavý et al. had constructed their dataset with the same unequal eye that narrative phylogeneticists use to build scenarios.

One route via which unexamined assumptions entered cladistic analyses was by scoring characters for higher-level terminal taxa based on assumed ground patterns of primitive character states rather than carefully inferred ancestral states (Prendini, 2001). For instance, Zrzavý et al. (1998) scored the presence of a coelom in priapulids, but this is unlikely to be a primitive character state as coeloms are only known from a single, meiofaunal priapulid species (Jenner, 2001: 737). For other characters, ground patterns reflected assumptions about the evolutionary process that were unsupported by observations about organisms. Examples include scoring the presence of an orthogonal (rope-ladder-like) nervous system in taxa that lack this, such as nematomorphs and priapulids, based on the implicit assumption that these taxa lost an orthogon (Jenner, 2004b: 301); coding a character that defines the gonocoels of molluscs and arthropods as secondarily reduced coeloms, which reflects the previously common assumption that these taxa evolved from annelid-like ancestors with more developed coeloms (see Chapter 9) (Jenner, 2004c: 80); and scoring chordates as having a trimeric body plan, which reflects the previously widespread view that they evolved from ancestors with three sets of coeloms, a body plan configuration still seen in echinoderms and hemichordates (Jenner, 2001: 732). Even when the systematic relationships within a terminal taxon are known, character variation is often understood too poorly to allow rigorous ancestral state reconstruction. Therefore, although many, or perhaps even most, ancestral assumptions about higher-level taxa may well capture character states that are truly primitive, uncertainty inevitably remains.

Ancestral assumptions were further reflected in the choice and resolution of terminal taxa. For example, in Nielsen's 2001 morphological cladistic analysis of the animal phyla, Sipuncula was included as a separate phylum, and resolved as a sister group to a clade containing molluscs, annelids, and panarthropods (Nielsen, 2001). Sipunculans are now understood to be a subclade of annelids (Struck et al., 2011), but this is impossible to discover without splitting annelids up into constituent subtaxa and including data in the analysis relevant for sorting them. Another common assumption that was present in Nielsen's 2001 analysis was that acoelomorphs were platyhelminths, rather than a separate lineage of worms, which they are now understood to be. Assumptions are inevitable in phylogenetic analyses, and these assumptions about sipunculans and acoelomorphs were reasonable at the time. But such assumptions affect both the resolution of phylogenetic relationships and the nature of inferred evolutionary events. Phylogenetic analyses of supraspecific taxa are therefore often ensnared in a web of assumptions that only lower-level phylogenetic analyses can resolve. Indeed, the sheer size of the challenge of constructing ancestral-assumption-free datasets was a major factor in the demise of deep morphological metazoan cladistics in the first decade of the twentieth century (Giribet and Edgecombe, 2020).

Another sign of the persistence of narrative elements in cladistic analyses was the appearance of cousins of HAM, the Hypothetical Ancestral Mollusc that we encountered in Chapter 9, as hypothetical ancestors to root trees. For instance, the first morphological cladistic analysis of relationships within Sipuncula was rooted with HAS, a Hypothetical Ancestral Sipunculan (Cutler and Gibbs, 1985), while HAN, or Hypothetical Ancestral Nauplius, was used to root a tree of barnacle nauplius larvae (Newman and Arnold, 2001). Although these were subjective constructs, at least they were constructed carefully by taxonomic experts. True chimeras emerged as the "all-zero" hypothetical ancestors represented entirely by zeros in the large cladistic data matrices that encompassed all Metazoa. The all-zero hypothetical ancestor that Fred Schram used to root his 2001 tree conjures an impossible mutant composed of mutually incompatible character states, while the information content of the one that roots the analysis of Zrzavý et al. (1998) is equal to that of a vacuum. Although metazoan cladistic analyses published around the turn of the twenty-first century still contained some unwarranted narrative elements, they also signaled the death of narrative phylogenetics. Its unmistakable headstone was the ubiquity of absence character states in data matrices.

The vast majority of morphological characters used in metazoan cladistic analyses were coded as binary absence/presence characters (Jenner, 2002, 2004b). Although presence character states were generally carefully delineated, absence states were typically treated as a default that could be scored for any taxon lacking the presence state, irrespective of its morphology. As a result, the zeros scored for absence states in datasets reveal absolutely nothing about the morphology of the taxa involved. This introduces interpretational difficulties when characters reverse to undefined absence states, and it can introduce spurious clade support when such reversals group taxa based on nonhomologous morphologies. For example, one of the four unambiguous synapomorphies supporting the monophyly of Bilateria in the phylogenetic analysis of Nielsen (2001) is the reversal of a character that scores the presence of an adult brain derived from, or associated with, the larval apical organ. But this reversal groups taxa on the basis of clearly nonhomologous nervous system configurations, such as the lack of an adult brain but presence of a larval apical organ (echinoderms), and the presence of an adult brain but lack of larvae with an apical organ (chaetognaths).

It also brings into sharp focus the fundamental difference in explanatory strategy between narrative phylogenetics and modern phylogenetic analyses. As we have seen in the preceding chapters, traditional narrative phylogenetic explanations involve linking antecedent and descendant states into continuous lineages. The present is thereby explained in terms of the past, which makes narrative explanations historical explanations. This explanatory strategy stalls when antecedent states are not specified. The synapomorphies that support the branching pattern of trees cannot reach back into the past without the careful specification of plesiomorphic (primitive) character states. Synapomorphies that emerge from empirically empty absence states cannot be captured in an evolutionary narrative of cause and effect. Although synapomorphies provide evidence for the phylogenetic relationships of taxa, without identifiable plesiomorphies they are mute about the past. In other words, characters without specified plesiomorphic states can provide evidence for the synchronic explanation of sister group hypotheses between collateral relatives, but they lack the diachronic explanatory power of historical hypotheses.

Truly novel characters, or as Günter Wagner has called them, type I novelties (Wagner, 2014, 2015), therefore pose a problem for narrative phylogenetic explanations. Narrative explanations only have purchase on characters that have identifiable ancestral character states, which Wagner classifies as type II novelties. Wagner calls insect wings "the paradigm" of a type I novelty (Wagner, 2014: 126) and tetrapod limbs, which evolved from fish fins, a type II novelty. Type I novelties fall outside the scope of transformational homology, at least when one restricts oneself to a single level of the somatic hierarchy (for some of the fascinating intricacies of this topic see, for example, Havstad et al., 2015, and Wagner, 2014).[1] Novel characters of both types can therefore provide evidence for inferring phylogenetic relationships between taxa, but only type II novelties fall within the explanatory purview of narrative phylogenetics. This is why evolutionary morphologists put so much emphasis on the principle of no *de novo* origins, as we saw in Chapter 5, for without it there was nothing for them to explain.

Cladistics removed storytelling from phylogenetics by removing considerations of the evolutionary process from the reconstruction of trees. However, this did not extinguish narrative phylogenetics altogether. The engine that drives molecular phylogenetic analyses is narrative, although the narrator isn't human, and no one is listening to the story. Molecular phylogenies are based upon stories of nucleotide and amino acid evolution spun by substitution models. The probability calculus of substitution events that phylogenetic software performs is, epistemologically speaking, the equivalent of traditional speculative storytelling. Both approaches produce narratives about the evolutionary process based on what a substitution model or a biologist considers to have been the most likely events that can explain the observed data. Field et al. (1988) didn't produce the first molecular

---

[1] Recent research suggests that insect wings are not pure type I novelties and that they incorporate transformed ancestral morphological structures (Bruce and Patel, 2020).

metazoan phylogeny by speculating about the evolution of coeloms and cleavage types and nervous systems and larvae. Instead, they delegated storytelling to the Jukes-Cantor substitution model. The model took an alignment of less than 1,500 nucleotides of 18S rRNA data and told a story about the substitutions that could have happened to produce this particular alignment, while assuming that the base frequencies and mutation rates remained equal at all times. The trees based on this simplistic story were then used by the authors to make inferences about the evolution of animal body plans, not by mere speculation of course, but by performing ancestral state reconstruction with "the canon of parsimony" (Field et al., 1988: 751). What could be more objective than that?

10.2 Cutting Corners with Pruned Trees

How powerful could the canon of parsimony be, when most of the animal phyla were lacking from the trees of Field et al? How reliable could their conclusion be that spiral cleavage is primitive for bilaterians, when most phyla without spiral cleavage were absent from the trees? And how trustworthy could the conclusion be that coelomates were monophyletic, when most noncoelomates were excluded? None of the four trees that Field et al. produced correctly reconstructed deep metazoan divergences. This insight is, of course, produced in hindsight. Authors can only work with the data they have at hand, but reconstructing ancestral states based on incomplete trees introduces the same biases that had long made narrative scenarios so subjective. Incomplete sampling of relevant taxa was a common feature of the first generation of molecular metazoan phylogenies, and any evolutionary inferences derived from them should therefore be approached with caution. But not everyone was cautious. In some cases, authors seemed to have simply forgotten the existence of relevant taxa.

Hausdorf (2000), for instance, published a phylogenetic analysis of deep metazoan relationships based on several nuclear protein-coding gene sequences in the leading journal *Systematic Biology*. Based on his tree, he concluded not only that spiral cleavage was primitive for bilaterians, but also that his results were inconsistent with a monophyletic clade Spiralia. How his tree provides any support for either conclusion is beyond me. His tree included five animals: a platyhelminth, a nematode, an arthropod, a mouse, and a human. He rooted his tree with a plant. The flatworm was the sister group of the remaining animals and represents the only taxon with spiral cleavage. In contrast to Hausdorf's conclusion, the optimization of cleavage type at the base of his bilaterian clade is ambiguous. And there is nothing in his tree that is inconsistent with a clade Spiralia, although other than a flatworm, no spiralians were included. Hausdorf used his tree to offer some further speculations about the nature and timing of animal evolution, but with the vast majority of animal phyla excluded, one wonders where he got the confidence to offer these views.

Explicit caveats about the incompleteness of early molecular trees were infrequent, and pruned trees were frequently biased in precisely the direction needed to confirm the conclusions drawn from them. Balavoine (1998), for instance, surveyed the molecular phylogenetic literature and drew a consensus tree with 18 taxa about which he noted that

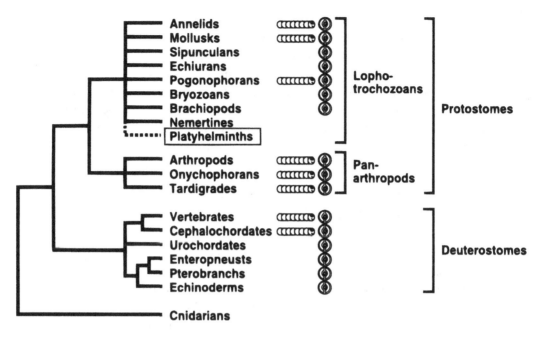

Figure 10.1 The heavily pruned and unresolved phylogeny used by Balavoine to claim that platyhelminths secondarily lack coeloms and segmentation. However, since this tree lacks 10 noncoelomate and nonsegmented bilaterian phyla known at the time, it fails as a phylogenetic framework for evolutionary inference
(after figure 3 in Balavoine, 1998, by permission of Oxford University Press).

"[m]ost of them are coelomates, which of course, can best be explained by the fact that the ancestor was already a coelomate" (Balavoine, 1998: 854) (Figure 10.1). He therefore concluded that platyhelminths were probably secondarily noncoelomate. Indeed, this may well be so, but with the exception of platyhelminths and tardigrades, no noncoelomate bilaterian phyla were included in his consensus tree. Worse, Balavoine leveraged the lack of resolution in his tree to argue that Urbilateria had not only been a coelomate creature, but a segmented one. To take advantage of such biases to boost one's conclusions is a tactic straight from the handbook of narrative phylogenetics. To some this tactic had great allure. Ten years after Balavoine's paper, Gudo and Syed (2008) used an almost identical pruned and unresolved tree to likewise argue for a segmented, coelomate Urbilateria.

A decade earlier Peterson et al. (1997) published the fascinating set-aside cell hypothesis, according to which life cycles with maximal indirect development, in which the adult develops from cells set aside during the larval stage, are primitive for bilaterians. Supporting their theory with a tree, Peterson et al. (1997: 628) pointed out that it "indicates 14 different phyla, 11 of which... contain some maximally indirect-developing species." Although that may sound convincing, I wonder what happened to the bilaterian phyla that do not show any traces of maximal indirect development, of which there are at least nine missing from the tree. Since the authors provided no information about the evidence used to produce this tree, it is hard to escape the conclusion that they were simply engaged in the age-old pastime of speculative evolutionary storytelling.

The use of evidently incomplete trees to boost hypotheses now seems to be a thing of the past. The standards by which phylogenetic and comparative analyses are judged today during peer review are much higher than they were just two decades ago. In the past, the use of pruned and sometimes unresolved trees could afford authors a shortcut to their conclusions, although this may not have been a conscious strategy. But as we will see in the remainder of this chapter, some storytelling shortcuts have been used much more deliberately.

10.3 Wringing Water from Stones

A core component of the conceptual revolution that cladistics brought to phylogenetics was to clarify the role that fossils could play in shedding light on ancestors. Fossils could no longer simply be declared ancestors without making sure that they were not collateral relatives instead. Nevertheless, while being carried by the flow of narrative, scientists sometimes still casually apply this label, for instance, calling the stem-arthropods *Kerygmachela* and *Anomalocaris* "fossil ancestors" (Ortega-Hernández et al., 2019: 157). Although this is conceptually untidy, it causes little harm in literature intended for professionals. Damage can be done, however, when scientists use such imprecise and misleading language deliberately in press releases and interviews to boost the impact of their work. When the fossil primate *Darwinius masillae* was described in 2009 and affectionately named Ida, the scientists and journalists involved engaged in a frenzied bout of Idalatry, where the little creature was crowned as a human ancestor (Jenner, 2009). The paper that presented Ida to the world (Franzen et al., 2009) didn't include any phylogenetic analysis, and the authors were quickly called out for making unsubstantiated and misleading claims about Ida's ability to shed light on our ancestry. Part of the surprising defense of Ida's describers was to claim that the language they were using to describe her in the media – as being on our ancestral line, as being a member of the ancestors, as being a representative of an ancestral group, as being the first link to all humans (Jenner, 2009) – was meant to be more understandable to the general public than using more scientific terms. Indeed, fossils are frequently labeled ancestors in the popular press, but rather than making things more understandable, the cavalier use of the ancestral label actually fuels misunderstanding in the general public.

Although it has long been unacceptable to simply declare fossils ancestors without the support of a phylogenetic analysis, there is a refugium in the scientific literature where this strategy survives. Some authors consider finding the oldest fossil of X as evidence that X is ancestral. This type of argument accords phylogenetic significance to the vagaries of fossil preservation. For example, in 2004 the ill-fated Precambrian fossil *Vernanimalcula* was described as the earliest known bilaterian, possessing beautifully preserved coeloms (Chen et al., 2004). Before this interpretation was debunked (Bengtson et al., 2012), Schmidt-Rhaesa (2007: 167) noted, with little precision, that the great age of these fossils made it "likely that coeloms are quite ancestral in bilaterians." When Li et al. (1998) described the oldest known sponge fossils with preserved cellular details and assigned them to demosponges, they concluded that this implied "that the ancestral form in sponges lies among the demosponges" (Li et al., 1998: 881). Likewise, when Zhang and Dong (2015) described the oldest known animal larva as a cnidarian-grade creature, they concluded that it implies indirect development is primitive for Metazoa, even though cnidarians are not even the earliest diverging animal lineage. And when Chen et al. (2007) described the oldest known ctenophore fossil as lacking tentacles, they inferred from this that the last common ancestor of ctenophores was atentaculate as well. This ancestral assumption was subsequently challenged by Fu et al. (2019), who concluded that ctenophores are primitively tentaculate because another early Cambrian ctenophore fossil does have tentacles. Without accompanying phylogenetic analyses, the age of fossils cannot reveal their primitiveness.

A final example of attempting to wring water from stones is the reasoning offered by Kevin Peterson and his colleagues (2005: 44) that "the known fossil record suggests that the benthos is the primitive site of animal evolution given that the earliest known bilaterian macrofossil, *Kimberella*, is a benthic animal." Without context, and as part of a narrative strategy to convince readers, this sounds eminently reasonable. However, according to data cited in their paper, the earliest occurrence of *Kimberella* in the rock record postdates the estimated origin of Bilateria by 24 million years and the origin of animals by 109 million years. One should certainly be lenient toward error bars on divergence time estimates, but when time's arrow is stretched this far, it is liable to break. Yet although the age of fossils is no precise proxy for how primitive their characters are, a broad correlation of some sort between these features does of course exist because derived character states evolve from primitive character states, and the longer a lineage has evolved, the greater the opportunity for evolutionary change will have been. Indeed, a broad correlation between the geological age and the perceived primitiveness of fossils has long been at the core of the paleontological method of phylogenetics (see Chapter 7). The intuition that older fossils are, on average, better guides to plesiomorphies than younger fossils is therefore not based in a conceptual error. However, the widely used ancestral shortcuts discussed in the next section are phylogenetic faux pas.

10.4 Ancestral Adjectives and Using Trees as Predictors

Ernst Haeckel (1891: 506) called amphioxus "the true Urvertebrate" and "that ancient, typical paradigm of the original vertebrate organisation." Adjectival flourishes like this record a long tradition of evolutionists using evocative language to advocate the ancestral nature of selected organisms. The term "lower" became the most common label to indicate the assumed primitiveness of organisms. As we saw in Chapter 9, early-nineteenth-century morphologists used the terms "lower" and "higher" as an index to place organisms along the scale of being based on their relative degree of morphological complexity. Evolutionists adopted this terminology on the assumption that evolution generally proceeds from simple to complex. Although this terminology is outdated and conceptually deflated, it continues to be with us (Rigato and Minelli, 2013). The vocabulary used to evoke an ancestral aura around favored organisms today additionally includes such terms as "ancient," "early," "archaic," "prototypical," "classic," "basal," "early branching," "generalized," and "living fossil." These words are "For Sale" signs and what they sell is ancestors. They linearize or temporalize the relationships between collateral relatives and invite readers to commit the sin of the scala.

For example, in a paper on one of the most expensive aquarium fishes in the world, the Asian arowana (*Scleropages formosus*), Yue et al. (2006: 1) give it phylogenetic labels befitting its ornamental exclusivity. Based on their phylogenetic analysis they call it a "basal teleost," which belongs to "an ancestral teleost lineage" that is "located near the baseline of the teleost tree." But when you consult their tree, you will see that this beautiful fish is just as basal as the common carp and conger eel and less basal with respect to the ancestral teleost node than rainbow trout and moray eel. The authors' use of these ancestral labels is just hype, but they are certainly not alone in pimping their favorite species.

When we glance upon a zebrafish, a cypriniform teleost model organism that is officially known as *Danio rerio*, its investigators tell us that we are observing an "early vertebrate," "an ancient vertebrate," "a basal vertebrate" (Koduru et al., 2006: 133; Nolan et al., 2006: 144; Froehlich et al., 2014: 51354). To apply these words to a living species in the carp family renders them meaningless, but the ubiquity of the word "basal" suggests it has special significance. When you search the recent literature in the Web of Science, you learn that students of the freshwater cnidarian *Hydra* claim to be working on a basal metazoan. Researchers working on corals, sea anemones, sponges, ctenophores, placozoans, and platyhelminths claim the same thing. Likewise, those dedicating their attention to lampreys, cartilaginous fishes such as the elephant shark, ray-finned fishes such as zebrafish, and amphibians such as *Xenopus*, all assure the reader that these organisms are basal vertebrates. Scientists studying the sea squirt *Ciona* claim to work on a basal chordate, a basal vertebrate, a protochordate, and a proto-vertebrate, while those working on amphioxus, sea urchins, enteropneusts, and *Xenoturbella* highlight that these are basal deuterostomes. The crown of 'basal insect' is worn by crickets, cockroaches, hemipterans, aphids, firebrats, and silverfish.

That one label can be attached to such diverse casts of taxa suggests it is imprecise and imprecisely used. The conceptual problems associated with basal taxa have been outlined before (Krell and Cranston, 2004; Crisp and Cook, 2005; Jenner, 2006; Jenner and Wills, 2007; Bronzati, 2017), but the continued use of this label diagnoses a widespread and persistent misunderstanding of a crucial topic in evolutionary biology that deserves treatment here. The issue is the relationship between the phylogenetic position of taxa and the nature of their characters. It is widely thought that the phylogenetic position of a taxon has at least some predictive value about the nature of its character states. Kuntner and Agnarsson (2009: 193) wrote that "[p]hylogenies are underutilised, powerful predictors of traits in unstudied species," and they illustrate this by successfully predicting web-architectures and behaviors in unstudied spider species. Predictions such as these are made possible because, broadly speaking, more closely related taxa have more characters in common. An unstudied taxon can be predicted to have a character state that is present in its close relatives, but this prediction is uncertain because the character may have changed along its lineage. Trees can thus be used as imperfect predictors of character states for unstudied taxa located within clades with a known distribution of character states. This predictive power relies on evolutionary change not having occurred.

In contrast, it is impossible to predict the nature of character states for taxa located in trees without a known distribution of character states, or for taxa falling outside a clade with a known distribution of character states. Such predictions would require one to know beforehand when and where character change occurred. The phylogenetic position of a taxon is of no help in such guesswork. Yet, it is precisely this kind of guesswork that authors are engaged in when they brandish labels like "early branching," "early diverging," or "basal" for their taxa. What they are searching for is taxa that retain ancestral character states.

Many authors think that the key to inferring ancestral traits lies in basal taxa. When I was doing a postdoc at the University of Cambridge, my friend Cassandra Extavour and her supervisor Michael Akam were trying to reconstruct the evolution of metazoan germ line specification. In a landmark review paper on this topic, they remarked that "[w]ithin each of these clades, the relationships between phyla are poorly resolved, so at present it is not easy to predict which phyla are most likely to retain ancestral characteristics" (Extavour and Akam, 2003: 5875). Such predictions are rooted in the widespread idea that "more basal lineages within a clade have (by definition) fewer derived characters, and more characters retaining ancestral states that may be generalized to other groups" (Bolker, 1995: 454). With a basal species, "one inevitably learns something about probable basal states of the character in question" (Wilkins, 2002: 520). Those "taxa basal in the lineage are those with predominately ancestral characters" (Wake, 2003: 211). And Hölldobler and Wilson (2009: 183) wrote about "species that are phylogenetically more basal ('primitive')."

These quotes show that biologists with very different backgrounds and specializations accept the idea of a close correlation between the phylogenetic positions of taxa and the nature of their character states. Indeed, some consider this correlation to be so tight that they use language that fuses these two aspects. Caron et al. (2013: 506) speak of taxa with a "primitive position" in trees, which Debernardi and Serrelli (2013: 7) call "an ancestral phylogenetic position," while others write about the "derived position" of taxa (Telford and Budd, 2011: R966; Schrödl, 2014: 161). Others call the sister taxa of species-rich clades "plesiomorphic clades" (Yeates, 1995: 345; Poore, 2005: 19). Is there theoretical or empirical support for these views?

Phylogenetic theory provides no basis for them. In a tree of extant taxa, all terminals are the tips of evolving lineages that have evolved for the same amount of time. Phylogenetic position just records the pattern of lineage splitting events in a tree. Nodes or terminal taxa can be labeled as more or less basal with respect to a common ancestral node in the tree, but in the absence of character state information about these taxa, this phylogenetic information has no predictive value about whether they are more or less likely to have retained plesiomorphic character states. Even when the tree is unbalanced and fully pectinate (comb-like), and character states are known for all taxa except the earliest diverging one, one cannot predict what character state it is likely to have. Early diverging or basal lineages, which branch off from the other lineages closer to the root of a clade, continue to evolve just like the other lineages, and their characters may have diverged just as much or more as those of any of the lineages that separated later in time.

Does evolutionary theory provide some rationale for thinking that basal taxa are most likely to retain primitive character states? It does. If evolutionary change is in some way punctuational and associated with lineage splitting or speciation events, then taxa may be likely to have more derived character states when there are more nodes located along their lineage. This idea fits with the theory of punctuated equilibrium, for instance, which proposes that morphological divergence of taxa is concentrated in geologically brief bursts associated with speciation (Gould, 2002b: 779–780). But even if evolution happens like this, it still doesn't necessarily turn phylogenetic position into a reliable predictor of a taxon's character states. If most evolutionary change in lineages accumulates during times not associated with branching speciation, there may not be any correlation between the number of speciation events along a lineage and the amount of evolution that it has accumulated.

For instance, Pagel et al. (2006) analyzed 122 molecular datasets for plants, animals, and fungi and discovered a strong punctuational, speciation-associated, signal of sequence divergence that accounted for 22 percent of DNA evolution. But this means that 78 percent of sequence evolution was not associated with speciation and accumulated during times when lineages didn't split. This turns the problem into an empirical issue that needs to be assessed for individual taxa and characters. Although there is no space here to review this interesting literature, the jury is still out. For some taxa and characters, there is evidence for punctuational evolution, while for others there isn't (Pagel et al., 2006; Bromham et al., 2015; Voje et al., 2020). What is clear is that substantial evolutionary change, both molecular and morphological, can accumulate in lineages without any connection to speciation (Pagel et al., 2006; Voje, 2016). In any case, unless a character of interest evolves in a punctuational mode, there is no reason to expect that the phylogenetic position of unstudied taxa is informative in any way about whether they are likely to have retained ancestral character states.

Phylogenetic position is not a useful predictor of ancestral character states, even when labels like "basal," "basally branching," and "early diverging" are used precisely. But in the vast majority of cases they are not. If the ability of phylogenetic position to predict the probability of character change along a tree depends on the number of speciation events along lineages, it is essential to count these accurately. This requirement is met in a fully sampled and fully resolved tree of relevant species. But this ideal situation is rarely met. Trees often include supraspecific taxa, which collapse multiple nodes into one. For instance, McGregor et al. (2008: 487) claimed that "[c]helicerates are basally branching arthropods and therefore their phylogenetic position is especially good for understanding the evolution of development in this phylum." However, their paper was written to promote the use of two spider species as models for studies in development and evolution. Both are nested deeply in the phylogeny of spiders and are therefore considerably less basal with respect to the common ancestral node of arthropods than insinuated. The lack of fossils in trees of extant taxa creates further unresolvable ambiguities in the use of these labels. When stem group fossils are introduced into the picture, chelicerates suddenly look a good deal less basal than when only living taxa are included (Edgecombe, 2020).

Why, then, is it so widely thought that phylogenetic position is in some way predictive of the probability that a taxon will retain ancestral character states? Why are these labels still so often used? I think there are two factors in play, one relating to constraints on the language we use when we talk about evolution, and one relating to a poorly appreciated aspect of the mechanics of ancestral state reconstruction. First, modern evolutionists know that extant organisms do not line up along a *scala naturae*. Yet, the language many of us use nevertheless seems to suggests the sin of the scala. As I pointed out in Chapter 4, and will discuss in more detail in Chapter 11, an important reason for this is that we use the taxic language of systematics to talk about lineage evolution. In the absence of a vocabulary expressly designed to facilitate precise communication about lineages, we are forced to use the language of taxonomy, which was created for the realm of collateral relatives. Ambiguities and imprecision inevitably result. A telltale sign of this is the common error of temporalizing sister taxa by labelling one as basal. This conceptual slip is exceedingly common, and spending one's entire career thinking about fossils and phylogenies is not guaranteed to confer immunity.

For instance, in their book on the Cambrian explosion, eminent paleobiologists Doug Erwin and James Valentine (2013: 298) write that *Trichoplax* may represent "the most basal living epitheliozoan," when their tree shows it is the sister taxon to the other epitheliozoans. A few pages later, they write that acoels "have appeared in recent phylogenies as representing the deepest bilaterian branch among living forms... preceding the branch at the protostome/deuterostome [last common ancestor]" (Erwin and Valentine, 2013: 304). But their tree shows that the lineages of acoels and the remaining bilaterians are sister taxa, and these lineages reach back into the evolutionary past equally deeply.

Other authors reveal their unconscious urge to temporalize relationships by talking about taxa being the "basal sister group" to their sister taxon (Neiber et al., 2011: 3; Gauthier et al., 2012: 32; Kutschera et al., 2015: 21). Another sign of the mismatch between our taxic language and the evolving lineages we wish to talk about is to label terminal taxa "intermediate" between other taxa. Telford et al. (2015: R881), for instance, refer to xenacoelomorphs as "an early branching clade intermediate between diploblasts... and all other bilaterians." The hypothesis they are referring to places xenacoelomorphs as sister to the other bilaterians and different diploblast lineages as successive sister taxa to this clade. Indeed, to facilitate discussion of animal body plan evolution, I have used the intermediate taxa label myself to indicate that mapping of these taxa's character states allows the inference of true intermediate character states along ancestral lineages (Jenner, 2014).

These labels are used as narrative shortcuts to facilitate thinking and talking about evolving lineages. But in transplanting terminology that fits the linear realm of descent to the branching realm of collateral relatives, it also sows the seeds of misunderstanding. So do statements that an early diverging living lineage, like amphioxus, is "the closest extant relative to the stem chordate" (Garcia-Fernàndez and Benito-Gutiérrez, 2009: 665) or that nereidid polychaetes, like *Platynereis dumerilii* (see Chapter 9), are "most closely related to stem annelids" (Winchell et al., 2010). Sometimes the urge to use linearizing language to assist storytelling is so strong that clades seem to collapse into paraphyly. Valentine and Marshall (2015: 33) wrote about the "ancestry of sponges in Choanoflagellata," even though choanoflagellates and animals are monophyletic sister taxa, because it suited their story of tracing animal ancestry back to choanoflagellate-like ancestors. Although the use of linearizing language is a phylogenetic faux pas, it is an understandable sin when composing evolutionary narratives about lineages using the language of systematics, which is the only available language for doing so.

The second, and probably prime, reason why modern evolutionists continue to think that more basal taxa are more likely to retain primitive character states is epistemological. Ancestral state reconstruction discovers character evolution by finding character state differences between sister taxa in a tree, and probabilistic algorithms can supplement inferences based on this observable evidence with suggested invisible character changes based on evolutionary models. Consequently, because each node in a tree offers a chance to sample character state differences between sister taxa, the probability of discovering character evolution along a lineage is positively correlated with the number of nodes along it. Since more basal taxa are separated from a given ancestral node by fewer nodes than less basal taxa, this may introduce an artifact known as the *node density artifact* (Fitch and Beintema, 1990; Venditti et al., 2006). It can cause the amount of inferred character evolution along more basal lineages to be underestimated. Although originally described and best understood for molecular data, the node density effect can affect phenotypic data as well, but this subject remains poorly studied and, for phenotypic data at least, poorly appreciated (Jenner, 2006; Jenner and Wills, 2007). My own unpublished research suggests that it can significantly affect the inference of phenotypic evolution along morphological trees for a diversity of taxa and characters. It can give the false impression that more basal taxa have more conservative phenotypes, a problem that clearly deserves further research.

Even if the node density effect is not in play, it is easy to get the false impression that more basal taxa are more likely to retain plesiomorphies than less basal taxa. Character change can occur everywhere along a tree. Especially in a pectinate tree, chances are that most character change will be located some distance away from the root of the tree, even if change is randomly assigned to its branches. Given the imbalance of such trees, it is easy to get the impression that the taxon-poor basal lineage is more conservative than its more taxon-rich sister lineage and that would be a correct impression when they are compared as sister lineages. But from this comparison one should not conclude that the basal lineage is more conservative than any of the individual lineages comprising its sister clade.

For instance, in a pectinate (comb-like) tree of ten extant species, character evolution has the same probability of occurring in each of the 10 lineages as each brackets exactly the same amount of evolutionary time. However, the chance that character evolution occurs in each of the eight supraspecific clades in the tree is higher than for each of their sister taxa, which are represented only by single species, because these clades encompass a greater total of evolutionary time. Such imbalances in the diversity of sister taxa are ubiquitous in trees, even when they are not fully pectinate. Such trees give us the *correct* impression that different amounts of character evolution have happened in sister lineages of unequal size, because this involves a taxic comparison of differently sized clades. Yet, they simultaneously give us the *wrong* impression that more basally diverging lineages are more likely to retain ancestral character states than individual lineages that diverge higher up in the tree, because all individual lineages in the tree are equivalent with respect to the amount of evolutionary time that was available for character evolution to occur. Proper lineage thinking can protect us from this flawed expectation. Of course, in pectinate trees, basal taxa will have a strong effect upon ancestral states reconstructed at the base of the tree. But there is no reason whatsoever to think that they are more likely than any other taxa to have retained these ancestral states.

The paradox of basal extant taxa is this. It is precisely the basal, unbranched lineages that have such strong effects on the reconstruction of ancestral states at the basal node of the tree that one should most mistrust. The less branched a lineage is, the less chance there is to infer character change along it, and the more likely it is that derived character states are mistaken for primitive states. It is in this sense that there is a tension between the epistemic power of basal taxa and the veracity of the evidence they provide for ancestral state reconstruction. One should therefore be especially wary of very old basal unbranched lineages, as their character states may sing a siren song of ancestral promise that may turn out to be a lie when their long branch is broken up by the addition of more taxa.

Summing up, the widespread expectation that the phylogenetic position of an extant taxon is in any way informative about whether it is likely to have retained ancestral traits is baseless. It is refuted by phylogenetic theory and proper lineage thinking, and empirical research into the importance of punctuational evolution has so far not provided convincing support for turning trees into predictors. The continued uncritical and imprecise use of labels like "basal" diagnoses flawed thinking. Diogo et al. (2015: 518) have argued that advice to avoid using language like this shows "how being too politically correct could be counterproductive and even go against basic common sense." They argue that calling one of two sister taxa basal "clearly helps communication between phylogeneticists, evolutionary biologists, and developmental biologists" and "allows other researchers to make biological predictions and focus their biological questions" (Diogo et al., 2015: 518). I hope that I have convinced you that this is dangerous advice.

Yet, my argument goes against the grain of the intuitions of many experienced biologists. Some years ago, I addressed this topic when teaching a summer school in evolutionary developmental biology (evo-devo) for PhD students, postdocs, and established researchers. One of the students, who also happens to be one of my friends and colleagues at the Natural History Museum, disagreed passionately with me in class. He kept telling me that I was wrong. I don't think so, Pete. But as the next chapter will show, taxic distortions of lineage thinking are disconcertingly common.

11

Taxic Distortions of Lineage Thinking

Let's start with a little catechism about lineage thinking to outline what lineages are, what they do in evolution, and how they relate to taxa. Taxa are bookkeeping constructs that allow biologists to communicate about segments of life's genealogical nexus. Taxa are not the fundamental units of evolution. Lineages are (De Queiroz, 1999, 2005). Each lineage is a single, independent, and unbranched ancestor–descendant continuum. They are the basic units of evolutionary descent. Zooming in and out of life's genealogical nexus, taxa can be seen to be lineages, composites of lineages, and parts of lineages, ranging from population-level lineages up to supraspecific taxa. Vertebrata, for instance, is a supraspecific lineage that contains many species-level lineages, and it is part of the larger lineage of Chordata, which is itself part of the larger lineage Metazoa. The only taxon that is a complete lineage is monophyletic life as a whole.

New lineages and new taxa evolve when lineages split, but also when lineages permanently fuse, for instance through hybridization. Lineages are diachronic entities that traverse the systematic hierarchy of taxa. If a taxon is evolving, it is a lineage, or part of a lineage. If evolution is happening within a supraspecific taxon, it is lineages that are evolving.

Supraspecific taxa, like monophyletic clades, as well as para- and polyphyletic groups, are groups of two or more lineages. Of these, only monophyletic clades are lineages. Para- and polyphyletic groups are abstractions that lack the individuality of lineages. Segments of a lineage are related by ancestor-descendant relationships. Different lineages are related as collateral relatives in a type of taxic relationship that we call systematic or phylogenetic relationships. As soon as you think about branched taxic relationships, you have left the realm of pure ancestor–descendant relationships, and you are dealing with lineages composed of two or more sublineages.

Because supraspecific taxa always include collateral relatives, they cannot be interpreted as ancestors, although they can contain ancestral lineages to a taxon or trait of interest. Gaining epistemic access to these ancestral lineages requires comparative observations in the realm of collateral relatives. The resulting inferences allow one to trace the linear pathways of descent of every trait and taxon of interest, unless lineages have permanently fused, in which case the history of descent is branching.

These are the basics of lineage thinking. They are the thread of Ariadne that will keep your thinking straight when you ponder the relationship between taxonomy/systematics –

the discipline concerned with the discovery and documentation of the diversity of life and its effective communication – and evolutionary biology and phylogenetics – the disciplines concerned with understanding evolution. Forget or ignore these basics, and you will assuredly get lost in the maze. This chapter discusses those who have, accidentally or deliberately, gotten lost.

11.1 Gould's Blind Spot and His Denial of Linear Descent

As we saw in previous chapters, narrative phylogeneticists have produced a rich literature of stories that imaginatively reconstruct the evolution of lineages. Their scenarios are infused with a priori evolutionary intuitions and hypothetical ancestors deliberately equipped with precursors to traits of interest, and their speculations typically do not take all relevant systematic relationships into account. Today this type of evolutionary storytelling is rightly rejected as solid science. We all agree that the scientific approach to reconstructing phylogeny is rooted in a comprehensive systematic framework, where all relevant taxa, and only relevant taxa, provide the data that feed into our inferences. The direct speculative imagining of linear ancestor–descendant lineages has been replaced by their indirect inference based on the branching patterns of collateral relatives. Surprisingly, we will see that influential scientists and educators have taken this to mean that evolutionary descent isn't linear and that linear evolutionary stories should no longer be told.

The linear march of progress from apish ancestor to modern human is simultaneously one of the most recognizable and most reviled evolutionary icons. Its flagship is artist Rudolph Zallinger's four and half page foldout in a section of Francis Clark Howell's 1965 book *Early Man*, which is titled "The Road to Homo Sapiens." It depicts in 15 drawings a series of reconstructions of fossil primates marching left to right and culminating in modern man. The foldout's title and the fact that all creatures are drawn facing and walking to the right invites readers to interpret it as a depiction of a series of human ancestors. However, the accompanying text makes it clear that it wasn't designed to be read in this way. Nevertheless, this image has inspired countless parodies to depict the evolution of anything from Homer Simpson to ISIS terrorists.

Despite its ubiquity in popular culture, this linear imagery is universally rejected as an accurate depiction of evolution. In his book on the history of evolutionary imagery, David Archibald (2009: 571–572) expresses scientific consensus when he writes that "this iconography has helped perpetuate the misrepresentation of evolution as a straight-line process or ladder." This critical view is echoed in countless articles, books, and websites. For example, in an online article titled "Evolution doesn't proceed in a straight line – so why draw it that way?," two systematists and a science communicator write that "these images bother us because they misrepresent how the process of evolution really works – and run the risk of reinforcing the public's misconceptions" (https://theconversation.com/evolution-doesnt-proceed-in-a-straight-line-so-why-draw-it-that-way-109401; last accessed September 24, 2021). Archibald considers this imagery to be a barely disguised temporalized version of

Figure 11.1 A version of the iconic linear march of human evolution. It is widely condemned by the scientific community as a misleading picture of evolution. However, if the images represent hypothetical ancestors in a lineage rather than collateral relatives, it accurately depicts the shape of evolutionary descent
(after an original by José-Manuel Benitos, modified by M. Garde, and licensed under a Creative Commons Attribution-Share Alike 3.0 Unported license: https://commons.wikimedia.org/wiki/File:Human_evolution_scheme.svg; last accessed February 12, 2022).

the *scala naturae,* the idea that all of God's creations can be arranged in a single, static, ascending chain of being (Figure 11.1).

In his book, Archibald (2014: 19) called the enduring stain of the *scala naturae* on evolutionary thinking "Steve Gould's Bane." Stephen Jay Gould was the most widely read writer on evolution of the second half of the twentieth century. He fought an influential battle against the linear depiction of evolution in many of his works, from his first popular book in 1977 titled *Ever since Darwin*, to his final professional tome in 2002 titled *The Structure of Evolutionary Theory*. Gould dubbed the mistaken impression that evolution is linear "Life's Little Joke" because it is best illustrated with taxa that were decimated by extinction to just a single surviving genus or species. The resulting iconography is a ladder from past to present, with horse and human evolution as classic examples. In reality, Gould argued, evolution branches and produces a bush-like genealogy, and "we can linearize a bush only if it maintains but one surviving twig that we can falsely place at the summit of a ladder" (Gould, 1991a: 181).

Gould's wasn't the only influential voice to lament linear evolutionary imagery. Evolutionary biologist Robert O'Hara coined the now ubiquitous phrase "tree thinking" for what he thought was the antidote for this conceptual sin (O'Hara, 1988, 1992, 1997). He conceived tree thinking as a phylogenetic counterpart to Ernst Mayr's "population thinking," which Mayr claimed was one of Darwin's conceptual triumphs over pre-Darwinian typological thinking. Tree thinking expresses the fundamental truths that evolution is endlessly branching and is not predictable, goal-directed, or arranged along a main line that culminates in us. Proper tree thinking, O'Hara argued, rejects linear evolutionary narratives because the evolutionary chronicle is branching, not linear.

The writings of Gould and O'Hara have led the charge against the linear conception and depiction of evolution for decades, inspiring a literature that warns lay, student, and professional audiences of the dangers of the linear evolutionary straitjacket (O'Hara, 1988, 1992, 1997; Gould, 1995a; Gregory, 2008; Omland et al., 2008; Mead, 2009; Catley et al., 2010; Matuk and Uttal, 2011; Baum and Smith, 2013; Matuk and Uttal, 2020). But all is not well (Jenner, 2018). The logic of life's little joke is fundamentally flawed. Just as Ernst Mayr summoned population thinking to do battle with a foe that turned out to be largely smoke and mirrors (Amundson, 2005; Winsor, 2006), O'Hara has misused tree thinking as a weapon against what turns out to be an imaginary enemy.

Gould was the premier popularizer of evolutionary biology and paleontology in the second half of twentieth century. Few things gave me more intellectual pleasure as a biology student than reading Gould's essays and books. I'm not sure if I would have ended up working in a natural history museum and pursuing a career in evolutionary biology had my enthusiasm not been kindled by Gould's writings. It is therefore with mixed feelings that I have to disagree so strongly with Gould's thinking here. Since the iconography of evolution was one of his favorite topics, his works have special significance. Gould first called what he thought was a conceptual error "life's little joke" in an eponymous 1987 essay in *Natural History* magazine, which is reprinted in *Bully for Brontosaurus* (Gould, 1991a: 168-181). Gould writes that modern horses are represented today by several species in the genus *Equus*, but the fossil record shows that the horse tree was once much more diverse, with many species having gone extinct. Gould accepts the earliest member of the horse lineage, *Hyracotherium*, as the ancestor of all horses, and wrote:

> To be sure, *Hyracotherium* is the base of the trunk (as now known), and *Equus* is the surviving twig. We can, therefore, draw a pathway of connection from a common beginning to a lone result. But the lineage of modern horses is a twisted and tortuous excursion from one branch to another, a path more devious than the road marked by Ariadne's thread from the Minotaur at the center to the edge of our culture's most famous labyrinth. Most important, the path proceeds not by continuous transformation but by lateral stepping. (Gould, 1991a: 175)

Gould paints an incredibly misleading picture of evolution here. Evolutionary lineages are not twisted, tortuous, and labyrinthine. If the pathway from *Hyracotherium* to *Equus* is an ancestor–descendant lineage, it is straight in the sense of simply tracking the arrow of time. Tracing the phylogenetic path between the root of the horse tree and *Equus* in either direction involves navigating a number of turns along internal branches of the tree, but such bends are just pictorial constraints of drawing a tree. All internal branches between each terminal twig and the root of a tree form an unbroken, unbranched lineage of ancestors and continuous evolutionary transformation through time. In the absence of processes that cause the horizontal transfer of genetic information between lineages, such as hybridization, horizontal gene transfer, or symbiogenesis, evolutionary descent is linear (I ask readers to keep this important caveat in mind). Without these processes, which Gould doesn't discuss, evolution by lateral stepping is nonsense, as is his conclusion that lineages trace "just a series of indirect paths to every twig that ever graced the periphery of the bush"

(Gould, 1991a: 175), or the bizarre suggestion that a lineage "hops from branch to branch through a phylogenetic bush" (Gould, 1991a: 180). His conclusion that the last surviving lineage in a mostly extinct clade records the "last gasp of a richer ancestry" (Gould, 1989: 35) confuses the branching realm of collateral relatives with the linear realm of ancestry. Lineages are direct connections between ancestors and descendants. Evolution runs straight through lineages from root to tip, with diverging lineages embarking on their own independent journeys.

By incorrectly claiming that evolution is irreducibly branched, Gould creates a false antithesis between the linear and branching aspects of evolution. He accused fellow professionals who did not recognize this flawed conceptual divide of laboring within a conceptual straitjacket. Gould claimed that so-called "friends of linearity" (Gould, 1996: 67), such as paleontologist George Gaylord Simpson, "held a lifelong commitment to the predominant role of evolution by transformational change within populations rather than by accumulation across numerous events of discrete, branching speciation [because they] could not entirely let go of biases imposed by the metaphor of the ladder" (Gould, 1991a: 178). But these viewpoints aren't in opposition. Tracing evolution along a lineage by inferring the transformation of traits or by counting the number of lineage splitting events that it traverses are two sides of the same coin. The linearity of descent and the splitting of diverging lineages go hand in hand, always. Bushier clades don't just contain more collateral relatives, they also contain more linear lineages of ancestors and descendants. Gould seemed to have failed to understand that you can't have one without the other.

"Evolution rarely proceeds by the transformation of a single population from one stage to the next. Such an evolutionary style, technically called *anagenesis*, would permit a ladder, a chain, or some similar metaphor of linearity to serve as a proper icon of change" (Gould, 1996: 62–63; italics in original). But without the lateral transfer of genetic information between lineages, this is precisely how *all* evolution happens! The ticking of evolutionary time stretches unbroken populations into ancestor–descendant lineages. Instead, according to Gould, "evolution proceeds by an elaborate and complex series of branching events, or episodes of speciation (technically called *cladogenesis*, or branch making)" (Gould, 1996: 63; italics in original). But cladogenesis isn't the opposite of anagenetic evolution happening in time-extended populations. The branching of lineages just produces more lineages. That Gould failed to see this point across the span of his entire career is deeply puzzling.

In his 1988 Presidential Address to the Paleontological Society, Gould (1988: 319) wrote, "Professionals, of course, recognize the bushiness of equid and hominid trees, but still view the lone survivors as end-products of a coherent anagenetic sequence *within* the bush" [italics in original]. But of course. It couldn't be otherwise. Every evolutionary endpoint is the unique culmination of changes that have accumulated *within* its lineage after it diverged from its nearest evolutionary relatives. All evolutionary change is anagenetic in this sense because all evolution happens within lineages. Speciation just produces more lineages, while reticulate evolution fuses them. Without anagenetic change within lineages, branching speciation would only produce identical twins. His ambitious attempt to achieve "a radical reformulation" (Gould, 2002b: 78) of macroevolutionary theory by reconceptualizing evolutionary trends "as products of the differential success of certain kinds of species, rather than as adaptive anagenesis of lineages" (Gould, 2002b: 78) is flawed because trends cannot emerge from the differential elimination of identical taxa.

Yet Gould insisted on seeing evolutionary linearity and branching as conflicting viewpoints and the linear view as flawed. In an essay on human evolution, Gould writes that "research on hominid history has stressed one primary theme above all others: The bush gets bushier and bushier" (Gould, 1998: 206). For Gould this signals a "fundamental change from the linear to the bushy view of our evolutionary history" (Gould, 1998: 205). A failure to relinquish the linear view of human evolution, Gould argued, reflects a continuing "disabling cultural bias" (Gould, 1998: 201), which consists of "our persisting preference for viewing history as a tale of linear progress" (Gould, 1998: 210). He may well be correct in rejecting as simplistic a strictly linear view of the history of human cultures, but this insight cannot support an argument against a linear view of evolution. The branched totality of an evolutionary tree consists of the combined linear flows of evolutionary change along its various shared and independent parallel lineages. The discovery that a given group of organisms is more diverse than previously thought cannot be ammunition against a linear conception of evolution. The bushier the bush, the more numerous the linear lineages that compose it. When Gould wrote that "[s]peciation has replaced linearity as the dominant theme" (Gould, 2002b: 909) for human evolution as a result of new fossil discoveries, he presented this as a fundamental change in our understanding of how evolution happens. It isn't. Two floating logs arriving at a branched river delta will not suddenly cease their linear journey, no matter how many branches there are. They will simply continue to float down one of the branches, although their paths may now diverge.

One of the insights that Gould frequently showcased in his writings is that what he called "conceptual locks" (Gould, 1988: 319) are often as important as barriers to scientific understanding as a dearth of empirical evidence. His own failure to see the conceptual harmony of the linear and branching views of evolution provides an unwittingly powerful illustration of this insight. And it didn't just lure him to false conclusions about the evolution of primates and horses. In an essay on the famous ostracoderm fish *Cephalaspis*, Gould compared how its relationships to other taxa were interpreted in the late nineteenth and early twentieth centuries by a nonevolutionist, Louis Agassiz, and two evolutionists, William Patten and Erik Stensiö. Agassiz presented a diagram of fish relationships with taxa depicted as spindles against the background of a geological chart, with the spindles more or less converging on each other further back in time, but never quite connecting, as Agassiz denied their common ancestry. Patten, in contrast, drew a strangely shaped phylogeny in which ostracoderms were placed as an intermediate group between the other vertebrates and their arachnid ancestors. Finally, Stensiö presented a cladogram-like diagram that placed *Cephalaspis* nearer to lamprey than to hagfish.

There is much that can be said about the conceptual differences underpinning these diagrams, but concluding, as Gould does, that Agassiz's scheme differs from Patten's because "Agassiz based his vision on differentiation (radiation of numerous lineages from common points of origin), while Patten embraced linear progress" (Gould, 1993: 435) misses the mark completely. As Gould himself points out, Agassiz's independently created spindles don't touch, so there are no truly common points of origin. Even though Patten's peculiar view of evolution centered upon a conviction that there was a main line of evolution marked out by extraordinary morphological and behavioral progress culminating in us, his diagram is a branching tree with diverging lineages. To then conclude that Stensiö's scheme is actually "closer to Agassiz's pre-evolutionary version than to Patten's supposed improvement" (Gould, 1993: 436) is to force a conceptual affinity where none exists. Patten's diagram clearly focuses on the phylogenetic flow of body plans along lineages of ancestors and descendants, while Stensiö's tree is an austere depiction of the branching relationships of collateral relatives, but both are fundamentally united as depictions of related evolutionary lineages.

Gould fails here and elsewhere to explain that the issue is epistemological. The linear pathways of evolution are populated by empirically inaccessible ancestors. All observable evidence that can be used to trace evolution is located at the terminal twigs of evolutionary trees, either because the taxa involved truly are terminal and cannot be arranged in ancestor–descendant series or because ancestral fossils have not been recognized as such. The great challenge facing phylogeneticists is to follow the linear course of evolution's empirically inaccessible highways by exploring the observable ends of its diverging byways. Our inability to line up observable evidence does not negate the linearity of descent. Yet in his writings Gould never explains that the seeming replacement of a linear by a branching view of evolution is due to the complex relationship between reality and the epistemological tools with which we study it.

Gould argued that our tendency to see evolution as linear is the unfortunate result of our cultural preference for linear stories. This is wrong. Our intuition that evolution is linear is correct. The irony of life's little joke is that the decimation of a bush until only a single species lineage remains lays bare the bedrock truth of evolution. The linearity of descent just chimes perfectly with our penchant for linear stories. Gould illustrated life's little joke with the evolution of horses and humans in a number of his popular and professional books, including *Bully for Brontosaurus, Wonderful Life, Full House, Leonardo's Mountain of Clams and the Diet of Worms,* and *The Structure of Evolutionary Theory* (Gould, 1977a: 56–62; 1989: 27–43; 1991a: 168–181; 1996: 57–73; 1998: 197–212; 2002b: 905–912). Sadly, all reveal that Gould failed to grasp this irony.

It is puzzling that Gould didn't seem to understand the linearity of descent or that the linear and branching aspects of evolution are not in conflict. To explain why a professional evolutionist would fail to grasp these basic truths one would have to posit the presence of a pretty powerful mental blinker, and I think there is one. I think Gould was blinded by his desire to erect an autonomous macroevolutionary theory that didn't just flow smoothly from the fundamental precepts of microevolutionary theory. Gould wanted to shift the prime engine of evolutionary change from processes operating in populations to the level of the birth and death of species. In a section of *The Structure of Evolutionary Theory* titled "The Grand Analogy: A Speciational Basis for Macroevolution," Gould wrote:

> if species act as stable units of geological scales, then evolutionary trends—the fundamental phenomenon of macroevolution—could be conceptualized as results of a 'higher order'

selection upon a pool of speciational events that might occur at random with respect to the direction of a trend. In such a case, the role of species in a trend would become directly comparable with the classical status of organisms as units of change within a population under natural selection. (Gould, 2002b: 715)

From the time that he and Niles Eldredge proposed the theory of punctuated equilibrium in the early 1970s, Gould had been convinced that most evolutionary change occurred in association with speciation events or cladogenesis. Gould thought that the amount of evolutionary change that accumulates within lineages during times of unbranching evolution (anagenesis or phyletic evolution)[1] was small in comparison to the amount of change that happens during the splitting of lineages. This is an interesting topic, but Gould somehow seemed to think that when a lineage splits, evolution's linearity is replaced by branching. It isn't. Speciation splits lineages; it doesn't fracture them. Evolution continues to happen within unbroken lineages. Lineages snake their way along stretches of anagenesis and through bottlenecks of cladogenesis without ever breaking or changing into another lineage. But no matter the importance of species making and breaking as a factor in explaining macroevolutionary trends, the linearity of descent indissolubly links the realms of micro- and macroevolution. Gould failed to see that, without linear evolution, his macroevolutionary engine would grind to a halt as surely as microevolution would.

11.2 Tree Thinking and the Rejection of Linear Narratives

Gould wasn't the only influential voice that railed loudly against the linearity of evolution. In his papers on the narrative nature of evolutionary biology, Robert O'Hara (O'Hara, 1988, 1992, 1997) coined the now ubiquitous phrase "tree thinking" to help banish the linear evolutionary narratives that are still pervasive in papers, books, and museum exhibitions. O'Hara considers linear evolutionary stories to be the rank remains of the *scala naturae*. He thinks that tree thinking can help erase this conceptual stain. For O'Hara tree thinking is to look at evolution through the lens of cladistics, with a focus on the branched relationships between collateral relatives. Although O'Hara doesn't deny the linearity of evolutionary descent, he nevertheless thinks, like Gould, that the impression that evolution is linear is an artifact caused by forcing a branched reality into a linear mold.

In the same year that Gould denied the linearity of lineages in his Presidential Address to the Paleontological Society, O'Hara argued that the linearity of our evolutionary stories is a flaw deliberately perpetrated by biologists so they can produce "Artificial closure" (O'Hara, 1988: 152) for the benefit of storytelling. O'Hara (1992: 143) wrote that the "research chronicle of the systematist differs in an important respect from that of the historian,

---

[1] Anagenesis has different definitions in the literature (Vaux et al., 2016; Allman, 2017). Gould contrasted anagenesis, as evolution along unbranched lineages, with cladogenesis, or evolution associated with speciation. However, as discussed earlier, he was inconsistent in his definition of anagenesis, as he also defined it as evolution happening in a single time-extended population. He seemed not to realize that in the absence of horizontal transfer of genetic material, this is how all evolution happens.

however, in that the chronicle of evolutionary history is strongly branched." Since evolutionary trees branch, O'Hara called for biologists to "free themselves from the ontogenetic view of evolution, and from linear evolutionary narratives. The evolutionary narratives of the future must branch and take their readers down any chosen thread of the evolutionary tree" (O'Hara, 1988: 153).

O'Hara misses a fundamental point in advocating this viewpoint. He accepts the tenets of the individuality thesis, as developed by Michael Ghiselin, David Hull, and others (Hull, 1975; Ghiselin, 1997), according to which species and clades are ontologically individuals that have a concrete reality in space and time. Such individuals can therefore function as central subjects in evolutionary narratives and provide these with unity and continuity through time. The problem is that O'Hara thinks that clades are the lowest-level central subjects of evolutionary narratives (O'Hara, 1988: 152) and, since clades are branching, so must our narratives be. To illustrate this, O'Hara (1994: 15) wrote that if you were to grab a phylogenetic tree "at any point, and cut immediately below your grip – below in the sense of toward the root – the chunk of the tree in your hand would by definition be a clade." Since clades are branched, the chunk in your hand would be a branched entity. That is certainly true, unless the tree has species as terminal taxa, in which case you could be holding a single species in your hand.

Clades, however, are not the most fundamental ontological individuals that can be central subjects of evolutionary narratives. O'Hara forgets that clades are bundles of independently evolving lineages, each of which can be the subject of its own story (Hull, 1975, 1989; Ghiselin, 1997; De Queiroz, 1999; Dupré, 2017; Dupré and Nicholson, 2018). It is therefore perfectly legitimate to construct evolutionary scenarios for linear genealogical segments and to interpret the transformation of traits along these segments to be analogous to the ontogenetic changes of individual organisms, a notion that O'Hara (1988: 153) labels an "absurdity."

Far from it being an absurdity, Haeckel was already on philosophically solid ground in the nineteenth century when he conceived the linear paths of ontogenesis and phylogenesis to reflect the parallels that exist between the developmental history of individual organisms and the evolutionary history of the lineages they belong to (see Chapter 4). O'Hara, however, is not alone in failing to properly come to terms with the relationship between the linearity of ontogeny and the branching of evolutionary lineages. In his widely cited paper on how Haeckel built his trees, Dayrat (2003) incorrectly argues that Haeckel's trees were not branching Darwinian diagrams because he reconstructed them with the linear view of evolution that was anchored in his biogenetic law. Following Dayrat, Ragan (2009: 21) asked "how could ontogeny (a linear process in a single organism) recapitulate phylogeny, unless phylogeny too was linear?", a criticism echoed by Scott Gilbert (1991: 843). The answer is that the ontogeny of an organism can only provide clues about its own, linear ancestry. As we saw in Chapter 4, for Haeckel ontogeny could recapitulate phylogeny because both represent linear processes in time: the development of an individual organism and the history of its evolutionary lineage. Using the biogenetic law as his guide, Haeckel reconstructed the phylogenetic history of a focal lineage, with other lineages branching off at the points where their ontogenies diverged. So although Dayrat is correct in concluding that Haeckel viewed his branching trees through the linear lens of the biogenetic law, this linearity does not reflect the *scala naturae*.

Proper tree thinking will help you avoid the sin of the scala. The sin of the scala is to line up collateral relatives as if they were a series of ancestors and descendants. This inappropriate linearity results when evolutionary trees are read across their tips. In Chapter 7 I presented several striking examples of this strategy. However, the fundamentally branched nature of the relationships between collateral relatives is underpinned by linear lineages of ancestors, each of which terminates in an extant or extinct taxon. These branched and linear realms are distinct but inextricably linked. The branched realm of the systematic relationships between sister groups is synchronic and has no time dimension. In contract, the realm of the linear relationships between ancestors and descendant is diachronic and traverses time. By adopting a narrow cladistic focus on the synchronic realm of systematic relationships, O'Hara seems to have lost sight of the diachronic realm of evolution.

O'Hara (1992) identified four strategies, which he called narrative devices, that scientists use to abstract linear stories from evolution's branching chronicle. They are (1) the linear sequencing of contemporary taxa in text and pictures, (2) the pruning of side branches from a focal lineage, (3) the naming of paraphyletic taxa so that they can be used as stages in a linear narrative, and (4) collapsing parts of a tree to highlight a focal lineage. Sandvik (2009) added a fifth: (5) arranging terminal taxa from left to right on a tree to suggest a reading direction. Although these narrative devices can be used to focus attention on any lineage in a phylogeny, O'Hara and Sandvik rightly point out that they have most often been used to zoom in on our own. They also claim that these narrative devices "distort evolutionary history by imposing a trend on it" (Sandvik, 2009: 427), and falsely suggest that "the evolutionary process underlying the historical account had been aiming for the focal taxon" (Sandvik, 2009: 427). These narrative tools "conflict with the underlying chronicle of evolution, which is not linear, but branched" (O'Hara, 1992: 135).

I agree that narrative devices one and five invite people to commit the sin of the scala. But the remaining narrative devices do not distort evolutionary history by making it appear inappropriately linear or redolent of teleology. The pruning and collapsing of clades that are side branches of a focal lineage are perfectly legitimate ways to focus attention. Epistemologically speaking, the evolutionary lineage of any trait, clade, or taxon of interest can only come into focus with the help of observations on collateral relatives, but as these lineages evolve independently, their stories can be told independently. I will discuss the issue of paraphyletic taxa in the final section of this chapter.

To prevent confusing or misleading readers when depicting ancestral lineages, it is crucial to specify what kind of organisms images depict. Gould was embarrassed with the cover design of the Dutch translation of *Ever since Darwin* because it showed a linear march of five hominids that became increasingly modern human-like, the very image, Gould wrote, that his "books are dedicated to debunking" (Gould, 1989: 30). But the image (figure 1.5 in Gould 1989) does not mention, nor does Gould make clear, whether the four hominids depicted to the left of *Homo sapiens* are fossil species or reconstructed or imagined human ancestors. The same ambiguity is present in a version of Rudolph Zallinger's original march of progress that is reproduced in Archibald's book on evolutionary imagery (Archibald, 2014: 20). Archibald (2014: 19) writes that this "kind of representation fuels profound misunderstandings by the general public of how evolution operates," without explaining anywhere whether the depicted creatures are reconstructed ancestors or observed taxa. Only in the latter case may the sin of the scala have been committed.

11.3 The Influence of Gould's Blind Spot and the Cladistic Blindfold

It is perhaps ironic that as soon as the branching iconography of evolution took root, it legitimized the tracing of its linear pathways. Frustratingly, the empirically inaccessible realm of evolutionary process can only be accessed indirectly via the observable evidence preserved in the realm of the systematic relationships of collateral relatives. Yet the invisibility of the realm of lineal descent doesn't mean that it doesn't exist or that it is in conflict with the branching realm of systematics. The challenge is how to properly integrate taxic and lineage thinking. In his copious writings on the topic, Gould never seems to realize that the tension between the linear and branching aspects of evolution is only apparent, and it relates to the epistemic challenge of recognizing ancestors. Naive tree thinking has likewise become a trap for the unwary. The cladistic approach to phylogenetics accords epistemic priority to the discovery of systematic pattern over the inference of phylogenetic process. The advent of cladistics allowed biologists to infer the systematic relationships between collateral relatives while staying empirically grounded in observable evidence and without having to engage with the epistemic hinterland of invisible ancestors and unobservable evolutionary processes. However, it seems that many have adopted such a shallow cladistic focus on sister groups relationships that it has caused a kind of evolutionary process blindness.

Only by invoking the cladistic blindfold can I understand why an eminent biologist like the late Rudy Raff would think that our "expectations of a linear pattern of evolution distorted the picture [of horse evolution] to make it seem that there had been a majestic 50-million-year evolutionary procession to the modern horse" (Raff, 1996: 25) or why in a paper for science teachers educational specialist Esther Van Dijk and philosopher of science Thomas Reydon conclude that a bushy horse tree implies that "there is no linear evolution towards modern horses with progress consisting in increasing body size" (Van Dijk and Reydon, 2010: 660). There assuredly was a majestic procession, even though the lineage leading to *Equus* is just one of many lineages in the tree and even though the tangible evidence for horse evolution cannot be simply lined up along a single lineage. This doesn't imply, of course, that character evolution was necessarily unidirectional across a lineage's lifespan. As the examples that follow will show, the pernicious influence of Gould's blind spot and the cladistic blindfold go a long way toward explaining why the linear view of evolution has been so widely, ruthlessly, and relentlessly sacrificed on the altar of the branching bush.

Paleoanthropology provides a revealing example of the impact of Gould's thinking. Ian Tattersall, a paleoanthropologist at the American Museum of Natural History in New York, points out that Gould's views have been influential in paleoanthropology. He wrote that Gould's "tireless advocacy of the idea that human phylogeny presents us with a 'bush' rather than with a 'ladder' introduced into paleoanthropological thought a powerful and compelling metaphor that continues to gather momentum" (Tattersall, 2013: 115). Indeed, as Tattersall points out, Gould had already developed this argument in the 1970s. He approvingly cites from a 1976 essay where Gould rejects the view that "'ladders' (evolution as a continuous sequence of ancestors and descendants) do not represent the path of evolution" (Tattersall, 2013: 119). But I don't know what evolutionary descent is, if not a continuous sequence of ancestors and descendants.

Nature editor Henry Gee has also drunk deeply from the Gouldian spring. His book *Deep Time* is a diatribe in which he urges his readers to reject the lunacy of seeing human evolution as linear:

> The conventional portrait of human evolution – and, indeed, of the history of life – tends to be one of ancestors and descendants... The conventional linear view easily becomes a story in which the features of humanity are acquired in a sequence that can be discerned retrospectively... New fossil discoveries are fitted into this pre-existing story. We call these new discoveries 'missing links', as if the chain of ancestry and descent were a real object for our contemplation, and not what it really is, a complete human invention created after the fact, shaped to accord with human prejudices. (Gee, 2000: 31-32)

The one nugget of insight in this quote – that fossils should not be uncritically slotted into lineages – is buried in a perversely warped view of evolutionary inference and reality. Are general readers, for whom Gee wrote his book, served well by being told that evolution does not flow along lineages of ancestors and descendants, that chains of ancestry are fantasies rather than inferences, and that we have not evolved our characteristic traits sequentially?

In his enjoyable book on the history of paleoanthropology, *The Strange Case of the Rickety Cossack*, Tattersall repeated the flawed argument that Gould called life's little joke when he wrote that "To members of a single species that dominates the world today, it may seem natural to reconstruct our biological history by simple extrapolation back in time; but as we will see, this perspective of ours is only made possible by some highly unusual circumstances" (Tattersall, 2015: 94). Tattersall is not referring here to the complexities introduced by the hybridization of hominin lineages, such as modern humans and Neanderthals. For Tattersall, the realization that the human evolutionary tree has numerous branches apparently negates the fact that our ancestral lineage extends straight back in time. He wrote that the seeds for this insight were planted in the late 1960s when he studied lemurs in Madagascar: "the whole lemur fauna loudly blares diversity at you" (Tattersall, 2015: 9). Realizing that this diversity of species was typical for the evolutionary history of hominids as well, Tattersall felt this refuted the view of evolution as a "linear, perfecting process" (Tattersall, 2015: 9), and that "this pattern of events entirely changes our perspective on how we became the highly unusual creatures we are" (Tattersall, 2015: 9). Instead of being "the burnished product of incremental improvement over the eons. . . we are one particular outcome of an active process of experimentation with the evidently many ways there are to be a hominid" (Tattersall, 2015: 9). This, it will be clear by now, is a false dichotomy, for both are, in all probability, true. Henry Gee likewise echoed life's little joke when he wrote that "I wonder how much of our view of evolution as linear and progressive is conditioned by our solitude; whether we might see things differently were *Homo sapiens* just one of a family of coexisting species like wapiti, wallabies or warblers" (Gee, 2000: 221). Why should we? The linearity of a taxon's descent is unaffected by how many evolutionary cousins it has.

By presenting the linear and branching interpretations of evolution as opposing viewpoints, these authors faithfully echo Gould's misleading message. Tattersall is of course completely right that arranging all fossil hominid remains in a *single* line leading to *Homo sapiens* is grossly misleading, despite its widespread acceptance throughout much of the twentieth century (Tattersall, 2000, 2013, 2015). This is a proper criticism of linear evolutionary thinking. But his arguments also set up a false dichotomy. Yes, our independently evolving lineage is just one experiment of hominid evolution, but this provides no rationale for rejecting the notion that evolution has proceeded via a linear sequence of incremental changes along our own, or any other, hominid lineage. Major evolutionary innovations, including the origin of bipedalism, increases in brain size, the use and fashioning of complex tools, and the acquisition of language still record steps that can be plotted at different points along our lineage in the hominid tree. Rejecting the linearity of evolutionary descent because of the bushiness of trees is a non sequitur. They are inextricably linked as two sides of the same coin. Indeed, the bushier a clade is, the more confidently we can sketch the details of the linear descent of its component lineages.

It is worrying that the confusion between the linear realm of descent and the branching realm of systematic relationships is particularly rife in literature written for popular and educational audiences. It is here that the influence of Gould's blind spot and the cladistic blindfold is particularly pernicious. In a paper titled "The great chain of being is still here" published in the journal *Evolution: Education and Outreach*, Rigato and Minelli (2013: 2) deplored that "evolution is indeed often perceived as a linear, progressive process rather than as a story of unceasing branching. . . This misleading progressionism is scientifically undefensible [sic]." Progressionism is indeed a problematic notion and evolutionary trees do indeed branch, but portraying the linear and branching aspects of evolution as somehow being in opposition, and citing Gould's and O'Hara's works in support of this argument, is equally indefensible.

A paper by Ryan Gregory in the same journal, titled "Understanding evolutionary trees," likewise incorporates life's little joke and O'Hara's tree thinking in the lesson it is trying to teach (Gregory, 2008). It contains a cladogram that shows the relationships between groups of living primates (Figure 11.2), on the basis of which Gregory concludes that we did not evolve from monkey ancestors "and no sane biologist suggests otherwise" (2008: 132). Yet, his tree suggests we did evolve from monkeys because monkeys are paraphyletic with respect to apes and humans. Of course our monkey ancestors are extinct species that don't belong to the living lineages of New World or Old World monkeys. By omitting this

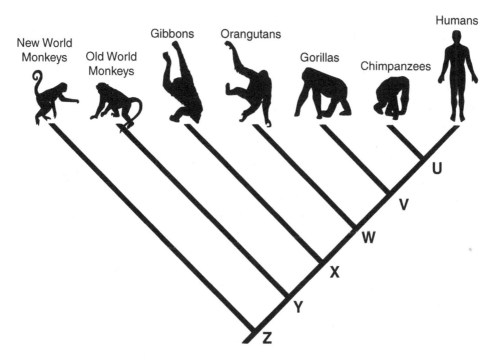

Figure 11.2 Evolutionary tree depicting the phylogenetic relationships between humans and their nearest primate relatives. Because monkeys are paraphyletic, the internal branch between Z and Y represents a lineage of extinct monkeys, which is part of our ancestral lineage (after figure 14 from Gregory, 2008, reprinted by permission Springer, Copyright, 2008).

linguistic detail, which carries a lot of conceptual baggage, Gregory creates a misleading paradox between his text and figure. The denial that we evolved from monkeys is a widespread canard in the popular and educational literature, and it is typically presented with the same lack of linguistic and conceptual precision. A representative example is an evolution quiz and accompanying article published on the website of the BBC on 25 September 2018 (www.bbc.co.uk/news/science-environment-45564594; last accessed September 27, 2021). It tells the reader that we didn't evolve from monkeys or apes, but instead share common ancestors with them, without noting that these now extinct ancestors would be classified as monkeys and apes.

In the acclaimed book, *Evolution. What the Fossils Say and Why it Matters*, which he dedicated to Stephen Jay Gould and Niles Eldredge, paleontologist Donald Prothero wrote that "the tendency to put things into simple linear order is a common metaphor for evolution – and also one of its greatest misperceptions... *Evolution is a bush, not a ladder!*" (italics in original) (Prothero, 2007: 125). This allows Prothero (2007: 347) to conclude that thinking "we come from monkeys is simply wrong", and that to believe otherwise would be "to live in a web of lies," despite presenting a primate tree identical to the one in Gregory (2008), and concluding earlier in the book that we still "have the genes for the long tails of our monkey ancestors" (Prothero, 2007: 98). Again, in a paper in the journal *Science*

Education, Halverson et al. (2011: 799) present as evidence of the poor understanding of evolution by students a study that showed that "many students believed humans directly evolved from monkeys," without explaining that this is indeed true, but that these ancestors were not extant but extinct monkeys.

The rejection of the concept of missing links is a telltale sign of a failure to understand or make clear the distinction between the linear and branching realms of evolution. Gould (1998: 202) labeled missing links a "hoary and clichéd concept," while two other paleontologists referred to them as an "archaic expression but also an outmoded approach to studying macroevolution" (Padian and Angielczyk, 2007: 197). Henry Gee, who was trained as a paleontologist, wrote an entire book "to explain why terms such as 'missing link' encapsulate more than a century of error in thinking about evolution, particularly of human beings. They reinforce a monstrous view of evolution whose function is to cement our own self-regard as the imagined pinnacle of creation" (Gee, 2013: x). Science writer and fossil boffin Riley Black – who was the resident paleontologist for the 2015 blockbuster movie *Jurassic World* – wrote an online article on the concept of missing links in 2018 for the website of the Smithsonian National Museum of Natural History (www.smithsonianmag.com/science-nature/whats-missing-link-180968327/; last accessed September 27, 2021). When she interviewed Smithsonian's fossil mammal curator Nicholas Pyenson and anthropologist Briana Pobiner, they agreed that the linear view of evolution is flawed and that the "chain metaphor that 'missing link' implies would have us looking for straight lines, when the reality of evolution is much more discursive." Donald Prothero wrote that in any case "there is no such thing as a missing link" (Prothero, 2007: 82) – paleontologist Gishlick (2003: 41) dismisses the concept in exactly the same words – because evolution is properly conceived of as a branching rather than a linear process. Two papers on how to teach evolution present precisely the same argument for why "paleontologists are not searching for missing links" (Mead, 2009: 313; Pobiner, 2016: 238). Likewise, a team of paleoanthropologists and paleontologists call the concept of missing links an "erroneous metaphor" promulgated by "Darwin and his less cautious contemporaries and intellectual descendants" (White et al., 2015: 4877). The existence of transitional taxa, a concept identical to missing links, has been similarly dismissed (Reisz and Müller, 2004; Bronzati, 2017).

These authors are right in one important sense. Paleontologists generally don't search for missing links and transitional forms. But this is not because evolution doesn't flow along linear lineages or because the concept is flawed. Missing links, in the sense of direct lineal ancestors, can in principle be found (Foote, 1996) and have indeed been found (Tsai and Fordyce, 2015; Carr et al., 2017; Parins-Fukuchi et al., 2019). Even Ian Tattersall, a prominent advocate for the branching view of human evolution, admits some fossils in directly ancestral positions in his trees (Figure 11.3). Apart from the great imperfection of the fossil record, this search is often futile because many fossils are discovered not to be lineal ancestors to living taxa, and even when they are, it is difficult to recognize them.

We shouldn't forget the truly amazing fact that every living species today represents the leading edge of an evolving lineage that extends back in time continuously for billions of years to the origin of life itself. The scope for discovering the organisms that are the missing links in these immense chains of ancestry should therefore be, in theory, substantial (Foote,

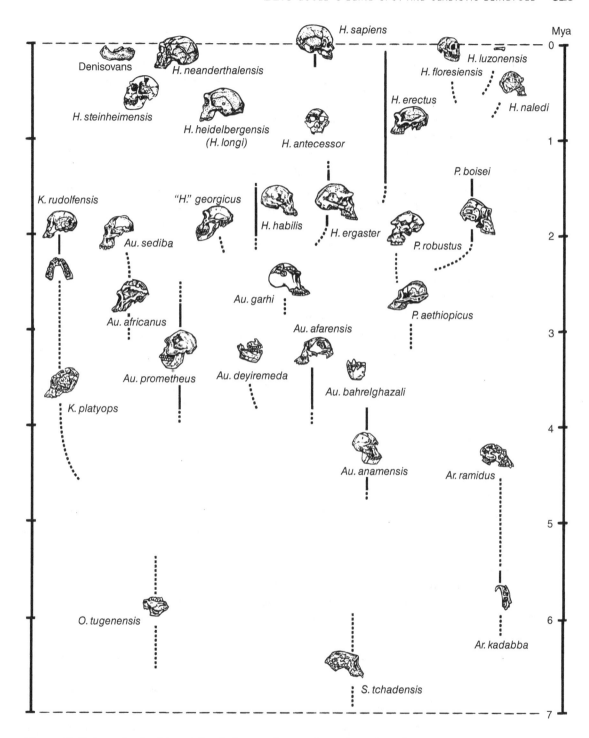

Figure 11.3 A synthesis of hominin phylogeny that combines branching lineages and fossils in ancestral positions
(by Ian Tattersall, 2021, with permission).

1996). I am truly flabbergasted that these fossil experts instead present missing links as a flawed concept rather than a tough epistemological challenge and one, when it is occasionally met, that is achieved with the unique evidence from their own discipline. Their views potently illustrate the dangers of flawed lineage thinking and the cladistic blindfold that stipulates that all taxa are terminal taxa.

The nonsensical concepts of "collateral ancestors" and "collateral ancestry" provide another illustration of how the cladistic blindfold can obscure the deep linearity of descent. These flawed views have, again, been advocated chiefly by paleontologists (Gishlick, 2003; Padian and Angielczyk, 2007; Prothero, 2007; Mead, 2009), but also, and especially worryingly, in books written for broader audiences on topics such as human evolution and creationism (Scott, 2004; Ayala and Cela-Conde, 2018). Prothero (2007: 82) explains that a lineal ancestor lies in the direct line of descent of a taxon, whereas a collateral ancestor shares an ancestor with it, but is not directly ancestral to it. He uses *Archaeopteryx* as an example of a collateral ancestor, which while not lying on the direct line of descent to modern birds, does show transitional traits. But this makes a collateral ancestor simply a collateral relative, and relatives aren't ancestors. Relatives are side branches of a focal lineage, and although their traits may provide clues for inferring the likely traits of ancestors along the focal lineage, in terms of ancestry they are literally beside the point. Labeling relatives as ancestors muddles the distinction between the realms of systematic pattern and evolutionary process, and it obscures the central epistemic challenge of phylogenetics: how to see the ghostly contours of ancestors in the distorted reflections of their descendants.

A failure to make this conceptual distinction crystal clear is a God's gift to creationists, as Ernst Haeckel and Konrad Lorenz noted in the late nineteenth and early twentieth centuries. By rightly denying that fossils such as *Archaeopteryx* are lineal ancestors, creationists try to undermine the reality of macroevolution. The standard response of scientists is to claim that creationists are confused about the distinction between lineal and collateral ancestors (Gishlick, 2003; Padian and Angielczyk, 2007). I don't think they are. Creationists understand the difference between ancestors and relatives as well as they do the distinction between parents and cousins. The problem is more fundamental than that. Creationists refuse to admit that inference can play a decisive role in arguments supporting the reality of evolution. Even if they understand the epistemic constraint under which evolutionists labor, they will only be swayed by observable evidence, which is profoundly ironic considering their overall worldview. If that observable evidence sits on a side branch, creationists simply dismiss it. Padian and Angielczyk (2007: 223) accuse a prominent creationist of seeing phylogeny as a *scala naturae* and of being confused about the distinction of "lineal and collateral ancestry." It doesn't help our fight against creationist nonsense if our own view is equally muddled.

Processes such as introgressive hybridization and horizontal gene transfer lead to the fusion of evolutionary lineages, and any affected taxa and traits therefore do not have linear evolutionary histories. For instance, when you trace the evolution of non-African human genomes back in time you find that the lineage splits into two around 50,000 to 60,000 years ago, which reflects the interbreeding of Neandertals and early modern humans. These ancestral strands rejoin again almost 600,000 years ago, reflecting the common ancestry of modern humans and Neandertals (Mendez et al., 2016; Dannemann et al., 2017; Prüfer and al., 2017; Bergström et al., 2021). But split or looping ancestries can only be produced by the fusion of lineages. Without this, evolutionary ancestries are straight. Yet some scientists disagree.

Biologists Todd Oakley and Sabrina Pankey published a paper on eye evolution in the journal *Evolution: Education and Outreach* aimed at educational audiences, where they approvingly cite both Gould's and O'Hara's tree thinking as the necessary remedy for linear evolutionary thinking (Oakley and Pankey, 2008). They correctly warn readers against the sin of the scala – lining up the visual systems of extant invertebrates in a direct ancestor-descendant series. But they then criticize the linear view of eye evolution, which they claim prevailed from Darwin to the present time, because it apparently prevents one from appreciating that the constituent parts of complex structures such as eyes can evolve in an integrated manner. They wrote, "Eyes and their integrated sets of components (mainly proteins encoded by genes) do not have linear evolutionary histories; instead their histories are branching histories. Examining eye evolution with a 'tree thinking' perspective illustrates how components of currently integrated structures have differing histories, assembled from other systems over evolutionary time" (Oakley and Pankey, 2008: 392). Indeed, gene duplication and co-option have likely played important roles in eye evolution, but in the absence of the lateral exchange of genetic information between lineages, these processes create converging linear ancestral tracks that culminate in single, complex integrated structures.

Oakley and Pankey (2008: 394) are concerned that "[s]tudents of biology at all levels, but especially those less experienced, commonly have a strong tendency to view evolution as a linear series of events." This shouldn't be a reason for concern because that's precisely what evolution is when you trace individual lineages. The branching of evolution becomes an irreducible reality only when you try to trace the evolution along multiple diverging lineages simultaneously. Nevertheless, Oakley and Pankey's misplaced worry expresses a broad consensus among educators concerned with developing the evolutionary understanding of high school and college students. The Evolution and the Nature of Science Institutes (https://ensiweb.bio.indiana.edu; last accessed September 27, 2021) is dedicated to improving the teaching of evolution in high schools in the United States and its website provides materials for the benefit of teachers. ENSI vigorously promotes cladistics and tree thinking, but in doing so they deny the linearity of descent: "As teachers, we often encounter the **misconception** that **ancestral** relationships are **lineal**, when in fact family trees and phylogenies are **branched**" [bold in original] (https://web.archive.org/web/20170708163434/http://www.indiana.edu/~ensiweb/phylo.%20tree.html; last accessed August 20, 2021). It would be a miracle if students and teachers aren't confused by this statement.

The reader is then urged to look at a cartoon drawn by paleontologist Matthew Bonnan (Figure 11.4) that contrasts linear and branching depictions of evolution. Bonnan's cartoon is great for helping novices appreciate how the branching nature of evolution relates to common ancestors. But it also illustrates the tension between the realms of systematic pattern and evolutionary process. It points out that a linear progression of a fish →

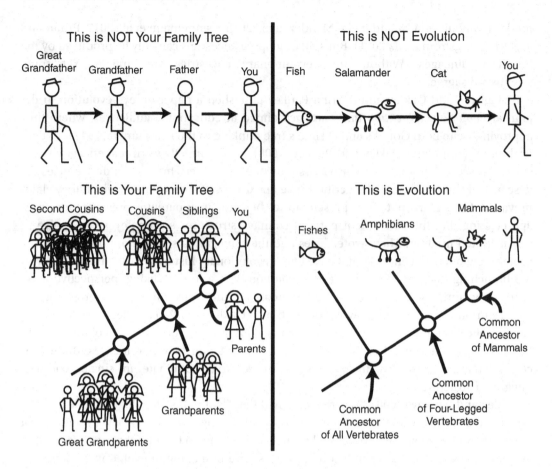

Figure 11.4 A cartoon drawn by paleontologist Matthew Bonnan to explain what evolution is. It made the top 10 in the 2010 Stick Figure Science contest run by Florida Citizens for Science (www.flascience.org/ss2010top10.html; with permission from Matthew Bonnan).

salamander → cat → human is not a proper depiction of evolution. Instead it shows a tree, titled "This is Evolution," in which these taxa form successive sister groups, with arrows highlighting the ancestral nodes. This tree nicely illustrates taxic thinking, but adding a few arrows along the internal branches and up to the terminal taxa could enhance its ability to illustrate lineage thinking as well, by revealing the direction of evolutionary time. This could help clarify the relationship between systematic pattern and evolutionary process.

A series of papers coauthored by biologist Kefyn Catley and cognitive psychologist Laura Novick cite and faithfully echo Gould's and O'Hara's thinking about lineages and trees, resulting in the by now familiar flaws (Catley and Novick, 2008; Catley et al., 2010; Novick et al., 2011). The root of the problem with these papers is their extreme taxic outlook. They view evolution exclusively from a taxic perspective. They claim that new species can arise either by cladogenesis (branching of lineages) or anagenesis, which they define as "one taxon turning into another over time" without branching (Catley and Novick, 2008: 984). They write that "[d]iversification and radiation of species within clades can be achieved only by branching or cladogenic events. If new species only arise by anagenesis, then there would never be a net increase in the total number of species; and while change within a single species is well documented, one species changing into another species is not" (Catley and Novick, 2008: 984).

These authors view the evolutionary process exclusively in terms of the taxonomic bookkeeping of its products. To establish the anagenetic origin of species is problematic because it is difficult to determine species boundaries within a continuous, unbranching lineage. But this epistemological challenge does not diminish the reality of anagenetic evolutionary change along unbranching lineages. Like Gould, these authors fail to realize that, without anagenetic evolution, cladogenesis would just produce identical twins. Their failure to grasp the central importance of linear evolution has caused Catley and Novick to offer ill-conceived and damaging advice to teachers and to misjudge the evolutionary understanding of students. They advise writers of school and university textbooks to avoid putting labels along internal branches of evolutionary diagrams, either to indicate taxa, or even the origin of characters because, they argue, that would incorrectly suggest that taxa evolve into each other. Catley et al.'s (2010) paper in the *Journal of Research in Science Teaching* presents a deeply misleading perspective on the proper depiction of evolutionary relationships. They consider anagenetic change and evolutionary diagrams with taxa located on internal branches as "inappropriate" (Catley et al., 2010: 1) (Figure 11.5). Similarly, although they "recognize the value of teaching the concept of common ancestry, because it is impossible to know which taxon was the common ancestor of any given clade,

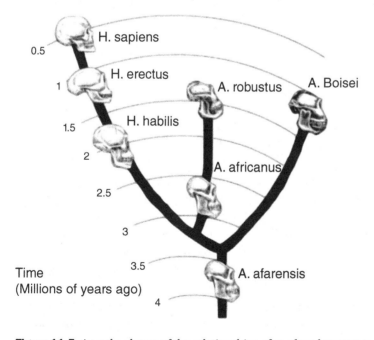

Figure 11.5 A textbook tree of the relationships of modern humans to fossil relatives that Catley et al. (2010: 1) reject for depicting "an inappropriate (anagenic) conception of evolutionary history" (after figure 3 in Catley et al., 2010, by permission of John Wiley and Sons).

its inclusion [in evolutionary diagrams] seems disingenuous at best and bad science at worst" (Catley and Novick, 2008: 985). This is the cladistic blindfold at its most extreme. Surely the difficulty of inferring the nature of ancestors can be no reason for airbrushing ancestry out of evolutionary diagrams.

These authors' failures to understand the distinction between the reality of evolutionary descent and how to get epistemic access to it causes problems when they judge students' understanding of evolution. Catley et al. (2010) think that students have a less sophisticated understanding of evolution when they use process phrases, such as "evolved into/from," "evolved to become," or "became," rather than pattern phrases that cast evolution in terms of taxa being related due to sharing common ancestors. Both are appropriate, depending on whether one focuses on the relationships between terminal taxa or the processes of evolutionary descent that they help reveal. By seeing evolution strictly in terms of taxic relationships, Catley et al. seem to be blind to the linearity of descent. The complete absence of lineage thinking is apparent when they write that use of the terms ancestor and descendant "suggest[s] a grasp of the notion that any given taxon arose, not as a result of some other taxon becoming it, but due to an ancestral population split that gave rise to two separate taxa" (Catley et al., 2010: 872).

Catley et al. (2010) and Novick et al. (2011) further express concerns that, when students are shown evolutionary diagrams that show taxa along internal branches, such as fossils in a hominid tree or supraphyletic groups in Haeckel's *Pedigree of Man*, they interpret these anagenetically. But that shouldn't be a reason for concern. Instead of explaining the great epistemological challenge of recognizing fossils as ancestors and warning students not to interpret paraphyletic groups as ancestors, they conclude that "[l]inear depictions of historical events are simplistic because they suggest a single, straight-line path from a starting point to an ending point" (Novick et al., 2011: 555–556). That is not simplistic. It is the simple truth.

The fact that these views are common in the educational literature is worrying. What makes the alarm bells really ring is that practicing scientists cite these papers approvingly. For example, paleontologist and horse evolution expert Bruce MacFadden and his colleagues published a paper in the journal *Evolution: Education and Outreach* that cites and endorses the papers by Catley, Novick, and their colleagues (Catley and Novick, 2008; Catley et al., 2010; Novick et al., 2011). They write that "instead of a linear sequence in which ancestral species evolve directly into their descendants, the evolutionary tree of horses is bushy" (MacFadden et al., 2012: 31). Here again we meet the false dichotomy between the branching and linear aspects of evolution promoted by other paleontologists like Gould, Prothero, Tattersall, and others. Yes, the phylogeny of horses is bushy, but it is a bush of lineages, within each of which ancestors are tied linearly to their descendants, from root to tip.

MacFadden and his colleagues did a survey of museum exhibitions about horse evolution to determine the prevalence of linear and branching components. They diagnosed exhibitions in which fossils are lined up along internal branches of phylogenies as orthogenetic, and they found that orthogenetic components are present in most exhibitions they surveyed. They consider this a big problem. If these exhibitions do indeed incorrectly slot fossils into ancestral positions in lineages, they would paint an incorrect evolutionary picture. But MacFadden and his colleagues coded *all* linear arrangements of fossils that don't overlap in geological time as orthogenetic. That cannot be correct, because in theory fossils can and sometimes do represent lineal ancestors rather than collateral relatives without this arrangement being orthogenetic (see Chapter 7 for a fuller discussion of orthogenetic thinking). Indeed, MacFadden's own horse phylogeny in the same paper (figure 7 in MacFadden et al., 2012) positioned several fossil species along internal branches. Their allergy to linear thinking is so strong that they even label the idea that we evolved from monkeys and apes as orthogenetic (MacFadden et al., 2012: 36)! It is yet another example of authors becoming blind to evolutionary lineages by overzealous taxic thinking.

11.4 The Importance of Lineage Thinking

Writers such as Gould have often attacked untenable evolutionary notions, such as predictable unidirectional progress, goal-directedness, anthropocentrism, and evolution following a main line, by taking aim at the bull's eye of evolutionary linearity. But linearity is the wrong target. Committing the sin of the scala isn't wrong because it involves linear thinking, but because it lines up the wrong entities. Students who are asked to depict the relationships of a set of contemporaneous taxa, even imaginary ones, often draw diagrams that connect the dots of extant diversity directly, often in a linear fashion, and students asked to draw the first image of evolution that comes to mind also most often draw a linear series of forms (Halverson et al., 2011; Matuk and Uttal, 2011, 2020). Is this a pernicious result of the stranglehold of folk theories of evolution and our desire to tell linear stories, as claimed by Matuk and Uttall (2011, 2020), who have done extensive research on undergraduates' understanding of cladograms? Or might this reflect the students' correct intuition that lines of descent are linear? Evolution tracks the arrow of time. How linear descent relates to the branched relationships of collateral relatives that are produced by speciation is much more difficult to grasp. It is important to realize that the struggles of students and laypeople today to understand tree thinking recapitulate the history of attempts by evolutionists to achieve the same thing. Many biologists and paleontologists produced more or less scala-like diagrams well into the twentieth century, lining up fossils or paraphyletic taxa to trace the pathways of evolutionary descent. Proper tree thinking doesn't come easy.

If even professional biologists, paleontologists, anthropologists, and philosophers today don't get the relationship between the realms of systematic relationships and evolutionary descent right, we should not be surprised that students and the broader public continue to struggle with this issue. Naive tree thinking creates a cladistic blindfold, where a myopic focus on the products of evolution causes a kind of process blindness. Proper tree thinking should go hand in hand with proper lineage thinking. Tree thinking shows how the realm of evolutionary descent relates to the patterns of relationships between collateral relatives. It is the bridge between taxic and lineage thinking. The central point to emphasize to students is that evolution is a process. It produces a pattern of split lineages, in which collateral relatives make up the branched realm of systematic relationships. But to achieve this, teachers must first remind themselves of this fundamental truth, lest their teaching derails from the beginning. On the first page of the preface of their book *Tree Thinking. An Introduction to Phylogenetic Biology* Baum and Smith wrote:

> [t]ree thinking runs counter to standard perceptions of evolution in popular culture. We do not know why it should be so, but we have learned from working with thousands of students that, without contrary training, people tend to have a one-dimensional and progressive view of evolution. We tend to tell evolution as a story with a beginning, a middle, and an end (Baum and Smith, 2013: xv).

By this point, readers will agree with me that evolution is not *a* story. There are millions of stories, each with a beginning, a middle, and an end. Baum and Smith have forgotten this basic truth. Some, however, reject this truth categorically.

11.5 Paraphyly, Ancestors, and Taxic Extremism

One of the triumphs of cladistic over traditional systematic thinking is the rejection of paraphyletic taxa. Paraphyletic taxa are incomplete groups, artifacts created by the deliberate or accidental exclusion of parts of monophyletic groups: invertebrates without vertebrates, dinosaurs without birds, apes without humans. Discovering and resolving paraphyletic taxa into monophyletic clades therefore represents a clear criterion by which to track progress in systematic research. But paraphyletic taxa have retained a toehold in the literature as narrative devices that facilitate the telling of evolutionary stories. This, according to tree thinkers, is a serious problem (O'Hara, 1988, 1992; Sandvik, 2008, 2009). Sandvik (2009: 428) considers their use a "distortion of the evolutionary past." O'Hara (1988: 152) thinks that authors use paraphyletic groups to force artificial closure on evolutionary stories. As Frost et al. (2006: 12) put it in the then largest phylogenetic analysis of amphibians ever published:

> A serious impediment in amphibian biology, and systematics generally, [is] the willingness of many taxonomists to embrace, if only tacitly, paraphyletic groupings... Recognizing paraphyletic groups is a way of describing trees in a linear way for the purpose of telling *great* stories and providing favored characters a starring role. Because we think that storytelling reflects a very deep element of human communication, many systematists, as normal storytelling humans, are unwilling to discard paraphyly. (italics in original)

How much of a faux pas is the use of paraphyletic taxa in evolutionary storytelling? Reconstructing the history of lineages necessarily involves acts of evolutionary storytelling. Since lineages are the central subjects of such narratives, how do you talk about evolution in terms of taxa, the only observable subjects of systematics? As we saw in previous chapters, since the founding of phylogenetics, biologists and paleontologists have marked out the backbones of their scenarios with hypothetical ancestors and with supraspecific paraphyletic ancestral taxa. These ancestral taxa can include fossil and living species or be

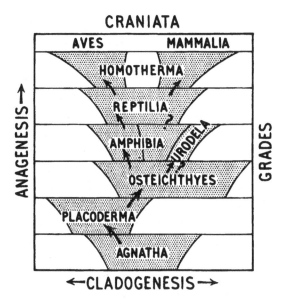

Figure 11.6 An evolutionary tree of vertebrates with paraphyletic taxa, drawn by Julian Huxley (after figure 2 from Huxley, 1957, reprinted by permission from Springer, Copyright, 1957).

wholly hypothetical (see Figure 5.4, for instance). Today we recognize that advocating paraphyletic taxa is bad systematics, but it didn't become an official sin until the concept was fully developed by Hennig and his successors in the 1960s and 1970s (Nelson, 1971; Vanderlaan et al., 2013). Before that time, the pervasive use of paraphyletic taxa was an understandable result of evolutionists trying to communicate evolutionary ideas in the language of systematics. I will argue that it still is.

Adolf Naef, an important early-twentieth-century reformer of systematics and phylogenetics (see Chapter 6), recognized stem groups as assemblages of taxa that were close relatives of the ancestors of a focal taxon.[2] He realized that these stem groups hinted at a phylogenetic truth while telling a systematic lie. Put in modern terms, Naef knew that a monophyletic taxon and its paraphyletic stem group couldn't have the same taxonomic rank. To create a proper, fully resolved tree, stem groups would need to be dissolved into their constituent taxa, until each was grouped together with its closest relative as sister taxa of equivalent rank (Naef, 1919: 49–51). But because they contained pointers to evolutionary descent, Naef accepted these systematically problematic paraphyletic stem groups. Paraphyletic ancestral taxa permeate the writings of narrative phylogeneticists because they are the only way in which lineage thinking can be expressed in taxic terms (Figure 11.6).

---

[2] Naef was an evolutionist concerned with keeping the empirical study of comparative morphology separate from phylogenetic speculations based upon it (a hallmark of post-evolutionary idealistic morphology, or what Naef called systematic morphology). Although he wrote the section referred to here in terms of systematic morphology and types, I have given it a phylogenetic interpretation.

The trouble started when paraphyletic taxa were rightly recognized as taxonomic artifacts without diagnostic traits. Because such ancestral taxa aren't real things, cladistic logic rejects them, and cladists have therefore long denounced the hunt for ancestors as a conceptually flawed and empirically empty exercise (Nelson, 1970; Engelmann and Wiley, 1977; Patterson, 1978, 1981, 2002; Eldredge and Cracraft, 1980; Forey, 1982, 2004; Smith, 1994: 126; Sandvik, 2008: 48; Williams and Ebach, 2008: 18, 2009, 2020; Podani, 2019: 130). With the exception of a minority of misguided workers who continue to defend the legitimacy of paraphyletic taxa (Quinn, 2017), we all chime in with this refrain. But it creates a puzzle. If paraphyletic taxa are anathema, can we no longer talk about fishes, animals, monkeys, apes, crustaceans, invertebrates, protists, reptiles, and dinosaurs? How can we discuss evolution as "taxon X evolved from Y" if Y thereby disappears as a nonentity? Didn't birds evolve from dinosaurs, and insects from crustaceans, and we from monkeys? Of course. The problem is taxic thinking in the realm of lineages.

Gareth Nelson (1994: 126) expressed the issue succinctly: "taxa do not exist as biological phenomena related as ancestor and descendant." This means that it is impossible to accurately express ideas about lineages in terms of the taxic language of systematics. Phylogenetic explanations are lineage explanations for which an appropriate language is lacking. Evolutionists attempt to talk about lineages with taxic concepts, such as paraphyletic taxa, but the result is imprecise. We didn't evolve from monkeys, but from a specific lineage of monkeys. Birds didn't evolve from dinosaurs, but from one lineage of theropod dinosaurs. These ancestral lineages don't coincide with any specific, named taxa. The only taxa with which lineages precisely coincide are monophyletic taxa, and only these can be properly recognized with specific names. The monkey lineage that evolved into humans, however, is a paraphyletic lineage that can be traced back into the tree of primates and beyond.

When you look at Figure 11.2, you can see that this lineage runs from humans down to node Z, with nodes U to X representing hypothetical ancestors that would be classified as apes and nodes Y and Z representing hypothetical ancestors that would be classified as monkeys. This lineage represents a single filament of biodiversity that successively joins the lineages of other primates, other mammals, other vertebrates, and so on. It is unique to humans and doesn't coincide with any particular named taxon. Instead, if we trace it back in time it traverses the taxonomic hierarchy as it climbs from the species *Homo sapiens* to the genus *Homo*, to the tribe Hominini (node U and its descendants), to the subfamily Homininae (node V and its descendants), to the family Hominidae (node W and its descendants), etc. Unless it encounters paraphyletic taxa, this lineage will not cross the boundaries of taxa of equal rank. For example, it will not cross from the genus *Homo* into the genus *Australopithecus*, but merely join its lineage.

Lineages fall outside the bookkeeping language of systematics. This is what makes it so challenging to accurately communicate ideas about evolutionary descent. When paraphyletic taxa enter the dialogue, it doesn't necessarily mean that a conceptual sin is in progress. The series of terminal taxa one encounters when descending the branches of a tree from a focal trait or taxon, like ourselves - chimps → other apes → monkeys → other primates →

other mammals – represents a kind of index of the empirically accessible evidence that allows one to infer character state evolution along that lineage. This index exists irrespective of whether these diverging taxa are classified as artificial paraphyletic groups or neat monophyletic taxa. What about the paraphyletic nature of lineages themselves? Isn't that a problem? For example, the evolutionary lineage that connects humans to the last common ancestor of mammals is paraphyletic in that a large segment of it is also shared by other lineages that lead to other species of mammals. Doesn't that mean that it is subject to the same criticism as paraphyletic taxa? No, it doesn't.

Paraphyletic taxa are artificial groups of collateral relatives that do not uniquely share a common history because they exclude taxa with the same common ancestry. This is not so for paraphyletic lineages. A lineage represents a hypothetical ancestor–descendant series unique to a particular trait or taxon (Figure 11.2). It includes all relevant hypothetical ancestors from a particular point in the phylogeny, excluding nothing relevant and including nothing irrelevant. Monophyletic and paraphyletic lineages are identical in this respect. For example, when you trace the lineage of a focal taxon down a phylogeny, you are tracing a monophyletic lineage up to the point where you meet the lineage leading to its sister taxon. From that point onward, the lineage you are tracing back in time is paraphyletic as you deliberately exclude the lineages leading to the sister taxon and any other taxa that diverged lower down the tree. Taken as a whole, this paraphyletic lineage is unique to your focal taxon, as it doesn't include irrelevant history and doesn't exclude any relevant history. It is not an artifact, but represents a segment of a larger lineage that is individuated by a unique evolutionary history that traces back to the origin of life and that is marked out by a unique series of character state transformations.

In short, other than representing an index of observable evidence with which inferences can be made about lineages, paraphyletic taxa have nothing directly to do with ancestors and ancestry. Neither do monophyletic taxa, or any other kinds of groupings of collateral relatives. Nevertheless, as we already saw earlier in this chapter, crude taxic thinking frequently distorts lineage thinking, even in careful scientists. When Phil Donoghue (2005) reviewed the concept of stem groups, he highlighted their paraphyly as problematic. He rejected Peter Ax's stem lineage concept[3] as even worse because "this concept is even more paraphyletic than the stem group... Clearly, Ax's stem lineage, despite its distinct and precise definition, is a stem group in all but name" (Donoghue, 2005: 555). It isn't. Lineages aren't groups, and Ax understood this. He rejected stem groups as valid taxa because of their paraphyly (Ax, 1985: 286). He proposed his stem lineage concept precisely to solve this theoretical issue. As I explained earlier, the paraphyly of lineages is not a conceptual problem, unlike the paraphyly of taxa. Ax's stem lineage concept is a useful tool for lineage thinking. Donoghue admits this, but then goes in search of "an alternative term that would convey the meaning of a stem group, but without the implication of a taxonomic group" (Donoghue, 2005: 555). Donoghue misses the point that Ax tailored his stem lineage

---

[3] According to Ax, a stem lineage comprises one or more stem species that are directly ancestral to a species or clade and which are evolutionary species that form an ancestor-descent lineage.

concept precisely for this end. Instead, Donoghue advocates following an emerging trend in the literature and simply referring to "stems" or "stem-(crown taxon name)." This is perfectly fine, but it shows that lineages, the fundamental entities of evolution, are not as prominent in the consciousness of researchers as they should be.

The lack of lineage thinking is prominent in the literature where cladists evaluate the work of their predecessors. For example, Peter Forey, Colin Patterson, and Pascal Tassy offered their views on early fossil-based phylogenies, comparing those drawn in the nineteenth century by the French paleontologist Jean Albert Gaudry and the German paleontologist Franz Hilgendorf (see Chapter 7) on the one hand, with those of Ernst Haeckel, T. H. Huxley, and the twentieth-century trees of Alfred Romer on the other hand (Patterson, 1981; Forey, 2004; Tassy, 2011). They laud Gaudry's and Hilgendorf's trees as "a faithful depiction of evolutionary theory in which species give rise to species" (Forey, 2004: 152). They accuse the other authors of advocating supraspecific taxa as ancestors, which are "units of systematics rather than evolution" (Patterson, 1981: 204), and Haeckel, Huxley, and Romer therefore fall "into the trap" (Tassy, 2011: 95) of paraphyly. They miss the point that these precladistic evolutionists were lineage thinkers. They used ancestral taxa so they could talk about lineages in the vocabulary of systematics, the only language available to them. These judgments further miss the point that Gaudry's trees don't just depict transitions between species, but also transitions between genera, as the genus *Hipparion* is paraphyletic with respect to the genus *Equus*. Patterson (1981: 204) admits that earlier authors shouldn't be judged anachronistically, but he and many other cladists have been so caught up in taxic thinking that they have failed to see that the use of paraphyletic ancestral taxa by lineage thinkers is often little more than sloppy systematic bookkeeping, rather than a logical fallacy. I will close this chapter with a few words about the most extreme form of taxic thinking, which issues from the pen of pattern cladists.

Both the defenders and the critics of the idea that paraphyletic taxa can be interpreted as ancestors have the wrong target. Paraphyletic taxa are a concept in the realm of sister group relationships between collateral relatives, not the realm of lineages and ancestor-descendant relationships. Yet the observable evidence in the realm of systematics can give indirect, inferential access to the realm of evolutionary descent. It is this transition that transforms systematics into phylogenetics and makes evolutionary biology a historical science. Pattern cladists don't like that. They adhere strictly to the epistemic primacy of systematic pattern over evolutionary process and adopt a position of stern agnosticism with respect to the transformations involved in evolutionary descent. They are pure taxic thinkers. For them lineage thinking is anathema. Indeed, they don't consider themselves to be phylogeneticists. And yet, they try and adjudicate issues of evolutionary descent while wearing a thick cladistic blindfold.

My friends and colleagues Dave and Malte are pattern cladists. They recently condensed years of thinking and writing in a book called *Cladistics. A Guide to Biological Classification* (Williams and Ebach, 2020). They titled their introductory chapter "Carving nature at its joints, or why birds are not dinosaurs and men are not apes." The question they intend to address is not whether women are apes, but whether these statements make any sense. To lineage thinkers they make immediate sense. To pattern cladists they are nonsense. Fifteen chapters later we learn that because dinosaurs and apes are artificial paraphyletic taxa, humans aren't apes, birds aren't dinosaurs, and neither did we evolve from apes, nor birds from dinosaurs (Williams and Ebach, 2020: 376–381). Indeed, one cannot argue with these conclusions when they are spelled out like this in the taxic language of systematics. But Dave and Malte aren't creationists. They are evolutionists. They know that evolutionary descent flows along lineages. Questions about the ancestry of taxa can be posed in the language of systematics, but they can only be answered with the concepts of lineage thinking. When they cite a biologist who gives the correct answer, they cannot follow: "Note how the text moves from observation ('crocodiles are more closely related to birds than they are to lizards') to interpretation ('Birds are, in fact, descended from a lineage of theropod dinosaurs'), confusing interpretation for data, as if it was evidence" (Williams and Ebach, 2020: 379).

This is a strange conclusion, and not just for saying that relationships are observed rather than inferred. Interpretation, or better, inference, is not the same as data or evidence, but is based upon them. All scientists know this. That birds are a lineage of dinosaurs (the proper phrasing if dinosaurs are a clade that includes birds) or have evolved from dinosaurs (if dinosaurs are considered a paraphyletic taxon of equal rank to birds) are inferences supported by available evidence. But pattern cladists refuse to engage in evolutionary inference. And like cladists of other stripes, they have for decades tried to refute statements about ancestors and lineages by claiming these are paraphyletic nonentities (e.g. Williams et al., 1996: 131).

Pattern cladist Andrew Brower (2014: 110) considers interpreting common ancestors as real things as a form of phylogenetic fundamentalism. Maybe so. Lineage thinking is indeed based on assuming the existence of things that can typically only be inferred, not observed. But reading Dave and Malte's book you are told that pattern cladists are able to observe homology and relationships rather than infer them (Williams and Ebach, 2020: 99, 204, 418), that data matrices are unnecessary constructs (Williams and Ebach, 2020: 225), that testing homology conjectures in the light of other characters using numerical techniques is a "perverse" practice that should defer to the "judgments of what is and is not a character" made by taxonomists (Williams and Ebach, 2020: 224); that cladograms should only be read from top to bottom (Williams and Ebach, 2020: 380), that "the explanatory side of historical biology has little to do with… understanding evidence" (Williams and Ebach, 2020: 351), and that interpreting a tree as "an evolutionary 'event-o-gram'" (Williams and Ebach, 2020: 126) is a myth. Readers have to decide for themselves what views they think are more fundamentalist: believing that comparative research reveals an unseen world of evolving lineages, which evolutionary theory tells us must exist, or maintaining an unmovable agnosticism about evolutionary descent, no matter how robust the bedrock of our background knowledge about evolution now is.

I have had many enjoyable discussions with Dave and Malte, often over pints of beer. But I've always come away thinking that in the end, I just don't understand them. I suspect the feeling is mutual. I now realize why. Pattern cladists are pure taxic thinkers. I am a lineage thinker. So are the vast majority of our colleagues, and all those who came before us who tried to capture fragments of life's ancestry in their mind's eye. I regard Dave and Malte's determination to stay strictly cheek to cheek with empirical evidence with a mixture of pity and awe. I admire them for working strictly from first principles and adhering to a philosophical point of view they have carefully worked out. But I'm sad for them that they don't dare to append any evolutionary interpretations to the fruits of their labor. I think their evolutionary agnosticism will always remain a point of principle, rather than a diminishing quantity of doubt that will eventually dissipate. I don't think they will ever free their minds to probe the uncertain outlines of evolutionary inferences that may be drawn from their data. I think they will always remain systematists only, afraid or unwilling to wonder what might lie beyond what their eyes can see.

12

Making Sense with Stories

12.1 Evolutionary Storytelling Is Inescapable

In 2013 paleontologist Michael Novacek, Provost of Science at the American Museum of Natural History in New York, was interviewed for the *Los Angeles Times* about a *Science* paper he had coauthored on the evolution of placental mammals (O'leary et al., 2013). Novacek and his colleagues had performed a phylogenetic analysis of a huge morphological matrix, which scored 4,541 characters for 86 living and fossil mammal species. Mapping these traits on their tree, they reconstructed the hypothetical placental mammal ancestor in exquisite detail, which artist Carl Buell captured in a beautiful lifelike drawing. Asked about this slender-snouted and fluffy-tailed insectivore, Novacek remarked that "[t]his is fairly novel, to reconstruct an ancestor. I don't know of any other" (www.latimes.com/science/la-sci-sn-placental-mammal-ancestor-20130207-story.html; last accessed September 29, 2021). This would have been a surprising statement even if it had been uttered a century earlier.

As this book has shown, throngs of hypothetical ancestors march through the pages of the evolutionary literature, including ones remarkably similar to the creature reconstructed by Novacek and his colleagues (Figure 12.1). They form the epistemic backbones of countless evolutionary narratives that try to provide lineage explanations for the origins of traits and taxa. Many hypothetical ancestors, especially in the early phylogenetic literature, were designed specifically to maximize their precursor potential, and thereby the explanatory power of the scenarios in which they acted as central subjects. Today, hypothetical ancestors are visualized far less often and exist mostly as disembodied clouds of character states optimized along the branches of trees. The *Los Angeles Times* interview ended with this: "'People who think this organism is cute have asked me for a name,' Novacek said. 'I tell them we can't give it a Latin name, because it never really existed. It doesn't have a nickname either. It's just the hypothetical placental mammal ancestor.'" Lay readers would have been better served by an explanation that it is of course perfectly possible that something very much like it did, in fact, exist. Indeed, under a realist interpretation of science, it is precisely the discovery, through inference, of hypothetical ancestors that is a major goal of phylogenetics. These creatures carry the entire weight of phylogenetic explanations on their tiny hypothetical shoulders.

Figure 12.1 Adolf Naef's conception of the general type of marsupials. As an idealistic morphologist, he conceived of this type as a thinking tool that could help him understand how forms could theoretically be derived from one another. But it could also be given a materialistic interpretation, for example, as a possible marsupial ancestor
(after figure 3 in Levit and Meister, 2006, reprinted by permission from Springer, Copyright, 2006).

When I told Dave Williams, currently outgoing President of the Systematics Association, that I was writing a book about ancestors, he said: "That will be a short book then, Ron." Dave is a pattern cladist who doesn't much care to speculate about ancestors, so if he would have written this book, you would have finished it hundreds of pages earlier. But critical remarks about the scientific status of evolutionary storytelling are as old as the discipline itself. In what is a typically dismissive sentence in a thoroughly scathing book, Henry Gee, editor at *Nature*, writes that "Deep Time can never support narratives of evolution" (Gee, 2000: 4). Gee claims that cause and effect are torn asunder in the depths of geological time. His book *Deep Time. Cladistics, the Revolution in Evolution*, is a diatribe against phylogenetics and paleontology as historical sciences. It puts flesh on the bones of Michael Lynch's glib dismissal of evolutionary storytelling with which I started this book. Gee's main message is that "we want to think of the history of life as a story, but that is precisely what we cannot do" (Gee, 2000: 4). Gee tells us that "most palaeontologists do not think of the history of life in terms of scenarios or narratives, and that they rejected the story-telling mode of evolutionary history as unscientific more than thirty years ago" (Gee, 2000: 5).

Gee is right about one thing here. The direct storytelling approach to evolutionary biology, which I have called narrative phylogenetics, can indeed be unscientific when relevant systematic relationships and conflicting evidence are ignored. Judged in the light of modern phylogenetics, this problem afflicts a large part of the earlier evolutionary literature. However, Gee is wrong to suggest that we mustn't think of the history of life as a story and that most biologists and paleontologists don't think about evolution in terms of stories. We do all the time. We have no choice. Lineage thinking is narrative thinking. Every time we wonder about what may have happened to an evolving lineage, we engage in narrative thinking. The methods we use to infer evolution, such as parsimony-based cladistic analysis, may not be narrative themselves, but if we wish to use them to shed light on what happened in evolution, we can only express our insights in narrative form. Telling more or less informed stories is our only route to making sense of evolution, and sketching such historical narratives is a major aim of evolutionary biology (Hull, 1975; Richards, 1992b; Ghiselin, 1997; Winther, 2008, 2011; Currie and Sterelny, 2017; Ereshefsky and Turner, 2020; Otto and Rosales, 2020). Gee knows this, of course. He may deride linear evolutionary narratives and prefer to wear the cladistic blindfold, but in his own writings and the many papers he has allowed into the pages of *Nature*, all kinds of evolutionary stories are told.

Sir Peter Medawar remarked that "[t]he most interesting and exciting of all intellectual problems is how the imagination is harnessed for the performance of scientific work, so that the steam, instead of blowing off in picturesque clouds and rattling the lid of the kettle is now made to turn a wheel" (Medawar, 1996: xv). My book illustrates the important role that the imagination plays in evolutionary storytelling. Medawar's ears perceived rather more rattling of lids here than the smooth swoosh of spinning wheels: "In the heyday of evolutionary biology there was thought to be an obligation upon entomologists to trace out, as far as possible, all the lines of evolution within insects, but most modern entomologists have given up this activity as tiresome and fruitless: nothing of any importance turns on the allocation of one ancestry rather than another" (Medawar and Medawar, 1977: 28). During my foray into the phylogenetic literature, I have heard plenty of lids rattling myself. Yet the imagination remains an indispensable tool in evolutionary thinking, even though evolutionary storytelling has changed over the past century and a half, largely due to the spread of modern phylogenetics.

Trees now firmly constrain speculation. Fossils are generally no longer simply anointed as ancestors, and hypothetical ancestors have become the products of phylogenetic inference rather than deliberate design. Evolutionary intuitions now generally take a back seat to the elucidation of systematic pattern, but they continue to play a central role in the judging of trees and the construction of scenarios, even though many of these narrative evaluations are never written down. What hasn't changed is that the power of phylogenetic hypotheses to explain the evolution of form is rooted in the ability to trace back traits to ancestral states.

While the first generations of evolutionary morphologists and embryologists were restricted to the level of the phenotype, biologists can now trace strands of homology across different levels of organization. When the trail of homology runs cold on the level of morphology, the continuity of descent may be traced further on the levels of genes and regulatory networks and the interactions between their products (Shubin et al., 2009; Tschopp and Tabin, 2017; DiFrisco and Jaeger, 2019, 2021; Linz et al., 2020). The realm of what has been called deep homology (Shubin et al., 2009) has dramatically expanded the size of our storytelling canvas.

The subjective element cannot be expelled from evolutionary thinking. Sophisticated algorithms and models may be used to reconstruct phylogenetic trees and infer ancestral character states, but our imagination and evolutionary intuitions always come into play to animate and judge scenarios. A few examples will suffice to show how our conclusions continue to show the footprints of our assumptions.

My retired Natural History Museum colleague Andrew Smith and two other echinoderm experts performed phylogenetic analyses to reconstruct the evolution of pedicellariae, which are small appendages located on the tests of sea urchins. Their trees suggested that venom-delivering pedicellariae may have evolved several times. However, they concluded that "this seems unlikely, given the similarity" between the pedicellariae in different groups (Coppard et al., 2012: 135). They therefore preferred a less parsimonious result and proposed homology of venomous pedicellariae across sea urchins, implying multiple losses of this trait. They may well be correct. The intuition that convergent losses are more likely than convergent gains of complex characters is an enduring and widespread intuition. It is, for instance, at the center of the debate about whether stick insects have re-evolved wings after they were lost (Whiting et al., 2003; Trueman et al., 2004; Whiting and Whiting, 2004). Using one's evolutionary intuitions to prefer a result that is suboptimal when judged on the basis of a phylogenetic optimality criterion shows that speculative storytelling is alive and well.

Hervé Philippe and his colleagues performed a phylogenetic analysis of the distribution of microRNAs to resolve the position of acoels and related taxa in the metazoan tree. They found that the most parsimonious tree "rather implausibly implies non-monophyly of Acoela" (Philippe et al., 2011: 257), so they accepted a less parsimonious tree without explaining why they thought that acoels should be monophyletic. In contrast, in their study of the evolution of bat echolocation, Nojiri et al. (2021) carefully outlined the evolutionary intuitions that made them reject the phylogenetic interpretation best supported by their analyses. Their model-based ancestral state reconstruction suggests that the last common ancestor of bats was most likely a laryngeal echolocator and that the echolocation systems diverged in the two main echolocating lineages. However, Taro Nojiri and his colleagues instead offer two reasons for why they prefer the hypothesis that laryngeal echolocation evolved convergently in these lineages. First, they expect that if laryngeal echolocation is homologous in the two main lineages, the development and growth of the echolocation system should only show "minimal differences" (Nojiri et al., 2021: 1355). Second, they find it unlikely that laryngeal echolocation can be lost without a trace, as their tree suggests happened in Old World fruit bats. These intuitions may well be correct, but what are they based upon?

First, the expectation that only limited phenotypic differences will accrue in lineages after their divergence has no basis. The laryngeal echolocating bat lineages diverged about 60 million years ago. The observed developmental differences between these lineages are compatible both with hypotheses of descent with modification and convergent evolution. Yet the authors only consider them as evidence for the latter. Second, they also conclude that approximately 30 million years is not long enough for the Old World fruit bats to have lost laryngeal echolocation without leaving vestigial traces. They may be right. Their reason for thinking this is that, after the evolutionary loss of other structures in other taxa such as hindlimbs in dolphins and teeth in mice, vestigial structures have been retained for longer periods of time. The important point is that these authors use these evolutionary instincts to reject the hypothesis that descent with modification explains their observations, a hypothesis that is in fact supported by their tree-based ancestral state reconstruction. Instead, they use narrative reasoning to tell an evolutionary story that makes more sense to them.

Telling competing evolutionary stories will always remain an important component of evolutionary thinking. The mind of the scientist is not a *tabula rasa* upon which results are simply inscribed. No matter how sophisticated our methods, or voluminous our data, we will, and I think should, always ask ourselves: does this make sense? This question is the first step on the path to increased understanding, and it is the spark that triggers new research. The phylogenomic hypothesis that the morphologically complex comb jellies diverged from the remaining animals before sponges did (Dunn et al., 2008) went against the grain of zoological dogma. This shakeup of entrenched ideas continues to drive research more than a decade later (Halanych, 2015; Nielsen, 2019; Kapli and Telford, 2020; Li et al., 2021). When Pyron and Burbrink (2014) published a phylogenetic reconstruction of reproductive mode in squamates that suggested that egg laying had evolved multiple times in live-bearing lineages, it caused much head scratching among experts. The ensuing thoughts were published in a special issue of the *Journal of Experimental Zoology B (Molecular and Developmental Evolution)* (issue 6 of volume 6 in 2015). The authors in this issue used genetic, physiological, ecological, developmental, and anatomical arguments to argue that Pyron and Burbrink's results were unconvincing and that egg laying should only rarely re-evolve in live-bearing lineages, if at all. By selectively focusing on different bits of evidence, these authors tried to make sense of the surprising results by offering complex, subjective, and speculative arguments for what they thought is and isn't likely to happen during squamate evolution. The seeds of scientific progress may sprout from the clash of such stories. However, most of today's evolutionary stories aren't told to be heard, and they aren't told by humans.

12.2 Leveraging Stories with Stories

The commonest stories today are about the evolution of molecular sequences, and they are told by substitution models. Their evolutionary intuitions are called parameters, and they specify those aspects of sequence evolution that the models consider to be most important for spinning their stories. Based on these assumptions and some input data, such as a sequence alignment, the substitution model constructs a scenario of sequence evolution that best matches the input data to a tree with a certain topology and branch lengths. But most humans who put these stochastic storytellers to work aren't interested in scenarios of sequence evolution. Instead, they try to leverage these untold stories to tell other stories, and this is where things get interesting.

It is the norm today to reconstruct ancestral states with the help of such probabilistic models (Pagel, 1994, 1999; Schluter et al., 1997; Cunningham et al., 1998; Pagel et al., 2004; Cusimano and Renner, 2014; Joy et al., 2016). Some of the earliest uses of these models were for reconstructing hypothetical ancestral sequences (Yang et al., 1995; Koshi and Goldstein, 1996). But that is not what is of greatest interest to many biologists today. Instead, we want to understand phenotypic evolution, and we use the scenarios of sequence evolution spun by these models to do this. Specifically, we use the tree topologies and molecular branch lengths produced with substitution models to inform the reconstruction of phenotypic ancestral states. If the substitution model fits the input data and if the evolutionary history of the sequences matches that of the traits one wishes to study – a nontrivial assumption that can be invalidated, for instance, by mixing paralogous sequences – then it makes eminent sense to use the topology of the resulting trees to inform the reconstruction of phenotypic evolution. But it is the use of branch lengths where ancestral state reconstruction with probabilistic models deviates most sharply from that based on parsimony.

Branch lengths are thought to be relevant for reconstructing ancestral states because time, or a proxy of time, may be considered a measure for the opportunity of trait evolution. Accordingly, on a tree that is not time-calibrated, evolutionary change can be thought to be more likely along longer branches than shorter ones because more sequence evolution has happened along them. Because this information is considered, ancestral state reconstruction with probabilistic models can differ substantially from parsimony optimization of traits on the same topology because parsimony mapping doesn't take branch lengths into account. The proxy for time produced by substitution models is the number of substitutions inferred along the lineages in the tree. Why would these be relevant for informing the reconstruction of phenotypic evolution? In his seminal papers on ancestral state reconstruction, Mark Pagel explained the rationale: "Genetic distances uncorrected for molecular clock assumptions may be the best estimate of the 'opportunity for change' in the sense of indicating the total amount of evolution between pairs of species" (Pagel, 1994: 42). Put differently, "Branch lengths in units of genetic divergence may usefully record the underlying opportunity for trait evolution" (Pagel et al., 2004: 675), as these are "units of the 'opportunity for selection'" (Pagel, 1999: 613).

The molecular loci that are used to provide the branch lengths that inform the probabilistic reconstruction of phenotypic ancestral states typically have nothing directly to do with these traits, which probably evolved under a completely different selective regime. I therefore find it hard to understand why the carefully calculated distribution of branch lengths produced by a substitution model chosen for its fit to a specific molecular dataset would provide any reliable information about the opportunity for evolution of an unrelated phenotypic trait along those branches. Yet the use of probabilistic ancestral state reconstruction of phenotypic characters is de rigueur in the literature. There exists, of course, a positive correlation of some sort between sequence evolution and phenotypic evolution, but the genotype-phenotype map is complex. In my view, the idea that molecular branch lengths are informative for reconstructing the ancestral states of unrelated phenotypic characters is a modern evolutionary intuition. I see little theoretical justification for this practice, unless the sequences are directly involved in the development, and therefore the individuality and evolvability, of the phenotypic trait. The modularity of development, which confers a degree of individuality upon characters, implies that the evolution of characters will be insulated or buffered to a substantial degree from sequence evolution that happens at unrelated loci.

Research on the correlation of rates of molecular and phenotypic evolution is fascinating and ongoing. Broad correlations have been detected in some taxa, but not in others. Clear evidence that these rates are generally correlated sufficiently across trees so that one can be used to help infer the other is lacking (Bromham et al., 2002; Seligmann, 2010; Janecka et al., 2012; O'Reilly et al., 2015; Yu et al., 2021). This is why many researchers try to avoid this assumption by transforming these branch lengths of genetic divergence into branch lengths that estimate relative or absolute time. They do this by filtering them through the assumptions of relaxed molecular clocks and then using the resulting chronograms or timetrees for ancestral state reconstruction. Although this removes the simplistic assumption of a strict correlation between sequence and phenotypic evolution, it still leaves a transformed form of this assumption in place, which is added to a complex mix of other assumptions that together make the interpretation of timetrees fiendishly challenging (Bromham, 2019; Carruthers and Scotland, 2021).

The goal of using probabilistic models and molecular branch lengths is to improve the reconstruction of phenotypic ancestral states, and maybe they do. But these methods also require one to accept a number of questionable, and often poorly articulated, assumptions about the relationship between molecular and phenotypic evolution. They force one to make difficult choices about exactly which methods, data, and trees to use to infer ancestral states, which are subjective choices that can have dramatic effects on the resulting evolutionary stories (Litsios and Salamin, 2012; Cusimano and Renner, 2014; McCann et al., 2016). In view of the uncertainty of these assumptions, I doubt whether the ancestral state probability pie charts that model-based methods produce add any useful measure of precision or nuance. When I'm trying to interpret these little diagrams plotted on trees, I find it impossible to disentangle the effects of evidence versus assumptions.

12.3 Optimality Criteria versus Storytelling

What do we want our preferred phylogenetic hypotheses to represent? Do we wish them to reflect the congruence or coherence of available evidence, without any necessary reference to external reality? Or do we hope they point us to features of natural entities and processes that exist independent of us as interpreters? The first sentiment fits what is referred to as a coherence theory of truth, whereas the second marks a correspondence theory of truth (Ghiselin, 1997: 165; Rieppel, 2007; Brower, 2016: 90). Does this matter? I think it does. Different phylogenetic optimality criteria receive their justification under different theories. The principle of parsimony is rooted in the coherence theory of truth. We don't apply this criterion to infer trees and map character states because we think that the process of evolution follows the dictates of parsimony. It is a methodological premise that corresponds to Ockham's razor, which tries to avoid positing unnecessary hypotheses about unobserved things or events. Here, the principle of parsimony makes no explicit ontological claims.

The justification for model-based probabilistic methods is very different and is rooted in the correspondence theory of truth. The parameters of the probabilistic models that fuel maximum likelihood and Bayesian analyses are designed to capture aspects of the evolutionary process as accurately as possible. By incorporating these specific ontological claims, these models produce hypotheses about the evolution, relationships, and character states of lineages that, if the models accurately reflect reality, correspond to what really happened. This doesn't mean that all probabilistic methods tell you what you should believe – Bayesian inference does, likelihood doesn't (Sober, 2008) – but it does mean that if you use parsimony to build trees and infer ancestral character states, a certain philosophical dissonance emerges when you then use these results as a basis for your beliefs about evolution. A certain feeling of unease may arise as a result of the implicit assumptions that your acceptance of the most parsimonious solution may entail. Let me illustrate this with two examples.

Martin-Durán et al. (2018) investigated the phylogenetic distribution of bilaterian nerve cords and the genes involved in their dorsoventral patterning. They found that, although medially condensed nerve cords and the expression of these patterning genes co-occur in arthropods, annelids, and vertebrates, they are decoupled in other taxa. Evaluating whether this means that medially condensed nerve cords and their dorsoventral patterning mechanisms evolved once, followed by losses in various taxa, or whether they evolved convergently in the three taxa where they co-occur, the authors prefer the second hypothesis because they claim it is more parsimonious. The implicit assumption underpinning this preference is that the gain of nerve cords and the gain of the coordinated expression of patterning genes are weighed the same as the loss of these features. How believable is such a weighting scheme? If you are not concerned with whether your hypothesis corresponds to evolutionary history, you don't have to worry about this. But if you are, it becomes important to ask if and how you can assign relative probabilities to the various evolutionary events involved so that a quantitative optimality criterion can objectively chose between them. If the loss of nerve cords and/or the expression of their dorsoventral patterning genes is much easier than evolving them, the arithmetic of parsimony can change dramatically. The modelling of developmental dynamics is one promising route to understand the relative probabilities of different events (DiFrisco and Jaeger, 2021). Without a way to assess these probabilities, how can you be confident that your parsimony analysis offers the best possible approximation of the evolutionary events you are interested in?

A second example concerns the evolution of venom in squamate reptiles. Two camps exist in the venom research community: one believes that venom evolved once at the base of the squamatan clade Toxicofera (Fry et al., 2013), and the other believes that venoms evolved multiple times independently in different groups of lizards and snakes (Mackessy and Saviola, 2016). Both hypotheses have been defended as being the most parsimonious one (Hargreaves et al., 2015; Mackessy and Saviola, 2016). However, without some kind of probability calculus for the gain and loss of the components that make up integrated venom systems, do we really want to give parsimony the authority to decide the issue for us? The factors that are relevant for understanding the evolvability of venom systems are many and complex (Jackson and Fry, 2016; Jackson et al., 2019), and these have to be considered if we want to try to fit our hypotheses to what may have happened during evolutionary history.

Parsimony is a neat and tidy epistemic tool that keeps inferences as close as possible to observable evidence. But it does not concern itself with what may have happened during evolution, and it remains mute about the question of whether the optimal hypothesis is the one that is closest to the truth. It is an optimality criterion that is deliberately agnostic about external reality. I think that its poverty as a phylogenetic tool becomes apparent when you try to understand the evolution of complex phenotypes that consist of many parts that covary imperfectly. The epistemic strength of parsimony analysis lies in its refusal to incorporate uncertain background knowledge because it may be wrong. Its epistemic weakness is this same refusal if there is good reason to believe that this background knowledge is correct.

Likelihood and Bayesian methods implement probabilistic models that try to leverage assumptions about the evolutionary process to aid evolutionary inference. The parameters of these models specify background knowledge about how they "think" the real world works, and by relying on these more or less corroborated intuitions, these models produce narrative accounts of evolutionary history in essentially the same manner as human storytellers do. At this point, we could dive into the philosophical abyss that opens up when asking whether reality is knowable at all, but we will not. I just want to close by emphasizing that if you want to know what happened in evolution, and you are incorporating into this endeavor aspects of what you think you already know, your approach is narrative and you are engaged in evolutionary storytelling. This is the epistemic core that connects the first generations of evolutionary biologists to comparative biologists today. Perhaps the most notable difference is that we now use computers to do most of our storytelling for us.

References

Abel, O. (1912). *Grundzüge der Palaeobiologie der Wirbeltiere*. Stuttgart: E. Schweizerbart'shce Verlagsbuchhandlung (Erwin Nägele).

(1914). Paläontologie und Paläozoologie. Pages 303–395 in R. Hertwig, R. v. Wettstein, eds. *Abstammungslehre, Systematik, Paläontologie, Biogeographie*. Leipzig: Druck und Verlag Von B. G. Teubner.

(1918). Methoden und Ziele der Paläobiologie. *Die Naturwissenschaften* 6:497–502.

(1919). *Die Stämme der Wirbeltiere*. Berlin: Vereinigung wissenschaftlicher Verleger.

(1920). *Lehrbuch der Paläozoologie*. Jena: Verlag von Gustav Fischer.

(1929). Das biologische Trägheitsgesetz. *Palaeontologische Zeitschrift* 11:7–17.

Adamowicz, S. J., Purvis, A., Wills, M. A. (2008). Increasing morphological complexity in multiple parallel lineages of the Crustacea. *Proceedings of the National Academy of Sciences USA* 105:4786–4791.

Adoutte, A., Balavoine, G., Lartillot, N., de Rosa, R. (1999). Animal evolution: the end of the intermediate taxa? *Trends in Genetics* 15:104–108.

Adoutte, A., Balavoine, G., Lartillot, N., et al. (2000). The new animal phylogeny: reliability and implications. *Proceedings of the National Academy of Sciences USA* 97:4453–4456.

Agassiz, A. (1874). *Revision of the Echini, Part IV*. Illustrated Catalogue of the Museum of Comparative Zoology at Harvard College. Cambridge, MA: University Press, Welch, Bigelow, & Co. 7.

(1876). Haeckel's Gastraea theory. *The American Naturalist* 10:73–75.

(1880). Paleontological and embryological development. *Science* 1:142–149.

Agassiz, L. (1962). *Essay on Classification*. Cambridge, MA: The Belknap Press of Harvard University Press.

Allen, G. E. (1981). Morphology and twentieth-century biology: a response. *Journal of the History of Biology* 14:159–176.

Allmon, W. D. (2017). Species, lineages, splitting, and divergence: why we still need "anagenesis" and "cladogenesis." *Biological Journal of the Linnean Society* 120:474–479.

(2020). Invertebrate paleontology and evolutionary thinking in the US and Britain, 1860–1940. *Journal of the History of Biology* 53:423–450.

Almeida, W. O., Christoffersen, M. L., Amorim, D. S., Garraffoni, A. R. S., Silva, G. S. (2003). Polychaeta, Annelida, and Articulata are not monophyletic: articulating the Metameria (Metazoa: Coelomata). *Revista Brasileira de Zoologia* 20:23–57.

Amundson, R. (1998). Typology reconsidered: two doctrines on the history of evolutionary biology. *Biology & Philosophy* 13:153–177.

(2002). Phylogenetic reconstruction then and now. *Biology & Philosophy* 17:679–694.

(2005). *The Changing Role of the Embryo in Evolutionary Thought. Roots of Evo-Devo*. Cambridge: Cambridge University Press.

Anderson, D. T. (1973). *Embryology and Phylogeny of Annelids and Arthropods*. Oxford: Pergamon Press.

(1982). Origins and relationships among the animal phyla. *Proceedings of the Linnean Society of New South Wales* 106:151–166.

Andersson, K. I., Norman, D., Werdelin, L. (2011). Sabretoothed carnivores and the killing of large prey. *PLoS ONE* 6:e24971.

Andrade, S. C. S., Novo, M., Kawauchi, G. Y., et al. (2015). Articulating "Archiannelids": phylogenomics and annelid relationships, with emphasis on meiofaunal taxa. *Molecular Biology and Evolution* 32:2860–2875.

Andrews, E. A. (1885). Affinities of annelids to vertebrates. *The American Naturalist* 19:767–774.

Anonymous. (1889). Sketch of Pierre Belon. *The Popular Science Monthly* 34:692–697.

Appel, T. A. (1980). Henri de Blainville and the animal series: a nineteenth-century chain of being. *Journal of the History of Biology* 13:291–319.

(1987). *The Cuvier-Geoffroy Debate. French Biology in the Decades before Darwin*. Oxford: Oxford University Press.

Arber, A. (1950). *The Natural Philosophy of Plant Form*. Cambridge: Cambridge University Press.

Archibald, J. D. (2009). Edward Hitchcock's pre-Darwinian (1840) "Tree of Life." *Journal of the History of Biology* 42:561–592.

(2014). *Aristotle's Ladder, Darwin's Tree. The Evolution of Visual Metaphors for Biological Order*. New York: Columbia University Press.

Arenas-Mena, C. (2010). Indirect development, transdifferentiation and the macroregulatory evolution of metazoans. *Philosophical Transactions of the Royal Society B: Biological Sciences* 365:653–669.

Arendt, D. (2018). Hox genes and body segmentation. *Science* 361:1310–1311.

Arendt, D., Benito-Gutierrez, E., Brunet, T., Marlow, H. (2015). Gastric pouches and the mucociliary sole: setting the stage for nervous system evolution. *Philosophical Transactions of the Royal Society B: Biological Sciences* 370:20150286.

Arendt, D., Nübler-Jung, K. (1994). Inversion of dorsoventral axis? *Nature* 371:26.

(1997). Dorsal or ventral: similarities in fate maps and gastrulation patterns in annelids, arthropods and chordates. *Mechanisms of Development* 61:7–21.

Arthur, W. (2011). *Evolution. A Developmental Approach*. Oxford: Wiley-Blackwell.

(2014). *Evolving Animals. The Story of Our Kingdom*. Cambridge: Cambridge University Press.

(2021). *Understanding Evo-Devo*. Cambridge: Cambridge University Press.

Asma, S. T. (1996). *Following Form and Function. A Philosophical Archaeology of Life Science*. Evanston, IL: Northwestern University Press.

Averof, M., Akam, M. (1995). Hox genes and the diversification of insect and crustacean body plans. *Nature* 376:420–423.

Ax, P. (1985). Stem species and the stem lineage concept. *Cladistics* 1:279–287.

(1989). Basic phylogenetic systematization of the Metazoa. Pages 229–245 in B. Fernholm, K. Bremer, H. Jörnvall, eds. *The Hierarchy of Life*. Amsterdam: Excerpta Medica/Elsevier.

(1999). *Das System der Metazoa II. Ein Lehrbuch der phylogenetischen Systematik*. Stuttgart: Gustav Fischer Verlag.

Ayala, F. J., Cela-Conde, C. J. (2018). *Processes in Human Evolution. The Journey from Early Hominins to Neanderthals and Modern Humans*. Oxford: Oxford University Press.

Baguñà, J., Martinez, P., Paps, J., Riutort, M. (2008). Back in time: a new systematic proposal for the Bilateria. *Philosophical Transactions of the Royal Society B: Biological Sciences* 363:1481–1491.

Baguñà, J., Riutort, M. (2004). The dawn of bilaterian animals: the case of acoelomorph flatworms. *BioEssays* 26:1046–1057.

Bailey, I. W. (1944). The development of vessels in angiosperms and its significance in morphological research. *American Journal of Botany* 31:421–428.

Balavoine, G. (1998). Are Platyhelminthes coelomates without a coelom? An argument based on the evolution of Hox genes. *American Zoologist* 38:843–858.

Balavoine, G., Adoutte, A. (2003). The segmented *Urbilateria*: a testable scenario. *Integrative and Comparative Biology* 43:137–147.

Balfour, F. M. (1880a). Larval forms: their nature, origin, and affinities. *Quarterly Journal of Microscopical Science* 20:381–407.

(1880b). On the structure and homologies of the germinal layers of the embryo. *Quarterly Journal of Microscopical Science* 2:247–273.

(1880c). *A Treatise on Comparative Embryology. Volume I.* London: Macmillan.

(1881). *A Treatise in Comparative Embryology. Volume II.* London: Macmillan.

(1883). The anatomy and development of *Peripatus capensis*. *Quarterly Journal of Microscopical Science* s2-23:213–259.

(1885). *A Treatise on Comparative Embryology. Volume I.* London: Macmillan.

Ballard, J. W. O., Olsen, G. J., Faith, D. P., et al. (1992). Evidence form 12S ribosomal RNA sequences that onychophorans are modified arthropods. *Science* 258:1345–1348.

Barnes, R. D. (1968). *Invertebrate Zoology, 2nd ed.* Philadelphia: W. B. Saunders.

(1974). *Invertebrate Zoology, 3rd ed.* Philadelphia: W. B. Saunders.

Bartolomaeus, T., Purschke, G., Hausen, H. (2005). Polychaete phylogeny based on morphological data - a comparison of current attempts. *Hydrobiologia* 535:341–356.

Bateson, W. (1886). The ancestry of the Chordata. *Quarterly Journal of Microscopical Science* 26:535–571.

Bather, F. A. (1920). *Fossils and Life. Report of the British Association for the Advancement of Science 1920 (88th Meeting)*:61–86.

Baum, D. A., Smith, S. D. (2013). *Tree Thinking. An Introduction to Phylogenetic Biology.* Greenwood Village, CO: Roberts and Company Publishers.

Beard, J. (1889). Some annelidan affinities in the ontogeny of the vertebrate nervous system. *Nature* 39:259–261.

Beklemishev, W. N. (1969a). *Principles of Comparative Anatomy of Invertebrates. Volume 1. Promorphology.* Chicago: The University of Chicago Press.

(1969b). *Principles of Comparative Anatomy of Invertebrates. Volume 2. Organology.* Chicago: The University of Chicago Press.

Bengtson, S., Cunningham, J. A., Yin, C. Y., Donoghue, P. C. J. (2012). A merciful death for the "earliest bilaterian," *Vernanimalcula. Evolution & Development* 14:421–427.

Benson, K. R. (1979). *William Keith Brooks (1848–1908): A Case Study in Morphology and the Development of American Biology. Unpublished PhD thesis.* Oregon State University.

Bentham, G. (1862). Anniversary meeting. *Proceedings of the Linnean Society of London* 74: lxvi–lxxxiii.

(1863). Anniversary meeting. *Proceedings of the Linnean Society of London* 75:xi–xxix.

(1873). Notes on the classification, history, and geographical distribution of Compositae. *Journal of the Linnean Society of London, Botany* 13:335–577.

Bergström, A., Stringer, C., Hajdinjak, M., Scerri, E. M., Skoglund, P. (2021). Origins of modern human ancestry. *Nature* 590:229–237.

Bergström, J., Hou, X.-G. (2003). Arthropod origins. *Bulletin of Geosciences* 78:323–334.

(2005). Early Palaeozoic non-lamellipedian arthropods. Pages 73–94 in S. Koenemann, R. A. Jenner, eds. *Crustacea and Arthropod Relationships. Festschrift for Frederick R. Schram*. Boca Raton, FL: CRC Press.

Bergström, J., Naumann, W. W., Viehweg, J., Martí-Mus, M. (1998). Conodonts, calcichordates and the origin of vertebrates. *Mitteilungen aus dem Museum für Naturkunde in Berlin, Geowissenschaftlich Reihe* 1:81–92.

Bernard, H. M. (1892). *The Apodidae. A Morphological Study*. London: Macmillan.

(1898). A new reading for the annulate ancestry of the Vertebrata. *Natural Science* 13:17–30.

Bidder, G. P. (1941). Obituaries. Mr. W. H. Caldwell. *Nature* 148:557–559.

Bitsch, C., Bitsch, J. (2004). Phylogenetic relationships of basal hexapods among the mandibulate arthropods: a cladistic analysis based on comparative morphological characters. *Zoologica Scripta* 33:511–550.

Blais, C., Archibald, J. M. (2021). The past, present and future of the tree of life. *Current Biology* 31: R314–R321.

Bleidorn, C. (2019). Recent progress in reconstructing lophotrochozoan (spiralian) phylogeny. *Organisms Diversity & Evolution* 19:557–566.

Bock, W. (1959). Preadaptation and multiple evolutionary pathways. *Evolution* 13:194–211.

(2007). Explanations in evolutionary theory. *Journal of Zoological Systematics and Evolutionary Research* 45:89–103.

Bode, H. R., Steele, R. E. (1989). Phylogeny and molecular data. *Science* 243:549–550.

Bolker, J. A. (1995). Model systems in developmental biology. *BioEssays* 17:451–455.

Bonik, K., Gutmann, W., Peters, D. (1977). Optimierung und Ökonomisierung im Kontext von Evolutionstheorie und phylogenetischer Rekonstruktion. *Acta Biotheoretica* 26:75–119.

Boulenger, G. A. (1907). A revision of the African silurid fishes of the subfamily *Clariinae. Proceedings of the Zoological Society of London* 2:1062–1097.

Bowler, P. J. (1989). *Evolution. The History of an Idea*. Berkeley: University of California Press.

(1992a). *The Eclipse of Darwinism. Anti-Darwinian Evolution Theories in the Decades around 1900*. Baltimore: The Johns Hopkins University Press.

(1992b). *The Fontana History of the Environmental Sciences*. London: Fontana Press.

(1994). Are the Arthropoda a natural group? An episode in the history of evolutionary biology. *Journal of the History of Biology* 27:177–213.

(1996). *Life's Splendid Drama. Evolutionary Biology and the Reconstruction of Life's Ancestry, 1860–1940*. Chicago: The University of Chicago Press.

(2000). Philosophy, instinct, intuition: what motivates the scientist in search of a theory? *Biology & Philosophy* 15:93–101.

(2013). *Darwin Deleted. Imagining a World without Darwin*. Chicago: University of Chicago Press.

(2021). *Progress Unchained. Ideas of Evolution, Human History and the Future*. Cambridge: Cambridge University Press.

Bowler, P. J., Morus, I. R. (2005). *Making Modern Science. A Historical Survey*. Chicago: The University of Chicago Press.

Bracken-Grissom, H. D., Cannon, M. E., Cabezas, P., et al. (2013). A comprehensive and integrative reconstruction of evolutionary history for Anomura (Crustacea: Decapoda). *BMC Evolutionary Biology* 13:1–1.

Brady, R. H. (1984). The causal dimension of Goethe's morphology. *Journal of Social and Biological Structures* 7:325–344.

(1985). On the independence of systematics. *Cladistics* 1:113–126.

(1994). Pattern description, process explanation, and the history of morphological sciences. Pages 7-31 in L. Grande, O. Rieppel, eds. *Interpreting the Hierarchy of Nature. From Systematic Patterns to Evolutionary Process Theories*. San Diego: Academic Press.

Breidbach, O. (2003). Post-Haeckelian comparative biology – Adolf Naef's idealistic morphology. *Theory in Biosciences* 122:174-193.

(2006). *Goethes Metamorphosenlehre*. Munich: Wilhelm Fink Verlag.

Breidbach, O., Ghiselin, M. T. (2002). Lorenz Oken and *Naturphilosophie* in Jena, Paris and London. *History and Philosophy of the Life Sciences* 24:219-247.

(2006). Baroque classification: a missing chapter in the history of systematics. *Annals of the History and Philosophy of Biology* 11:1-30.

Brinkman, P. D. (2010). Charles Darwin's *Beagle* voyage, fossil vertebrate succession, and "The Gradual Birth & Death of Species." *Journal of the History of Biology* 43:363-399.

Bromham, L. (2019). Six impossible things before breakfast: assumptions, models, and belief in molecular dating. *Trends in Ecology & Evolution* 34:962-962.

Bromham, L., Hua, X., Lanfear, R., Cowman, P. F. (2015). Exploring the relationships between mutation rates, life history, genome size, environment, and species richness in flowering plants. *American Naturalist* 185:507-524.

Bromham, L., Woolfit, M., Lee, M. S. Y., Rambaut, A. (2002). Testing the relationships between morphological and molecular rates of change along phylogenies. *Evolution* 56:1921-1930.

Bronzati, M. (2017). Should the terms "basal taxon" and "transitional taxon" be extinguished from cladistic studies with extinct organisms? *Palaentologia Electronica* 20.2.3E:1-12.

Brooks, W. K. (1882a). *Lucifer: a study in morphology*. London: W. Bowyer.

(1882b). *Speculative zoölogy*. *Popular Science Monthly* 22:195-204.

(1883). *Speculative zoölogy*. *Popular Science Monthly* 22:364-380.

(1893a). *The Genus Salpa*. Baltimore: The Johns Hopkins Press.

(1893b). Salpa and its relation to the evolution of life. Johns Hopkins University. *Studies from the Biological Laboratory* 5:129-211.

(1908). Biographical memoir of Alpheus Hyatt. 1839-1902. *Biographical Memoirs of the National Academy of Sciences* 6:311-325.

Brower, A. V. Z. (2014). Communing with the ancestors. *Cladistics* 30:107-111.

(2015). Paraphylophily. *Cladistics* 31:575-578.

(2016). Are we all cladists? Pages 88-114 in D. M. Williams, M. Schmitt, Q. D. Wheeler, eds. *The Future of Phylogenetic Systematics. The Legacy of Willin Hennig*. Cambridge: Cambridge University Press.

Browne, J. (1995). *Charles Darwin. Voyaging. Volume I of a Biography*. London: Pimlico.

Bruce, H. S., Patel, N. H. (2020). Knockout of crustacean leg patterning genes suggests that insect wings and body walls evolved from ancient leg segments. *Nature Ecology & Evolution* 4:1703-1712.

Bruce, R. (2016). Hennig, Løvtrup, evolution and biology. Pages 344-355 in D. Williams, M. Schmitt, Q. Wheeler, eds. *The Future of Phylogenetic Systematics: The Legacy of Willi Hennig*. Cambridge: Cambridge University Press.

Brunet, T., King, N. (2017). The origin of animal multicellularity and cell differentiation. *Developmental Cell* 43:124-140.

Brunet, T., Lauri, A., Arendt, D. (2015). Did the notochord evolve from an ancient axial muscle? The axochord hypothesis. *BioEssays* 37:836-850.

Brusca, R. C., Brusca, G. J. (1990). *Invertebrates*. Sunderland, MA: Sinauer Associates.

(2003). *Invertebrates, 2nd ed*. Sunderland, MA: Sinauer Associates.

Bryant, H. N. (1995). The threefold parallelism of Agassiz and Haeckel, and polarity determination in phylogenetic systematics. *Biology & Philosophy* 10:197–217.

Budd, G. E., Jensen, S. (2015). The origin of the animals and a "Savannah" hypothesis for early bilaterian evolution. *Biological Reviews of the Cambridge Philosophical Society* 92:446–473.

Bullock, T. H., Horridge, G. A. (1965). *Structure and Function in the Nervous Systems of Invertebrates.* San Francisco: W. H. Freeman and Company.

Burkhardt, R. W. J. (1995). *The Spirit of System. Lamarck and Evolutionary Biology.* Cambridge, MA: Harvard University Press.

Butler, S. (1922). *Luck, or Cunning?* London: Jonathan Cape.

Bütschli, O. (1876). Über die Bedeutung der Entwickelungsgeschichte für die Stammesgeschichte der Thiere. *Bericht über die Senckenbergische naturforschende Gesellschaft*:61–74.

   (1884). *Bemerkungen zur Gastraeatheorie. Morphologisches Jahrbuch* 9:415–427.

   (1921). *Vorlesungen über vergleichende Anatomie.* Berlin: Verlag von Julius Springer.

Caianiello, S. (2015). Succession of functions, from Darwin to Dohrn. *History and Philosophy of the Life Sciences* 36:335–345.

Calcott, B. (2009). Lineage explanations: explaining how biological mechanisms change. *The British Journal for the Philosophy of Science* 60:51–78.

Caldwell, W. H. (1885). Blastopore, mesoderm and metameric segmentation. *Quarterly Journal of Microscopical Science* 25:15–28.

Calman, W. T. (1904). On the classification of the Crustacea Malacostraca. *Annals and Magazine of Natural History* 7:144–158.

   (1909). *A Treatise on Zoology. Part VII. Appendiculata, Third Fascicle, Crustacea.* London: Adam and Charles Black.

   (1936). The origin of insects. *Proceedings of the Linnean Society of London* 148:193–204.

Camardi, G. (2001). Richard Owen, morphology and evolution. *Journal of the History of Biology* 34:481–515.

Cameron, C. B., Garey, J. R., Swalla, B. J. (2000). Evolution of the chordate body plan: new insights from phylogenetic analyses of deuterostome phyla. *Proceedings of the National Academy of Sciences USA* 97:4469–4474.

Cammarata, M., Pagliara, P. (2018). Elie Metchnikoff and the multidisciplinary link novelty among zoology, embryology and innate immunity. *Invertebrate Survival Journal* 15:2340–2239.

Caporael, L. (1994). Of myth and science: origin stories and evolutionary scenarios. *Social Science Information* 33:9–23.

Carignan, M. (2003). Analogical reasoning in Victorian historical epistemology. *Journal of the History of Ideas* 64:445–464.

Caron, J.-B., Conway Morris, S., Cameron, C. B. (2013). Tubicolous enteropneusts from the Cambrian period. *Nature* 495:503–506.

Carr, T. D., Varricchio, D. J., Sedlmayr, J. C., Roberts, E. M., Moore, J. R. (2017). A new tyrannosaur with evidence for anagenesis and crocodile-like facial sensory system. *Scientific Reports* 7:44942.

Carroll, S. B. (2005). *Endless Forms Most Beautiful. The New Science of Evo Devo and the Making of the Animal Kingdom.* New York: W. W. Norton and Company.

Carruthers, T., Scotland, R. W. (2021). Uncertainty in divergence time estimation. *Systematic Biology* 70:855–861.

Casane, D., Laurenti, P. (2013). Why coelacanths are not "living fossils." *BioEssays* 35:332–338.

Caspari, R., Wolpoff, M. H. (2012). The Dubois Syndrome. *History and Philosophy of the Life Sciences* 34:33–42.

Cassini, II. (1826). *Opuscules Phytologiques.* Paris: Levrault.

Catley, K. M., Novick, L. R. (2008). Seeing the wood for the trees: an analysis of evolutionary diagrams in biology textbooks. *Bioscience* 58:976–987.

Catley, K. M., Novick, L. R., Shade, C. K. (2010). Interpreting evolutionary diagrams: when topology and process conflict. *Journal of Research in Science Teaching* 47:861–882.

Cavaillon, J.-M., Legout, S. (2016). Centenary of the death of Elie Metchnikoff: a visionary and an outstanding team leader. *Microbes and Infection* 18:577–594.

Cavalier-Smith, T. (1998). A revised six-kingdom system of life. *Biological Reviews of the Cambridge Philosophical Society* 73:203–266.

(2017). Origin of animal multicellularity: precursors, causes, consequences-the choanoflagellate/ sponge transition, neurogenesis and the Cambrian explosion. *Philosophical Transactions of the Royal Society B: Biological Sciences* 372:20150476.

Chakrabarty, P. (2010). The transitioning state of systematic ichthyology. *Copeia* 2010:513–515.

Chen, J.-Y., Schopf, J. W., Bottjer, D. J., et al. (2007). Raman spectra of a Lower Cambrian ctenophore embryo from southwestern Shaanxi, China. *Proceedings of the National Academy of Sciences USA* 104:6289–6292.

Chen, J. Y., Bottjer, D. J., Oliveri, P., et al. (2004). Small bilaterian fossils from 40 to 55 million years before the Cambrian. *Science* 305:218–222.

Chen, X., Li, Q., Wang, J., et al. (2009). Identification and characterization of novel amphioxus microRNAs by Solexa sequencing. *Genome Biology* 10:R78.

Chipman, A. D., Edgecombe, G. D. (2019). Developing an integrated understanding of the evolution of arthropod segmentation using fossils and evo-devo. *Proceedings of the Royal Society B-Biological Sciences* 286:20191881.

Christen, R., Ratto, A., Baroin, A., et al. (1991a). An analysis of the origin of metazoans, using comparisons of partial sequences of the 28S RNA, reveals an early emergence of triploblasts. *The EMBO Journal* 10:499–503.

(1991b). Origin of metazoans. A phylogeny deduced from sequences of the 28S ribosomal RNA. Pages 1–9 in A. M. Simonetta, S. Conway Morris, eds. *The Early Evolution of Metazoa and the Significance of Problematic Taxa.* Cambridge: Cambridge University Press.

Christoffersen, M. L., Araújo-de-Almeida, E. (1994). A phylogenetic framework of the Enterocoela (Metameria: Coelomata). *Revista Nordestina De Biologia* 9:173–208.

Churchill, F. B. (1989). The guts of the matter. Infusoria from Ehrenberg to Bütschli: 1838–1876. *Journal of the History of Biology* 22:189–213.

Clark, R. B. (1964). *Dynamics in Metazoan Evolution.* Oxford: Clarendon Press.

Cloud Jr., P. E. (1968). Pre-metazoan evolution and the origins of the Metazoa. Pages 1–72 in E. T. Drake, ed. *Evolution and Environment.* New Haven, CT: Yale University Press.

Cohen, C. (2017). "How nationality influences Opinion": Darwinism and palaeontology in France (1859–1914). *Studies in History and Philosophy of Biological & Biomedical Science* 66:8–17.

Coleman, W. (1973). Limits of the recapitulation theory: Carl Friedrich Kielmeyer's critique of the presumed parallelism of earth history, ontogeny, and the present order of organisms. *Isis* 62:341–350.

(1976). Morphology between type concept and descent theory. *Journal of the History of Medicine and Allied Sciences* 31:149–175.

Conway Morris, S. (2000). Evolution: bringing molecules into the fold. *Cell* 100:1–11.

(2003). *Life's Solution. Inevitable Humans in a Lonely Universe.* Cambridge: Cambridge University Press.

(2005). Aliens like us? *Astronomy & Geophysics* 46:4.24–4.26.

(2009). The predictability of evolution: glimpses into a post-Darwinian world. *Naturwissenschaften* 96:1313–1337.

Cope, E. D. (1896). *The Primary Factors of Organic Evolution*. Chicago: The Open Court Publishing Company.

Coppard, S. E., Kroh, A., Smith, A. B. (2012). The evolution of pedicellariae in echinoids: an arms race against pests and parasites. *Acta Zoologica* 93:125–148.

Cracraft, J. (1981). The use of functional and adaptive criteria in phylogenetic systematics. *American Zoologist* 21:21–36.

(2005). Phylogeny and evo-devo: characters, homology, and the historical analysis of the evolution of development. *Zoology* 108:345–356.

Crampton, G. (1916). The phylogenetic origin and the nature of the wings of insects according to the paranotal theory. *Journal of the New York Entomological Society* 24:1–39.

Crampton, G. C. (1921). The phylogenetic origin of the mandibles of insects and their arthropodan relatives – a contribution to the study of the evolution of the Arthropoda. *Journal of the New York Entomological Society* 29:63–100.

(1938). The interrelationships and lines of descent of living insects. *Psyche* 45:165–181.

Crisp, M. D., Cook, L. G. (2005). Do early branching lineages signify ancestral traits? *Trends in Ecology & Evolution* 20:122–128.

Cronquist, A. (1956). Significance of orthogenesis in angiosperm taxonomy. *The Southwestern Naturalist* 1:97–99.

Crow, W. B. (1926). Phylogeny and the natural system. *Journal of Genetics* 17:88–155.

Crowson, R. A. (1982). Computers versus imagination in the reconstruction of phylogeny. Pages 245–255 in K. A. Joysey, A. E. Friday, eds. *Problems of Phylogenetic Reconstruction*. London: Academic Press.

Cunningham, C. W., Omland, K. E., Oakley, T. H. (1998). Reconstructing ancestral character states: a critical reappraisal. *Trends in Ecology & Evolution* 13:361–366.

Currie, A., Sterelny, K. (2017). In defence of story-telling. *Studies in History and Philosophy of Science* 62:14–21.

Cusimano, N., Renner, S. S. (2014). Ultrametric trees or phylograms for ancestral state reconstruction: does it matter? *Taxon* 63:721–726.

Cutler, E. B., Gibbs, P. E. (1985). A phylogenetic analysis of higher taxa in the phylum Sipuncula. *Systematic Zoology* 34:162–173.

D'Hombres, E. (2012). The "division of physiological labour": the birth, life and death of a concept. *Journal of the History of Biology* 45:3–31.

(2016). The Darwinian muddle on the division of labour: an attempt at clarification. *History and Philosophy of the Life Sciences* 38:1–22.

Dahl, E. (1983). Alternatives in malacostracan evolution. *Memoirs of the Australian Museum* 18:1–5.

Daly, M., Brugler, M. R., Cartwright, P., et al. (2007). The phylum Cnidaria: a review of phylogenetic patterns and diversity 300 years after Linnaeus. *Zootaxa* 1668:127–182.

Dannemann, M., Prüfer, K., Kelso, J. (2017). Functional implications of Neandertal introgression in modern humans. *Genome Biology* 18:61.

Darwin, C. (1859). *On the Origin of Species by Means of Natural Selection, or the Preservation of Favored Races in the Struggle for Pife (Penguin Facsimile of First Edition)*. London: Penguin Books.

(1862). *On the Various Contrivances by which British and Foreign Orchids are Fertilised by Insects, and on the Good Effects of Intercrossing*. London: John Murray.

(1876). *On the Origin of Species by Means of Natural Selection, or the Preservation of Favored Races in the Struggle for Life. Sixth Edition, with Additions and Corrections to 1872*. London: John Murray.

(1901). *The Descent of Man and Selection in Relation to Sex*. London: John Murray.

Dawkins, R. (2004). *The Ancestor's Tale. A Pilgrimage to the Dawn of Evolution*. Boston: Houghton Mifflin Company.

Dayrat, B. (2003). The roots of phylogeny: how did Haeckel build his trees? *Systematic Biology* 52:515–527.

de Beer, G. R. (1940). *Embryos and Ancestors*. Oxford: Clarendon Press.

(1954). The evolution of Metazoa. Pages 24–33 in J. Huxley, A. C. Hardy, E. B. Ford, eds. *Evolution as a Process*. London: George Allen & Unwin Ltd.

(1958). *Embryos and Ancestors*. Oxford: Clarendon Press.

de Pinna, M. C. C. (1991). Concepts and tests of homology in the cladistic paradigm. *Cladistics* 7:367–394.

De Queiroz, K. (1999). The general lineage concept of species and the defining properties of the species category. Pages 49–89 in R. A. Wilson, ed. *Species. New Interdisciplinary Essays*. Cambridge, MA: MIT Press.

(2005). Different species problems and their resolution. *BioEssays* 27:1263–1269.

De Robertis, E. M., Sasai, Y. (1996). A common plan for dorsoventral patterning in Bilateria. *Nature* 380:37–40.

Dean, B. (1912). Orthogenesis in the egg capsules of *Chimaera*. *Bulletin of the American Museum of Natural History* 31:35–40.

Debernardi, M., Serrelli, E. (2013). From bacteria to Saint Francis to Gaia in the symbiotic view of evolution. *Evolution: Education and Outreach* 6:4.

Delsman, H. C. (1922). *The Ancestry of Vertebrates. As a Means of Understanding the Principal Features of their Structure and Development*. Amersfoort: Valkhoff & Co.

Desmond, A. (1982). *Archetypes and Ancestors. Palaeontology in Victorian London 1850–1875*. Chicago: The University of Chicago Press.

(1985). The making of institutional zoology in London 1822–1836: part I. *History of Science* 23:153–185.

(1989). *The Politics of Evolution. Morphology, Medicine, and Reform in Radical London*. Chicago: The University of Chicago Press.

(1994). *Huxley: The Devil's Disciple*. London: Michael Joseph.

(1997). *Huxley: Evolution's High Priest*. London: Michael Joseph.

Desmond, A., Moore, J. (2009). *Darwin*. London: Penguin Books.

Dewel, R. A. (2000). Colonial origin for Eumetazoa: major morphological transitions and the origin of bilaterian complexity. *Journal of Morphology* 243:35–74.

Dewel, R. A., Dewel, W. C., McKinney, F. K. (2001). Diversification of the Metazoa: Ediacarans, colonies, and the origin of eumetazoan complexity by nested modularity. *Historical Biology* 15:193–218.

Di Gregorio, M. A. (2005). *From Here to Eternity. Ernst Haeckel and Scientific Faith*. Göttingen: Vandenhoeck & Ruprecht.

DiFrisco, J., Jaeger, J. (2019). Beyond networks: mechanism and process in evo-devo. *Biology & Philosophy* 34:54.

(2021). Homology of process: developmental dynamics in comparative biology. *Interface Focus* 11:20210007.

Diogo, R., Ziermann, J. M., Linde-Medina, M. (2015). Is evolutionary biology becoming too politically correct? A reflection on the scala naturae, phylogenetically basal clades, anatomically plesiomorphic taxa, and "lower" animals. *Biological Reviews of the Cambridge Philosophical Society* 90:502–521.

Dobzhansky, T. (1965). Mendelism, Darwinism, and evolutionism. *Proceedings of the American Philosophical Society* 109:205–215.

Dohrn, A. (1875). *Der Ursprung der Wirbelthiere und das Princip des Functionswechsels. Genealogische Skizzen.* Leipzig: Wilhelm Engelmann.

Dominici, S., Eldredge, N. (2010). Brocchi, Darwin, and transmutation: phylogenetics and paleontology at the dawn of evolutionary biology. *Evolution: Education and Outreach* 3:576–584.

Donoghue, P. C. J. (2005). Saving the stem group – a contradiction in terms? *Paleobiology* 31:553–558.

Doolittle, W. F., Bapteste, E. (2007). Pattern pluralism and the Tree of Life hypothesis. *Proceedings of the National Academy of Sciences USA* 104:2043–2049.

Duerden, J. E. (1919). Crossing the North African and South African ostrich. *Journal of Genetics* 8:155–198.

Dunn, C. W., Hejnol, A., Matus, D. Q., et al. (2008). Broad phylogenomic sampling improves the resolution of the animal tree of life. *Nature* 452:745–749.

Dunn, C. W., Leys, S. P., Haddock, S. H. D. (2015). The hidden biology of sponges and ctenophores. *Trends in Ecology & Evolution* 30:282–291.

Dupré, J. (2017). *The metaphysics of evolution. Interface Focus* 7:20160148.

Dupré, J., Nicholson, D. J. (2018). Towards a processual philosophy of biology in D. J. Nicholson, J. Dupré, eds. *Everything Flows. Towards a Processual Philosophy of Biology.* Oxford: Oxford University Press.

Ebach, M. C. (2005). *Anschauung* and the archetype. The role of Goethe's delicate empiricism in comparative biology. *Janus Head* 8:254–270.

Eckermann, J. P. (1908). *Goethes Gespräche mit J. P. Eckermann.* Leipzig: Insel-Verlag.

Edgecombe, G. D. (2020). Arthropod origins: integrating paleontological and molecular evidence. *Annual Review of Ecology, Evolution, and Systematics* 51:1–25.

Edgecombe, G. D., Giribet, G., Dunn, C. W., et al. (2011). Higher-level metazoan relationships: recent progress and remaining questions. *Organisms, Diversity & Evolution* 11:151–172.

Eernisse, D. J., Albert, J. S., Anderson, F. E. (1992). Annelida and arthropoda are not sister taxa: a phylogenetic analysis of spiralian metazoan morphology. *Systematic Biology* 41:305–330.

Eigen, E. A. (1997). Overcoming first impressions: Georges Cuvier's types. *Journal of the History of Biology* 30:179–209.

Eigenmann, C. H. (1917–1929). The American Characidae. memoirs of the Museum of Comparative Zoology 42:1–558.

Eisig, H. (1878). Der Nebendarm der Capitelliden und seine Homologa. *Zoologischer Anzeiger* 1:148–152.

Eldredge, N. (2010). How systematics became "phylogenetic." *Evolution: Education and Outreach* 3:491–494.

Eldredge, N., Cracraft, J. (1980). *Phylogenetic Patterns and the Evolutionary Process. Method and Theory in Comparative Biology.* New York: Columbia University Press.

Eldredge, N., Novacek, M. J. (1985). Systematics and paleobiology. *Paleobiology* 11:65–74.

Engel, M. S., Kristensen, N. P. (2013). A history of entomological classification. *Annual Review of Entomology* 58:585–607.

Engelbrecht, D. v. Z. (1967). The annelid ancestry of the chordates and the origin of the chordate central nervous system and the notochord. *Journal of Zoological Systematics and Evolutionary Research* 7:18–30.

(1971). The phylogenetic origin of the lateral eyes and the optic chiasma of vertebrates. *Journal of Zoological Systematics and Evolutionary Research* 9:30–48.

Engelmann, G. F., Wiley, E. O. (1977). The place of ancestor-descendant relationships in phylogeny reconstruction. *Systematic Zoology* 26:1–11.

Ereshefsky, M., Turner, D. (2020). Historicity and explanation. *Studies in the History and Philosophy of Science* 80:47–55.

Ereskovsky, A. V., Dondua, A. K. (2006). The problem of germ layers in sponges (Porifera) and some issues concerning early metazoan evolution. *Zoologischer Anzeiger* 245:65–76.

Erives, A., Fritzsch, B. (2020). A screen for gene paralogies delineating evolutionary branching order of early Metazoa. G3: Genes, Genomes, *Genetics* 10:811.

Erwin, D. H., Valentine, J. W. (2013). *The Cambrian Explosion. The Construction of Animal Biodiversity.* Greenwood Village, CO: Roberts and Company.

Extavour, C. G., Akam, M. (2003). Mechanisms of germ cell specification across the metazoans: epigenesis and preformation. *Development* 130:5869–5884.

Farber, P. L. (1976). The type-concept in zoology during the first half of the nineteenth century. *Journal of the History of Biology* 9:93–119.

Farley, J. (1974). The initial reactions of French biologists to Darwin's *Origin of Species*. *Journal of the History of Biology* 7:275–300.

Fauchald, K. (1974). Polychaete phylogeny: a problem in protostome evolution. *Systematic Zoology* 23:493–506.

Felsenstein, J. (2004). *Inferring Phylogenies.* Sunderland, MA: Sinauer Associates.

Fernández, R., Laumer, C. E., Vahtera, V., et al. (2014). Evaluating topological conflict in centipede phylogeny using transcriptomic data sets. *Molecular Biology and Evolution* 31:1500–1513.

Ferrari, F. D. (2010). Morphology, development, and sequence. *Journal of Crustacean Biology* 30:767–769.

Ferrari, F. D., Fornshell, J., Vagelli, A. A., Ivanenko, V. N., Dahms, H.-U. (2011). Early post-embryonic development of marine chelicerates and crustaceans with a nauplius. *Crustaceana* 84:869–893.

Ferrier, D. E. K. (2012). Evolutionary crossroads in developmental biology: annelids. *Development* 139:2643–2653.

Field, K. G., Olsen, G., Lane, D. J., et al. (1988). Molecular phylogeny of the animal kingdom. *Science* 239:748–753.

Field, K. G., Olsen, G. J., Giovannoni, S. J., et al. (1989). Phylogeny and molecular data. *Science* 243:550–551.

Figueirido, B., Lautenschlager, S., Pérez-Ramos, A., Van Valkenburgh, B. (2018). Distinct predatory behaviors in scimitar- and dirk-toothed sabertooth cats. *Current Biology* 28:3260–3266.

Fisher, D. C. (2008). Stratocladistics: intergrating temporal data and character data in phylogenetic inference. *Annual Review of Ecology, Evolution and Systematics* 39:365–385.

Fisler, M., Lecointre, G. (2013). Categorizing ideas about trees: a tree of trees. *PLoS ONE* 8:e68814.

Fitch, W. M., Beintema, J. J. (1990). Correcting parsimonious trees for unseen nucleotide substitutions: the effect of dense branching as exemplified by ribonuclease. *Molecular Biology and Evolution* 7:438–443.

Fitzhugh, K. (2006). The abduction of phylogenetic hypotheses. *Zootaxa* 1145:1–110.

Fokin, S. I. (2013). Otto Bütschli (1848–1920): where we will genuflect? *Protistology* 8:22–35.

Foote, M. (1996). On the probability of ancestors in the fossil record. *Paleobiology* 22:141–151.

Forey, P. L. (1982). Neontological analysis versus palaeontological stories. Pages 119–157 in K. A. Joysey, A. E. Friday, eds. *Problems of Phylogenetic Reconstruction.* London: Academic Press. (2004). *Systematics and paleontology.* Pages 149–180. *Milestones in Systematics.* Boca Raton, FL: CRC Press.

Fortey, R. A., Thomas, R. H. (1993). The case of the velvet worm. *Nature* 361:205–206.

Foster, M., Sedgwick, A., eds. (1885). *The Works of Francis Maitland Balfour. Volume I.* London: Macmillan.

Franzen, J., Gingerich, P., Habersetzer, J., et al. (2009). Complete primate skeleton from the Middle Eocene of Messel in Germany: morphology and paleobiology. *PLoS ONE* 4:e5723.

Froehlich, J. M., Seiliez, I., Gabillard, J. C., Biga, P. R. (2014). Preparation of primary myogenic precursor cell/myoblast cultures from basal vertebrate lineages. *Journal of Visualized Experiments* 86:51354.

Frost, D. R., Grant, T., Faivovich, J., et al. (2006). The amphibian tree of life. *Bulletin of the American Museum of Natural History* 297:8–370.

Fry, B., Undheim, E., Ali, S., et al. (2013). Squeezers and leaf-cutters: differential diversification and degeneration of the venom system in toxicoferan reptiles. *Molecular & Cellular Proteomics* 12:1881–1899.

Fryer, G. (1992). The origin of the Crustacea. *Acta Zoologica* 73:273–286.

(1996). Reflections on arthropod evolution. *Biological Journal of the Linnean Society* 58:1–55.

Fu, D.-J., Tong, G., Dai, T., et al. (2019). The Qingjiang biota – A Burgess Shale–type fossil Lagerstätte from the early Cambrian of South China. *Science* 363:1338–1342.

Fusco, G., Minelli, A. (2013). Arthropod segmentation and tagmosis. Pages 197–221 in A. Minelli, G. Boxshall, G. Fusco, eds. *Arthropod Biology and Evolution.* Berlin: Springer-Verlag.

Galera, A. (2021). Etienne Geoffroy Saint-Hilaire and the first embryological evolutionary model on the origin of vertebrates. *Journal of the History of Biology* 54:229–245.

Garcia-Fernàndez, J., Benito-Gutiérrez, E. (2009). It's a long way from amphioxus: descendants of the earliest chordate. *BioEssays* 31:665–675.

Garey, J. R. (2002). The lesser-known protostome taxa: an introduction and a tribute to Robert P. Higgins. *Integrative and Comparative Biology* 42:611–618.

Garstang, W. (1928). The morphology of the Tunicata, and its bearing on the phylogeny of the Chordata. *Quarterly Journal of Microscopical Science* 72:51–187.

Gąsiorowski, L., C., A., Janssen, R., et al. (2021). Molecular evidence for a single origin of ultrafiltration-based excretory organs. *Current Biology* 31:1–10.

Gaskell, W. H. (1906). On the origin of vertebrates, deduced from the study of ammocoetes. *Journal of Anatomy and Physiology* 40:305–317.

(1908). *The Origin of Vertebrates.* London: Longmans, Green, and Co.

(1910). Origin of the vertebrates. *Proceedings of the Linnean Society of London* 122:46–50.

Gauthier, J. A., Kearney, M., Maisano, J. A., Rieppel, O., Behlke, A. D. B. (2012). Assembling the squamate tree of life: perspectives from the phenotype and the fossil record. *Bulletin of the Peabody Museum of Natural History* 53:3–308.

Gee, H. (2000). *Deep Time. Cladistics, the Revolution in Evolution.* London: Fourth Estate.

(2013). *The Accidental Species. Misunderstandings of Human Evolution.* Chicago: The University of Chicago Press.

(2018). *Across the Bridge. Understanding the Origin of the Vertebrates.* Chicago: University of Chicago Press.

Gegenbaur, C. (1859). *Grundzüge der vergleichenden Anatomie.* Leipzig: Verlag von Wilhelm Engelmann.

(1870). *Grundzüge der vergleichenden Anatomie. Zweite umgearbeitete Auflage.* Leipzig: Verlag von Wilhelm Engelmann.

(1876). Die Stellung und Bedeutung der Morphologie. *Morphologisches Jahrbuch* 1:1–19.

(1878). *Elements of Comparative Anatomy.* London: Macmillan.

Geoffroy Saint-Hilaire, E. (1830). *Principes de philosophie zoologique.* Paris: Pichon et Didier.

Ghiselin, M. T. (1969). *The Triumph of the Darwinian Method.* Chicago: The University of Chicago Press.

(1972). Models in phylogeny. Pages 130–145 in T. J. M. Schopf, ed. *Models in Paleobiology.* San Francisco: Freeman, Cooper & Company.

(1976). Two Darwins: history versus criticism. *Journal of the History of Biology* 9:121–132.

(1988). The origin of molluscs in the light of molecular evidence. *Oxford Surveys in Evolutionary Biology* 5:66–95.

(1991). Classical and molecular phylogenetics. *Bolletino di Zoologia* 58:289–294.

(1994). The origin of vertebrates and the principle of succession of functions. Genealogical sketches by Anton Dohrn, 1875. An English translation from the German, introduction and bibliography. *History and Philosophy of the Life Sciences* 16:3–96.

(1995). Darwin, progress, and economic principles. *Evolution* 49:1029–1037.

(1996). Charles Darwin, Fritz Müller, Anton Dohrn, and the origin of evolutionary physiological anatomy. *Memorie della Società Italiana di Scienze Naturali e del Museo Civico di Storia Naturale di Milano* 22:49–58.

(1997). *Metaphysics and the Origin of Species.* Albany: State University of New York Press.

(2000a). The assimilation of Darwinism in systematic biology. Pages 265–281 in A. Minelli, S. Casellato, eds. *Giovanni Canestrini: Zoologist and Darwinist.* Venezia: Inst. Veneto Sci. Lett. ed Arti.

(2000b). The founders of morphology as alchemists. Pages 39–49 in M. T. Ghiselin, A. E. Leviton, eds. *Cultures and Institutions of Natural History: Essays in the History and Philosophy of Science.* San Francisco: California Academy of Sciences.

(2003). Carl Gegenbaur versus Anton Dohrn. *Theory in Biosciences* 122:142–147.

(2005a). The Darwinian revolution as viewed by a philosophical biologist. *Journal of the History of Biology* 38:123–136.

(2005b). Homology as a relation of correspondence between parts of individuals. *Theory in Biosciences* 24:91–103.

(2016). Homology, convergence and parallelism. *Philosophical Transactions of the Royal Society B: Biological Sciences* 371:20150035.

Ghiselin, M. T., Groeben, C. (1997). Elias Metschnikoff, Anton Dohrn, and the metazoan common ancestor. *Journal of the History of Biology* 30:211–228.

Gilbert, S. F. (1991). *Developmental Biology.* Sunderland, MA: Sinauer Associates.

Gingerich, P. D. (1979). The stratophenetic approach to phylogeny reconstruction in vertebrate paleontology. Pages 41–77 in J. Cracraft, N. Eldredge, eds. *Phylogenetic Analyses and Paleontology.* New York: Columbia University Press.

Giribet, G. (2008). Assembling the lophotrochozoan (=spiralian) tree of life. *Philosophical Transactions of the Royal Society B: Biological Sciences* 363:1513–1522.

(2009). On velvet worms and caterpillars: science, fiction, or science fiction? *Proceedings of the National Academy of Sciences USA* 106:E131.

Giribet, G., Distel, D. L., Polz, M., Sterrer, W., Wheeler, W. C. (2000). Triploblastic relationships with emphasis on the acoelomates and the position of Gnathostomulida, Cycliophora, Plathelminthes, and Chaetognatha: a combined approach of 18S rDNA sequences and morphology. *Systematic Biology* 49:539–562.

Giribet, G., Edgecombe, G. D. (2020). *The Invertebrate Tree of Life.* Princeton: Princeton University Press.

Giribet, G., Okusu, A., Lindgren, A. R., et al. (2006). Evidence for a clade composed of molluscs with serially repeated structures: monoplacophorans are related to chitons. *Proceedings of the National Academy of Sciences USA* 103:7723–7728.

Gishlick, A. D. (2003). Icons of evolution? Why much of what Jonathan Wells writes about evolution is wrong. *National Centre for Science Education*:1–64.

Glaubrecht, M. (2012). Franz Hilgendorf's dissertation "Beiträge zur Kenntnis des Süåwasserkalks von Steinheim" from 1863: transcription and description of the first Darwinian interpretation of transmutation. *Zoosystematics and Evolution* 88:231–259.

Gliboff, S. (2007). H.G. Bronn and the history of nature. *Journal of the History of Biology* 40:259–294.
(2008). *H. G. Bronn, Ernst Haeckel, and the Origins of German Darwinism.* Cambridge, MA: MIT Press.

Göbbel, L., Schultka, R. (2003). Meckel the Younger and his epistemology of organic form: morphology in the pre-Gegenbaurian age. *Theory in Biosciences* 122:127–141.

Goldschmidt, R. (1920). Otto Bütschli 1848–1920. *Naturwissenschaften* 8:543–549.

Goodrich, E. S. (1931). The scientific work of Edwin Ray Lankester. *The Quarterly Journal of Microscopical Science* s2-74:363–382.

Gordon, S. (2016). Elie Metchnikoff, the man and the myth. *Journal of Innate Immunity* 8:223–227.

Gould, S. J. (1974). The origin and function of "bizarre" structures: antler size and skull size in the "Irish elk," *Megaloceros giganteus. Evolution* 28:191–220.
(1977a). *Ever Since Darwin. Reflections in Natural History.* London: Penguin Books.
(1977b). *Ontogeny and Phylogeny.* Cambridge, MA: The Belknap Press of Harvard University Press.
(1980). *The Panda's Thumb. More Reflections in Natural History.* New York: W. W. Norton and Company.
(1986). Geoffroy and the homeobox. Pages 205–218 in H. C. Slavkin, ed. *Progress in Developmental Biology, Part A.* New York: Alan R. Liss, Inc.
(1988). Trends as changes in variance: a new slant on progress and directionality in evolution. *Journal of Paleontology.* 62:319–329.
(1989). *Wonderful Life. The Burgess Shale and the Nature of History.* London: Penguin Books.
(1991a). *Bully for Brontosaurus. Reflections in Natural History.* London: Penguin Books.
(1991b). *Time's Arrow, Time's Cycle. Myth and Metaphor in the Discovery of Geological Time.* London: Penguin Books.
(1993). *Eight Little Piggies. Reflections in Natural History.* London: Penguin Books.
(1995a). Ladders and cones: constraining evolution by canonical icons. Pages 37–67 in R. B. Silvers, ed. *Hidden Histories of Science.* New York: The New York Review of Books.
(1995b). Redrafting the tree of life. *Proceedings of the American Philosophical Society* 141:30–54.
(1996). *Full House. The Spread of Excellence from Plato to Darwin.* New York: Harmony Books.
(1998). *Leonardo's Mountain of Clams and the Diet of Worms. Essays on Natural History.* New York: Harmony Books.
(2000). *The Lying Stones of Marrakech. Penultimate Reflections in Natural History.* New York: Harmony Books.
(2002a). *I Have Landed. Splashes and Reflections in Natural History.* London: Vintage.
(2002b). *The Structure of Evolutionary Theory.* Cambridge, MA: The Belknap Press of Harvard University Press.

Gould, S. J., Lewontin, R. C. (1979). The spandrels of San Marco and the Panglossian paradigm: a critique of the adaptationist programme. *Proceedings of the Royal Society B: Biological Sciences* 205:581–598.

Gourko, H., Williamson, D. I., Tauber, A. I., eds. (2000). *The Evolutionary Biology Papers of Elie Metchnikoff* Dordrecht: Kluwer Academic Publishers.

Grant, E. (2007). *A History of Natural Ohilosophy. From the Ancient World to the Nineteenth Century.* Cambridge: Cambridge University Press.

Grant, R. E. (1826). On the structure and nature of the Spongilla friabilis. *Edinburgh Philosophical Journal* 14:270–284.

Grant, T. (2019). Outgroup sampling in phylogenetics: severity of test and successive outgroup expansion. *Journal of Zoological Systematics and Evolutionary Research* 57:748–763.

Grasshoff, M., Gudo, M. (2002). The origin of Metazoa and the main evolutionary lineages of the animal Kingdom: the gallertoid hypothesis in the light of modern research. *Senckenbergiana lethaea* 82:295-314.

Gregory, T. R. (2008). Understanding evolutionary trees. *Evolution: Education and Outreach* 1:121-137.

Grehan, J., Ainsworth, R. (1985). Orthogenesis and evolution. *Systematic Zoology* 34:174-192.

Grell, K. G. (1979). *Die Gastraea-Theorie. Medizinhistorisches Journal* 14:275-291.

Grene, M. (2001). Darwin, Cuvier and Geoffroy: comments and questions. *History and Philosophy of the Life Sciences* 23:187-211.

Griesemer, J. R. (1996). Some concepts of historical science. *Memorie della Società Italiana di Scienze Naturali e del Museo Civico di Storia Naturale di Milano* 27:60-69.

Groeben, C. (1985). Anton Dohrn: the statesman of Darwinism: to commemorate the 75th anniversart of the death of Anton Dohrn. *Biological Bulletin* 168 (Supplement):4-25.

Groeben, C., Oppenheimer, J. M. (1993). Karl Ernst von Baer (1792-1876), Anton Dohrn (1840-1909): correspondence. *Transactions of the American Philosophical Society* 83:1-156.

Gudo, M. (2005). An evolutionary scenario for the origin of pentaradial echinoderms – implications from the hydraulic principles of form determination. *Acta Biotheoretica* 53:191-216.

Gudo, M., Syed, T. (2008). *100 years of Deuterostomia (Grobben, 1908): cladogenetic and anagenetic relations within the Notoneuralia domain. ArXiv*:0811.2189v0811.

Gutmann, W. F. (1966). Coelomgliederung, Myomerie und die Frage der Vertebraten-Antezedenten. *Zeitschrift für zoologische Systematik und Evolutionsforschung* 4:13-57.

(1981). Relationships between invertebrate phyla based on functional-mechanical analysis of the hydrostatic skeleton. *American Zoologist* 21:63-81.

Hadži, J. (1953). An attempt to reconstruct the system of animal classification. *Systematic Zoology* 2:145-154.

(1958). Zur Diskussion über die Abstammung der Eumetazoen. *Zoologischer Anzeiger Supplement* 21:169-179.

(1963). *The Evolution of the Metazoa.* Oxford: Pergamon Press.

Haeckel, E. (1866a). *Generelle Morphologie der Organismen: allgemeine Grundzüge der organischen Formen-Wissenschaft, mechanisch begründet durch die von Charles Darwin reformirte Descendenz-Theorie. Bd. 1, Allgemeine Anatomie der Organismen.* Berlin: Verlag von Georg Reimer.

(1866b). *Generelle Morphologie der Organismen: allgemeine Grundzüge der organischen Formen-Wissenschaft, mechanisch begründet durch die von Charles Darwin reformirte Descendenz-Theorie. Bd. 2, Allgemeine Entwickelungsgeschichte der Organismen.* Berlin: Verlag von Georg Reimer.

(1870). *Studien über Moneren und andere Protisten nebst einer Rede über Entwicklungsgang und Aufgabe der Zoologie.* Leipzig: Verlag von Wilhelm Engelmann.

(1872). *Die Kalkschwämme. Erster Band (Genereller Theil). Biologie der Kalkschwämme.* Berlin: Verlag von Georg Reimer.

(1874). The Gastrea-theory, the phylogenetic classification of the animal kingdom and the homology of the germ-lamellae. *Quarterly Journal of Microscopical Science* 14:142-165, 223-247.

(1875). Ziel und Wege der heutigen Entwicklungsgeschichte. *Jenaische Zeitschrift für Naturwissenschaft* 10 (Supplement):1-99.

(1877). *Studien zur Gastraea-Theorie.* Jena: Verlag von Hermann Dufft.

(1885). Ursprung und Entwicklung der thierischen Gewebe. Ein histogenetischer Beitrag zur Gastraea-Theorie. *Jenaische Zeitschrift für Medicin und Naturwissenschaft* 18:206-276.

(1889). *Natürliche Schöpfungs-Geschichte.* Berlin: Druck und Verlag von Georg Reimer.

(1891). *Anthropogenie oder Entwicklungsgeschichte der Menschen. Keimes- und Stammes-Geschichte.* Leipzig: Verlag von Wilhelm Engelmann.

(1894). *Systematische Phylogenie der Protisten und Pflanzen.* Berlin: Verlag von Georg Reimer.

(1895). *Systematische Phylogenie der Wirbelthiere (Vertebrata).* Berlin: Verlag von Georg Reimer.

(1896). *Systematische Phylogenie der Wirbellosen Thiere (Invertebrata).* Berlin: Verlag von Georg Reimer.

(1908). *Unsere Ahnenreihe (Progonotaxis hominis).* Jena: Verlag von Gustav Fischer.

(1919). *Die Welträtsel. Gemeinverständliche Studien über monistische Philosophie. Elfte verbesserte Auflage.* Stuttgart: Alfred Kröner Verlag.

(2008). *Art Forms in Nature.* Munich: Prestel Verlag.

Halanych, K. M. (2015). The ctenophore lineage is older than sponges? That cannot be right! Or can it? *Journal of Experimental Biology* 218:592–597.

(2016). How our view of animal phylogeny was reshaped by molecular approaches: lessons learned. *Organisms Diversity & Evolution* 16:319–328.

Hall, B. K. (1999). The paradoxical platypus. *Bioscience* 49:211–218.

(2005). Betrayed by *Balanoglossus*: William Bateson's rejection of evolutionary embryology as the basis for understanding evolution. *Journal of Experimental Zoology B (Molecular and Developmental Evolution)* 304B:1–17.

(2007). Tapping many sources: the adventitious roots of evo-devo in the nineteenth century. Pages 467–497 in M. D. Laubichler, J. Maienschein, eds. *From Embryology to Evo-Devo: A History of Developmental Evolution.* Cambridge, MA: MIT Press.

Hallam, A. (1988). The contribution of palaeontology to systematics and evolution. Pages 128–147 in D. L. Hawksworth, ed. *Prospects in Systematics.* Oxford: Oxford University Press.

Halverson, K. L., Pires, C. J., Abell, S. K. (2011). Exploring the complexity of tree thinking expertise in an undergraduate systematics course. *Science Education* 95:794–823.

Hamilton, A., ed. (2014). *The Evolution of Phylogenetic Systematics.* Berkeley: University of California Press.

Hand, C. (1959). On the origin and phylogeny of the coelenterates. *Systematic Zoology* 8:191–202.

Handlirsch, A. (1908). *Die Fossilen Insekten und die Phylogenie der Rezenten Formen.* Leipzig: Verlag von Wilhelm Engelmann.

Hanson, E. D. (1958). On the origin of the Eumetazoa. *Systematic Zoology* 7:16–47.

(1963). Homologies and the ciliate origin of the Eumetazoa. Pages 7–22 in E. C. Dougherty, Z. Norwood Brown, E. D. Hanson, W. D. Hartman, eds. *The Lower Metazoa. Comparative Biology and Phylogeny.* Berkeley: University of California Press.

(1972). *Animal Diversity.* Englewood Cliffs, NJ: Prentice-Hall.

(1977). *The Origin and Early Evolution of Animals.* Middletown: Wesleyan University Press.

(1981). *Understanding Evolution.* New York: Oxford University Press.

Hargreaves, A. D., Tucker, A. S., Mulley, J. F. (2015). A critique of the toxicoferan hypothesis. Pages 69–86 in P. Gopalakrishnakone, A. Malhotra, eds. *Evolution of Venomous Animals and Their Toxins.* Dordrecht: Springer Netherlands.

Härlin, M. (1998). Tree-thinking and nemertean systematics, with a systematization of the Eureptantia. *Hydrobiologia* 365:33–46.

Hart, M. W., Grosberg, R. K. (2009). Caterpillars did not evolve from onychophorans by hybridogenesis. *Proceedings of the National Academy of Sciences USA* 106:19906–19909.

Hartman, W. D. (1963). A critique of the enterocoele theory. Pages 55–77 in E. C. Dougherty, ed. *The Lower Metazoa. Comparative Biology and Phylogeny.* Berkeley: University of California Press.

Hartnoll, R. (2016). In remembrance of Donald Williamson (6 January 1922 – 29 January 2016), planktologist, carcinologist, and evolutionist: a metamorphosis. *Journal of Crustacean Biology* 36:408–413.

Harvey, L. A. (1961). Speculations on ancestry and evolution. *Science Progress* 49:111–121.

Haszprunar, G. (1996). The Mollusca: coelomate turbellarians or mesenchymate annelids? Pages 1–28 in J. Taylor, ed. *Origin and Evolutionary Radiation of the Mollusca*. Oxford: Oxford University Press.

Haszprunar, G., Wanninger, A. (2012). Molluscs. *Current Biology* 22:R510–R514.

Hatschek, B. (1878). *Studien über die Entwicklungsgeschichte der Anneliden. Ein Beitrag zur Morphologie der Bilaterien. Arbeiten aus dem Zoologischen Institut der Universität Wien und der Zoologischen Station in Triest* 1:1–128.

(1888). *Lehrbuch der Zoologie. Eine Morphologische Übersicht des Thierreiches zur Einführung in das Studium dieser Wissenschaft*. Jena: Verlag von Gustav Fischer.

Hausdorf, B. (2000). Early evolution of the Bilateria. *Systematic Biology* 49:130–142.

Havstad, J. C., Assis, L. C., Rieppel, O. (2015). The semaphorontic view of homology. *Journal of Experimental Zoology B (Molecular and Developmental Evolution)* 324:578–587.

Heider, K. v. (1914). Phylogenie der Wirbellosen. Pages 453–529 in R. Hertwig, R. v. Wettstein, eds. *Die Kultur der Gegenwart. Teil 3. Abteilung 4. Band 4. Abstammungslehre, Systematik, Paläontologie, Biogeographie*. Leipzig: Druck und Verlag von B. G. Teubner.

Hejnol, A., Lowe, C. J. (2015). Embracing the comparative approach: how robust phylogenies and broader developmental sampling impacts the understanding of nervous system evolution. *Philosophical Transactions of the Royal Society B: Biological Sciences* 370:20150045.

Hejnol, A., Martín-Durán, J. M. (2015). Getting to the bottom of anal evolution. *Zoologischer Anzeiger* 256:61–74.

Hejnol, A., Martindale, M. Q. (2009). The mouth, the anus and the blastopore—open questions about questionable openings. Pages 33–40 in M. J. Telford, D. T. J. Littlewood, eds. *Animal Evolution. Genomes, Fossils, and Trees*. Oxford: Oxford University Press.

Hennig, W. (1965). Phylogenetic systematics. *Annual Review of Entomology* 10:97–116.

(1966). *Phylogenetic Systematics*. Urbana, IL: University of Illinois Press.

Hertwig, O., Hertwig, R. (1882). Die Coelomtheorie. Versuch einder Erklärung des mittleren Keimblattes. *Jenaische Zeitschrift für Naturwissenschaft* 15:1–150.

Hessler, R. R. (1983). A defense of the caridoid facies; wherein the early evolution of the Eumalacostraca is discussed. Pages 145–164 in F. R. Schram, ed. *Crustacean Phylogeny*. Rotterdam: A. A. Balkema.

Hessler, R. R., Newman, W. A. (1975). A trilobitomorph origin for the Crustacea. *Fossils and Strata* 4:437–459.

Hilgendorf, F. (1881). Die neu erschienene Schrift "The genesis of the tertiary species of Planorbis at Steinheim" by A. Hyatt. *Sitzungsberichte der Gesellschaft naturforschender Freunde Berlin* 95–100.

Hitchens, C. (2007). *God Is Not Great. The Case Against Religion*. London: Atlantic Books.

Holland, N. D. (2003). Early central nervous system evolution: an era of skin brains? *Nature Reviews Neuroscience* 4:1–11.

(2011). Walter Garstang: a retrospective. *Theory in Biosciences* 130:247–258.

Holland, N. D., Holland, L. Z., Holland, P. W. H. (2015). Scenarios for the making of vertebrates. *Nature* 520:450–455.

Holland, P. W. H. (2010). From genomes to morphology: a view from amphioxus. *Acta Zoologica* 91:81–86.

Hölldobler, B., Wilson, E. O. (2009). *The Superorganism. The Beauty, Elegance, and Strangeness of Insect Societies*. New York: W. W. Norton & Company.

Hopwood, N. (2006). Pictures of evolution and charges fraud. Ernst Haeckel's embryological illustrations. *Isis* 97:260–301.

(2015a). The cult of amphioxus in German Darwinism; or, our gelatinous ancestors in Naples' blue and balmy bay. *History and Philosophy of the Life Sciences* 36:371–393.

(2015b). *Haeckel's Embryos. Images, Evolution, and Fraud.* Chicago: The University of Chicago Press.

Hossfeld, U., Olsson, L. (2003). The road from Haeckel: the Jena tradition in evolutionary morphology and the origins of "evo-devo." *Biology & Philosophy* 18:285-307.

(2005). The history of the homology concept and the "Phylogenetisches Symposium." *Theory in Biosciences* 124:243-253.

Hossfeld, U., Porges, K., Levit, G. S., Olsson, L., Watts, E. (2019). Ernst Haeckel's embryology in biology textbooks in the German Democratic Republic, 1951-1988. *Theory in Biosciences* 138:31-48.

Hubrecht, A. (1905). The Gastrulation of the vertebrates. *Quarterly Journal of Microscopical Science* 49:403-419.

(1908). Early ontogenetic phenomena in mammals and their bearing on our interpretation of the phylogeny of the vertebrates. *The Quarterly Journal of Microscopical Science, New Series*, 53:1-181.

Hubrecht, A. A. W. (1883). On the ancestral form of the Chordata. *Quarterly Journal of Microscopical Science* 23:349-368.

(1887). The relation of the Nemertea to the Vertebrata. *Quarterly Journal of Microscopical Science* 27:605-644.

Hui, J. H. L., Raible, F., Korchagina, N., et al. (2009). Features of the ancestral bilaterian inferred from *Platynereis dumerilii* ParaHox genes. *BMC Biology* 7:43.

Hull, D. L. (1975). Central subjects and historical narratives. *History and Theory* 14:253-274.

(1989). *The Metaphysics of Evolution.* New York: State University of New York Press.

Huxley, J. (1957). The three types of evolutionary process. *Nature* 180:454-455.

(1958). The evolutionary process. Pages 9-33 in J. Huxley, A. C. Hardy, E. B. Ford, eds. *Evolution as a Process.* New York: Collier Books.

Huxley, J., Hardy, A. C., Ford, E. B., eds. (1958). *Evolution as a Process* New York: Collier Books.

Huxley, T. H. (1849). On the anatomy and the affinities of the family of the Medusae. *Philosophical Transactions of the Royal Society of London* 139:413-434.

(1850). Notes on medusae and polypes. *The Annals and Magazine of Natural History, Including Zoology, Botany, and Geology* 6:66-67.

(1853). On the morphology of the cephalous Mollusca, as illustrated by the anatomy of certain Heteropoda and Pteropoda collected during the voyage of H. M. S. "Rattlesnake" in 1846-50. *Philosophical Transactions of the Royal Society of London* 143:29-65.

(1858). On the theory of the vertebrate skull. *Proceedings of the Royal Society of London* 9:381-457.

(1863). *Evidence as to Man's Place in Nature.* New York: D. Appleton and Co.

(1868). On the animals which are most nearly intermediate between birds and reptiles. *The Popular Science Review* 7:237-247.

(1870). The anniversary address of the president. *Quarterly Journal of the Geological Society London* 18:xxix-lxiv.

(1875). On the classification of the animal kingdom. *Quarterly Journal of Microscopical Science* 15:52-56.

(1888). *A Manual of the Anatomy of Invertebrated Animals.* New York: D. Appleton and Company.

Hyatt, A. (1866). On the parallelism between the different stages of life in the individual and those in the entire group of the molluscous order Tetrabranchiata. *Memoirs of the Boston Society of Natural History* 1:193-209.

(1880). The genesis of the Tertiary species of Planorbis at Steinheim. *Anniversary Memoirs of the Boston Society of Natural History.* Boston: Boston Society of Natural History 1-114.

Hyman, L. H. (1940). *The Invertebrates. Protozoa through Ctenophora.* New York: McGraw-Hill.

(1951). *The Invertebrates. Platyhelminthes and Rhynchocoela. The Acoelomate Bilateria. Volume II.* New York: McGraw-Hill.

(1959). *The Invertebrates. Smaller Coelomate Groups. Chaetognatha, Hemichordata, Pogonophora, Phoronida, Ectoprocta, Brachipoda [sic], Sipunculida. The Coelomate Bilateria. Volume V.* New York: McGraw-Hill.

Irie, N., Satoh, N., Kuratani, S. (2018). The phylum Vertebrata: a case for zoological recognition. *Zoological Letters* 4:32.

Ivanov, A. V. (1988). On the early evolution of the Bilateria. *Fortschritte der Zoologie* 36:349–352.

Ivanova-Kazas, O. M. (2008). Origin of Chordata and the "Upside-Down Theory." *Russian Journal of Marine Biology* 34:391–402.

(2013). Origin of arthropods and of the clades of Ecdysozoa. *Russian Journal Developmental Biology* 44:221–231.

(2015). The secondary mouth and its phylogenetic significance. *Russian Journal of Marine Biology* 41:83–93.

Jackson, T., Fry, B. (2016). A tricky trait: applying the fruits of the "function debate" in the philosophy of biology to the "venom debate" in the science of toxinology. *Toxins* 8:263.

Jackson, T. N. M., Jouanne, H., Vidal, N. (2019). Snake venom in context: neglected clades and concepts. *Frontiers in Ecology and Evolution* 7:332.

Jacobs, N. X. (1989). From unit to unity: protozoology, cell theory, and the new concept of life. *Journal of the History of Biology* 22:215–242.

Jägersten, G. (1955). On the early phylogeny of the Metazoa. The Bilaterogastrea Theory. *Zoologiska Bidrag Uppsala* 30:321–354.

Jahn, I. (2002). Das „Meckel-Serres-Gesetz", sein Ursprung und seine Beziehung zu Evolutionstheorien des 19. *Jahrhunderts. Annals of Anatomy* 184:509–517.

Janecka, J., Chowdhary, B., Murphy, W. (2012). Exploring the correlations between sequence evolution rate and phenotypic divergence across the mammalian tree provides insights into adaptive evolution. *Journal of Biosciences* 37:897–909.

Janssen, R., Budd, G. E. (2017). Investigation of endoderm marker-genes during gastrulation and gut-development in the velvet worm *Euperipatoides kanangrensis*. *Developmental Biology* 427:155–164.

Janssen, R., Jörgensen, M., Lagebro, L., Budd, G. E. (2015). Fate and nature of the onychophoran mouth–anus furrow and its contribution to the blastopore. *Proceedings of the Royal Society B* 282:20142628.

Janz, H. (1999). Hilgendorf's planorbid tree – the first introduction of Darwin's theory of transmutation into palaeontology. *Paleontological Research* 3:287–293.

Jefferies, R. P. S. (1988). How to characterize the Echinodermata – some implications of the sister-group relationship between echinoderms and chordates. Pages 3–12 in C. R. C. Paul, A. B. Smith, eds. *Echinoderm Phylogeny and Evolutionary Biology*. Oxford: Clarendon Press.

Jenkins, B. (2016). Neptunism and transformism: Robert Jameson and other evolutionary theorists in early nineteenth-century Scotland. *Journal of the History of Biology* 49:527–557.

Jenner, R. A. (1999). Metazoan phylogeny as a tool in evolutionary biology: current problems and discrepancies in application. *Belgian Journal of Zoology* 129:245–262.

(2000). Evolution of animal body plans: the role of metazoan phylogeny at the interface between pattern and process. *Evolution & Development* 2:208–221.

(2001). Bilaterian phylogeny and uncritical recycling of morphological data sets. *Systematic Biology* 50:730–742.

(2002). Boolean logic and character state identity: pitfalls of character coding in metazoan cladistics. *Contributions to Zoology* 71:67–91.

(2003). Unleashing the force of cladistics? Metazoan phylogenetics and hypothesis testing. *Integrative and Comparative Biology* 43:207–218.

(2004a). Libbie Henrietta Hyman (1888–1969): from developmental mechanics to the evolution of animal body plans. *Journal of Experimental Zoology B (Molecular and Developmental Evolution)* 5:413–423.

(2004b). The scientific status of metazoan cladistics: why current research practice must change. *Zoologica Scripta* 33:293–310.

(2004c). Towards a phylogeny of the Metazoa: evaluating alternative phylogenetic positions of Platyhelminthes, Nemertea, and Gnathostomulida, with a critical reappraisal of cladistic characters. *Contributions to Zoology* 73:3–163.

(2004d). When molecules and morphology clash: reconciling conflicting phylogenies of the Metazoa by considering secondary character loss. *Evolution & Development* 6:372–378.

(2005a). Foiling vertebrate inversion with the humble nemertean. *Palaeontological Association Newsletter* 58:32–39.

(2005b). Meeting a nemertean nemesis. *Palaeontological Association Newsletter* 59:37–42.

(2006). Unburdening evo-devo: ancestral attractions, model organisms, and basal baloney. *Evolution & Development* 216:385–394.

(2009). *Idalatry. Palaeontological Association Newsletter* 71:94–102.

(2014). Macroevolution of animal body plans: is there science after the tree? *Bioscience* 64:653–664.

(2018). Evolution is linear: debunking life's little joke. *BioEssays* 40:doi: 10.1002/bies.201700196.

(2019). The origin of evolutionary storytelling. Pages 357–368 in G. Fusco, ed. *Perspectives on Evolutionary and Developmental Biology. Essays for Alessandro Minelli*. Padova: Padova University Press.

Jenner, R. A., Scholtz, G. (2005). Playing another round of metazoan phylogenetics: historical epistemology, sensitivity analysis, and the position of Arthropoda within the Metazoa on the basis of morphology. Pages 355–385 in S. Koenemann, R. A. Jenner, eds. *Crustacea and Arthropod Relationships*. Boca Raton, FL: CRC Press.

Jenner, R. A., Schram, F. R. (1999). The grand game of metazoan phylogeny: rules and strategies. *Biological Reviews of the Cambridge Philosophical Society* 74:121–142.

Jenner, R. A., Wills, M. A. (2007). The choice of model organisms in evo-devo. *Nature Reviews Genetics* 8:311–319.

Jepsen, G. L. (1949). Selection, "orthogenesis," and the fossil record. *Proceedings of the American Philosophical Society* 93:479–500.

Jockusch, E. L. (2017). Developmental and evolutionary perspectives on the origin and diversification of arthropod appendages. *Integrative and Comparative Biology* 57:533–545.

Jones, W. T., Feldmann, R. M., Schram, F. R., Schweitzer, C. E., Maguire, E. P. (2016). The proof is in the pouch: *Tealliocaris* is a peracarid. *Palaeodiversity* 9:75–88.

Joy, J. B., Liang, R. H., McCloskey, R. M., Nguyen, T., Poon, A. F. Y. (2016). *Ancestral reconstruction. PLoS Computational Biology* 12:e1004763.

Kapli, P., Telford, M. J. (2020). Topology-dependent asymmetry in systematic errors affects phylogenetic placement of Ctenophora and Xenacoelomorpha. Science *Advances* 6:eabc5162.

Kappers, C. U. A. (1921). On structural laws in the nervous system: the principles of neurobiotaxis. *Brain* 44:125–149.

(1927). On neurobiotaxis. A pschycical law in the structure of the nervous system. *Acta Psychiatrica Scandinavica* 2:118–145.

Kemp, T. S. (1988). Haemothermia or Archosauria? The interrelationships of mammals, birds and crocodiles. *Zoological Journal of the Linnean Society* 92:67–104.

King, N., Rokas, A. (2017). Embracing uncertainty in reconstructing early animal evolution. *Current Biology* 27:R1081–R1088.

Kingsley, J. S. (1885). Notes on the embryology of *Limulus*. *Quarterly Journal of Microscopical Science* 2:521–576.

Kingsolver, J. G., Pfennig, D. W. (2004). Individual-level selection as a cause of Cope's rule of phyletic size increase. *Evolution* 58:1608–1612.

Kleinenberg, N. (1886). Die Entstehung des Annelids aus der Larve von *Lopadorhynchus*, nebst Bemerkungen über die Entwicklung anderer Polychaeten. *Zeitschrift für wissenschaftliche Zoologie* 44:1–227.

Kluge, A. G. (2007). Completing the neo-Darwinian synthesis with an event criterion. *Cladistics* 23:613–633.

(2009). Explanation and falsification in phylogenetic inference: exercises in Popperian philosophy. *Acta Biotheoretica* 57:171–186.

Kocot, k. M., Poustka, A. J., Stöger, I., Halanych, K. M., Schrödl, M. (2020). New data from Monoplacophora and a carefully-curated dataset resolve molluscan relationships. *Scientific Reports* 10:101.

Koduru, S., Vegiraju, S. R., Nadimpalli, S. K., et al. (2006). The early vertebrate *Danio rerio* Mr 46000 mannose-6-phosphate receptor: biochemical and functional characterisation. *Development, Genes and Evolution* 216:133–143.

Kolchinsky, E., Levit, G. S. (2019). The reception of Haeckel in pre-revolutionary Russia and his impact on evolutionary theory. *Theory in Biosciences* 138:73–88.

Kolchinsky, E. I. (2019). Russian editions of E. Haeckel's works and the evolution of their perception. *Theory in Biosciences* 138:49–71.

Korschelt, E., Heider, K. (1899). *Text-book of the Embryology of Invertebrates. Vol. III. Arachnida, Pentastomidae, Pantopoda, Tardigrada, Onychophora, Myriopoda, Insects.* London: Swan Sonnenschein and Co.

Koshi, J. M., Goldstein, R. A. (1996). Probabilistic reconstruction of ancestral protein sequences. *Journal of Molecular Evolution* 42:313–320.

Kotikova, E. A., Mamkaev, Y. V. (2006). Artemii Vasil'evich Ivanov, an outstanding zoologist-evolutionist (to the 100-anniversary). *Journal of Evolutionary Biochemistry and Physiology* 42:226–229.

Krell, F.-T., Cranston, P. S. (2004). Which side of the tree is more basal? *Systematic Entomology* 29:279–281.

Kuntner, M., Agnarsson, I. (2009). Phylogeny accurately predicts behaviour in Indian Ocean Clitaetra spiders (Araneae : Nephilidae). *Invertebrate Systematics* 23:193–204.

Kusnezov, N. (1959). Die allgemeinen Gesetze der organischen Evolution. *Acta Biotheoretica* 13:47–86.

Kutschera, U. (2007). Palaeobiology: the origin and evolution of a scientific discipline. *Trends in Ecology & Evolution* 22:172–173.

(2011). From the scala naturae to the symbiogenetic and dynamic tree of life. *Biology Direct* 6:33.

Kutschera, V., Maas, A., Mayer, G., Waloszek, D. (2015). Calcitic sclerites at base of malacostracan pleopods (Crustacea) – part of a coxa. *BMC Evolutionary Biology* 15:117.

Kuznetsov, A. N. (2012). Five longitudes in chordate body. *Theoretical Biology Forum* 105:21–35.

Lacalli, T. C. (1996). Dorsoventral axis inversion: a phylogenetic perspective. *BioEssays* 18:251–254.

(1997). The nature and origin of deuterostomes: some unresolved issues. *Invertebrate Biology* 116:363–370.

(1999). Tunicate tails, stolons, and the origin of the vertebrate trunk. *Biological Reviews of the Cambridge Philosophical Society* 74:177–198.

(2010). The emergence of the chordate body plan: some puzzles and problems. *Acta Zoologica* 91:4–10.

Landau, M. (1984). Human evolution as narrative: have hero myths and folktales influenced our interpretations of the evolutionary past? *American Scientist* 72:262–268.

(1991). *Narratives of Human Evolution.* New Haven, CT: Yale University Press.

Lankester, E. R. (1873). On the primitive cell-layers of the embryo as the basis of genealogical classification of animals, and on the origin of vascular and lymph systems. *The Annals and Magazine of Natural History* 11:321–338.

(1875). Dohrn on the origin of the Vertebrata and on the principle of succession of functions. *Nature* 12:479–481.

(1876). An account of Professor Haeckel's recent additions to the Gastraea Theory. *Quarterly Journal of Microscopical Science* s2-16:51–66.

(1877). Notes on the embryology and classification of the animal kingdom: comprising a revision of speculations relative to the origin and significance of the germ-layers. *Quarterly Journal of Microscopical Science* 17:399–454.

(1880). *Degeneration. A Chapter in Darwinism.* London: Macmillan.

(1881). Limulus an Arachnid. *Quarterly Journal of Microscopical Science* 21:504–548, 609–649.

Lankester, E. R., ed. (1900). *A Treatise on Zoology. Part II. The Porifera and Coelenterata.* London: Adam and Charles Black.

(1904). The structure and classification of the Arthropoda. *Quarterly Journal of Microscopical Science* 47:523–582.

Laumer, C. E., Fernández, R., Lemer, S., et al. (2019). Revisiting metazoan phylogeny with genomic sampling of all phyla. *Proceedings of the Royal Society B: Biological Sciences* 286:20190831.

Laumer, C. E., Gruber-Vodicka, H., Hadfield, M. G., et al. (2018). Support for a clade of Placozoa and Cnidaria in genes with minimal compositional bias. *Elife* 7:e36278.

Laundon, D., Larson, B. T., McDonald, K., King, N., Burkhardt, P. (2019). The architecture of cell differentiation in choanoflagellates and sponge choanocytes. *PLoS Biology* 17:e3000226.

Le Guyader, H. (2004). *Geoffroy Saint-Hilaire. A Visionary Naturalist.* Chicago: The University of Chicago Press.

Lee, M. S. Y., Doughty, P. (1997). The relationship between evolutionary theory and phylogenetic analysis. *Biological Reviews of the Cambridge Philosophical Society* 72:471–495.

Lefèvre, W. (2001). Jean Baptiste Lamarck. Pages 176–201 in I. Jahn, M. Schmitt, eds. *Darwin & Co.* Munich: Verlag C. H. Beck.

Lemche, H., Wingstrand, K. G. (1959). The anatomy of *Neopilina galatheae* Lemche, 1957. *Galathea Report* 3:1–71.

Lenoir, T. (1981). The Göttingen school and the development of transcendental Naturphilosophie in the Romantic era. *Studies in the History of Biology* 5:111–205.

(1987). The eternal laws of form: morphotypes and the conditions of existence in Goethe's biological thought. Pages 17–28 in F. Amrine, F. J. Zucker, H. Wheeler, eds. *Goethe and the Sciences: A Reappraisal.* Dordrecht: D. Reidel.

Leroi, A. M. (2014). *The Lagoon. How Aristotle Invented Science.* London: Bloomsbury Circus.

Lester, J., Bowler, P. J. (1995). *E. Ray Lankester and the Making of Modern British Biology.* Oxford: British Society for the History of Science.

Levit, G. S. (2007). The roots of evo-devo in Russia: is there a characteristic "Russian Tradition"? *Theory in Biosciences* 126:131–148.

Levit, G. S., Hossfeld, U., Olsson, L. (2004). The integration of Darwinism and evolutionary morphology: Alexej Nikolajevich Swertzoff (1866-1936) and the developmental basis of evolutionary change. *Journal of Experimental Zoology B (Molecular and Developmental Evolution)* 302B:343-354.

(2006). From the "modern synthesis" to cybernetics: Ivan Ivanovich Schmalhausen (1884-1963) and his research program for a synthesis of evolutionary and developmental biology. *Journal of Experimental Zoology B (Molecular and Developmental Evolution)* 306B:89-106.

(2014). The Darwinian revolution in Germany: from evolutionary morphology to the modern synthesis. *Endeavour* 38:268-279.

Levit, G. S., Meister, K. (2006). The history of essentialism vs. Ernst Mayr's "Essentialism Story": a case study of German idealistic morphology. *Theory in Biosciences* 124:281-307.

Levit, G. S., Olsson, L. (2006). "Evolution on rails": mechanisms and levels of orthogenesis. *Annals for the History and Philosophy of Biology* 11:97-136.

Lewes, G. H. (1871). *The History of Philosophy from Thales to Comte, Volume I.* London: Longmans, Green.

Lewis, J. G. E. (1981). *The Biology of Centipedes.* Cambridge: Cambridge University Press.

Leys, S. P., Eerkes-Medrano, D. (2005). Gastrulation in calcareous sponges: in search of Haeckel's Gastraea. *Integrative and Comparative Biology* 45:342-351.

Li, C.-W., Chen, J.-Y., Hua, T.-E. (1998). Precambrian sponges with cellular structures. *Science* 279:879-882.

Li, Y., Shen, X.-X., Evans, B., Dunn, C. W., Rokas, A. (2021). Rooting the animal tree of life. *Molecular Biology and Evolution* 38:4322-4333.

Lightner, J. K., Cserhati, M. (2019). The uniqueness of humans is clearly demonstrated by the gene-content statistical baraminology method. *Creation Research Society Quarterly Journal* 55:132-141.

Lindberg, D. R., Ghiselin, M. T. (2003). Fact, theory and tradition in the study of molluscan origins. *Proceedings of the California Academy of Sciences* 54:663-686.

Linz, D. M., Hu, Y., Moczek, A. P. (2020). From descent with modification to the origins of novelty. *Zoology* 143:125836.

Lipscomb, D. (1992). Parsimony, homology and the analysis of multistate characters. *Cladistics* 8:45-65.

Litsios, G., Salamin, N. (2012). Effects of phylogenetic signal on ancestral state reconstruction. *Systematic Biology* 61:533-538.

Lorenz, K. (1941). Vergleichende Bewegungsstudien an Anatinen. *Journal für Ornithologie* 89:194-294.

Lovejoy, A. O. (1936). *The Great Chain of Being. A Study of the History of an Idea.* Cambridge, MA: Harvard University Press.

Lozano-Fernandez, J., Giacomelli, M., Fleming, J., et al. (2019). Pancrustacean evolution illuminated by taxon-rich genomic-scale data sets with an expanded remipede sampling. *Genome Biology and Evolution* 11:2055-2070.

Lynch, M. (2006). The frailty of adaptive hypotheses for the origins of organismal complexity. *Proceedings of the National Academy of Sciences USA* 104:8597-8604.

Lyons, S. (1993). Thomas Huxley: fossils, persistence, and the argument from design. *Journal of the History of Biology* 26:545-569.

MacBride, E. W. (1895). Sedgwick's theory of the embryonic phase of ontogeny as an aid to phylogenetic theory. *Quarterly Journal of Microscopical Science* s237:325-342.

(1909). The formation of the layers in amphioxus and its bearing on the interpretation of the early ontogenetic processes in other vertebrates. *Quarterly Journal of Microscopical Science* 54:279-345.

(1914). *Text-book of Embryology. Vol. I. Invertebrata.* London: Macmillan.

(1929a). Mimicry. *Nature* 123:712–713.

(1929b). Natural selection. *Nature* 124:225.

MacFadden, B. J., Oviedo, L. H., Seymour, G. M., Ellis, S. (2012). Fossil horses, orthogenesis, and communicating evolution in museums. *Evolution: Education and Outreach* 5:29–37.

Macfarlane, J. M. (1918). *The Causes and Course of Organic Evolution. A Study in Bioenergics.* New York: Macmillan.

Mackessy, S. P., Saviola, A. J. (2016). Understanding biological roles of venoms among the Caenophidia: the importance of rear-fanged snakes. *Integrative and Comparative Biology* 56:1004–1021.

MacLeay, W. S. (1819–1821). *Horae Entomologicae or Essays on the Annulose Animals.* London: S. Bagster.

Mah, J. L., Christensen-Dalsgaard, K. K., Leys, S. P. (2014). Choanoflagellate and choanocyte collar-flagellar systems and the assumption of homology. *Evolution & Development* 16:25–37.

Maienschein, J. (1994). "It's a long way from *Amphioxus*" Anton Dohrn and late nineteenth century debates about vertebrate origins. *History and Philosophy of the Life Sciences* 16:465–478.

(2000). Competing epistemologies and developmental biology. Pages 122–1137 in R. Creath, J. Maienschein, eds. *Biology and Epistemology.* Cambridge: Cambridge University Press.

Malakhov, V. V. (2013). A revolution in zoology: new concepts of the metazoan system and phylogeny. *Herald of the Russian Academy of Sciences* 83:123–127.

Malakhov, V. V., Bogomolova, E. V., Kuzmina, T. V., Temereva, E. N. (2019). Evolution of metazoan life cycles and the origin of pelagic larvae. *Russian Journal of Developmental Biology* 50:303–316.

Mallatt, J., Chen, J. Y. (2003). Fossil sister group of craniates: predicted and found. *Journal of Morphology* 258:1–31.

Manton, S. M. (1973). Arthropod phylogeny – a modern synthesis. *Journal of the Zoological Society of London* 171:111–130.

Manuel, M. (2009). Early evolution of symmetry and polarity in metazoan body plans. *Comptes Rendus Biologies* 332:184–209.

Marcus, E. (1958). On the evolution of the animal phyla. *Quarterly Review of Biology* 33:24–58.

Margulis, L., Chapman, M., Guerrero, R., Hall, J. (2006). The last eukaryotic common ancestor (LECA): acquisition of cytoskeletal motility from aerotolerant spirochetes in the Proterozoic Eon. *Proceedings of the National Academy of Sciences USA* 103:13080–13085.

Marlétaz, F., Peijnenburg, K. T. C. A., Goto, T., Satoh, N., Rokhsar, D. (2019). A new spiralian phylogeny places the enigmatic arrow worms among gnathiferans. *Current Biology* 29:312–318.

Martin, J. W., Davis, G. E. (2001). An updated classification of the Recent Crustacea. *Natural History Museum of Los Angeles County Science Series* 39:1–124.

Martín-Durán, J. M., Hejnol, A. (2021). A developmental perspective on the evolution of the nervous system. *Developmental Biology* 475:181–192.

Martín-Durán, J. M., Pang, K., Børve, A., et al. (2018). Convergent evolution of bilaterian nerve cords. *Nature* 553:45–50.

Martín-Durán, J. M., Passamaneck, Y. J., Martindale, M. Q., Hejnol, A. (2017). The developmental basis for the recurrent evolution of deuterostomy and protostomy. *Nature Ecology & Evolution* 1:0005.

Martindale, M. Q., Hejnol, A. (2009). A developmental perspective: changes in the position of the blastopore during bilaterian evolution. *Developmental Cell* 17:162–174.

Martynov, A. V. (2012). Ontogeny, systematics, and phylogenetics: perspectives of future synthesis and a new model of the evolution of Bilateria. *Biology Bulletin of the Russian Academy of Sciences* 39:393–401.

Maslin, T. P. (1952). Morphological criteria of phyletic relationships. *Systematic Zoology* 1:49–70.

Masterman, A. T. (1897a). On the Diplochorda. *Quarterly Journal of Microscopical Science* 40:281–338.

(1897b). On the structure of *Actinotrocha* considered in relation to the suggested chordate affinities of *Phoronis*. *Proceedings of Royal Society of Edinburgh* 21:129–136.

(1897c). Preliminary note on the structure and affinities of *Phoronis*. *Proceedings of Royal Society of Edinburgh* 21:59–71.

(1898). On the theory of archimeric segmentation and its bearing upon the phyletic classification of the *Coelomata*. *Proceedings of Royal Society of Edinburgh* 22:270–310.

Matuk, C., Uttal, D. H. (2011). Narrative spaces in the representation and understanding of evolution. Pages 119–144 in K. S. Rosengren, S. K. Brem, E. M. Evans, G. M. Sinatra, eds. *Evolution Challenges. Integrating Research and Practice in Teaching and Learning About Evolution*. Oxford: Oxford University Press.

(2020). The effects of invention and recontextualization on representing and reasoning with trees of life. *Research in Science Education* 50:1991–2033.

Mayr, E. (1969). *Principles of Systematic Zoology*. New York: McGraw-Hill.

(1971). *Populations, Species, and Evolution*. Cambridge, MA: The Belknap Press of Harvard University Press.

(1972). Lamarck revisited. *Journal of the History of Biology* 5:55–94.

(1982). *The Growth of Biological Thought. Diversity, Evolution, and Inheritance*. Cambridge, MA: The Belknap Press of Harvard University Press.

(1988). *Towards a New Philosophy of Biology. Observations of an Evolutionist*. Cambridge, MA: Harvard University Press.

Mayr, E., Bock, W. J. (2002). Classifications and other ordering systems. *Journal of Zoological Systematics and Evolutionary Research* 40:169–194.

McCann, J., Schneeweiss, G. M., Stuessy, T. F., Villasenor, J. L., Weiss-Schneeweiss, H. (2016). The impact of reconstruction methods, phylogenetic uncertainty and branch lengths on inference of chromosome number evolution in American saisies (*Melampodium*, Asteraceae). *PLoS ONE* 11: e0162299.

McGregor, A. P., Hilbrant, M., Pechmann, M., et al. (2008). *Cupiennius salei* and *Achaearanea tepidariorum*: spider models for investigating evolution and development. *BioEssays* 30:487–498.

McLaughlin, P. A., Lemaitre, R. (1997). Carcinization in the Anomura – fact or fiction? I. Evidence form adult morphology. *Contributions to Zoology* 67:79–123.

McLaughlin, P. A., Lemaitre, R., Tudge, C. C. (2004). Carcinization in the Anomura – fact or fiction? II. Evidence from larval, megalopal and early juvenile morphology. *Contributions to Zoology* 73:165–205.

McMurrich, J. P. (1891). The Gastraea Theory and its Successors. Pages 79–106. *Biological lectures delivered at the Marine Biological Laboratory of Wood's Holl in the summer session of 1890*. Boston: Ginn & Company.

McShea, D. W. (1991). Complexity and evolution: what everybody knows. *Biology & Philosophy* 6:303–324.

(2016). Three trends in the history of life: an evolutionary syndrome. *Evolutionary Biology* 43:531–542.

Mead, L. S. (2009). Transforming our thinking about transitional forms. *Evolution: Education and Outreach* 2:310–314.

Medawar, P. (1996). *The Strange Case of the Spotted Mice and Other Classic Essays in science*. Oxford: Oxford University Press.

Medawar, P. B., Medawar, J. S. (1977). *The Life Science. Current Ideas of Biology*. London: Wildwood House Ltd.

Meglitsch, P. A. (1967). *Invertebrate Zoology*. New York: Oxford University Press.

Meglitsch, P. A., Schram, F. R. (1991). *Invertebrate Zoology*. Oxford: Oxford University Press.

Mendez, F. L., Poznik, G. D., Castellano, S., Bustamante, C. D. (2016). The divergence of Neandertal and modern human Y chromosomes. *The American Journal of Human Genetics* 98:728–734.

Metchnikoff, O. (1921). *Life of Elie Metchnikoff 1845–1916*. London: Constable and Company Ltd.

Metschnikoff, E. (1874). Zur Entwickelungsgeschichte der Kalkschwämme. *Zeitschrift für wissenschaftliche Zoologie* 24:1–14.

(1882). Vergleichend-embryologische Studien. 3. Über die Gastrula einiger Metazoen. *Zeitschrift für wissenschaftliche Zoologie* 37:286–313.

(1883). Untersuchungen über die intracelluläre Verdauung bei wirbellosen Thieren. *Arbeiten aus dem Zoologischen Institut der Universität Wien und der Zoologischen Station in Triest* 5:141–168.

(1886). *Embryologische Studien an Medusen. Ein Beitrag zur Genealogie der Primitiv-Organe*. Vienna: Alfred Hölder.

Meyrick, E. (1895). *A Handbook of British Lepidoptera*. London: Macmillan.

Mikhailov, K. V., Konstantinova, A. V., Nikitin, M. A., et al. (2009). The origin of Metazoa: a transition from temporal to spatial cell differentiation. *BioEssays* 31:758–768.

Minelli, A. (2009). *Perspectives in Animal Phylogeny and Evolution*. Oxford: Oxford University Press.

(2016). Tracing homologies in an ever-changing world. *Rivista di estetica* 62:40–55.

Minot, C. S. (1897). Cephalic homologies. A contribution to the determination of the ancestry of vertebrates. *The American Naturalist* 31:927–943.

Mivart, S. G. (1874). *Man and Apes. An Exposition of Structural Resemblances and Differences Bearing upon Questions of Affinity and Origin*. New York: D. Appleton & Co.

Moczek, A. P., Sears, K. E., Stollewerk, A., et al. (2015). The significance and scope of evolutionary developmental biology: a vision for the 21st century. *Evolution & Development* 17:198–219.

Moore, J., Willmer, P. (1997). Convergent evolution in invertebrates. *Biological Reviews of the Cambridge Philosophical Society* 72:1–60.

Morgan, T. H. (1895). A study of metamerism. *Quarterly Journal of Microscopical Science* 2:395–476.

Moroz, L. L., Kohn, A. B. (2016). Independent origins of neurons and synapses: insights from ctenophores. *Philosophical Transactions of the Royal Society B: Biological Sciences* 371:201500241.

Morrison, D. A. (2014). Is the Tree of Life the best metaphor, model, or heuristic for phylogenetics? *Systematic Biology* 63:628–638.

(2015). [Review of] Aristotle's Ladder, Darwin's Tree. *Systematic Biology* 64:892–895.

(2016). Genealogies: pedigrees and phylogenies are reticulating networks not just divergent trees. *Evolutionary Biology* 43:456–473.

Moseley, H. N. (1874). On the structure and development of *Peripatus capensis*. *Philosophical Transactions of the Royal Society* 164:757–782.

Moseley, H. N., Sedgwick, A. (1882). Note on a discovery, as yet unpublished, by the late Professor F. M. Balfour, concerning the existence of a blastopore, and on the origin of the mesoblast in the embryo of *Peripatus Capensis*. *Proceedings of the Royal Society of London* 24:390–393.

Müller, F. (1869). *Facts and Arguments for Darwin*. London: John Murray.

Müller, I. (1973). Der "Hydriot" Nikolai Kleinenberg oder: Spekulation und Beobachtung. *Medizinhistorisches Journal* 8:131–153.

Naef, A. (1919). *Idealistische Morphologie und Phylogenetik (zur Methodik der Systematischen Morphologie)*. Jena: Verlag von Gustav Fischer.

(1928). *Cephalopoda. Embryology. Part I, Vol. II*. Washington D. C.: Smithsonian Institution Libraries.

Nakanishi, N., Sogabe, S., Degnan, B. M. (2014). Evolutionary origin of gastrulation: insights from sponge development. *BMC Biology* 12:26.

Nardi, F., Spinsanti, G., Boore, J. L., et al. (2003). Hexapod origins: monophyletic or paraphyletic? *Science* 299:1887–1889.

Negrisolo, E., Minelli, A., Valle, G. (2004). The mitochondrial genome of the house centipede *Scutigera* and the monophyly versus paraphyly of myriapods. *Molecular Biology and Evolution* 21:770–780.

Neiber, M. T., Hartke, T. R., Stemme, T., et al. (2011). Global biodiversity and phylogenetic evaluation of Remipedia (Crustacea). *PLoS ONE* 6:e19627.

Nelson, G. (1994). Homology and systematics. Pages 101–149 in B. K. Hall, ed. *Homology. The Hierarchical Basis of Comparative biology*. San Diego: Academic Press.

   (2004). Cladistics: its arrested development in D. M. Williams, P. L. Forey, eds. *Milestones in Systematics*. Boca Raton: CRC Press.

Nelson, G., Platnick, N. (1981). *Systematics and Biogeography. Cladistics and vicariance*. New York: Columbia University Press.

Nelson, G. J. (1970). Outline of a theory of comparative biology. *Systematic Zoology* 19:373–384.

   (1971). Paraphyly and polyphyly: redefinitions. *Systematic Zoology* 20:471–472.

Newell, N. D. (1959). The nature of the fossil record. *Proceedings of the American Philosophical Society* 103:264–285.

Newman, W. A. (2005). Origin of the Ostracoda and their maxillopodan and hexapodan affinities. *Hydrobiologia* 538:1–21.

Newman, W. A., Arnold, R. (2001). Prospectus on larval cirriped setation formulae, revisited. *Journal of Crustacean Biology* 21:56–77.

Nezlin, L. P. (2010). The golden age of comparative morphology: laser scanning microscopy and neurogenesis in trochophore animals. *Russian Journal of Developmental Biology* 41:381–390.

Nielsen, C. (1985). Animal phylogeny in the light of the trochaea theory. *Biological Journal of the Linnean Society* 25:243–299.

   (1987). Structure and function of metazoan ciliary bands and their phylogenetic significance. *Acta Zoologica* 68:205–262.

   (2001). *Animal Evolution. Interrelationships of the Living Phyla*. Oxford: Oxford University Press.

   (2003). Proposing a solution to the Articulata-Ecdysozoa controversy. *Zoologica Scripta* 32:475–482.

   (2008). Six major steps in animal evolution: are we derived sponge larvae? *Evolution & Development* 10:241–257.

   (2012). *Animal Evolution. Interrelationships of the Living Phyla, 3rd ed.* Oxford: Oxford University Press.

   (2013). Life cycle evolution: was the eumetazoan ancestor a holopelagic, planktotrophic gastraea? *BMC Evolutionary Biology* 13:1–1.

   (2017). Evolution of deuterostomy – and origin of the chordates. *Biological Reviews of the Cambridge Philosophical Society* 92:316–325.

   (2019). Early animal evolution: a morphologist's view. *Royal Society Open Science* 6:190638.

Nielsen, C., Brunet, T., Arendt, D. (2018). Evolution of the bilaterian mouth and anus. *Nature Ecology & Evolution* 2:1358–1376.

Nielsen, C., Nørrevang, A. (1985). The trochaea theory-an example of life cycle phylogeny. Pages 28–41 in S. Conway Morris, J. D. George, R. Gibson, H. M. Platt, eds. *The Origin and Relationships of Lower Invertebrate Groups*. Oxford: Oxford University Press.

Nielsen, C., Scharff, N., Eibye-Jacobsen, D. (1996). Cladistic analyses of the animal kingdom. *Biological Journal of the Linnean Society* 57:385–410.

Nieuwenhuys, R., Ten Donkelaar, H. J., Nicholson, C. (1998). *The Central Nervous System of Vertebrates. Volume 1*. Berlin: Springer-Verlag.

Nieuwkoop, P., Sutasurya, L. (1976). Embryological evidence for a possible polyphyletic origin of the recent amphibians. *Journal of Embryology and Experimental Morphology* 35:159–167.

Nixon, K. C., Carpenter, J. M. (1993). On outgroups. *Cladistics* 9:413–426.

Noever, C., Glenner, H. (2018). The origin of king crabs: hermit crab ancestry under the magnifying glass. *Zoological Journal of the Linnean Society* 182:300–318.

Nojiri, T., Wilson, L. A. B., Lopez-Aguirre, C., et al. (2021). Embryonic evidence uncovers convergent origins of laryngeal echolocation in bats. *Current Biology* 31:1353–1365.

Nolan, C. M., McCarthy, K., Eivers, E., Jirtle, R. L., Byrnes, L. (2006). Mannose 6-phosphate receptors in an ancient vertebrate, zebrafish. *Development, Genes and Evolution* 216:144–151.

Northcutt, R. G. (2012). Evolution of centralized nervous systems: two schools of evolutionary thought. *Proceedings of the National Academy of Sciences USA* 109:10626–10633.

Novick, L. R., Shade, C. K., Catley, K. M. (2011). Linear versus branching depictions of evolutionary history: implications for diagram design. *Topics in Cognitive Science* 3:536–559.

Nübler-Jung, K., Arendt, D. (1994). Is ventral in insects dorsal in vertebrates? A history of embryological arguments favouring axis inversion in chordate ancestors. *Roux's Archive for Developmental Biology* 203:357–366.

Nyhart, L. K. (1995). *Biology Takes Form. Animal Morphology and the German universities, 1800–1900.* Chicago: University of Chicago Press.

(2002). Learning from history: morphology's challenges in Germany ca. 1900. *Journal of Morphology* 252:2–14.

(2003). The importance of the "Gegenbaur School" for German morphology. *Theory in Biosciences* 122:162–173.

O'Hara, R. J. (1988). Homage to Clio, or, toward an historical philosophy for evolutionary biology. *Systematic Zoology* 37:142–155.

(1991). Representations of the natural system in the nineteenth century. *Biology & Philosophy* 6:255–274.

(1992). Telling the tree: narrative representation and the study of evolutionary history. *Biology & Philosophy* 7:135–160.

(1994). Evolutionary history and the species problem. *American Zoologist* 34:12–22.

(1997). Population thinking and tree thinking in systematics. *Zoologica Scripta* 26:323–329.

O'Leary, M. A., Bloch, J. I., Flynn, J. J., et al. (2013). The placental mammal ancestor and the post-K-Pg radiation of placentals. *Science* 339:662–667.

O'Reilly, J. E., dos Reis, M., Donoghue, P. C. J. (2015). Dating tips for divergence-time estimation. *Trends in Genetics* 31:637–650.

Oakley, T. H., Pankey, M. S. (2008). Opening the "Black Box": the genetic and biochemical basis of eye evolution. *Evolution: Education and Outreach* 1:390–402.

Oakley, T. H., Wolfe, J. M., Lindgren, A. R., Zaharoff, A. K. (2012). Phylotranscriptomics to bring the understudied into the fold: monophyletic Ostracoda, fossil placement, and pancrustacean phylogeny. *Molecular Biology and Evolution* 30:215–233.

Ochoa, C. (2021). Inertia, trend, and momentum reconsidered: G. G. Simpson—an orthogeneticist? Pages 261–290 in R. G. Delisle, ed. *Natural Selection. Revisiting its Explanatory Role in Evolutionary Biology.* Cham, Switzerland: Springer Nature Switzerland.

Ochoa, C., Barahona, A. (2014). *El jano de la morfología. De la homología a la homoplasia, historia, debates y evolución.* Mexico City: Universidad Nacional Autónoma de México and Centro de Estudios Filosóficos, Políticos y Sociales Vicente Lombardo Toledano.

364 REFERENCES

Ogilvie, B. W. (2006). *The Science of Describing. Natural History in Renaissance Europe.* Chicago: The University of Chicago Press.

Oikkonen, V. (2009). Narrating descent: popular science, evolutionary theory and gender politics. *Science as Culture* 18:1–21.

Olson, M. E. (2012). Linear trends in botanical systematics and the major trends of xylem evolution. *The Botanical Review* 78:154–183.

Olsson, L. (2007). A clash of traditions: the history of comparative and experimental embryology in Sweden as exemplified by the research of Gösta Jägersten and Sven Hörstadius. *Theory in Biosciences* 126:117–129.

Olsson, L., Levit, G. S., Hossfeld, U. (2017). The "Biogenetic Law" in zoology: from Ernst Haeckel's formulation to current approaches. *Theory in Biosciences* 136:19–29.

Omland, K. E., Cook, L. G., Crisp, M. D. (2008). Tree thinking for all biology: the problem with reading phylogenies as ladders of progress. *BioEssays* 30:854–867.

Ommundsen, A., Noever, C., Glenner, H. (2016). Caught in the act: phenotypic consequences of a recent shift in feeding strategy of the shark barnacle *Anelasma squalicola* (Lovén, 1844). *Zoomorphology* 135:51–65.

Oppenheimer, J. M. (1940). The non-specificity of the germ-layers. *The Quarterly Review of Biology* 15:1–27.

Ortega-Hernández, J., Janssen, R., Budd, G. E. (2019). The last common ancestor of Ecdysozoa had an adult terminal mouth. *Arthropod Structure & Development* 49:155–158.

Osborn, H. F. (1922). Orthogenesis as observed from paleontological evidence beginning in the year 1889. *The American Naturalist* 56:134–143.

Osigus, H.-J., Eitel, M., Schierwater, B. (2013). Chasing the urmetazoon: striking a blow for quality data? *Molecular Phylogenetics and Evolution* 66:551–557.

Ospovat, D. (1981). *The Development of Darwin's Theory. Natural History, Natural theology, and Natural Selection, 1838–1859.* Cambridge: Cambridge University Press.

Otto, S. P., Rosales, A. (2020). Theory in service of narratives in evolution and ecology. *The American Naturalist* 195:290–299.

Owen, R. (1848). *On the Archetype and Homologies of the Vertebrate Skeleton.* London: John van Voorst.

(1849). *On the Nature of Limbs. A Discourse.* London: John Van Voorst.

(1894). *The Life of Richard Owen, Volume II.* London: John Murray.

Padian, K. (2007). Richard Owen's Quadrophenia: The Pull of Opposing Forces in Victorian Cosmogony. PagesLIII–XCI in *On the Nature of Limbs: A Discourse.* Chicago: The University of Chicago Press.

Padian, K., Angielczyk, K. D. (2007). "Transitional forms" versus transtional features. Pages 197–230 in A. J. Petto, L. R. Godfrey, eds. *Scientists Confront Intelligent Design and Creationism.* New York: Norton.

Pagel, M. (1994). Detecting correlated evolution on phylogenies: a general method for the comparative analysis of discrete characters. *Proceedings of the Royal Society B: Biological Sciences* 255:37–45.

(1999). The maximum likelihood approach to reconstructing ancestral character states of discrete characters on phylogenies. *Systematic Biology* 48:612–622.

Pagel, M., Meade, A., Barker, D. (2004). Bayesian estimation of ancestral character states on phylogenies. *Systematic Biology* 53:673–684.

Pagel, M., Venditti, C., Meade, A. (2006). Large punctuational contribution of speciation to evolutionary divergence at the molecular level. *Science* 314:119–121.

Panchen, A. L. (1992). *Classification, Evolution, and the Nature of Biology*. Cambridge: Cambridge University Press.

(2001). Étienne Geoffory St.-Hilaire: father of "evo-devo"? *Evolution & Development* 3:41–46.

Pandey, A., Braun, E. L. (2020). Phylogenetic analyses of sites in different protein structural environments result in distinct placements of the metazoan root. *Biology* 9:64.

Parins-Fukuchi, C., Greiner, E., MacLatchy, L. M., Fisher, D. C. (2019). Phylogeny, ancestors, and anagenesis in the hominin fossil record. *Paleobiology* 45:378–393.

Parry, L., Tanner, A., Vinther, J. (2014). The origin of annelids. *Palaeontology* 57:1091–1103.

Patten, W. (1912). *The Evolution of the Vertebrates and Their Kin*. London: J. & A. Churchill.

(1920). *The Grand Strategy of Evolution. The Social Philosophy of a Biologist*. Boston: Richard G. Badger, The Gorham Press.

Patterson, C. (1978). Verifiability in systematics. Systematics *Zoology* 27:218–222.

(1981). Significance of fossils in determining evolutionary relationships. *Annual Review of Ecology and Systematics* 12:195–223.

(2002). Evolutionism and creationism. *The Linnean* 18:15–33.

Patterson, C., Williams, D. M., Gill, A. C. (2011). Adventures in the fish trade. *Zootaxa* 2946:118–136.

Penny, D. (2011). Darwin's Theory of Descent with Modification, versus the Biblical Tree of Life. *PloS Biology* 9:e1001096.

Perrard, A. (2020). Wasp waist and flight: convergent evolution in wasps reveals a link between wings and body shapes. *The American Naturalist* 195:181–191.

Peterson, K. J., Cameron, R. A., Davidson, E. H. (1997). Set-aside cells in maximal indirect development: evolutionary and developmental significance. *BioEssays* 19:623–631.

Peterson, K. J., Eernisse, D. J. (2001). Animal phylogeny and the ancestry of bilaterians: inferences from morphology and 18S rDNA gene sequences. *Evolution & Development* 3:170–205.

(2016). The phylogeny, evolutionary developmental biology, and paleobiology of the Deuterostomia: 25 years of new techniques, new discoveries, and new ideas. *Organisms Diversity & Evolution* 16:401–418.

Peterson, K. J., McPeek, M. A., Evans, D. A. D. (2005). Tempo and mode of early animal evolution: inferences from rocks, Hox, and molecular clocks. *Paleobiology* 31:36–55.

Philippe, H., Brinkmann, H., Copley, R. R., et al. (2011). Acoelomorph flatworms are deuterostomes related to *Xenoturbella*. *Nature* 470:255–258.

Pietsch, T. W. (2012). *Trees of Life. A Visual History of Evolution*. Baltimore: The Johns Hopkins University Press.

Pinto-Correia, C. (1997). *The Ovary of Eve. Egg and Sperm and Preformation*. Chicago: The University of Chicago Press.

Pobiner, B. (2016). Accepting, understanding, teaching, and learning (human) evolution: obstacles and opportunities. *American Journal of Physical Anthropology* 159:232–274.

Podani, J. (2019). The coral of life. *Evolutionary Biology* 46:123–144.

Podgorny, I. (2017). Manifest ambiguity: intermediate forms, variation, and mammal paleontology in Argentina, 1830–1880. *Studies in History and Philosophy of Biological and Biomedical Sciences* 66:27–36.

Poore, G. C. B. (2005). Peracarida: monophyly, relationships and evolutionary success. *Nauplius* 13:1–27.

Pozdnyakov, I. R., Sokolova, A. M., Ereskovsky, A. V., Karpov, S. A. (2017). Kinetid structure of choanoflagellates and choanocytes of sponges does not support their close relationship. *Protistology* 11:248–264.

Prendini, L. (2001). Species or supraspecific taxa as terminals in cladistic analysis? Groundplans versus examplars revisited. *Systematic Biology* 50:290–300.

Prothero, D. R. (2007). *Evolution. What the Fossils Say and Why it Matters*. New York: Columbia University Press.

Prüfer, K., et al., (2017). A high-coverage Neandertal genome from Vindija Cave in Croatia. *Science* 358:655–658.

Putten, K. v. (2020). Trees, coral, and seaweed: an interpretation of sketches found in Darwin's papers. *Journal of the History of Biology* 53:5–44.

Pyron, R. A., Burbrink, F. T. (2014). Early origin of viviparity and multiple reversions to oviparity in squamate reptiles. *Ecology Letters* 17:13–21.

Quammen, D. (2003). The man who knew too much: Stephen Jay Gould's opus posthumous. *Harper's Magazine* June:73–80.

Queiroz, K. d. (1988). Systematics and the Darwinian revolution. *Philosophy of Science* 55:238–259.

Quinn, A. (2017). When is a cladist not a cladist? *Biology & Philosophy* 32:581–598.

Raff, R. A. (1996). *The Shape of Life. Genes, Development, and the Evolution of Animal Form*. Chicago: The University of Chicago Press.

Ragan, M. A. (2009). Trees and networks before and after Darwin. *Biology Direct* 4:43.

Raia, P., Fortelius, M. (2013). Cope's Law of the Unspecialized, Cope's Rule, and weak directionality in evolution. *Evolutionary Ecology Research* 15:747–756.

Raineri, M. (2008). Old and new concepts in EvoDevo. Pages 95–114 in P. Pontarotti, ed. *Evolutionary Biology from Concept to Application*. Heidelberg: Springer-Verlag.

(2009). On some historical and theoretical foundations of the concept of chordates. *Theory in Biosciences* 128:53–73.

Rainger, R. (1981). The continuation of the morphological tradition: American paleontology, 1880–1910. *Journal of the History of Biology* 14:129–158.

(1985). Paleontology and philosophy: a critique. *Journal of the History of Biology* 18:267–287.

Rasser, M. W. (2013). Darwin's dilemma: the Steinheim snails' point of view. *Zoosystematics and Evolution* 89:13–20.

(2014). Evolution in isolation: the *Gyraulus* species flock from Miocene Lake Steinheim revisited. *Hydrobiologia* 739:7–24.

Raw, F. (1953). The external morphology of the trilobite and its significance. *Journal of Paleontology* 27:82–129.

(1960). Outline of a theory of origin of the vertebrate. *Journal of Paleontology* 34:497–539.

Redmond, A. K., McLysaght, A. (2021). Evidence for sponges as sister to all other animals from partitioned phylogenomics with mixture models and recoding. *Nature Communications* 12:1783.

Rees, David J., Noever, C., Høeg, Jens T., Ommundsen, A., Glenner, H. (2014). On the origin of a novel parasitic-feeding mode within suspension-feeding barnacles. *Current Biology* 24:1429–1434.

Regier, J. C., Shultz, J. W., Zwick, A., et al. (2010). Arthropod relationships revealed by phylogenomic analysis of nuclear protein-coding sequences. *Nature* 463:1079–1083.

Rehbock, P. F. (1975). Huxley, Haeckel, and the oceanographers: the case of *Bathybius haeckelii*. *Isis* 66:504–533.

(1983). *The Philosophical Naturalists. Themes in Early Nineteenth Century British Biology*. Madison, WI: The University of Wisconsin Press.

Reif, W.-E. (1983). Hilgendorf's (1863) dissertation on the Steinheim planorbids (Gastropoda; Miocene): the development of a phylogenetic research program for paleontology. *Paläontologische Zeitschrift* 57:7–20.

(1986). The search for a macroevolutionary theory in German paleontology. *Journal of the History of Biology* 19:79–130.

(2002). Evolution of organ systems: phylogeny, function and reciprocal illumination. *Senckenbergiana lethaea* 82:357–366.

Reisinger, E. (1972). Die Evolution des Orthogons der Spiralier und das Archicölomatenproblem. *Zeitschrift für zoologische Systematik und Evolutionsforschung* 10:1–43.

Reisz, R. R., Müller, J. (2004). The comparative method for evaluating fossil calibration dates: a reply to Hedges and Kumar. *Trends in Genetics* 20:596–597.

Remane, A. (1950). Entstehung der Metamerie der Wirbellosen. *Zoologischer Anzeiger Supplement* 14:16–23.

   (1956). *Die Grundlagen des Natürlichen Systems, der vergleichenden Anatomie und der Phylogenetik.* Leipzig: Akademische Verlagsgesellschaft Geest & Portig K.-G.

   (1963). The enterocoelic origin of the celom. Pages 78–90 in E. C. DOugherty, ed. *The Lower Metazoa. Comparative Biology and Phylogeny.* Berkeley, CA: University of California Press.

   (1967). Die Geschichte der Tiere. Pages 589–677 in G. Heberer, ed. *Die Evolution der Organismen,* vol. I. Jena: Gustav Fischer.

Rensch, B. (1960a). *Evolution above the Species Level.* New York: Columbia University Press.

   (1960b). The laws of evolution. Pages 95–116 in S. Tax, ed. *Evolution after Darwin. Volume 1. The Evolution of Life.* Chicago: The University of Chicago Press.

   (1991). *Das universale Weltbild.* Darmstadt: Wissenschaftliche Buchgesellschaft.

Reynolds, A., Hülsmann, N. (2008). Ernst Haeckel's discovery of *Magosphaera planula*: a vestige of metazoan origins? *History and Philosophy of the Life Sciences* 30:339–386.

Richards, R. J. (1992a). *The Meaning of Evolution. The Morphological Construction and Ideological Reconstruction of Darwin's Theory.* Chicago: The University of Chicago Press.

   (1992b). The structure of narrative explanation in history and biology. Pages 19–53 in M. H. Nitecki, D. V. Nitecki, eds. *History and Evolution.* New York: State University and New York Press.

   (2000). The epistemology of historical interpretation. Progressivity and recapitulation in Darwin's theory. Pages 64–88 in R. Creath, J. Maienschein, eds. *Biology and Epistemology.* Cambridge: Cambridge University Press.

   (2002). *The Romantic Conception of Life. Science and Philosophy in the Age of Goethe.* Chicago: The University of Chicago Press.

   (2006). Nature is the poetry of mind, or how Schelling solved Goethe's Kantian problems. Pages 27–50 in M. Friedman, A. Nordman, eds. *The Kantian Legacy in Nineteenth Century Science.* Cambridge, MA: MIT Press.

   (2008). *The Tragic Sense of Life. Ernst Haeckel and the Struggle over Evolutionary Thought.* Chicago: The University of Chicago Press.

   (2013). *Was Hitler a Darwinian? Disputed Questions in the History of Evolutionary Theory.* Chicago: The University of Chicago Press.

Richards, R. J., Ruse, M. (2016). *Debating Darwin.* Chicago: The University of Chicago Press.

Rieger, R. M. (1986). Über den Ursprung der Bilateria: die Bedeutung der Ultrastrukturforschung für ein neues Verstehen der Metazoenevolution. *Verhandlungen der Deutschen Zoologischen Gesellschaft* 79:31–50.

   (1994). The biphasic life cycle – a central theme of metazoan evolution. *American Zoologist* 34:484–491.

   (2003). The phenotypic transition from uni- to multicellular animals. Pages 247–258 in A. Legakis, S. Sfenthourakis, R. Polymeni, M. Thessalou-Legaki, eds. *The New Panorama of Animal Evolution.* Sofia: PENSOFT Publishers.

Rieger, R. M., Haszprunar, G., Schuchert, P. (1991). On the origin of the Bilateria: traditional views and recent alternative concepts. Pages 107–112 in A. M. Simonetta, S. Conway Morris, eds. *The Early Evolution of Metazoa and the Significance of Problematic Taxa.* Cambridge: Cambridge University Press.

Riegner, M. F. (2013). Ancestor of the new archetypal biology: Goethe's dynamic typology as a model for contemporary evolutionary developmental biology. *Studies in History and Philosophy of Biological & Biomedical Sciences* 44:735–744.

Rieppel, O. (1985). Ontogeny and the hierarchy of types. *Cladistics* 1:234–246.

(1986). Atomism, epigenesis, preformation and pre-existence: a clarification of terms and consequences. *Biological Journal of the Linnean Society* 28:331–341.

(1988a). *Fundamentals of Comparative Biology*. Basal: Birkhäuser Verlag.

(1988b). Louis Agassiz (1807–1973) and the reality of natural groups. *Biology & Philosophy* 3:29–47.

(2001a). Étienne Geoffroy Saint-Hilaire (1772–1844). Pages 157–175 in I. Jahn, M. Schmitt, eds. *Darwin & Co.* München: Verlag C. H. Beck.

(2001b). Preformationist and epigenetic biases in the history of the morphological character concept. Pages 57–75 in G. P. Wagner, ed. *The Character Concept in Evolutionary Biology*. San Diego: Academic Press.

(2006). On concept formation in systematics. *Cladistics* 22:474–492.

(2007). The nature of parsimony and instrumentalism in systematics. *Journal of Zoological Systematics and Evolutionary Research* 45:177–183.

(2010a). The series, the network, and the tree: changing metaphors of order in nature. *Biology & Philosophy* 25:475–496.

(2010b). Sinai Tschulok (1875–1945) – a pioneer of cladistics. *Cladistics* 26:103–111.

(2011a). Adolf Naef (1883–1949), systematic morphology and phylogenetics. *Journal of Zoological Systematics and Evolutionary Research* 50:2–13.

(2011b). Ernst Haeckel (1834–1919) and the monophyly of life. *Journal of Zoological Systematics and Evolutionary Research* 49:1–5.

(2011c). The Gegenbaur Transformation: a paradigm change in comparative biology. *Systematics and Biodiversity* 9:177–190.

(2011d). Species are individuals-the German tradition. *Cladistics* 27:629–645.

(2012a). The dark side of the moon. *Cladistics* 28:1–3.

(2012b). Othenio Abel (1875–1946) and "the phylogeny of the parts". *Cladistics* 29:328–335.

(2013a). Biological Individuals and natural kinds. *Biological Theory* 7:162–169.

(2013b). Othenio Abel (1875–1946): the rise and decline of paleobiology in German paleontology. *Historical Biology* 25:313–325.

(2016). *Phylogenetic Systematics. Haeckel to Hennig*. Boca Raton, FL: CRC Press.

(2020). Morphology and phylogeny. *Journal of the History of Biology* 53:217–230.

Rieppel, O., Kearney, M. (2002). Similarity. *Biological Journal of the Linnean Society* 75:59–82.

Rieppel, O., Williams, D. M., Ebach, M. C. (2013). Adolf Naef (1883–1949): on foundational concepts and principles of systematic morphology. *Journal of the History of Biology* 46:445–510.

Rigato, E., Minelli, A. (2013). The great chain of being is still here. *Evolution: Education and Outreach*. 6:18

Romer, A. S. (1946). The early evolution of fishes. *The Quarterly Review of Biology* 21:33–69.

(1964). *Vertebrate Paleontology*. Chicago: The University of Chicago Press.

(1972). The vertebrate as a dual animal – somatic and visceral. *Evolutionary Biology* 6:121–156.

Romero, A. (2001). Scientists prefer them blind: the history of hypogean fish research. *Environmental Biology of Fishes* 62:43–71.

Ros-Rocher, N., Perez-Posada, A., Leger, M. M., Ruiz-Trillo, I. (2021). The origin of animals: an ancestral reconstruction of the unicellular-to-multicellular transition. *Open Biology* 11:200359.

Rosen, D. E., Forey, P. L., Gardiner, B. G., Patterson, C. (1981). Lungfishes, tetrapods, paleontology, and plesiomorphy. *Bulletin of the American Museum of Natural History* 167:159–276.

Rouse, G. W., Fauchald, K. (1997). Cladistics and polychaetes. *Zoologica Scripta* 26:139–204.

Rouse, G. W., Pleijel, F. (2001). *Polychaetes*. Oxford: Oxford University Press.

Rudwick, M. J. S. (1985). *The Meaning of Fossils. Episodes in the History of Palaeontology*. Chicago: The University of Chicago Press.

(1992). *Scenes from Deep Time. Early Pictorial Representations of the Prehistoric World*. Chicago: The University of Chicago Press.

(1997). *Georges Cuvier, Fossil Bones, and Geological Catastrophes. New Translations & Interpretations of the Primary Texts*. Chicago: The University of Chicago Press.

(2005). *Bursting the Limits of time. The Reconstruction of Geohistory in the Age of Revolution*. Chicago: The University of Chicago Press.

(2008). *Worlds before Adam. The Reconstruction of Geohistory in the Age of Reform*. Chicago: The University of Chicago Press.

Ruiz-Trillo, I., de Mendoza, A. (2020). Towards understanding the origin of animal development. *Development* 147:dev192575.

Rupke, N. A. (1993). Richard Owen's vertebrate archetype. *Isis* 84:231–251.

(2005). Neither creation nor evolution: the third way in mid-nineteenth century thinking about the origin of species. *Annals of the History and Philosophy of Biology* 10:143–172.

(2009). *Richard Owen. Biology without Darwin, A Revised Edition*. London: The University of Chicago Press.

Ruppert, E. E., Fox, R. S., Barnes, R. D. (2004). *Invertebrate Zoology*. Belmont: Brooks/Cole.

Ruse, M. (1996). *Monad to Man. The Concept of Progress in Evolutionary Biology*. Cambridge, MA: Harvard University Press.

(2000). Darwin and the philosophers. Epistemological factors in the development and reception of the theory of the *Origin of Species*. Pages 3–26 in R. Creath, J. Maienschein, eds. *Biology and epistemology*. Cambridge: Cambridge University Press.

Russell, E. S. (1916). *Form and Function. A Contribution to the History of Animal Morphology*. London: John Murray.

Salvini-Plawen, L. v. (1978). On the origin and evolution of the lower Metazoa. *Zeitschrift für zoologische Systematik und Evolutionsforschung* 16:40–88.

(1980). Phylogenetischer Status und Bedeutung der mesenchymaten Bilateria. *Zoologische Jahrbücher, Abteilung für Anatomie und Ontogenie der Tiere* 103:354–373.

(1982). A paedomorphic origin of the oligomerous animals? *Zoologica Scripta* 11:77–81.

(1985). Early evolution and the primitive groups. *Mollusca* 10:59–150.

(1989). Mesoderm heterochrony and metamery in Chordata. *Fortschritte der Zoologie* 35:213–219.

(1998a). Morphologie: Haeckel's Gastraea-Theorie und ihre Folgen. *Stapfia* 56:147–168.

(1998b). The urochordate larva and archichordate organization: chordate origin and anagenesis revisited. *Journal of Zoological Systematics and Evolutionary Research* 36:129–145.

(2000). What is convergent/homoplastic in Pogonophora? *Journal of Zoological Systematics and Evolutionary Research* 38:133–147.

Salvini-Plawen, L. v., Steiner, G. (2014). The Testaria concept (Polyplacophora+ Conchifera) updated. *Journal of Natural History* 48:2751–2772.

Sanders, H. L. (1955). The Cephalocarida, a new subclass of Crustacea from Long Island Sound. *Proceedings of the National Academy of Sciences USA* 41:61–66.

(1957). The Cephalocarida and crustacean phylogeny. *Systematic Zoology* 6:112–128, 148.

Sandvik, H. (2008). Tree thinking cannot taken for granted: challenges for teaching phylogenetics. *Theory in Biosciences* 127:45–51.

(2009). Anthropocentrisms in cladograms. *Biology & Philosophy* 24:425–440.

Satoh, N., Rokhsar, D., Nishikawa, T. (2014). Chordate evolution and the three-phylum system. *Proceedings of the Royal Society B: Biological Sciences* 281:20141729.

Schaeffer, B., Hecht, M. K., Eldredge, N. (1972). Phylogeny and paleontology. *Evolutionary Biology* 6:31–46.

Schaffner, J. H. (1937). Examples of orthogenetic series in plants and animals. *The Ohio Journal of Science* 37:267–287.

Schechter, V. (1959). *Invertebrate Zoology.* Englewood Cliffs, NJ: Prentice-Hall.

Scheltema, A. H. (1993). Aplacophora as progenetic aculiferans and the coelomate origin of mollusks as the sister taxon of Sipuncula. *Biological Bulletin* 184:57–78.

Scheltema, A. H., Schander, C. (2006). Exoskeletons: tracing molluscan evolution. *Venus* 65:19–26.

Schierwater, B. (2005). My favorite animal, *Trichoplax adhaerens. BioEssays* 27:1294–1302.

Schierwater, B., de Jong, D., Desalle, R. (2009a). Placozoa and the evolution of Metazoa and intrasomatic cell differentiation. *The International Journal of Biochemistry & Cell Biology* 41:370–379.

Schierwater, B., Eitel, M., Jakob, W., et al. (2009b). Concatenated analysis sheds light on early metazoan evolution and fuels a modern "Urmetazoon" hypothesis. *PLoS Biology* 7:e20.

Schierwater, B., Eitel, M., Osigus, H.-J., et al. (2011). *Trichoplax* and Placozoa: one of the crucial keys to understanding metazoan evolution. Pages 289–326 in R. DeSalle, B. Schierwater, eds. *Key Transitions in Animal Evolution.* Boca Raton, FL: CRC Press.

Schiffer, P. H., Robertson, H. E., Telford, M. J. (2018). Orthonectids are highly degenerate annelid worms. *Current Biology* 28:1970–1974.

Schileyko, A. A., Pavlinov, I. J. (1997). A cladistic analysis of the order Scolopendromorpha (Chilopoda). *Entomologica Scandinavica* 51:33–40.

Schindewolf, O. H. (1927). Prinzipienfragen der biologischen Systematik. *Palaeontologische Zeitschrift* 9:122–169.

(1962). "Neue Systematik." *Palaeontologische Zeitschrift* 36:59–78.

(1993). *Basic Questions in Paleontology. Geologic Time, Organic Evolution, and Biological Systematics.* Chicago: The University of Chicago Press.

Schluter, D., Price, T., Mooers, A. O., Ludwig, D. (1997). Likelihood of ancestor states in adaptive radiation. *Evolution* 51:1699–1711.

Schmidt-Rhaesa, A. (2007). *The Evolution of Organ Systems.* Oxford: Oxford University Press.

Schmitt, S. (2004). *Histoire d'une Question Anatomique: La Répétition des Parties.* Paris: Publications Scientifiques du Muséum National d'Histoire Naturelle.

(2009). From physiology to classification: comparative anatomy and Vicq d'Azyr's plan of reform for life sciences and medicine (1774–1794). *Science in Context* 22:145–193.

(2017). Serial homology as a challenge to evolutionary theory. The repeated parts of organisms from idealistic morphology to evo-devo. Pages 317–347 in P. Huneman, D. Walsh, eds. *Challenging the Modern Synthesis. Adaptation, Development, and Inheritance.* Oxford: Oxford University Press.

Schram, F. R. (1978). Arthropods: a convergent phenomenon. *Fieldiana Geology* 39:61–108.

(1982). The fossil record and evolution of Crustacea. Pages 93–147 in L. G. Abele, ed. *The Biology of Crustacea, Volume 1.* New York: Academic Press.

(1983). Remipedia and crustacean phylogeny. Pages 23–28 in F. R. Schram, ed. *Crustacean Issues Vol. 1. Crustacean phylogeny.* Rotterdam: Balkema.

(1986). *Crustacea.* Oxford: Oxford University Press.

(1991). Cladistic analysis of metazoan phyla and the placement of fossil problematica. Pages 35–46 in A. M. Simonetta, S. Conway Morris, eds. *The Early Evolution of Metazoa and the Significance of Problematic Taxa.* Cambridge: Cambridge University Press.

Schram, F. R., Hof, C. H. J. (1998). Fossils and interrelationships of major crustacean groups. Pages 233–302 in G. D. Edgecombe, ed. *Arthropod Fossils and Phylogeny.* New York: Columbia University Press.

Schram, F. R., Koenemann, S. (2004). Are the crustaceans monophyletic? Pages 319-329 in J. Cracraft, M. J. Donoghue, eds. *Assembling the Tree of Life.* New York: Oxford University Press.

Schrödl, M. (2014). Time to say "Bye-bye Pulmonata"? *Spixiana* 37:161-164.

Schuchert, C. (1926). Biographical memoir of John Mason Clarke. *National Academy Biographical Memoirs* 12:183-244.

Schuchert, P. (1993). Phylogenetic analysis of the Cnidaria. *Zeitschrift für zoologische Systematik und Evolutionsforschung* 31:161-173.

Schulze, F. E. (1883). *Trichiplax adhaerens,* nov. gen., nov. spec. *Zoologischer Anzeiger* 6:92-97.

Schwalbe, G. (1899). Studien über *Pithecanthropus erectus* Dubois. *Zeitschrift für Morphologie und Anthropologie* 1:16-240.

(1906). *Studien zur Vorgeschichte des Menschen.* Stuttgart: Verlag der E. Schwiezerbartschen Verlagsbuchhandlung (E. Nägele).

Schwentner, M., Combosch, D. J., Pakes Nelson, J., Giribet, G. (2017). A phylogenomic solution to the origin of insects by resolving crustacean-hexapod relationships. *Current Biology* 27:1818-1824.

Schwentner, M., Richter, S., Rogers, D. C., Giribet, G. (2018). Tetraconatan phylogeny with special focus on Malacostraca and Branchiopoda: highlighting the strength of taxon-specific matrices in phylogenomics. *Proceedings of the Royal Society B: Biological Sciences* 285:20181524.

Scott, E. C. (2004). *Evolution vs. Creationism. An Introduction.* Berkeley: University of California Press.

Secord, J. (1991). Edinburgh Lamarckians: Robert Jameson and Robert E. Grant. *Journal of the History of Biology* 24:1-18.

Sedgwick, A. (1884). On the origin of metameric segmentation and some other morphological questions. *Quarterly Journal of Microscopical Science* 24:43-82.

(1885). The development of Peripatus Capensis. *Proceedings of the Royal Society of London* 38:354-361.

Seligmann, H. (2010). Positive correlations between molecular and morphological rates of evolution. *Journal of Theoretical Biology* 264:799-807.

Semper, C. (1875). Die Stammesverwandschaft der Wirbelthiere und Wirbellosen. *Arbeiten aus dem Zoologisch-Zootomischen Institut in Würzburg* 2:25-76.

Sewertzoff, A. N. (1929). Directions of evolution. *Acta Zoologica* 10:59-141.

(1931). *Morphologische Gesetzmässigkeiten der Evolution.* Jena: Verlag von Gustav Fischer.

(1932). Ueber die Bedeutung des Princips der Substitution und einiger anderer Principien in der Phylogenese. *Archivio Zoologico Italiano* 16:128-139.

Shipman, P. (2001). *The Man Who Found the Missing Link. The Extraordinary Life of Eugene Dubois.* New York: Simon & Schuster.

Shipman, P., Storm, P. (2002). Missing links: Eugène Dubois and the origins of paleoanthropology. *Evolutionary Anthropology* 11:108-116.

Shubin, N., Tabin, C., Carroll, S. (2009). Deep homology and the origins of evolutionary novelty. *Nature* 457:818-823.

Siewing, R. (1969). Diskussionsbeitrag zur Phylogenie der Coelomaten. *Zoologischer Anzeiger* 179:132-176.

(1972). Zur Deszendenz der Chordaten - Erwiderung und Versuch einer Geschichte der Archicoelomaten. *Zeitschrift für zoologische Systematik und Evolutionsforschung* 10:267-291.

(1976). Probleme und neuere Erkenntnisse in der Groåsystematik der Wirbellosen. *Verhandlungen der Deutschen Zoologischen Gesellschaft* 70:59-83.

(1980). *Das Archicoelomatenkonzept. Zool. Jahrb. Abt. Anat.* 103:439-482.

Simion, P., Philippe, H., Baurain, D., et al. (2017). A large and consistent phylogenomic dataset supports sponges as the sister group to all other animals. *Current Biology* 27:958-967.

Simpson, G. G. (1944). *Tempo and Mode in Evolution.* New York: Hafner Publishing Company.

(1953). *The Major Features of Evolution*. New York: Columbia University Press.

(1956). *The Meaning of Evolution. A Special Revised and Abridged Edition*. New York: Mentor Books.

(1959). Anatomy and morphology: classification and evolution: 1859 and 1959. *Proceedings of the American Philosophical Society* 103:286–306.

Siomava, N., Fuentes, J. S. M., Diogo, R. (2020). Deconstructing the long-standing a priori assumption that serial homology generally involves ancestral similarity followed by anatomical divergence. *Journal of Morphology* 281:1110–1132.

Sloan, P. R. (2003). Whewell's philosophy of discovery and the archetype of the vertebrate skeleton: the role of German philosophy of science in Richard Owen's biology. *Annals of Science* 60:39–61.

Smith, A. B. (1994). *Systematics and the Fossil Record. Documenting Evolutionary Patterns*. Oxford: Blackwell Scientific Publications.

Smith, J. P. (1900). The biogenetic law from the standpoint of paleontology. *The Journal of Geology* 8:413–425.

Snodgrass, R. E. (1935). *The Principles of Insect Morphology*. New York: McGraw-Hill.

(1938). Evolution of Annelida, Onychophora and Arthropoda. *Smithsonian Miscellaneous Collections* 97:1–159.

Sober, E. (2008). *Evidence and Evolution. The Logic Behind the Science*. Cambridge: Cambridge University Press.

(2009). Did Darwin write the *Origin* backwards? *Proceedings of the National Academy of Sciences USA* 106:10048–10055.

Sogabe, S., Hatleberg, W. L., Kocot, K. M., et al. (2019). Pluripotency and the origin of animal multicellularity. *Nature* 570:519–522.

Sollas, W. J. (1884). On the development of *Halisarca lobularis* (O. Schmidt). *Quarterly Journal of Microscopical Science* 24:603–621.

Sørensen, M. V., Funch, P., Willerslev, E., Hansen, A. J., Olesen, J. (2000). On the phylogeny of the Metazoa in the light of Cycliophora and Micrognathozoa. *Zoologischer Anzeiger* 239:297–318.

Sperling, E., Peterson, K., Pisani, D. (2009). Phylogenetic-signal dissection of nuclear housekeeping genes supports the paraphyly of sponges and the monophyly of Eumetazoa. *Molecular Biology and Evolution* 26:2261–2274.

Sperling, E. A., Pisani, D., Peterson, K. J. (2007). Poriferan paraphyly and its implications for Precambrian palaeobiology. *Geological Society, London, Special Publications* 286:355–368.

Stafleu, F. A. (1971). Lamarck: the birth of biology. *Taxon* 20:397–442.

Stamos, D. N. (2005). Pre-Darwinian taxonomy and essentialism – a reply to Mary Winsor. *Biology & Philosophy* 20:79–96.

Starunov, V. V., Dray, N., Belikova, E. V., et al. (2015). A metameric origin for the annelid pygidium? *BMC Evolutionary Biology* 15:25.

Steigerwald, J. (2002). Goethe's morphology: *Urphänomene* and aesthetic appraisal. *Journal of the History of Biology* 35:291–328.

Steinböck, O. (1958a). Schlusswort zur Diskussion Remane-Steinböck. *Zoologischer Anzeiger Supplement* 21:196–218.

(1958b). Zur Phylogenie der Gastrotrichen. *Zoologischer Anzeiger Supplement* 21:128–169.

(1963). Origin and affinities of the lower Metazoa. The "aceloid" ancestry of the Eumetazoa. Pages 40–54 in E. C. Dougherty, Z. Norwood Brown, E. D. Hanson, W. D. Hartman, eds. *The Lower Metazoa. Comparative Biology and Phylogeny*. Berkeley: University of California Press.

Steinmetz, P. R. H. (2019). A non-bilaterian perspective on the development and evolution of animal digestive systems. *Cell and Tissue Research* 377:321–339.

Stevens, P. F. (1994). *The Development of Biological Systematics. Antoine-Laurent de Jussieu, Nature and the Natural System*. New York: Columbia University Press.

Stoczkowski, W. (2002). *Explaining Human Origins. Myth, Imagination and Conjecture*. Cambridge: Cambridge University Press.

Strausfeld, N. J., Hirth, F. (2013). Homology versus convergence in resolving transphyletic correspondences of brain organization. *Brain Behavior and Evolution* 82:215-219.

Strickland, H. E. (1840). Observations upon the affinities and analogies of organized beings. *Magazine of Natural History* 4:219-226.

   (1841). On the true method of discovering the natural system in zoology and botany. *The Annals and Magazine of Natural History* 6:184-194.

Striedter, G. F. (2020). A history of ideas in evolutionary neuroscience. Pages 3-16 in J. H. Kaas, ed. *Evolutionary Neuroscience*. London: Academic Press.

Struck, T. H., Paul, C., Hill, N., et al. (2011). Phylogenomic analyses unravel annelid evolution. *Nature* 470:95-98.

Tamborini, M. (2017). The reception of Darwin in late nineteenth-century German paleontology as a case of pyrrhic victory. *Studies in History and Philosophy of Biological and Biomedical Sciences* 66:37-45.

Tassy, P. (2011). Trees before and after Darwin. *Journal of Zoological Systematics and Evolutionary Research* 49:89-101.

Tattersall, I. (2000). Paleoanthropology: the last half-century. *Evolutionary Anthropology* 9:2-16.

Tattersall, I. (2013). Stephen J. Gould's intellectual legacy to anthropology. Pages 115-127 in G. A. Danieli, A. Minelli, T. Pievani, eds. *Stephen J. Gould: The Scientific Legacy*. Milan: Springer Verlag.

   (2015). *The Strange Case of the Rickety Cossack. And Other Cautionary Tales from Human Evolution*. New York: Palgrave Macmillan.

Tattersall, I., Eldredge, N. (1976). Fact, theory, and fantasy in human paleontology. *American Scientist* 65:204-211.

Tauber, A. I. (2003). Metchnikoff and the phagocytosis theory. *Nature Reviews Molecular Cell Biology* 4:897-901.

Taylor, P. D. (2020). *Bryozoan Paleobiology*. Hoboken, NJ: Wiley-Blackwell.

Telford, M. J. (2006). Animal phylogeny. *Current Biology* 16:R981-R985.

   (2007). A single origin of the central nervous system? *Cell* 129:237-239.

Telford, M. J., Budd, G. E. (2011). Invertebrate evolution: bringing order to the molluscan chaos. *Current Biology* 21:R964-R966.

Telford, M. J., Budd, G. E., Philippe, H. (2015). Phylogenomic Insights into animal evolution. *Current Biology* 25:R876-R887.

Tessmar-Raible, K., Arendt, D. (2003). Emerging systems: between vertebrates and arthropods, the Lophotrochozoa. *Current Opinion in Genetics & Development* 13:331-340.

Theobald, D. L. (2010). A formal test of the theory of universal common ancestry. *Nature* 465:219-222.

Theunissen, B. (1989). *Eugène Dubois and the Ape-Man from Java. The History of the First "Missing Link" and Its Discoverer*. Dordrecht: Kluwer.

Tiegs, O. W., Manton, S. M. (1958). The evolution of the Arthropoda. *Biological Reviews of the Cambridge Philosophical Society* 33:255-333.

Trienes, R. (1989). Type concept revisited. A survey of German idealistic morphology in the first half of the twentieth century. *History and Philosophy of the Life Sciences* 11:23-42.

Trueman, J. W. H., Pfeil, B. E., Kelchner, S. A., Yeates, D. K. (2004). Did stick insects *really* regain their wings? *Systematic Entomology* 29:138-139.

Tsai, C.-H., Fordyce, R. E. (2015). Ancestor-descendant relationships in evolution: origin of the extant pygmy right whale, Caperea marginata. *Biology Letters* 11:20140875-20140875.

Tsang, L. M., Chan, T.-Y., Ahyong, S. T., Chu, K. H. (2011). Hermit to king, or hermit to all: multiple transitions to crab-like forms from hermit crab ancestors. *Systematic Biology* 60:616–629.

Tschopp, P., Tabin, C. J. (2017). Deep homology in the age of next-generation sequencing. *Philosophical Transactions of the Royal Society B: Biological Sciences* 372:20150475.

Tschulok, S. (1922). *Deszendenzlehre*. Jena: Gustav Fischer.

Tyndall, J. (1870). *The Scientific Use of the Imagination*. London: Longmans, Green, and Co.

Ulett, M. A. (2014). Making the case for orthogenesis: the popularization of definitely directed evolution (1890–1926). *Studies in History and Philosophy of Biological and Biomedical Sciences* 45:124–132.

Ulrich, W. (1972). Die Geschichte des Archicoelomatenbegriffs und die Archicoelomatennatur der Pogonophoren. *Zeitschrift für zoologische Systematik und Evolutionsforschung* 10:301–320.

Vagvolgyi, J. (1967). On the origin of molluscs, the coelom, and coelomic segmentation. *Systematic Zoology* 16:153–168.

Valentine, J. W. (2004). *On the Origin of Phyla*. Chicago: The University of Chicago Press.

Valentine, J. W., Marshall, C. R. (2015). Fossil and transcriptomic perspectives on the origins and success of metazoan multicellularity. Pages 31–46 in I. Ruiz-Trillo, A. M. Nedelcu, eds. *Evolutionary Transitions to Multicellular Life. Principles and Mechanisms*. Dordrecht: Springer.

Van Cleave, H. J. (1931). *Invertebrate Zoology*. New York: McGraw-Hill.

Van den Biggelaar, J. A. M., Edsinger-Gonzales, E., Schram, F. R. (2002). The improbability of dorso-ventral axis inversion during animal evolution, as presumed by Geoffroy-Saint Hilaire. *Contributions to Zoology* 71:29–36.

Van der Hammen, L. (1981). Type-concept, higher classification and evolution. *Acta Biotheoretica* 30:3–48.

Van Dijk, E. M., Reydon, T. A. C. (2010). A conceptual analysis of evolutionary theory for teacher education. *Science & Education* 19:655–677.

Vanderlaan, T., Ebach, M., Williams, D., Wilkins, J. (2013). Defining and redefining monophyly: Haeckel, Hennig, Ashlock, Nelson and the proliferation of definitions. *Australian Systematic Botany* 26:347–355.

Vaux, F., Trewick, S. A., Morgan-Richards, M. (2016). Lineages, splits and divergence challenge whether the terms anagenesis and cladogenesis are necessary. *Biological Journal of the Linnean Society* 117:165–176.

Venditti, C., Meade, A., Pagel, M. (2006). Detecting the node-density artifact in phylogeny reconstruction. *Systematic Biology* 55:637–643.

Vigors, N. A. (1825). Observations on the natural affinities that connect the orders and families of birds. *Transactions of the Linnean Society of London* 14:395–517.

Vikhanski, L. (2016). *Immunity. How Elie Metchnikoff Changed the Course of Modern Medicine*. Chicago: Chicago Review Press.

Vinther, J., Parry, L., Briggs, D. E. G., Roy, P. V. (2017). Ancestral morphology of crown-group molluscs revealed by a new Ordovician stem aculiferan. *Nature Publishing Group* 542:471–474.

Voje, K. L. (2016). Tempo does not correlate with mode in the fossil record. *Evolution* 70:2678–2689.

Voje, K. L., Di Martino, E., Porto, A. (2020). Revisiting a landmark study system: no evidence for a punctuated mode of evolution in *Metrarabdotos*. *American Naturalist* 195:899–917.

von Baer, K. E. (1828). *Über Entwicklungsgeschichte der Thiere. Beobachtung und Reflexion. Erster Theil*. Königsberg: Gebrüdern Bornträger.

(1876). *Reden gehalten in wissenschaftlichen Versammlungen und kleinere Aufsätze vermischten Inhalts, Zweiter Theil*. St. Petersburg: Verlag der Kaiserlichen Hofbuchhandlung, H. Schmitzdorff.

von Graff, L. (1891). *Die Organisation der Turbellaria Acoela*. Leipzig: Verlag von Wilhelm Engelmann.

Von Ihering, H. (1877). *Vergleichende Anatomie des Nervensystemes und Phylogenie der Mollusken.* Leipzig: Verlag von Wilhelm Engelmann.

von Reumont, B. M., Jenner, R. A., Wills, M. A., et al. (2012). Pancrustacean phylogeny in the light of new phylogenomic data: support for Remipedia as the sister group of Hexapoda. *Molecular Biology and Evolution* 29:1031–1045.

Vopalensky, P., Pergner, J., Liegertova, M., et al. (2012). Molecular analysis of the amphioxus frontal eye unravels the evolutionary origin of the retina and pigment cells of the vertebrate eye. *Proceedings of the National Academy of Sciences USA* 109:15383–15388.

Waagen, W. (1869). Die Formenreihe des Ammonites Subradiatus. Versuch einer Paläontologischen Monographie. *Geognostisch-Paläontologische Beiträge* 2:179–256.

Wägele, J. W., Erikson, T., Lockart, P., Misof, B. (1999). The Ecdysozoa: artifact or monophylum? *Journal of Zoological Systematics and Evolutionary Research* 37:211–223.

Wägele, J. W., Kück, P. (2013). Arthropod phylogeny and the origin of Tracheata (= Atelocerata) from Remipedia-like ancestors. Pages 285–341 in J. W. Wägele, T. Bartolomaeus, eds. *Deep Metazoan Phylogeny: The Backbone of the Tree of Life.* Berlin: De Gruyter.

Wägele, J. W., Misof, B. (2001). On quality of evidence in phylogeny reconstruction: a reply to Zrzavy's defence of the "Ecdysozoa" hypothesis. *Journal of Zoological Systematics and Evolutionary Research* 39:165–176.

Wagner, G. P. (2014). *Homology, Genes, and Evolutionary Innovation.* Princeton: Princeton University Press.

   (2015). Evolutionary innovations and novelties: let us get down to business! *Zoologischer Anzeiger* 256:75–81.

Wake, M. H. (2003). Reproductive modes, ontogenies, and the evolution of body form. *Animal Biology* 53:209–223.

Walcott, C. D. (1912). Middle Cambrian Branchiopoda, Malacostraca, Trilobita, and Merostomata. *Smithsonian Miscellaneous Collections* 57:145–228.

Wallace, A. R. (1855). On the law which has regulated the introduction of new species. *Annals and Magazine of Natural History 2nd Series* 16:184–196.

   (1856). Attempts at a natural arrangement of birds. *The Annals and Magazine of Natural History (Ser. 2)* 18:193–219.

   (1891). *Darwinism. An Exposition of the Theory of Natural Selection with Some of its Applications.* London: Macmillan.

Wanninger, A., Wollesen, T. (2015). Mollusca. Pages 103–153 in A. Wanninger, ed. *Evolutionary Developmental Biology of Invertebrates 2.* Vienna: Springer-Verlag.

   (2019). The evolution of molluscs. *Biological Reviews of the Cambridge Philosophical Society* 94:102–115.

Watling, L., Hof, C. H. J., Schram, F. R. (2000). The place of the Hoplocarida in the malacostracan pantheon. *Journal of Crustacean Biology* 20 special no. 2:1–11.

Watson, D. M. S. (1940). George Albert Boulenger, 1858–1937. *Biographical Memoirs of Fellows of the Royal Society* 3:13–17.

Weigert, A., Bleidorn, C. (2016). Current status of annelid phylogeny. *Organisms Diversity & Evolution* 16:345–365.

Wenderoth, H. (1986). Transepithelial cytophagy by Trichoplax adhaerens F. E. Schulze (Placozoa) feeding on yeast. *Zeitschrift für Naturforschung* 41c:343–347.

West, D. A. (2003). *Fritz Müller. A Naturalist in Brazil.* Blacksburg, VA: Pocahontas Press.

West-Eberhard, M. J. (2019). Modularity as a universal emergent property of biological traits. *Journal of Experimental Zoology Part B-Molecular and Developmental Evolution* 332:356–364.

Westheide, W. (1997). The direction of evolution within the Polychaeta. *Journal of Natural History* 31:1-15.

Westheide, W., McHugh, D., Purschke, G., Rouse, G. (1999). Systematization of the Annelida: different approaches. *Hydrobiologia* 402:291-307.

Wheeler, W. C. (2012). *Systematics: A Course of Lectures* Oxford:Wiley-Blackwell.

Whelan, N. V., Kocot, K. M., Moroz, T. P., et al. (2017). Ctenophore relationships and their placement as the sister group to all other animals. *Nature Ecology & Evolution* 1:1737-1746.

White, P. (2003). *Thomas Huxley: Making the "Man of Science"*. Cambridge: Cambridge University Press.

White, T. D., Lovejoy, C. O., Asfaw, B., Carlson, J. P., Suwa, G. (2015). Neither chimpanzee nor human, *Ardipithecus* reveals the surprising ancestry of both. *Proceedings of the National Academy of Sciences USA* 112:4877-4884.

Whiting, M., Whiting, A. S. (2004). Is wing recurrence *really* impossible?: a reply to Trueman et al. *Systematic Entomology* 29:140-141.

Whiting, M. F., Bradler, S., Maxwell, T. (2003). Loss and recovery of wings in stick insects. *Nature* 421:264-267.

Wiley, E. O., Siegel-Causey, D., Brooks, D. R., Funk, V. A. (1991). *The Compleat Cladist. A Primer of Phylogenetic Procedures*. Lawrence, KS: The University of Kansas Museum of Natural History.

Wilkins, A. S. (2002). *The Evolution of Developmental Pathways*. Sunderland, MA: Sinauer Associates.

Wilkins, J. S. (2009). *Species. A History of the Idea*. Berkeley: University of California Press.

Willey, A. (1894). *Amphioxus and the Ancestry of the Vertebrates*. New York: MacMillan.

(1896). On Ctenoplana. *Quarterly Journal of Microscopical Science* 39:323-342.

Williams, D., Ebach, M. (2009). What, exactly, is cladistics? Re-writing the history of systematics and biogeography. *Acta Biotheoretica* 57:249-268.

Williams, D., Schmitt, M., Wheeler, Q., eds. (2016). *The Future of Phylogenetic Systematics. The Legacy of Willi Hennig*. Cambridge: Cambridge University Press.

Williams, D. M., Ebach, M. C. (2004). The reform of palaeontology and the rise of biogeography – 25 years after "ontogeny, phylogeny, palaeontology and the biogenetic law" (Nelson, 1978). *Journal of Biogeography* 31:685-712.

(2008). *Foundations of Systematics and Biogeography*. New York: Springer.

(2020). *Cladistics. A Guide to Biological Classifcation*. Cambridge: Cambridge University Press.

Williams, D. M., Forey, P. L., eds. (2004). *Milestones in Systematics*. Boca Raton, FL: CRC Press.

Williams, D. M., Scotland, R. W., Humphries, C. J., Siebert, D. J. (1996). Confusion in philosophy: a comment on Williams (1992). *Synthese* 108:127-136.

Williams, J. B. (1996). Sessile lifestyle and origin of chordates. *New Zealand Journal of Zoology* 23:111-133.

Williamson, D. I. (2009). Caterpillars evolved from onychophorans by hybridogenesis. *Proceedings of the National Academy of Sciences USA* 106:19901-19905.

(2012). The origins of chordate larvae. *Cell & Developmental Biology* 1:1.

(2013). Larvae, lophophores and chimeras in classification. *Cell & Developmental Biology* 2:4.

(2014). The origin of barnacles (Thecostraca, Cirripedia). *Crustaceana* 87:755-765.

(2015). Rhizocephala revisited. *Crustaceana* 88:939-943.

Willmann, R. (2003). From Haeckel to Hennig: the early development of phylogenetics in German-speaking Europe. *Cladistics* 19:449-479.

Willmer, E. N. (1974). Nemertines as possible ancestors of the vertebrates. *Biological Reviews of the Cambridge Philosophical Society* 49:321-363.

Willmer, P. (1990). *Invertebrate Relationships. Patterns in Animal Evolution*. Cambridge: Cambridge University Press.

Willmer, P., Holland, P. W. H. (1991). Modern approaches to metazoan relationships. *Journal of Zoology* 224:689–694.

Wills, M. A. (1998). Crustacean disparity through the Phanerozoic: comparing morphological and stratigraphic data. *Biological Journal of the Linnean Society* 65:455–500.

Wilson, E. B. (1891). Some problems of annelid morphology. Pages 53–78. *Biological Lectures Delivered at the Marine Biological Laboratory of Wood's Holl in the Summer Session of 1890.* Boston: Ginn & Company.

Wilson, E. O. (1990). *Success and Dominance in Ecosystems: The Case of the Social Insects.* Oldendorf/Luhe: Ecology Institute.

Winchell, C. J., Valencia, J. E., Jacobs, D. K. (2010). Confocal analysis of nervous system architecture in direct-developing juveniles of *Neanthes arenaceodentata* (Annelida, Nereididae). *Frontiers in Zoology* 7:17.

Winsor, M. P. (1976). *Starfish, Jellyfish, and the Order of Life.* New Haven, CT: Yale University Press.

(1991). *Reading the Shape of Nature. Comparative Zoology at the Agassiz Museum.* Chicago: The University of Chicago Press.

(1995). The English debate on taxonomy and phylogeny, 1937–1940. *History and Philosophy of the Life Sciences* 17:227–252.

(2003). Non-essentialist methods in pre-Darwinian taxonomy. *Biology & Philosophy* 18:387–400.

(2006). The creation of the essentialism story: an exercise in metahistory. *History and Philosophy of the Life Sciences* 28:149–174.

Winther, R. G. (2008). Systemic Darwinism. *Proceedings of the National Academy of Sciences USA* 105:11833–11838.

(2011). Part-whole science. *Synthese* 178:397–427.

Witteveen, J. (2018). Typological thinking: then and now. *Journal of Experimental Zoology Part B-Molecular and Developmental Evolution* 330:123–131.

Woltereck, R. (1904). Beiträge zur praktischen Analyse der Polygordius-Entwicklung nach dem "Nordsee-" und dem "Mittelmeer-typus." *Archiv für Entwicklungsmechanik der Organismen* 18:377–403.

Wood, T. C. (2011). Using creation science to demonstrate evolution? Senter's strategy revisited. *Journal of Evolutionary Biology* 24:914–918.

Worsaae, K., Kerbl, A., Di Domenico, M., et al. (2021). Interstitial Annelida. *Diversity* 13:77.

Wyse Jackson, P. N., Maderson, P. F. A. (2014). James Edwin Duerden (1865–1937): zoological polymath. Pages 231–265 in P. N. Wyse Jackson, M. E. Spencer Jones, eds. *Annals of Bryozoology 4. Aspects of the history of research on bryzoans.* Dublin: International Bryozoology Association.

Yager, J. (1981). Remipedia, a new class of Crustacea from a marine cave in the Bahamas. *Journal of Crustacean Biology* 1:328–333.

Yajima, M. (2007). Franz Hilgendorf (1839–1904): introducer of evolutionary theory to Japan around 1873. *Geological Society, London, Special Publications* 287:389–393.

Yang, Z. H., Kumar, S., Nei, M. (1995). A new method of inference of ancestral nucleotide and amino acid sequences. *Genetics* 141:1641–1650.

Yeates, D. K. (1995). Groundplans and exemplars – paths to the tree of life. *Cladistics* 11:343–357.

Yu, Y. L., Zhang, C., Xu, X. (2021). Deep time diversity and the early radiations of birds. *Proceedings of the National Academy of Sciences USA* 118:e2019865118.

Yue, G., Liew, W. C., Orban, L. (2006). The complete mitochondrial genome of a basal teleost, the Asian arowana (*Scleropages formosus,* Osteoglossidae). *BMC Genomics* 7:242.

Zangerl, R. (1948). The methods of comparative anatomy and its contribution to the study of evolution. *Evolution* 2:351–374.

Zhang, H., Dong, X.-P. (2015). The oldest known larva and its implications for the plesiomorphy of metazoan development. *Science Bulletin* 60:1947–1953.

Zrzavý, J., Hypsa, V., Tietz, D. F. (2001). Myzostomida are not annelids: molecular and morphological support for a clade of animals with anterior sperm flagella. *Cladistics* 17:170–198.

Zrzavý, J., Mihulka, S., Kepka, P., Bezdek, A., Tietz, D. (1998). Phylogeny of the Metazoa based on morphological and 18S ribosomal DNA evidence. *Cladistics* 14:249–285.

Index

Abel, Othenio, 6, 158, 186, 206–208, 211–213, 215–216
absence/presence character coding, 286–287
acoelomorphs, 169, 286
acoels, 114, 230, 241, 254, 260
Adanson, Michel, 47
adaptive force, 31–32, 35, 266
affinity relationships, 4, 44, 49–59, 65, 67–68, 72–76,
    78–80, 103, 153–156
    circular, 50, 54, 56, 58, 75
    linear, 49, 73
    reticulating, 4, 42, 47, 49–50, 57, 66, 76, 79, 155–157
Agassiz, Alexander, 84–85, 235
Agassiz, Louis, 54, 58, 63, 84, 152–154, 189, 303–304
Ahnenreihen, 158–159, 207–208, 211, 213, 215
ammonoids, 183–184, 200, 213
Amphiblastula theory, 222, 235, 238, 240
amphioxus, 35, 97, 99, 114, 116–117, 134, 218, 222,
    291–292, 295
amphistomy, 6, 243–244, 246–248, 251–253, 275
anagenesis, 45, 302–303, 305, 316–318
analogy, 29, 53–58, 67
ancestral adjectives, 291–297
ancestral state reconstruction, 5, 126, 149–151, 177,
    205, 215, 240, 253, 273, 282, 285, 288, 295–296,
    330–333
Anelasma squalicola, 125–126
anemones, 243–246, 249, 251, 292
angiosperms, 100, 191, 214
animal types, 21, 30, 39, 48, 58, 68, 110, 128, 152–153,
    219, 259, 264–265, 273
Anpassungsreihen, 207–208
anthozoans, 251, 255, 259
anus, 169, 213, 221–222, 241, 244–245, 250–253, 260,
    262, 268
Arber, Agnes, 41–42
Archaeopteryx lithographica, 97, 114, 181, 208, 314
    Geoffroy's, 21–22
    Goethe's, 10–14
    Owen's, 29–35, 40, 267, 269

archetypes, 8–10, 39, 41–42, 63, 80, 86, 104, 129, 264,
    270, 279
archiannelids, 275
archicoelomate theory, 140, 246–248, 250
Arendt, Detlev, 252–253, 264, 280–281
Aristotle, 17, 23, 45, 48, 267
arthropods
    origin of, 139, 275–276
    phylogeny of, 148, 167–168, 170–171
    polyphyly of, 147–149, 173, 275
Articulata, 58, 107, 110, 139, 153, 219, 253, 264,
    273–274, 279
Articulata versus Ecdysozoa, 139
artificial classification, 46–47, 55, 59
Audouin, Jean Victor, 264, 266

Baer, Karl Ernst von, 58, 82–83, 127–128, 130,
    218–219
Bailey, Irving Widmer, 214–215
Balfour, Francis Maitland, 6, 137, 222–225, 227, 231,
    235–236, 244, 246. *See also* Amphiblastula theory
baraminology, 175–176
basal taxa, 7, 239, 292–297
Bateson, William, 84
Bather, Francis Arthur, 184, 197, 201
Bathybius haeckelii, 115–116
Bayesian inference, 167, 334–335
Beklemishev, Vladimir, 251, 254
Belon du Mans, Pierre, 24
Bentham, George, 75–77
Bilateria
    hypotheses for the origin of, 242–263
bilaterian nerve cords, 113, 134, 162, 164, 334–335
Bilaterogastraea, 254
biogenetic law, 5, 79, 88, 107, 110, 112, 120–121, 163,
    212, 219, 232, 234–235, 238, 258–259, 262, 306
Blainville, Henri de, 48, 182, 267
Blastaea, 110, 112, 116, 219, 233, 261, 283
blastocoel, 112, 221–222, 231

blastopore, 220–223
  slit-like, 244, 248, 252
blastula, 112, 219–220, 222, 225, 229, 231–232
Bonnet, Charles, 48, 60–62
Boulenger, George Albert, 205
branchiopods, 277–280
Brocchi, Giambattista, 80, 182
Bronn, Heinrich Georg, 86–87, 103, 117, 164–166, 267
Brooks, William Keith, 84–85, 143, 188, 208, 210
Buffon, Georges-Louis Leclerc, Comte de, 16, 19, 23, 63–64
Butler, Samuel, 64
Bütschli, Otto, 6, 231–237, 240. *See also* Plakula theory

Caldwell, William Hay, 246–247
Calman, William Thomas, 271–272
Candolle, Augustin-Pyramus de, 49
caridoid facies, 271–272
Carus, Carl Gustav, 86, 88, 264, 266
causal law, 21, 62, 166
Cavalier-Smith, Thomas, 255
central subjects, 1, 4–5, 122–123, 214, 306, 320, 327
Cephalaspis, 303
cephalocarids, 269, 277–280
Ceratodus, 190
Chain, 91
character conflict, 4, 49–50, 57, 59, 67, 69, 72, 79, 140–142, 154–158, 206, 211
character hitchhiking, 213–214, 248
character states
  distinguishing primitive and derived character states, 72, 155–158
choanoflagellates, 224, 229, 236, 296
Chordaea, 112
ciliate theory for metazoan origins, 6–7, 133, 256–263
cladistic blindfold, 215, 308–319, 324, 329
cladistics, 2, 7, 126, 151, 162, 262, 282, 287, 290, 305, 308, 315
cladogenesis, 116, 302, 305, 316
cladogram, 79, 103, 204, 263, 282–283, 303, 310, 319, 325
Coelomaea, 112–113
coeloms, 113, 132, 134, 139–140, 144, 170, 241, 244, 247, 249–251, 254, 268, 285, 288, 291
collar cells, 170, 175–178
collateral ancestors, 314
convergent evolution, 6, 35, 67, 70, 106, 147, 168, 173, 175, 177, 238, 330, 334
Conway Morris, Simon, 34–35, 97
Cope, Edward Drinker, 215, 270
Cope's law of the unspecialized, 163, 206, 215
Cope's rule of phyletic size increase, 6, 124, 215, 269
Crampton, Guy Chester, 128, 137, 159

Cronquist, Arthur, 191–194, 198
crossing of specializations, 206–208, 211, 213–214
crossopterygians, 138, 211
Crow, William Bernard, 213, 215
Cuvier, Georges, 16–19, 21–23, 29, 45, 48, 50, 58, 108, 147, 267
Cytaea, 112

Darwin, Charles
  apprehension about German translation of the *Origin*, 86
  consequences of metaphysical revolution, 39–42
  differences lack phylogenetic value, 175
  influence on Ernst Haeckel, 87
  judgement of Haeckel's phylogenetics, 103
  primitiveness of homonomous repetition of parts, 274
  speculative reconstruction of ancestors, 81–82, 205
  systematic practice, 65, 122
  theory of common descent, 4, 42, 65, 87, 157
  transformation of archetype into ancestor, 4, 9, 29, 33, 35–37
  tree of life metaphor, 5, 66, 72–75, 202
  work on *Anelasma squalicola*, 125
Darwin, Erasmus, 64
Darwinian revision of the natural system, 64–78
Darwinius masillae, 290
Daubenton, Louis Jean Marie, 24
De Beer, Gavin, 166, 256, 258, 261
de novo origins of novel traits, 5, 127–133, 162, 249, 260, 287
De Vries, Hugo, 184
Dean, Bashford, 192, 197
Depaea, 112, 219
descent with modification, 4, 6, 65–66, 70, 72, 94, 122, 131, 152–155, 170, 172, 178–179, 182, 200, 331
developmental systems drift, 178
differences
  misuse as phylogenetic evidence, 19, 148, 163, 169, 172–179, 255
direct phylogenetics, 151–152, 163, 242
divergence time estimates, 291
division of (physiological) labor, 163, 219–221, 228, 230, 238, 266–269, 274
Dobzhansky, Theodosius, 194
Dohrn, Anton, 18, 84, 125, 131, 133
  annelid theory of vertebrate origins, 131, 133–135, 142–143
  principle of no *de novo* origins of novel traits, 127–128, 131
  principle of the succession of functions, 131, 134
Dollo, Louis, 185, 206, 211
Dollo's law, 163, 206

dorsoventral axis inversion, 26–29, 136–137, 166
Dubois, Eugène, 114, 181, 209–210
Duerden, James, 197–198

echidna, 99, 113, 117
ectoderm, 112, 218–219, 221, 228, 230, 232–233, 257
Eigenmann, Carl, 70
elephant evolution, 158
endoderm, 112, 132, 218, 220–221, 228, 230, 232–233, 249, 254, 257
enterocoel theory, 132, 244–247, 250–251, 253–256, 259
enterocoely, 246–247, 250
essentialism story, 9
evolutionary intuitions, 5–6, 14, 117, 123, 126, 151–152, 160–172, 180, 203, 217, 220, 235, 237–238, 240, 255, 258, 263, 266–267, 269, 279–280, 299, 329–331
    free-living habit is primitive, 113–114
    homonomous segmentation is primitive, 269–272, 274–280
    parasitic habit is derived, 114, 119, 124, 126, 163
    sessile habit is derived, 114–115, 119, 124, 163, 255–257, 259
    simple to complex morphologies, 118–119, 153, 164, 237, 257, 261, 267–271, 274, 291
    small body size is primitive, 269
    subterranean habit is derived, 119
evolutionary laws, 6, 163–164, 166, 206, 269
evolutionary morphology, 4, 10, 44, 80, 83–84, 87, 131, 142, 152, 158, 164, 274
evolutionary novelties, 85, 128–131, 172, 238, 262
explanatory power of narrative phylogenetic hypotheses, 130–131, 133, 149, 159, 172, 217, 228, 242, 258, 262, 287, 327
extracellular digestion, 219, 229, 231, 237

Ferrari, Frank, 170–172
Formenreihen, 184, 201, 206–208
functional morphology, 126
    in phylogenetic reasoning, 146–149, 176
Fuxianhuia, 269

Gaskell, Walter, 133, 135–137
Gastraea, 5–6, 42, 99, 102, 106–107, 109–113, 116, 120–121, 135, 140, 152, 217, 219–222, 224–225, 229–230, 235–237, 241, 243, 245–246, 248–249, 251, 253–254, 260–262, 268, 283
Gastraea theory, 6, 83, 110, 217–225, 227–237, 239–242, 244, 275
gastrula, 110–112, 117, 121, 141, 218–221, 224–225, 234–235, 238, 248, 253
gastrulation, 141, 218–225, 228–231, 234, 236–237, 244, 251–253, 263, 275

Gaudry, Jean Albert, 184–185, 324
Gee, Henry, 159, 309
Gegenbaur, Carl, 18, 37, 40, 68, 80, 84, 114, 134, 141, 158, 212–213, 260, 267, 269
general homology, 29, 31, 36
Geoffroy Saint-Hilaire, Étienne, 4, 9–10, 13–30, 33, 39, 45, 48, 63, 131, 147, 175, 182
    principle of connections, 10, 17, 20, 22–25, 37
    principle of elective affinities, 20
    principle of the balancement of organs, 20, 23
    theory of analogues, 20, 23
    unity of composition, 18–21, 30, 131, 175
geotheories, 80
Gessner, Conrad, 46
Ghiselin, Michael, 4, 35–36, 39, 62, 132, 147, 227, 272–274, 306
Goethe, Johann Wolfgang von, 4, 9
Gould, Stephen Jay, 3, 7, 10, 26, 38, 100–101, 182, 187, 203, 261, 307, 319
    dismissal of concept of missing links, 312
    flawed lineage thinking, 300–305
    influence on paleoanthropological thought, 309–310
    influence on popular science and educational literature, 310–312, 315
    life's little joke, 300–301, 304, 309–310
Grant, Robert Edmond, 17, 64, 66, 267
Grell, Karl, 234
ground patterns of assumed primitive character states, 284–285
Gryphaea, 195
gut, origin of, 217–241

Haacke, Wilhelm, 188
Hadži, Jovan, 256–263
Haeckel, Ernst, 2, 5–6, 37, 79–80, 82–83, 86–122, 130, 134, 141–142, 154, 156, 164, 181, 208–211, 213, 216–222, 224–225, 227–232, 235–238, 243, 248–250, 253, 256, 259, 267, 291, 306–307, 314, 324. *See also* Gastraea; Gastraea theory
    defense of evolutionary speculation, 108–109
    distinction genealogy and phylogeny, 88–89
    hypothetical ancestors, 104–106, 109–117
    law of progress, 117
    lineage thinking, 99–104
    phylogenetic evidence, 106–108
    phylogenetic trees, 89–99
    recognition of paraphyletic taxa, 94–95
    views on history of morphology, 108
    views on progressive evolution, 117–120
Hallam, Anthony, 182
Hanson, Earl, 256–258, 260, 262
Hatschek, Berthold, 243, 274
Heider, Karl von, 262

Hennig, Willi, 2, 42, 79, 156–157, 204, 207, 215, 321
  auxiliary principle, 19, 175
hermit crab evolution, 146
Herschel, John, 40
Hertwig, Oscar, 244, 250
Hertwig, Richard, 185, 244
heterobathmy, 207
heteronomous segmentation, 274, 276
Hilgendorf, Franz, 6, 68, 185–188, 190, 201–202, 207, 212, 324
historical narratives, 3, 5, 123, 125
hominization scenarios, 242
Homo erectus, 114, 181, 209
Homo neanderthalensis, 209
homonomous repetition of parts is original condition, 31, 266, 274
homonomous segmentation. *See* evolutionary intuitions; homonomous segmentation is primitive
horizontal gene transfer, 301, 305, 314
horse evolution, 301, 308, 318–319
Hubrecht, Ambrosius, 84, 129, 133–135, 137, 149, 221
  nemertean theory for vertebrate origins, 134–135
Huxley, Julian, 258
Huxley, Thomas Henry, 15, 32, 37, 42, 54, 76–78, 82, 84, 103, 115, 181, 185, 190, 208, 217–218, 220, 223, 245, 250, 265, 270, 272, 324
Hyatt, Alpheus, 6, 186–190, 192, 197, 200–202, 206, 208, 213, 216
hybridization, 55, 70, 72, 163, 172, 298, 301, 309, 314
Hydra, 99, 115, 117, 292
Hyman, Libbie, 236, 241, 249, 251, 253–257, 259
hypothetical ancestors
  all-zero, 286
  annelid or annelid-like hypothetical ancestors, 7, 109, 127, 133–140, 142, 145, 166, 203, 263–281, 285
  as central subjects, 1, 4–5, 122–124, 327
  epistemic rise, 122–149
  generalized, 269–272, 274
  nauplius-like hypothetical ancestors, 275
  of placental mammals, 327
  precursor potential of, 5, 7, 133–140, 148, 241, 246, 260–263, 279, 327
  with modular body plans, 279
hypothetical ancestral mollusc, 272–273, 286

idealistic morphology, 8–9, 17–18, 22, 36, 39, 41–42, 66, 76, 80, 87, 143, 152, 154–155, 158–159, 201, 212, 216, 264, 272, 321
Ihering, Hermann von, 259, 262
imagination as phylogenetic tool, 5–6, 81–85, 108–110, 329
individuality thesis, 4, 62, 306

intercalary types, 208
intermediate forms, 11, 51, 53, 81, 126, 137, 144, 158, 181, 208, 210–212, 234, 243, 247–248, 255, 295, 303
intracellular digestion, 229–231
Ivanov, Artemii, 234, 254
Ivanova-Kazas, Olga, 139

Jefferies, Richard, 213
Jussieu, Antoine-Laurent de, 47, 50, 56, 155

Kappers, Ariëns, 166
Kielmeyer, Carl Friedrich, 23, 62
Kimberella, 291
king crabs, 146
Kleinenberg, Nikolaus, 84, 131, 248–249
Kompetenzkonflikt, 140, 142, 144, 262
Kovalevsky, Alexander, 219, 228
Kovalevsky, Vladimir, 184–185, 201, 207
kynoblast, 230

Lamarck, Jean Baptiste, 15, 38–39, 49–50, 56, 63–64, 66, 80, 90, 108, 156, 182, 267
language of systematics, 95, 295, 321–322, 325
Lankester, Edwin Ray, 6, 83, 129, 131, 141, 220–222, 227, 229, 232, 235–237, 243, 248, 256
  Planula theory, 220–222
Latreille, Pierre André, 264
law of perfection, 163–164
Lewes, George, 33
Leys, Sally, 176–177, 225
lineage explanations, 3, 123, 179, 322, 327
lineage thinking, 3–7, 38, 48, 95, 159, 179, 204, 212–216, 297, 329. *See also* Gould, Stephen Jay
  flawed, 298–326
  Haeckel's, 99–104
  the emergence of, 44–85
linear evolutionary imagery, 7, 45, 299–301
  rejection of, 298–326
linear evolutionary storytelling
  rejection of, 305–306
linear series of forms, types, or taxa, 6, 47–50, 59, 75, 90, 95, 99, 152–154, 158–159, 204, 216, 318–319
linear series of fossils, 181–190, 199–206
linearity of descent, 95, 99, 206, 215, 302, 304–305, 314–315, 318
linearizing language, 291–297
Lingula, 190
Linnaeus, Carolus, 47, 49, 58, 108
living fossil, 99, 113, 190, 239, 264, 280, 292
Lorenz, Konrad, 206, 214, 264, 314
lungfish, 106, 141, 190, 211, 246

MacBride, Ernest, 130, 132, 246, 256

MacFadden, Bruce, 318

Macfarlane, John Muirhead, 149, 260

MacLeay, William Sharp, 50–58, 73, 78

Magosphaera planula, 116, 233

Manton, Sydnie, 147–148, 173, 275

Margulis, Lynn, 171, 263

Masterman, Arthur, 133, 247, 250

maximum likelihood inference, 150, 167, 334–335

maximum parsimony inference, 150, 167, 177, 288, 329, 332, 334–335

Mayr, Ernst, 9, 44, 94, 129, 194, 199, 300
  population thinking, 9, 44, 300

McLaughlin, Patsy, 146

Meckel, Johann Friedrich, 23, 63, 152–153

Meckel-Serres law, 152

mesoderm, 113, 230, 247, 254

Metazoa
  hypotheses for the origin of, 217–241, 257–263

Metschnikoff, Elie, 6, 121, 141, 143, 223, 225–231, 234–238, 248–249, 253–254

microRNAs, 330

Milne-Edwards, Henri, 129, 266–267. *See also* division of (physiological) labor

missing links, 97, 114, 125, 168, 181, 209, 309
  rejection of the concept, 312–314

Mivart, St. George, 67–70

model organisms, 280–281

molluscs, origin of, 273–274

monoplacophorans, 169, 273

Moraea, 112

morpho-clines, 158

morphological seriation, 152–155

Moseley, Henry, 244, 248

mouth
  evolutionary origin, 221–222, 231–232, 243–246, 259–260
  slit-like, 249–251

Müller, Fritz, 79, 83, 87, 143
  hypothesis for evolutionary origin of rhizocephalans, 124–126

myzostomids, 284–285

Naef, Adolf, 18, 42, 143, 151, 154, 157, 159–160, 163, 213, 242, 321

naïve phylogenetics, 160

narrative devices, 100, 163, 214, 307, 320

narrative phylogenetics, 3, 7, 121
  hypothetical ancestors at epistemic core of, 122–149

narrative shortcuts, 7, 214, 282–297

natural classification, 47

natural history, 5, 18, 45–47, 59, 63, 65, 76, 200

natural system, 4, 21, 38, 46
  Darwinian revision of, 64–78
  realist interpretation of, 55–58
  temporal interpretation of, 58–64
  the shape of, 47–55

Naudin, Charles, 73

Nautilus, 190

Neanderthals, 209–211, 215, 309

Neopilina galathaea, 273

Neumayr, Melchior, 183, 200–201

neurobiotaxis, 166–167

Newell, Norman, 182

Nielsen, Claus, 139, 240, 251–253, 262, 283, 285–286

node density artifact, 296

Novacek, Michael, 327

O'Hara, Robert, 7, 300–301, 305–307, 310, 315–316

Oken, Lorenz, 264

onychophorans, 147–148, 168, 171, 173, 222, 248, 252, 275

orthogenesis, 188, 190–204, 206, 216, 268

orthogenetic series, 191–194, 199, 205–206

orthogon theory, 164

Osborn, Henry Fairfield, 191, 197, 270

Owen, Richard, 4, 9–10, 13, 18, 29–36, 39–40, 57, 264, 266–267, 274. *See also* archetype; Owen's homonomous repetition of parts is original condition, 31
  law of vegetative repetition, 32
  polarizing force, 31–34, 266

paedomorphosis, 256, 262

paleontological method, 6, 182, 291

Pander, Heinrich, 218

parallel evolution, 123, 130, 197, 214, 270

paraphyletic taxa, 7, 94–95, 99, 114, 159, 162, 215, 307, 310, 318–325

Parazoa, 176, 224, 235

Parenchymella theory. *See* Phagocytella theory

Patten, William, 133, 137, 144, 303–304

pattern cladists, 42, 66, 324–326

Perrier, Edmond, 278

persistent types, 190

Phagocytella theory, 225–232, 234–235, 237–238, 240, 242, 249

Phagocytellozoa, 234

phagocytoblast, 230

phagocytosis, 227, 234

phenomenal law, 21

phyletic life cycle, 189–190

phylogenetic laws. *See* evolutionary laws

phylogenetic optimality criteria, 2, 5, 133, 140, 150, 167, 330, 333–335

phylogenetic systematics, 2, 42, 156, 283

phylogenetics
  emergence of, 3, 5, 86–89
Pithecanthropus alali, 181
Pithecanthropus erectus, 97, 113, 181, 209, 211, 215
pituitary gland
  origin of, 133–135
placozoans, 168, 234, 239, 292
Plakula theory, 231–234, 239–240
Planaea, 110, 219
Planula theory. *See* Lankester, Edwin Ray
planuloid-acoeloid theory, 251, 254
Platynereis dumerilii, 264, 279–281
platypus, 99, 113, 117, 246
Polygordius, 252, 275
precursor potential of hypothetical ancestors
  as phylogenetic optimality criterion,
    5, 133, 140
predictive power of phylogenetic position, 7, 292–297
preformationism, 130
primary germ layers, 217, 219–220, 222, 232, 234–235,
    237, 262
  origin of, 217–241
principle of compensation, 20
progressive evolution, 94, 99–120, 210
Prospondylus. *See* Urvertebrate
Protascus, 243
Proterospongia (Protospongia) haeckeli,
    229, 236, 238
Prothelmis, 243
Prothero, Donald, 311–314, 318
Protracheata, 248
pruned phylogenetic trees
  as narrative shortcut, 280, 288–290
punctuated equilibrium, 294, 305

quinarianism, 50–57

Raw, Frank, 144–145
recapitulation, 6, 63, 79, 83, 87, 107, 112, 124, 144,
    189, 197, 201, 208, 212, 214, 216, 220, 223, 237,
    246, 248, 258, 262, 275
regressive evolution, 259
Remane, Adolf, 42, 142, 158, 163–164, 241, 253–255,
    257
  homology criteria, 158, 212
remipedes, 278, 280
Rensch, Bernard, 159, 163, 269
rhizocephalans, 172
  hypothesis for the origin of, 124–127
Robinet, Jean-Baptiste, 60–62
Romer, Alfred Sherwood, 195, 200, 270, 324
Russell, Edward Stuart, 10, 14, 22, 102, 130–131, 152,
    185

sabertooth cats, 195, 197–198, 200
Salvini-Plawen, Luitfried von, 159, 161–162, 251, 273,
    276
Savigny, Jules César, 264
scala naturae, 5, 38, 48–50, 59–60, 90–96, 99, 118,
    153–154, 159, 267, 295, 300, 305, 307, 310, 314
scale of being, 31, 38, 60, 266, 268, 291
Schaffner, John, 191–192, 198
Schelling, Friedrich Wilhelm Joseph, 62
Schierwater, Bernd, 239–240
Schindewolf, Otto, 185, 191–192, 194–197, 199, 206,
    213, 270
Schmalhausen, Ivan, 143
Schram, Frederick, 171, 173, 272, 278, 283, 286
Schulze, Franz Eilhard, 224–225, 233
Schwalbe, Gustav, 209–211, 213, 215
sea anemones. *See* anemones
Sedgwick, Adam, 244–252, 255
segmentation, 124, 135, 139, 170, 244, 247, 266,
    273–276, 279
serial homology, 24, 31, 266, 274
Sewertzoff, Alexei, 143, 256
Simpson, George Gaylord, 139, 194–196, 199, 203, 302
sin of the scala, 159, 199, 203, 206, 213, 215, 295,
    307–308, 315, 319
Snodgrass, Robert, 275–276
Sollas, William, 176, 224
special homology, 31, 36
species fixism, 55, 58
Sphenula, 262
squamates, 170, 331, 334
  evolution of egg laying, 331
  evolution of venom, 334–335
Steinböck, Otto, 142, 256–258, 260–263
Steinheim planorbid snails, 6, 68, 185–190, 201–202,
    208
stem groups, 90, 162, 273, 295, 321, 323
stem lineage concept, 323
stratocladistics, 183
stratophenetics, 183
Strickland, Hugh Edwin, 57–58, 75
structuralism, 29
Stufenreihen, 158, 207–208, 211, 213
substitution models, 66, 70, 283, 287, 331–332
substitution of organs, 132, 249
Swainson, William, 56–57
Sycon raphanus, 223–225
systematic hierarchy, 153–154, 189, 298

Tattersall, Ian, 210, 282, 309–310, 312, 318
taxic language of systematics, 7
taxic thinking, 44, 87, 94–95, 119, 128, 214, 308, 316,
    319, 322–325

teleological thinking, 5

threefold parallelism, 5, 63, 106–107, 115, 151–154, 189

Tiedemann, Friedrich, 63

timetrees, 333

transcendental morphology, 8, 17, 22–23, 36, 80, 89. *See also* idealistic morphology

transformation series, 157, 179, 212–213

transitional forms, 95, 114, 213, 234, 312. *See also* intermediate forms; missing links

tree of life metaphor, 70–73

tree thinking, 7, 44, 90, 204–205, 215, 300, 305–308, 310, 315, 319

Treviranus, Gottfried Reinhold, 63

Trichoplax adhaerens, 99, 231, 234, 236, 239, 268, 295

triconodonts, 99, 113

Trochaea theory, 140, 251, 262, 283

Trochophore theory, 274

Trochozoon, 275

Tschulok, Sinai, 42, 154–158, 213, 215

Tyndall, John, 82

typological thinking, 9, 179, 300

Unger, Franz, 181

unity of composition. *See* Geoffroy Saint-Hilaire, Étienne

unity of plan, 4, 8, 10–11, 14–29, 36, 41, 44, 48, 64, 80, 89, 105, 147, 264

unity of type. *See* unity of plan

Urbilateria, 8, 240, 273, 280–281, 289

Urcrustacean, 269, 277, 280

Urdarm, 237–238

Urmund, 221, 237

Urvertebrate, 99, 104, 109, 112, 116, 291

Vernanimalcula, 290

Vertebraea. *See* Urvertebrate

vertebral theory of the skull, 264

vertebrate archetype, 18, 120, 264, 266. *See also* archetype; Owen's

vertebrates, origin of, 133–137, 142–146, 149. *See also* Dohrn, Anton; Hubrecht, Ambrosius

Vicq d'Azyr, Félix, 24

Vigors, Nicholas, 53

volvocine algae, 232, 261

Waagen, Wilhelm, 183–184, 201, 212–213

Walcott, Charles Doolittle, 203

Wallace, Alfred Russell, 8, 44, 73–75

Whewell, William, 33, 40–41

Williamson, Donald, 171–172

Wilson, Edmund Beecher, 80, 84, 97, 246, 268

Wilson, Edward Osborne, 268

Windelband, Wilhelm, 34

xenacoelomorphs, 295

Xenoturbella, 169, 292

Zallinger, Rudolph, 299, 307

zebrafish, 292

Systematics Association Special Volumes

1. The New Systematics (1940)[a]
   Edited by J. S. Huxley (reprinted 1971)

2. Chemotaxonomy and Serotaxonomy (1968)[*]
   Edited by J. C. Hawkes

3. Data Processing in Biology and Geology (1971)[*]
   Edited by J. L. Cutbill

4. Scanning Electron Microscopy (1971)[*]
   Edited by V. H. Heywood

5. Taxonomy and Ecology (1973)[*]
   Edited by V. H. Heywood

6. The Changing Flora and Fauna of Britain (1974)[*]
   Edited by D. L. Hawksworth

7. Biological Identification with Computers (1975)[*]
   Edited by R. J. Pankhurst

8. Lichenology: Progress and Problems (1976)[*]
   Edited by D. H. Brown, D. L. Hawksworth and R. H. Bailey

9. Key Works to the Fauna and Flora of the British Isles and Northwestern Europe, fourth edition (1978)[*]
   Edited by G. J. Kerrich, D. L. Hawksworth and R. W. Sims

10. Modern Approaches to the Taxonomy of Red and Brown Algae (1978)[*]
    Edited by D. E. G. Irvine and J. H. Price

11. Biology and Systematics of Colonial Organisms (1979)[*]
    Edited by C. Larwood and B. R. Rosen

12. The Origin of Major Invertebrate Groups (1979)[*]
    Edited by M. R. House

13. Advances in Bryozoology (1979)[*]
    Edited by G. P. Larwood and M. B. Abbott

14. Bryophyte Systematics (1979)[*]
    Edited by G. C. S. Clarke and J. G. Duckett

15. The Terrestrial Environment and the Origin of Land Vertebrates (1980)[*]
    Edited by A. L. Panchen

16. Chemosystematics: Principles and Practice (1980)[*]
    Edited by F. A. Bisby, J. G. Vaughan and C. A. Wright

17. The Shore Environment: Methods and Ecosystems (two volumes) (1980)[*]
    Edited by J. H. Price, D. E. C. Irvine and W. F. Farnham

18. The Ammonoidea (1981)[*]
    Edited by M. R. House and J. R. Senior

19. Biosystematics of Social Insects (1981)[*]
    Edited by P. E. House and J.-L. Clement

20. Genome Evolution (1982)[*]
    Edited by G. A. Dover and R. B. Flavell

21. Problems of Phylogenetic Reconstruction (1982)[*]
    Edited by K. A. Joysey and A. E. Friday

22. Concepts in Nematode Systematics (1983)[*]
    Edited by A. R. Stone, H. M. Platt and L. F. Khalil

23. Evolution, Time and Space: The Emergence of the Biosphere (1983)[*]
    Edited by R. W. Sims, J. H. Price and P. E. S. Whalley

24. Protein Polymorphism: Adaptive and Taxonomic Significance (1983)[*]
    Edited by G. S. Oxford and D. Rollinson

25. Current Concepts in Plant Taxonomy (1983)[*]
    Edited by V. H. Heywood and D. M. Moore

26. Databases in Systematics (1984)[*]
    Edited by R. Allkin and F. A. Bisby

27. Systematics of the Green Algae (1984)[*]
    Edited by D. E. G. Irvine and D. M. John

28. The Origins and Relationships of Lower Invertebrates (1985)[‡]
    Edited by S. Conway Morris, J. D. George, R. Gibson and H. M. Platt

29. Infraspecific Classification of Wild and Cultivated Plants (1986)[‡]
    Edited by B. T. Styles

30. Biomineralization in Lower Plants and Animals (1986)[‡]
    Edited by B. S. C. Leadbeater and R. Riding

31. Systematic and Taxonomic Approaches in Palaeobotany (1986)[‡]
    Edited by R. A. Spicer and B. A. Thomas

32. Coevolution and Systematics (1986)[‡]
    Edited by A. R. Stone and D. L. Hawksworth

33. Key Works to the Fauna and Flora of the British Isles and Northwestern Europe, fifth edition (1988)[‡]
    Edited by R. W. Sims, P. Freeman and D. L. Hawksworth

34. Extinction and Survival in the Fossil Record (1988)[‡]
    Edited by G. P. Larwood

35. The Phylogeny and Classification of the Tetrapods (two volumes) (1988)[‡]
    Edited by M. J. Benton

36. Prospects in Systematics (1988)[‡]
    Edited by J. L. Hawksworth

37. Biosystematics of Haematophagous Insects (1988)[‡]
    Edited by M. W. Service

38. The Chromophyte Algae: Problems and Perspective (1989)[‡]
    Edited by J. C. Green, B. S.C. Leadbeater and W. L. Diver

39. Electrophoretic Studies on Agricultural Pests (1989)[‡]
    Edited by H. D. Loxdale and J. den Hollander

40. Evolution, Systematics and Fossil History of the Hamamelidae (two volumes) (1989)[‡]
    Edited by P. R. Crane and S. Blackmore

41. Scanning Electron Microscopy in Taxonomy and Functional Morphology (1990)[‡]
    Edited by D. Claugher

42. Major Evolutionary Radiations (1990)[‡]
    Edited by P. D. Taylor and G. P. Larwood

43. Tropical Lichens: Their Systematics, Conservation and Ecology (1991)[‡]
    Edited by G. J. Galloway

44. Pollen and Spores: Patterns and Diversification (1991)[‡]
    Edited by S. Blackmore and S. H. Barnes

45. The Biology of Free-Living Heterotrophic Flagellates (1991)[‡]
    Edited by D. J. Patterson and J. Larsen

46. Plant–Animal Interactions in the Marine Benthos (1992)[‡]
    Edited by D. M. John, S. J. Hawkins and J. H. Price

47. The Ammonoidea: Environment, Ecology and Evolutionary Change (1993)[‡]
    Edited by M. R. House

48. Designs for a Global Plant Species Information System (1993)[‡]
    Edited by F. A. Bisby, G. F. Russell and R. J. Pankhurst

49. Plant Galls: Organisms, Interactions, Populations (1994)[‡]
    Edited by M. A. J. Williams

50. Systematics and Conservation Evaluation (1994)[‡]
    Edited by P. L. Forey, C. J. Humphries and R. I. Vane-Wright

51. The Haptophyte Algae (1994)[‡]
    Edited by J. C. Green and B. S. C. Leadbeater

52. Models in Phylogeny Reconstruction (1994)[‡]
    Edited by R. Scotland, D. I. Siebert and D. M. Williams

53. The Ecology of Agricultural Pests: Biochemical Approaches (1996)[**]
    Edited by W. O. C. Symondson and J. E. Liddell

54. Species: The Units of Diversity (1997)[**]
    Edited by M. F. Claridge, H. A. Dawah and M. R. Wilson

55. Arthropod Relationships (1998)[**]
    Edited by R. A. Fortey and R. H. Thomas

56. Evolutionary Relationships among Protozoa (1998)[**]
    Edited by G. H. Coombs, K. Vickerman, M. A. Sleigh and A. Warren

57. Molecular Systematics and Plant Evolution (1999)[‡‡]
    Edited by P. M. Hollingsworth, R. M. Bateman and R. J. Gornall

58. Homology and Systematics (2000)[‡‡]
    Edited by R. Scotland and R. T. Pennington

59. The Flagellates: Unity, Diversity and Evolution (2000)[‡‡]
    Edited by B. S. C. Leadbeater and J. C. Green

60. Interrelationships of the Platyhelminthes (2001)[‡‡]
    Edited by D. T. J. Littlewood and R. A. Bray

61. Major Events in Early Vertebrate Evolution (2001)[‡‡]
    Edited by P. E. Ahlberg

62. The Changing Wildlife of Great Britain and Ireland (2001)[‡‡]
    Edited by D. L. Hawksworth

63. Brachiopods Past and Present (2001)[‡‡]
   Edited by H. Brunton, L. R. M. Cocks and S. L. Long

64. Morphology, Shape and Phylogeny (2002)[‡‡]
   Edited by N. MacLeod and P. L. Forey

65. Developmental Genetics and Plant Evolution (2002)[‡‡]
   Edited by Q. C. B. Cronk, R. M. Bateman and J. A. Hawkins

66. Telling the Evolutionary Time: Molecular Clocks and the Fossil Record (2003)[‡‡]
   Edited by P. C. J. Donoghue and M. P. Smith

67. Milestones in Systematics (2004)[‡‡]
   Edited by D. M. Williams and P. L. Forey

68. Organelles, Genomes and Eukaryote Phylogeny (2004)[‡‡]
   Edited by R. P. Hirt and D. S. Horner

69. Neotropical Savannas and Seasonally Dry Forests: Plant Diversity, Biogeography and Conservation (2006)[‡‡]
   Edited by R. T. Pennington, G. P. Lewis and J. A. Rattan

70. Biogeography in a Changing World (2006)[‡‡]
   Edited by M. C. Ebach and R. S. Tangney

71. Pleurocarpous Mosses: Systematics and Evolution (2006)[‡‡]
   Edited by A. E. Newton and R. S. Tangney

72. Reconstructing the Tree of Life: Taxonomy and Systematics of Species Rich Taxa (2006)[‡‡]
   Edited by T. R. Hodkinson and J. A. N. Parnell

73. Biodiversity Databases: Techniques, Politics, and Applications (2007)[‡‡]
   Edited by G. B. Curry and C. J. Humphries

74. Automated Taxon Identification in Systematics: Theory, Approaches and Applications (2007)[‡‡]
   Edited by N. MacLeod

75. Unravelling the Algae: The Past, Present, and Future of Algal Systematics (2008)[‡‡]
   Edited by J. Brodie and J. Lewis

76. The New Taxonomy (2008)[‡‡]
   Edited by Q. D. Wheeler

77. Palaeogeography and Palaeobiogeography: Biodiversity in Space and Time (2011)[‡‡]
   Edited by P. Upchurch, A. McGowan and C. Slater

[a]Published by Clarendon Press for the Systematics Association

[*]Published by Academic Press for the Systematics Association

[‡]Published by Oxford University Press for the Systematics Association

[**]Published by Chapman & Hall for the Systematics Association

[‡‡]Published by CRC Press for the Systematics Association

Printed in the United States
by Baker & Taylor Publisher Services